THE CADASTRAL MAP IN THE SERVICE OF THE STATE

THE CADASTRAL MAP IN THE SERVICE OF THE STATE

A History of Property Mapping

ROGER J. P. KAIN

and

ELIZABETH BAIGENT

THE UNIVERSITY OF CHICAGO PRESS

Chicago & London

The University of Chicago Press, Chicago 60637
The University of Chicago Press, Ltd., London

Printed in the United States of America
01 00 99 98 97 96 95 94 93 92 5 4 3 2 1
ISBN (cloth): 0–226–42261–5

⊗The paper used in this publication meets the minimum requirements of the American National Standard for Information Sciences—Permanence of Paper for Printed Library Materials, ANSI Z39.48–1984.

Library of Congress Cataloging-in-Publication Data

Kain, R. J. P. (Roger J. P.)
The cadastral map in the service of the state : a history of property mapping / Roger J. P. Kain and Elizabeth Baigent. p. cm.
Includes bibliographical references and index.
1. Real property—Maps—History. I. Baigent, Elizabeth.
II. Title.
GA109.5.K35 1992
333.3′022′3—dc20 92-10661
CIP

In memory of our fathers,

G. W. BAIGENT *and* P. A. KAIN

CONTENTS

3
THE NORDIC COUNTRIES
47

6
FRANCE
205

7
ENGLAND AND WALES
236

8
COLONIAL SETTLEMENT FROM EUROPE

9
CADASTRAL MAPS IN THE SERVICE OF THE STATE

FIGURES

PREFACE

In 1607 in his *Surveyor's Dialogue* John Norden sought to convince estate owners of the value of the cadastral map when he wrote: "A plot rightly drawne by true information, discribeth so the lively image of a mannor, and every branch and member of the same, as the lord sitting in his chayre, may see what he hath, and where and how he lyeth, and in whole use and occupation of every particular is upon suddaine view." Exactly three hundred years later, in 1907, the government of Egypt completed its national mapped cadaster, which was "a comprehensive stocktaking of Egypt's agricultural resources and auxiliaries which was indispensable for the development of her highly artificial economy in many ways other than its primary fiscal purpose." The extension of property mapping from private realms to state-sponsored surveys in the period 1607 to 1907 is the core concern of this book. In it we explore the origin, development, and elaboration of cadastral mapping by governments to help implement land taxation and reallocation policies and for recording property ownership and resource characteristics. It is thus with the *use* of cadastral maps by governments and public institutions, such as polder drainage authorities, that this book is mainly concerned.

Our study ranges in time from a brief recapitulation of early initiatives in classical antiquity and by individual landowners in medieval Europe, through to the later nineteenth century, by which time the cadastral map was in many areas an established, if not axiomatic, adjunct to effective government monitoring and control of land. The book ranges in area from Europe to those parts of the New World which were colonized by European powers; each of the territories discussed illustrates a particular aspect of the development of cadastral cartography. The Netherlands are chosen for their early adoption of cadastral maps and especially of printed maps. The Nordic countries show both strikingly early development in some areas and a complete lack of maps in others. The experience in Germany and the Austrian Habsburg lands shows the development of mapping especially in the eighteenth century in situations where political control was fragmented and contested. By contrast, in France and England the significant developments originated in the nineteenth rather than the eighteenth century. Cadastral maps played very different roles in the settlement of Europe's colonies, and we examine the experience of mapping in North America, Australia, New Zealand, and India. Together the chapters provide not a complete history of cadastral mapping but one which covers most of Europe and much of the New World, one broad enough to trace the emergence and development of cadastral mapping under a wide variety of physical, political, economic, and cultural circumstances. The comprehensiveness of the coverage allows us to go beyond most other writings on the cadastral map, which stop at description of individual maps or mapping projects, to look for *general* preconditions for mapping and insurmountable obstacles to it and to try to set the development of the cadastral map in the context of contemporary political and economic struggles in the areas concerned. Each chapter addresses the question of why maps were produced, or indeed not produced, in a specific area. In the conclusion we move toward a general answer to this question.

Cadastral maps are maps of properties. Their essential feature is that they identify property owners, usually by linking properties on a map to a written register in which details of the property, such as the owner's name and its area, are recorded. The state or public-sponsored cadasters with which this book is concerned are differentiated from private maps of individual domains by the fact that they record a number of properties within administrative units, from perhaps only one parish, as do English enclosure maps, to whole territories, as does the Napoleonic *cadastre* of France. Though some maps of individual properties are discussed in this book, they are included very much as forerunners of the state cadasters, which are our proper concern. Thus our definition of "cadaster" is at once narrower than in that cartographic literature which considers the private estate map as "cadastral" and much broader than the view of yet others who would restrict the appellation "cadastral" to taxation mapping alone.

Sixty years ago, Marc Bloch drew attention to cadastral maps as sources for reconstructing the past rural histories of European countries. We follow his example and review the mapping only of rural properties; the recording of town properties on maps was quite different, both ideologically and technically, and we have just begun a separate study of such maps. Though we assess the same types of surveys as Bloch and his collaborators, our focus is different. We are concerned with the relationships between cadastral mapping and contemporary society: we view cadastral maps as instruments for effecting state policies with respect to landed property and for exerting political and economic control over land. Some of the surveys that we discuss, such as the English and Welsh tithe surveys, the *ancien cadastre* of France, and the plats and notes of the United States Federal Land Survey, have a long record of use as sources of historical evidence. We hope also that this book may draw attention to the similar potential of the other surveys which we review that have been as yet little exploited by historians or historical geographers.

Our book has been some five years in the writing, during which time we have incurred many debts, which it is our pleasure to acknowledge. We have received financial support from the British Academy, the Universities of Exeter and Oxford, The Leverhulme Trust, St. Anne's College and Lady Margaret Hall, Oxford, the Deutsche Akademische Austauschdienst, Ernst and Young (Eastern Europe), and the Swedish Institute. Continued help and support has been provided by the staff of the School of Geography and the History Faculty Library, Oxford, those of the Bodleian Library, Oxford, and the libraries of the University of Exeter and University College London, Francis Herbert and staff at the Royal Geographical Society, and the staff of the British Library Map Room. Of colleagues in the Geography Department at Exeter, we should like to thank especially Rodney Fry for cartography and Andrew Teed for photography. Thanks are also due to the many scholars who advised us on the surveys and documents of their home countries or regional specialties: they are acknowledged by name at the end of each chapter. Roger Kain would also like to acknowledge the contribution of Hugh Prince, with whom he wrote *The Tithe Surveys of England and Wales* (1985), in which the mid-nineteenth century in Europe is characterized as "an age of cadastral surveys," an observation which provided the original idea for this history of cadastral mapping. We both wish to thank Penelope Kaiserlian and staff at the University of Chicago Press for all their interest in, and help with, this project. Especial thanks are owed to our editor, Craig Noll, whose knowledge and expertise have added much to the finished work.

It is perhaps unusual for one coauthor to thank the other, but Roger Kain would like to acknowledge that, although Elizabeth Baigent began as a paid employee on this project to research the history of cadastral mapping, in the writing of the book she has taken a full share. Final thanks are due to our respective spouses, Annmaree and Hugh, for their support and forbearance.

Exeter and Oxford
Michaelmas, 1991

1

ANTIQUITY TO CAPITALISM: *Decline and Renaissance of Cadastral Mapping*

1.1 The Cadastral Map in Antiquity

The technique of representing landed properties on a map was known and was used to a limited extent in ancient Mesopotamia and Egypt. In ancient Mesopotamia, from c. 2300 B.C., Babylonian scribes drew plans of properties and buildings on clay tablets, probably when land was sold or boundaries disputed, but little survives.[1] The Egyptians possessed instruments for measuring land, but to date the only maps of landed property which have been discovered are from the Ptolemaic period (305–30 B.C.). Land taxation was an important source of revenue, and its collection would have been complicated by the fact that virtually all of the cultivable land of Egypt was subject to annual inundation by the Nile, which made the delimitation of boundaries difficult. Yet Egyptian property registers contain not even a sketch map. A. F. Shore sees a parallel between this lack of maps in the face of seemingly potent influences and the ancient Egyptians' late adoption of alphabetic script. He contends that "the ancient Egyptians showed no greater predilection for change or development once a certain level of attainment had been achieved. Their principles of drawing contained all that was necessary for map-making, and they possessed the means and the bureaucracy for measuring, calculating and registering areas."[2]

Although there is plenty of landscape and archaeological evidence of systematic and regular urban and rural land divisions in Greek colonies, there is as yet no evidence that the survey work was mapped.[3] By contrast with Greece, the rulers of Rome had a well-developed sense of map consciousness, seeing cadastral maps as a means by which they could exert and maintain control over the land resources of their far-flung dominions, especially for regulating the granting of land and accounting for its revenues.[4] Roman land divisions were even more systematized than those of the Greek city-states, with grid streets dividing urban space into square *insulae* and with the cultivated countryside set out in square *centuriae*.[5] There is considerable landscape evidence of the former extent of centuriation around Carthage, in the Po valley, in Campania around Capua, in the colonies of Spain and France, and, most extensively, in present Tunisia. Most, but

Fig. 1.1 Fragment of the Orange Cadaster (Cadaster B). Source: Centre National de la Recherche Scientifique, Paris.

not all, centuriation was applied to *ager publicus*, state land acquired through conquest and assigned to individuals.[6]

The Roman historian Granius Liciniamus relates that the consul Publius Cornelius Lenticulus ordered what was to be the first cadastral map, 170–165 B.C., to reclaim for the state some lands in Campania which had been appropriated by private individuals. The resultant map incised in bronze was displayed in the atrium Libertatis in Rome. A law of 111 B.C. required that *agrimensores* (land surveyors) should complete *formae* (maps) or *tabulae* (registers) of any land in Italy that land commissioners had granted, assigned, or abandoned. From 78 B.C. there was a *tabularium* in Rome serving as a central archive for the deposit of such cadastral maps and related documents.[7]

A didactic "textbook," a collection of manuscripts illustrated with diagrammatic examples of varying dates but probably put together about the fourth century A.D. and known as the *Corpus Agrimensorum*, provides evidence of the way Roman surveyors used instruments such as the *groma* (a type of cross-staff) to set out lands and construct maps. Much of the *agrimensores'* work concerned the definition and plotting of property boundaries. Hyginus Gromaticus, one of its putative authors, tells how the land surveyor had to enter areas, names of occupiers, and other cadastral information on the maps, and that two copies of each *forma* were required, one for the *tabularium* in Rome and one to be kept locally. He envisaged each being cut in bronze, the seizure and melting down of which by invading peoples perhaps explains why not a single example has survived from the thousands of *formae* from the *tabularium* in Rome.

The locally held copy of the centuriated cadaster of the Roman colony of Arausio (Orange, southeastern France) was cut in stone, and some fragments of the map at a scale of about 1:6,000 covering an area from Carpentras to Orange and Montelimar in the lower Rhone valley have survived (figure 1.1) to provide an artefactual link to the *agrimensores*.[8] As in Campania, the object of this survey was to reclaim privately appropriated state lands. A monumental inscription of A.D. 77 reads: "The Emperor Vespasian, in the eighth year of his tribunician power, so as to restore the state lands which the Emperor Augustus had given to soldiers of Legion II Gallica, but which for some years had been occupied by private individuals, ordered a survey map [*forma*] to be set up with a record on each 'century' of the annual rental."[9]

This cadastral survey was part of a concerted attempt by Vespasian to restore state revenues, in this case by repossessing state lands and reimposing taxation formerly uncollected. These cadastral maps were veritable "tools of statecraft" which more specifically by the close of the Roman empire were widely used to underpin state property rights and revenues.[10]

1.2 Private Estate Maps: The Precursors of State Cadasters

It might have been thought that the cadastral map was such a powerful instrument of social and political control that, once invented, it would have been taken up avidly by successive rulers. Yet with the fall of Rome the use of maps to describe and record landed property was effectively discontinued. To the medieval mind, both individual and corporate, the proper way to describe properties was in written descriptions of the extent of land parcels and their topological relationships. So complete was the obliteration of map consciousness in feudal Europe that such private property maps as were produced in the medieval world cannot be seen in any sense as survivals of a tradition from antiquity. Property mapping in antiquity is not, therefore, part of a continuous history of the state-sponsored cadastral mapping that came to characterize European countries in the Enlightenment. The cadastral map was effectively in decline from its Roman apogee; not until the sixteenth and early seventeenth centuries in the Netherlands did government bodies again use maps to exert control over land. In the emergent capitalist societies of Renaissance Europe, where land became a commodity and power relations were expressed through control of the means of production, which included land, there was now clearly a reason for mapping properties—namely, as an aid to developing the new systems of exclusive rights to land.

By the time that many European governments were rediscovering the mapped cadaster in Enlightenment Europe, cadastral maps had become well-established adjuncts for private estate management, for example, in England, and were used in all the countries reviewed in this book to substantiate claims to land in property disputes. It is argued here and at greater length elsewhere that their compilation can in large part be explained by the way that landed society in Renaissance Europe came to apply both monetary and symbolic values to specific parcels of land as feudalism gave way to capitalism.[11] But changes in the new capitalist society thereafter did not lead to an inexorable and consistent increase in the use of the "new" medium of communicating local cadastral detail. D. H. Fletcher's careful study of the use of maps on the extensive and widely scattered estates of Christ Church, Oxford, in the seventeenth and eighteenth centuries concludes that map consciousness "did not develop uniformly or consistently over time."[12]

Professional land surveyors were crucially important in the capitalist economies in changing attitudes toward graphic representation of topographical information within society in general and its governing institutions in particular. The economic imperatives of emergent capitalism encouraged developments in the science of surveying, and authors of didactic treatises such as the English surveyors Radulph Agas, John Norden, and Valentine Leigh were ardent advocates of the recording of surveys on maps. As F. M. L. Thompson writes in his history of the surveying profession, "The spur to their activities was not the joy of intellectual rediscovery, but the gain to be made out of a changing economic and social situation."[13]

The English surveyor Radulph Agas provides a late sixteenth-century view of the utility of private estate maps in his didactic treatise *A Preparative to Platting of Landes . . .* (1596). These can, he says, define the situation of a manor and can locate the manor house, the various tenements, curtilages, barns, fields, stables, and cottages "in their full number measure and forme." A map may display chases, warrens, parks, woods, fields, closes, pastures, and "euery parcel of land lying within the boundes thereof, in their exact measure: fashion and quantitie." Boundaries and abuttals are "so wholly put downe, as no booke may bee comparable with the same." Agas notes that the type of boundary can be specified on a map, whether wall, pale, hedge, ditch, river, lane, or path, "the true placing whereof, bringeth perfection to the woorke, and may in time to come bee many waeis most necessarie and profitable." Furthermore, a map might be a tool for management: "Heere have you also every parcel ready measured, to all purposes: you may also see upon the same, how conveniently this or that ground may be layd to this or that messuage." Last, when perhaps arable land had been laid down to pasture, or when "bounders and meeres" were moved, or when names were changed, "notwithstanding their auncient and faire bookes, for the abuttals thereof: the surueigh by plat, suffereth no such inconvenience, but shall be for continuall evidence."[14] Agas was aware that a map had the great advantage over a terrier in that property boundaries could be identified for all time, even when the landscape of balks and hedges had changed. In short, this experienced Tudor surveyor advanced three main advantages for survey by plan rather than written cadaster alone: precision of location, efficiency of land management, and permanence of record.

The English surveyor John Norden was, like Agas, a committed advocate of property survey by maps. In his *Surveyor's Dialogue . . .* (1607), he posed the question: "Is not the fielde itselfe a goodly map for the lord to looke upon, better than a painted paper?" His reply was: "A plot rightly drawne by true information, discribeth so the lively image of a mannor, and every branch and member of the same, as the lord sitting in his chayre, may see what he hath, and where and how he lyeth, and in whole use and occupation of every particular is upon suddaine view."[15]

These and other similar didactic treatises notwithstanding, it is clear that most landed properties were managed in medieval and early modern Europe without the aid of maps. Land was bought and sold, farmed and exploited, surveyed for private individuals and on behalf of crowns and governments for valuations and tax assessments by written description alone. Paul Harvey's calculations of the number of estate maps surviving from medieval England can be adduced to support this reasoning. He finds only three English property maps surviving from the mid-twelfth century to the mid-fourteenth century. For each of the fifty years between 1350 and 1500, however, there are about ten survivals, but then no fewer than about two hundred maps are known from the period 1500–1550.[16] He interprets these figures as representing real change, rather than just differential survival from an unchanged population of maps.

There are several reasons why property maps were rarely drawn in the Middle Ages but were increasingly compiled during the Renaissance. First, it can be argued that communities would be concerned with boundaries only when rival claimants appeared on the scene and that clashes of interest over peripheral woods, pastures, and rights to water and marsh might surface only when the reservoir of unclaimed land was perceived to be nearing exhaustion.[17] Second, income from land in the medieval period was not calculated by reference to area-based quotients but derived from the possession of rights over specific tracts of land according to the custom of the manor. Measurement of the surface area of the constituent parts of manors was therefore irrelevant unless perhaps these were to be divided or enclosed. Third, the established boundaries of manors and estates were marked by topographical features, supplemented where needed by merestones and balks placed specifically for the purpose. The likelihood is that these marks had existed little changed within longest living memory and were presumed by the local inhabitants to be immutable. Their continued existence was regularly checked by perambulation, and viewing of the marks and their nature was such that they could be described adequately in words for unerring recognition by their local readership.[18] Fourth, a manorial extent, an enumeration and valuation of the assets and rentals of a manor, is concerned first and foremost with what the manor contained and not where the buildings, orchards, pastures, woods, and plowlands were situated—that is, with economics and not geography.[19] Some of these descriptions are very elaborate and describe the lands, parcel by parcel, by means of lists of bounds and abuttals; they are virtually written maps.[20] *Cartes parlantes* is the term de Dainville has coined for some of the fourteenth-century terriers from Languedoc.[21] These and others like them from across the breadth of medieval Europe were written out by authors who had yet to take the critical step toward the more succinct method of cartographic description.

The precise location of a piece of land on a map and the exact measurement of its area were probably beyond the technical competence of medieval surveyors.[22] The well-developed exposition of the application of geometry to land surveying in Richard Benese's *This Boke Sheweth the Maner of Measurynge . . .* (1537) would at most represent late medieval best practice on English manors. But parcels could be arranged in their correct topological relationships, thus allowing a written terrier to be made into a map. Well-known examples of such maps from England include the plan of lands of the Benedictine Abbey of Chertsey of c. 1432, of some of the strips in the open fields of Shouldham, Norfolk, 1441, and from Alsace, also from 1441, a map of the land owned by the former Benedictine Abbey of Honau in La Wantzenau, Honau, and Abertsheim.[23] None of these maps is drawn to scale. The Alsace map is a square divided into halves by the Rhine, but it shows in diagrammatic form all the information which might have been set down in words in a written extent: farms and names of tenants and freeholders; arable, meadow, pasture, and commons; and the ownership of various tracts of land.

The process by which one landowner experimented with such a graphic representation of land as he became conscious of the value of property maps has been convincingly demonstrated from the documentary record of two small manors in the English West Midlands.[24] John Archer, who owned the estates between 1472 and 1519, was closely involved in their management as well as with a number of commercial ventures on the Continent (as such, he perhaps was not an average fifteenth-century English squire). From an analysis of his family papers, Brian Roberts has elucidated the way Archer "was clearly feeling the need for a map and seems to have been taking the critical mental step between a written survey describing the locality of each piece of land and doing the same more accurately and simply by means of a map."[25] About the year 1500, when a survey of his manor was being compiled, Archer constructed a sketch map in schematic form to serve as an aide-mémoire of the spatial relationships of part of his land.

Many of the property maps produced in medieval times were designed to clarify and record points at issue in disputes about land boundaries. Disputes over land became endemic in many medieval societies, exacerbated by diminishing land resources consequent on the *grands défrichements,* the occupation and colonization of untilled lands between the eleventh and thirteenth centuries.[26] An example is the earliest extant local map of a community in the *département* of Tarn in France. Points in dispute c. 1314 between the bishop of Albi and the seigneur of Puygouzon concerning the boundary between the two communities were set out on a bird's-eye view of the property entitled *La carta pentha et vehuta de la sonhoria dalby depart dessa lo pont et fazen division am Puyggozo et autras parta.*[27] From a little earlier, 1224 to 1249, is a map copied into the English Kirkstead Psalter depicting the nine *vaccaria* (cow pastures) disputed by the sokes of Bolinbroke, Horncastle, and Scrivelsby in Lincolnshire.[28]

The use of such maps to marshal evidence in disputes about ownership of, or rights to, particular pieces of land increased during the European Renaissance.[29] The construction of a map (more properly a picture or a *figure des lieux*) to assist in settling disputes was advocated by Jean Imbert in the French editions of his legal text *Institutions forenses ou practique judiciaire* (1563 and editions to 1641). Imbert advised that judges should enlist the services of a *peintre* to produce a *figure.* Then "they should enquire of the parties if the picture was well done, and if it is agreed, the judge should question the parties to determine the disputed territory and the respective boundaries claimed."[30] Imbert's advice was reiterated in the seventeenth century in a text by Jean l'Hoste, who noted that judges often called for a map of disputed territory to help them make a fair decision.[31]

It is clear, then, that one category of large-scale property maps was that used in litigation over rights to land. It is also possible to argue that, as the demand for land increased, so an accurately surveyed estate map might ward off future attempts to pretend rights. One of the arguments that John Norden adduced in defense of the surveyor in his imaginary dialogue with a skeptical farmer was the deterrent effect of a property map. "Where a due and true survey is made and continued, there is peace mayntayned between the lord and his tenants; where, if all things rest between them confused, questions and quarrels arise, to the disturbance of both."[32] There is some evidence that, by the end of the sixteenth century in England at least, careful housekeeping required that maps of estates be compiled for just this reason: to be ready to combat a vexatious action. Peter Eden suggests that All Souls College, Oxford, may have embarked on such a program of surveying after problems with a lease caused by the fact that the property had not been properly surveyed at the time it was let.[33]

Property maps, in the surveyor Valentine Leigh's phrase, were constructed for the "profit" they might bring.[34] There can be no doubt that one element of that profit was reckoned in money, but a property map, as landed property itself, was invested with symbolic values as well. The ownership of a landed estate with its fields, woods, mansion, farms, and cottages was the entrée to landed society; an estate was "a little commonwealth" in its own right. A map pictured this cosmology and was a touchstone to the rights and privileges which possession of land brought in train. It can be considered, as J. B. Harley remarked, "a seigneurial emblem, asserting the lord of the manor's legal power within the rural society. For him, the map was one badge of his local authority. Family coats of arms added within the margins were certainly more for him than mere decoration, for the right to these heraldic emblems also incorporated an individual's right, rooted in the past, to the possession of land."[35]

1.3 Cadastral Maps in the Service of the State

Beginning in the sixteenth and seventeenth centuries, there was a fundamental shift in the development of cadastral maps from their use as inventories of private land toward their use by public authorities and ultimately state governments. They were used initially as instruments to effect specific measures, notably tax reform, but ultimately became more general tools for the accurate recording of information relating to individual land parcels. That the cadastral map was by the nineteenth century a widespread and widely valued instrument of government land management is well attested. What is less well known is that in the early modern period the cadastral map was a highly contentious instrument for the extension and consolidation of power, not just of the propertied individual, but of the nation-state and the capitalist system which underlies it.

It is this controversy and the struggles which surrounded the reintroduction and use by the state of the cadastral map which we explore in chapters 2 to 8. Chapter 9 draws together the histories of these European and colonial territories to suggest a chronology of cadastral mapping and to address the question of why this chronology unfolded as it did.

2

THE NORTHERN AND SOUTHERN NETHERLANDS

Of all the territories discussed in this book, the Netherlands, and particularly the northwestern polder areas, stand out by reason of their very early mapped cadastral surveys. The continuing battle to win and preserve land, as well as the high population density and the early development of capitalist production relations, encouraged in the Netherlanders a highly developed sense of the value of land and of maps as instruments for its effective allocation and management. The history of cadastral cartography in the Netherlands is fully as remarkable as the better-known histories of their military and topographical maps and navigation charts.[1]

The Netherlands consist of the northern provinces, the mainly Protestant Dutch area now the Netherlands, and the southern provinces, the Catholic Walloon and Flemish area now Belgium (figure 2.1).[2] Both were part of the medieval Holy Roman Empire. From 1477 the southern provinces and Holland were in the control of the Spanish Habsburgs, who by 1543 had also inherited the northern provinces. A period of religious and political repression by the Habsburgs precipitated the Eighty Years' War, or Dutch War of Independence (1568–1648), at the end of which the Spanish were forced to recognize the independence of the seven northern United Provinces of Drenthe, Friesland, Gelderland, Groningen, Holland, Overijssel, and Zeeland at the Peace of The Hague. Under the terms of the treaties which ended the Thirty Years' War in 1648, the United Provinces left the empire. During the seventeenth century the Dutch built up their trading empire after the establishment of the East and West India companies in 1602 and 1621 respectively. After the War of the Spanish Succession the Peace of Utrecht in 1713–14 gave control of the Spanish (southern) Netherlands to the Austrian Habsburgs. In 1795 the French annexed the southern Netherlands (Belgium), and the Batavian (Bataafsche) Republic was declared in the northern Netherlands. The northern Netherlands came increasingly under French control. The Unitarian party assumed power with French support in 1798, and in 1806 Louis Bonaparte (Lodewijk Napoleon), Napoleon's brother, was declared king of Holland. Then in 1810 the French annexed the country. In 1813 after the Battle of Leipzig the Netherlands were freed from

Fig. 2.1 The Netherlands: regions, provinces, and places noted in the text. Sources: Droysens 1886; Muir and Philip 1927; Shepherd 1930; *Westermanns Atlas zur Weltgeschichte* 1956; Kinder and Hilgermann 1974–78; Darby and Fullard 1978.

French control, and at the Peace of Vienna in 1815 the Kingdom of the United Netherlands, covering roughly present-day Netherlands and Belgium with Luxembourg, was established. The Catholic Walloons resented their subordination to the Dutch, and a separate Kingdom of Belgium was established between 1831 and 1839.[3] Since the northern and southern Netherlands are so culturally distinct and since the northern Netherlands have such an important cartographic history, the two countries are treated separately in this chapter. The cadastral history of the Netherlands under Austrian Habsburg rule is discussed here rather than in chapter 5 (the Austrian Habsburg lands), since Austrian Habsburg rule was very brief and left the Netherlanders considerable autonomy, especially in fiscal matters.

The Northern Netherlands

2.1 Dikes and Polders

The coast of North Holland originally extended to a line marked by the curve of the West Friesland Islands, such that in Roman times there was a large freshwater lake northeast of Amsterdam. From the eleventh to the thirteenth centuries a combination of rising sea level and sinking land led to flooding and general encroachment by the sea. Prehistoric *terpen* and *wieren* (artificial mounds on which settlements were made) are the earliest evidence of reclamation; construction of dikes, sluices, and polders began around A.D. 800 or even earlier. By the thirteenth century there was a succession of dikes along the seacoasts, and the major rivers and distributaries were confined within walls and levees. From the fourteenth century onward there was a change from defensive measures to offensive reclamation and drainage.[4] The process of reclamation meant that there were always areas of new settlement in the Netherlands, and from at least the eleventh century such settlement was often organized. Surveyors typically divided the land into regular parcels which were shared out among the new settlers. Such surveying was not usually accompanied by mapping, but it established the intimate ties between surveying and reclamation from a very early date.[5]

Polder Authorities

Complex organizations were required to create new land and to manage existing land and water installations. In the thirteenth century the counts of Holland and the bishops of Utrecht amalgamated many long-established local polder and drainage boards, or *waterschappen*, into larger units, the *hoogheemraadschappen*. Rhineland, Schieland, Delfland, and Amstelland all had *hoogheemraadschappen* by 1300, and by the close of the Middle Ages all the western lowlands were organized into polders under self-governing polder boards.[6]

The *waterschappen* in Holland consisted of the dike reeve, a representative of the count of Holland, and the *heemraden* or *dijkschepenen*, representatives of the inhabitants of the area. The *waterschappen*, although ultimately dependent on the central administration, exerted a remarkable degree of independence from it and control over the inhabitants of their area, since they possessed legal authority to execute necessary maintenance and repairs, to resolve conflicts of interest in the case of drainage and water rights, and to punish offenders who endangered the safety of the polder. As the size of the projects which they undertook increased and became ever more specialized and professional in the sixteenth and seventeenth centuries, so the amount of money they needed to raise through taxation increased. In return, taxpayers often demanded greater representation on the polder board.[7] Thus the polder boards became at once increasingly entrepreneurial and increasingly representative of power from below rather than above. This provoked conflict when Emperor Charles V, who was also count of Holland, tried to exert his authority over the boards.[8] What is of even more importance here is the considerable use of maps by the polder authorities.

The term *waterschapskaarten* (polder authority maps) covers a wide range of both topographical and cadastral maps.[9] The dike reeve and the *hoogheemraden* needed maps to carry out their administrative, water management, and fiscal duties, but at the same time, Amsterdam publishers began to include maps of polders in their atlases. Often the cartographic *content* of the atlas maps was the same as in the former group, but the *context* in which they were produced was entirely different.[10] In the following discussion maps are grouped according to the reasons for their production by the polder authority—namely, tax collection, litigation, administration, and publicity.

Whatever their prime purpose, many *waterschap* maps had an important secondary function in that they could be hung as wall maps to display the territory over which the authority had jurisdiction. When combined with the coats of arms of the authority and the dike reeve and with poems and allegorical engravings, they expressed the authority's power and status. The commissioning of important surveys and maps was a reflection of the status of an authority; the cost of producing and printing such maps could be recouped through dike taxes. The spectacular wall maps which survive show the emotive value and monetary worth that the Netherlanders ascribed to their intensively used and hard-won land.[11] The style of the early *waterschap* maps was heavily influenced by map painters and later by engravers, but authorities' demands that symbols of their power be included did not remove the onus on the surveyor to produce an accurate map. Decoration is confined to borders and does not intrude on the map itself.[12]

Maps and Dike Taxes

The upkeep of dikes and other works demanded great efforts of the population. Upkeep was originally the duty of those whose lands abutted directly on the dikes, but this system was replaced under the counts of Holland by a system of *dijkverdeling* whereby each landholder in the polder was responsible for the maintenance of a length of dike proportional to the size of his holding. This called for measurement of landholdings to determine their extent, which in turn required surveyors. However, *dijkverdeling* also proved unsatisfactory, as it took no account of the quality of a holding, only of its area. It did not preclude defective work or neglect, so one person's negligence could endanger the entire polder, and it was extremely difficult to apportion the maintenance of costly dams and sluices. Above all, some of the major dikes protected far wider areas than the individual polders to which they belonged, and the high cost of their upkeep outstripped the resources of the immediate neighborhood. As sea level rose and the effects of dike breaches became more widespread and destructive, it was realized that water defenses should be a common responsibility. From the late fifteenth century many dikes were declared a public charge. In North Holland the first defenses to be so designated were those between Schoorl and Petten, just north of Alkmaar, in 1477. In order to prevent a repetition in this region of the St. Elizabeth's Day floods of 1421, the cost of repairing dikes was made a charge upon the inhabitants of most of Kennemerland and West Friesland. The method of paying for dike repairs out of communally levied funds meant that dikes were repaired by specialist dike managers and farmers were freed to concentrate their attention on agriculture.[13]

In the sixteenth century dike taxes were often apportioned among the villages according to their size. Each village then shared its quota among the villagers. This produced many anomalies and inequities, and as the burden of taxation was sizable, the practice of revising apportionment after a thorough survey became more widespread. In 1533 Charles V, as count of Holland, ordered a cadastral survey in the northern quarter of Holland to reform the system by which the dike tax (known as the *Hondsbossche-contributie*) was levied. The results of the survey showed that the areas which the villages had been deemed to cover for the purposes of taxation were often considerably at variance with their real area. The value of survey for removing injustices was clear.[14]

The practice of commissioning surveys for tax collection became increasingly common in the sixteenth century.[15] In 1539–40 the *heemraadschap* of Rhineland commissioned Simon Meeuwszoon of Edam to survey the whole area under the polder authority.[16] Meeuwszoon's Rhineland survey, like that for the count of Holland, showed that the areas on which tax was levied were often far from representative.[17]

In Putten, an island to the southwest of Rotterdam composed of many polders each with its own dike, it was decided in 1617 to make an accurately surveyed and mapped cadaster to remove the inequalities in assessment of dike taxes. The need for reform had become apparent during the sixteenth century. Floods in the 1530s increased the level of taxation, which in turn brought dissatisfaction about apportionment to a head. Revisions made in 1580 and again at the turn of

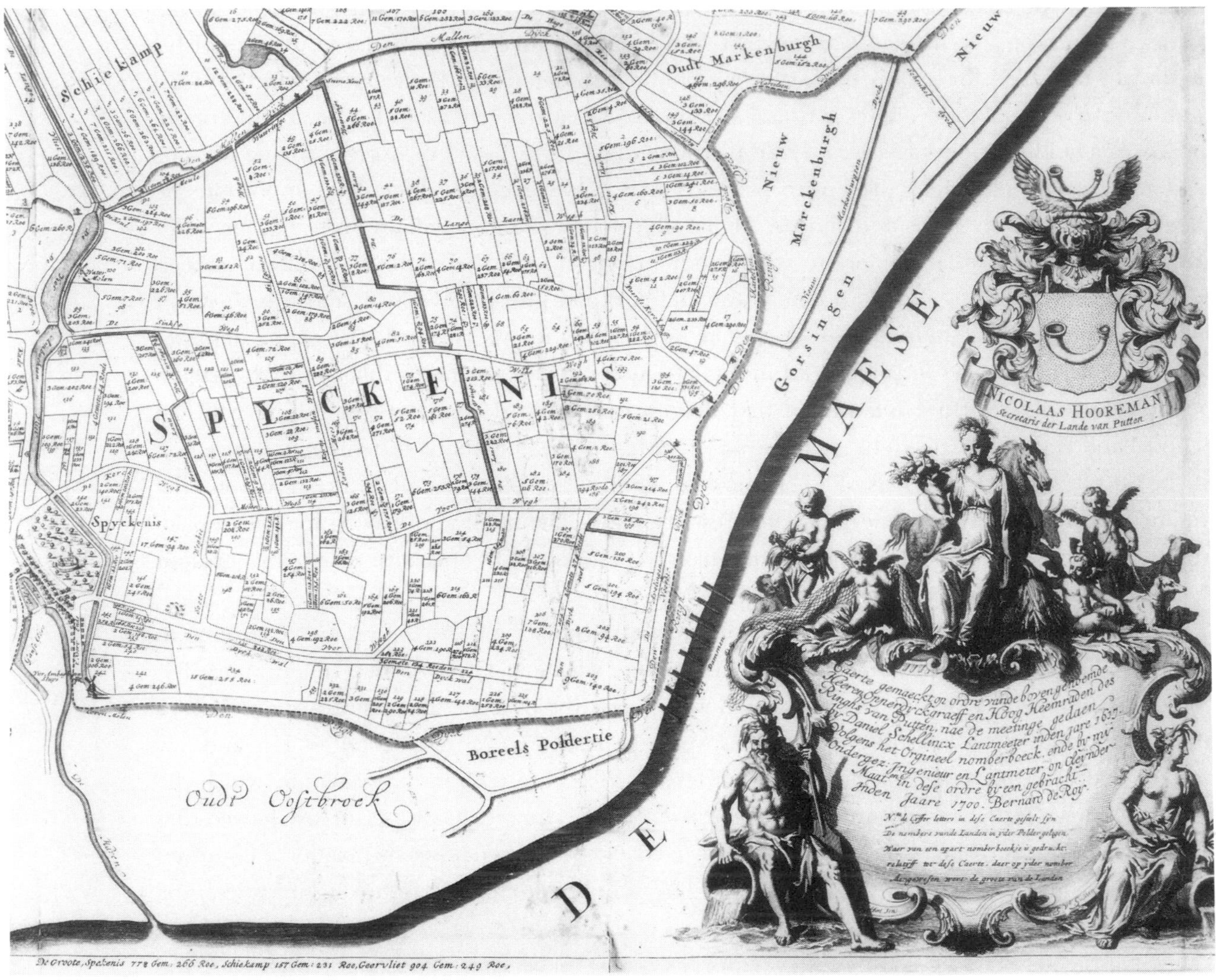

Fig. 2.2 Map commissioned by the *hoogheemraadschap* of the Ring of Putten, drawn in 1700 using the original survey and reference numbering of Daniel Schillincx of 1617. Source: Algemeen Rijksarchief, The Hague, VTH 2089.

the century failed to resolve the problem. This was in part because the reforms were based on adjustment of existing records using additional evidence from the *waarsmannen*, officials appointed by the sheriff with responsibility for assessing and collecting taxes in each polder. In 1617 the dike reeve and the *hoogheemraden* of Putten commissioned the surveyor Daniel Schillincx to make a cadaster which was to consist of written documents and cadastral maps. A general non-cadastral map was compiled from the 1:6,250 cadastral maps, and this map survives. Van der

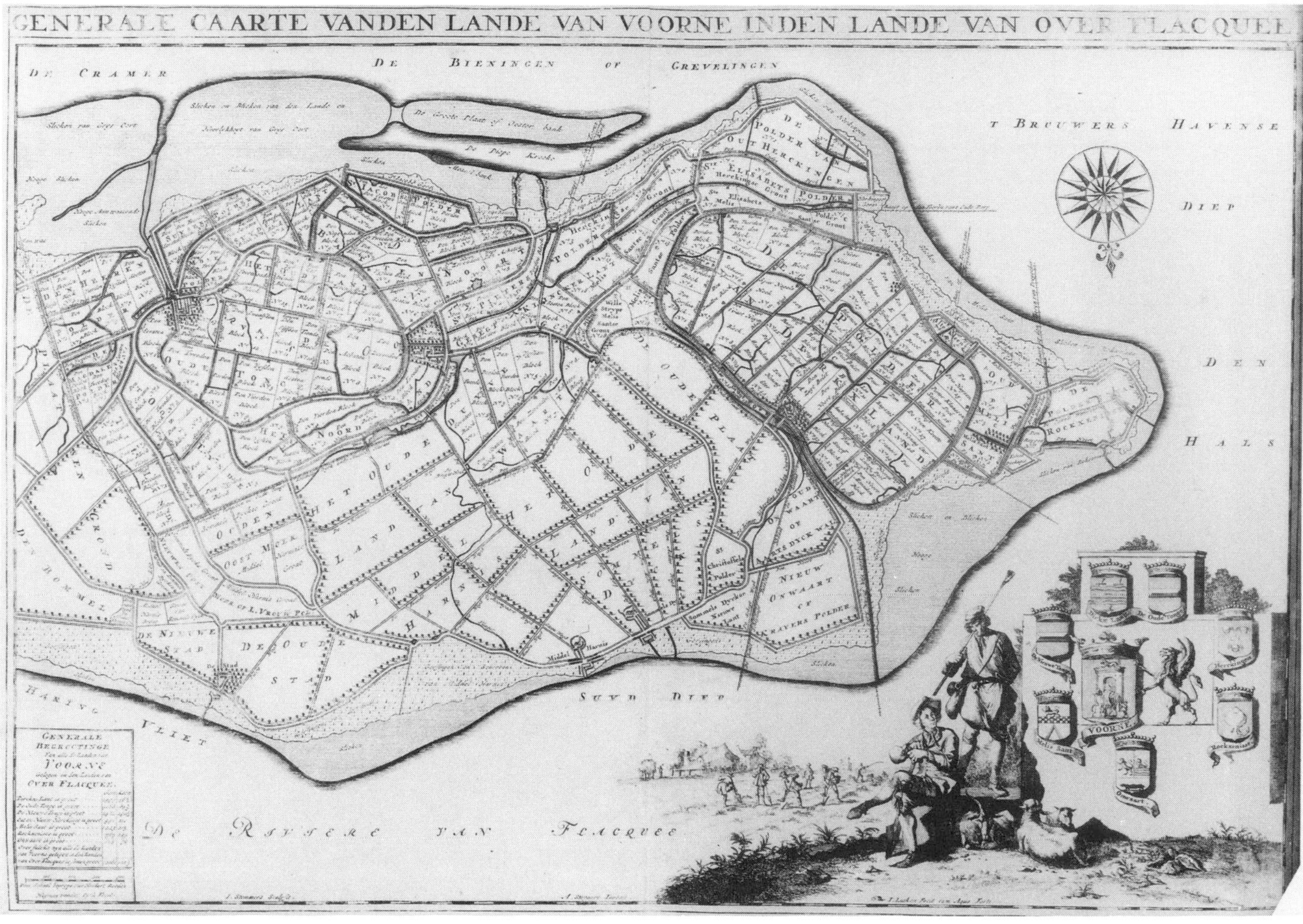

Fig. 2.3 Map of Voorne commissioned for administrative purposes and the collection of polder taxes, drawn by Heymann van Dijck between 1695 and 1701. The atlas consists of this general map at 1:20,000 and detailed maps at 1:5,000. Source: Algemeen Rijksarchief, The Hague, VTH E.

Gouw describes Schillincx's work as a "perfect cadaster"; it was in fact used for the collection of taxes in Putten for more than 250 years (figure 2.2). The new national cadaster of 1832 superseded it in some respects, but it continued to be used for the levying of taxes until 1862–67.[18] In nearby Voorne an atlas was commissioned for the collection of polder taxes and general administration. It was produced by Heymann van Dijck between 1695 and 1701 and consists of four general maps at scales of between 1:19,000 and 1:30,000 (figure 2.3) and detailed cadastral maps at a scale of 1:5,000.[19] Both the Putten and the Voorne maps are elaborately decorated with the coats of arms of the officials involved in polder administration (figures 2.4 and 2.5).

Fig. 2.4 Part of the frontispiece to the *verpondings* atlas of Voorne. The copperplate engraving is by Romein de Hoogh and was published in 1701. Source: Algemeen Rijksarchief, The Hague, VTH E.

Fig. 2.5 Coats of arms of members of the *hoogheemraadschap* of the Ring of Putten found on the map of Putte, 1700. Source: Algemeen Rijksarchief, The Hague, VTH 2089.

Polder Authorities and the Courts

Because the polder authorities were legally constituted bodies, it is not surprising that they became involved in legal disputes and that they used maps to resolve them. Tax collection inevitably provoked disputes, and many of the earliest *waterschap* maps were drawn for court cases about dike dues. The oldest map in the collection of the Rhineland polder authority is dated 1457 and is of the Spaarndame dike in North Holland. It was used in a court case about a dispute concerning upkeep responsibilities for the dike.[20] A map was drawn in 1545 to resolve a dispute about management of the Grooten polder near Zoeterwude to the southeast of Leiden. On this map, drawn at a scale of 1:6,000, Pieter Sluyter has shown each parcel of land, its owner, and its extent.[21] The production of legal maps by *waterschappen* is important in that it led to their employing surveyors regularly, rather than on an ad hoc basis, and to their interest in formalizing professional standards of surveying. The status of surveyors in the polder authorities was fixed in 1534 by Charles V, who required that all *waterschappen* engage surveyors as regular employees. Since the number of surveyors rose rapidly after 1550, polder authorities devised procedures for admitting "sworn surveyors" who had been proved professionally competent and who bound themselves to be impartial. Admission gave the surveyor legal status in the area covered by the body which examined him. Surveyors naturally wanted to be admitted by a body with wide jurisdiction, and hence admittance through one or more provincial bodies became the norm.[22]

Maps Used in the Administration of the Polders

By the sixteenth century, polder authorities regularly commissioned maps for use in the day-to-day running of the areas under their jurisdiction. It was common for the professions of surveyor and dike engineer, or surveyor and dike reeve, to be combined, and this indicates that the polder authorities were fully aware of the usefulness of surveying and mapping.[23] The famous surveyor-engineer Johan Sems, who, with Jan Pieterszoon Dou(w), wrote the surveying manual *Praktijck des Landmeterns* . . . (1600), was also the dike reeve of the area around Bunderneuland (Bondernieuwlandt) in East Friesland, now part of Germany, which had been impoldered in 1605. He drew several cadastral maps of the area between 1605 and 1628.[24]

These manuscript maps were suitable for the small Friesland authority, but as the *waterschappen* of Holland and western Utrecht became bigger, wealthier, and more powerful and complex, they began to commission printed maps. While Holland has nothing to rival the Swedish seventeenth-century *geometriska jordeböckerna* (see chapter 3.3) in terms of completeness or uniformity of conception, the fact that many Dutch maps were engraved and published gives them a unique position in European sixteenth- and seventeenth-century large-scale cartography.[25] There was a tradition of high quality reproduction of books, maps, and charts in the Netherlands by the seventeenth century, particularly in Amsterdam in the northern Netherlands. This publishing tradition was complemented by the need of many polder authorities for a large number of copies of cadastral maps. The administrative structure of the bigger *waterschappen* meant that tens or even hundreds of officials in an authority could work more efficiently if they each had access to maps. In addition, the success of reclamation schemes, whether organized by the *waterschappen* or by individual capitalists, depended on effective publicity, which again demanded the printing of maps. The larger polder authorities were often richer and hence could afford to undertake the comparatively expensive business of printing; especially in the west of the country the wealth of inhabitants ensured a market for such printed maps.[26]

Noncommercial bodies, such as towns or water or polder authorities, employed land surveyors who were responsible for the cartographic content of the maps, but they did not employ engravers or printers. Rather, they sent their surveys to commercial map producers who engraved the plates and printed the maps.[27] An example of cooperation between a polder authority and commercial publishers is that of the famous Cruquius map. In 1697 the *hoogheemraadschap* of Delfland employed the brothers Nicolaes and Jacobus Cruquius (or Kruikius), who were skilled and experienced land surveyors, to survey and map their territory. Work began in 1701. Jacobus died in 1706, and Nicolaes completed the surveying alone in 1709. Cruquius had a modern idea of cartography in that he aimed for technical perfection and wanted to exclude unnecessary decoration. He had to bow to the wishes of the dike reeve and council of the polder authority, however, who wanted their coats of arms and uplifting allegories on the map. Nicolaes Cruquius also organized the engraving and printing of the map. By 1712 a set of twenty-five sheet maps at a scale of 1:10,000 with additional title pages and an overview map at 1:45,000 were complete. This last was not very successful, as Cruquius tried to include so much information that it is almost illegible. The larger-scale maps show all property boundaries, but plots are not numbered or identified according to their owners. Although they are not cadastral maps in the strictest sense, it was intended that each landholder could recognize his property from the maps and that they could be used for the administration of tithes, *verponding*, and other polder dues and for settling boundary disputes.[28]

Maps as a Publicity Medium for Reclamation Schemes

Early reclamation projects were usually small and were often carried out by local people who reclaimed and improved dunes and mud flats belonging to the secular or ecclesiastical lord, who granted permission for land to be improved on condition that he received some of it as domain land. Schemes from as early as the eleventh century had involved surveyors: the regular, often rectangular land plots show that land had been carefully surveyed and parceled out.[29]

Early in the seventeenth century, reclamation activity shifted from *bedijkingen* (reclamation by dikes of coastal mud flats) to *droogmakerijen* (draining of shallow internal lakes under which were tracts of good-quality soils). This was made possible by changes in technology: the wind-driven water pumping mill, invented in the early fifteenth century, came into wide use only in the seventeenth century, notably in schemes devised by the famous engineer Jan Adrianszoon Leeghwater ("empty water"). Between 1607 and 1643 some twenty-seven lakes in Holland were drained and more than ninety square miles of land added to the area of Holland.[30] The first to be drained was the Beemster, southeast of Alkmaar, which was drained in 1612. This was followed by the draining of Purmer in 1622, Wormer in 1626, Heerhugowaard in 1630, and Schermer in 1635.

Ristow attributes this burst of lake-draining activity to the fact that the growing urban population created urgent demands for food and hence for more productive agricultural land. Since about 1550 there had been a steady growth of population combined with the destruction and uncertainty of food supplies during the war with Spain, as well as a decrease in the area of farmland resulting from the extension of peat workings. These circumstances produced a marked rise in food prices, compounded by generally high inflation.[31]

Draining cannot be explained adequately with reference to food prices alone. Those with the capital necessary to undertake land reclamation were untroubled by higher food prices. Rather, they viewed reclamation as a potentially profitable investment for their capital. The profitability of any reclamation scheme depended on the price of newly won land compared with the cost of winning it. The costs of reclaiming land consisted very largely of labor costs, which remained remarkably constant during the seventeenth and eighteenth centuries. Fluctuations in diking activity were thus mainly due to the availability of risk capital and of land suitable for reclamation. The seventeenth century was one of the most important periods of reclamation, since there was a surplus of capital and much suitable land was available. From the sixteenth century the nobility, who had taken the lead in reclamation activity, were replaced by wealthy urban merchants.[32] They had a surplus of capital from trade and manufacture and wanted to diversify their investments. Land still represented a fairly risk-free investment, especially by comparison with long-distance trade.[33]

The merchants' drainage projects were thoroughly capitalist undertakings. Commercial money was invested, and the plots created yielded commercial rents. Settlers on the new holdings were either involved in market-orientated farming to supply the towns or were merchants seeking seclusion from their urban businesses. Maps were used in the drainage projects to help in planning, to interest potential shareholders and buyers in the schemes, to allot the new land plots, and finally to give a pleasing representation of the land to decorate the walls of those who had invested in it. Surveyors even laid out formal pleasure gardens on the polders for wealthy merchants and made drawings and plans of the houses and gardens.[34]

Maps for such reclamation projects were often printed, as the object was to spread knowledge of schemes as widely as possible. An important example of a pre-seventeenth-century printed polder map is that by Adriaan Anthoniszoon (1541–1620) of Zijpe. The map, made on the order of the dike reeve Sebastiaen Craenhals, was printed in 1572 and shows Hontbos and Zijplant with some adjacent lands. It shows the dike system and land parcels and the extent of parcels is listed in an accompanying booklet. It is generally thought that the map served as an illustrated brochure to interest potential investors in the dike. The booklet describes the long process of land reclamation, which had been interrupted by destructive floods like that of All Saints' Day, 1570. The scheme for the new dike and reclamation as shown in the map won the approval of the prince in 1577; in 1597 the Zijpe was finally reclaimed.[35] A further printed cadastral map of Zijpe was made by Jan Dirkszoon Zoutman in 1665. It consists of six sheets at a scale of 1:10,000.[36] Dutch skills in engraving and printing helped make Adriaan Anthoniszoon's 1611 cadastral map of Wieringerwaard at 1:13,000 a useful publicity medium in an early seventeenth-century reclamation scheme. It was drawn while the authorities were coping with a dike break-through. Anthoniszoon was commissioned to survey the area, parcel out land, and print a map to interest buyers in the newly endiked land.[37]

Perhaps the most famous of all the polder schemes was the draining of the Beemster. Dirck van Oss, the administrator of the Dutch East India Company, was the chief instigator of the ambitious plan to convert the 7,000-hectare Beemster into farmland; he organized the financing of the project and engaged Leeghwater for the task. Work began in 1607 and, after setbacks in 1610, was completed by 1612. Noisome activities were excluded from the polder to make it a more pleasant place to live and thus attract high-rent activities.[38] In 1612 it was decided to publish the conditions under which land parcels on the Beemster were to be sold, together with a register of the plots and a map of the area showing all the plots as part of an application for a patent for the project from the States. The first edition was published in 1613 by Hessel Gherritszoon in Pascaert in Amsterdam,[39] and the map of the plots was drawn and published by Gherritszoon in Amsterdam c. 1632. A second edition appeared in 1696.[40]

There were some technical difficulties with the drainage of the Beemster, but financially it was a complete success. By 1640 there were 207 farmsteads on the polder, as well as 141 townsmen's houses and several more elaborate country houses. The polder had been divided into land parcels of varying sizes in an effort to create parcels suitable for all budgets. In the southeast corner close to Amsterdam very small square plots were laid out to encourage townspeople to build country houses and pleasure gardens on the small plots. Dirck van Oss himself had a fine house on the polder. Almost immediately after the polder fell dry, the 123 investors began receiving yearly rental payments which added up to more than a quarter of a million guilders, an annual rate of return of nearly 17 percent.[41]

The resultant landscape is revealed on Balthasar Floriszoon van Berckenrode's map of the Beemster (figure 2.6). Drawn in 1640 at 1:11,500, this map shows all the land plots, each of which is identified and numbered. Fockema Andreae considers it one of the finest of such maps and the apogee of early Dutch polder cartography.[42] It was engraved in Amsterdam in 1644 and published c. 1646. The printed map is very elegantly presented with a Latin poem which extols the virtues of the Beemster and a border of coats of arms; these do not intrude on to the body of the map, however, which is uncluttered and easily read. Extra copies were printed as early as 1660. In 1721 new coats of arms were added to the map, and it was republished.[43]

The success of the Beemster reclamation encouraged further schemes. For example, Dirck van Oss with other wealthy Amsterdammers and the *burgemeesters* (mayors) of Edam and Monnikendam combined to finance the drainage of the Purmer lake in 1617. The scheme was a suc-

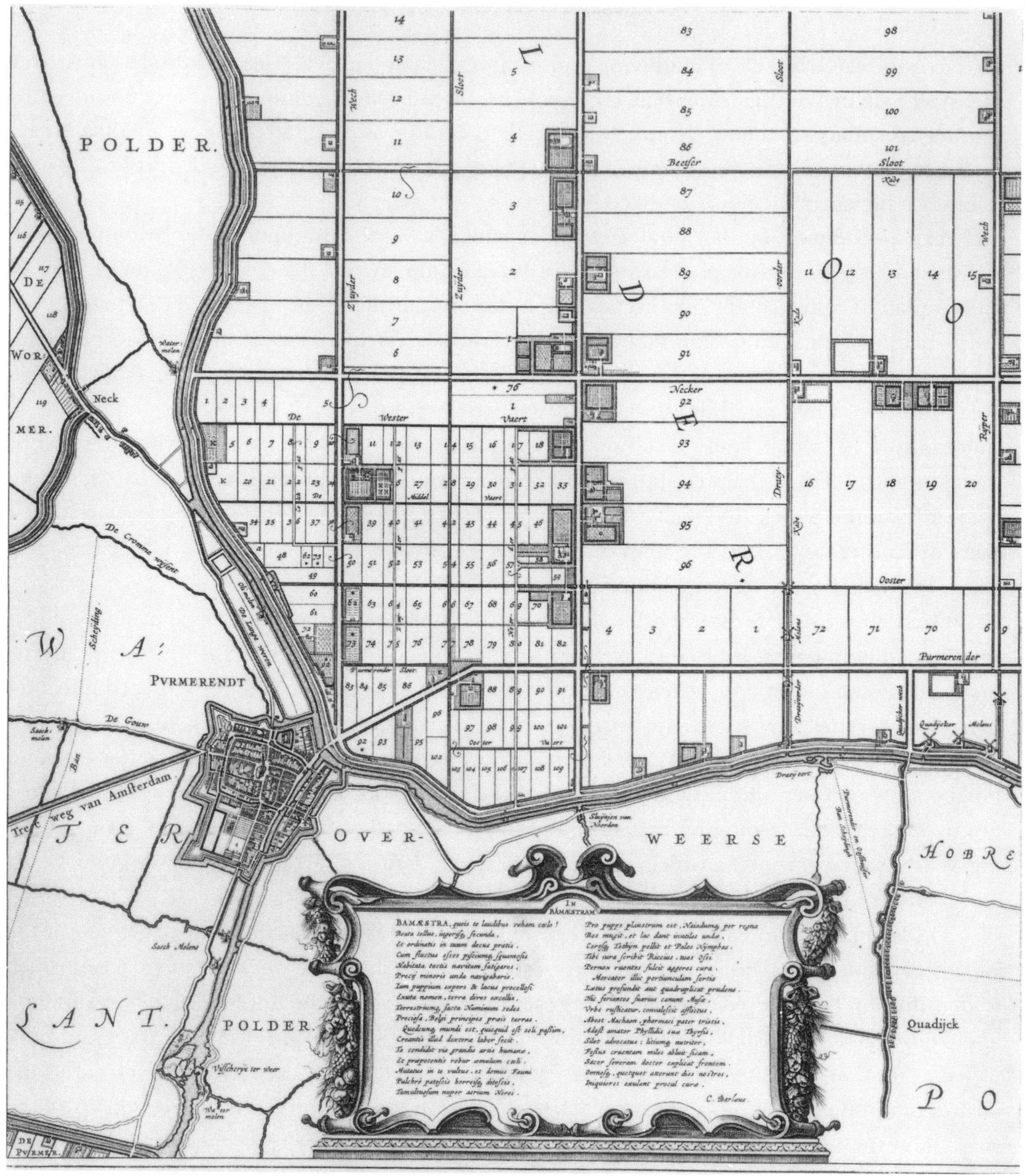

Fig. 2.6 Map of the Beemster drawn in 1640 by Balthasar Floriszoon van Berckenrode and engraved in 1644 by Daniel van Breen at a scale of 1:11,500. Note the small plots for pleasure houses in the part of the polder nearest the town and the larger farms at a distance. Source: de Vries 1983. Photo: Algemeen Rijksarchief, The Hague.

cess, but the draining of the Heerhugowaard which followed in 1625 was not so successful, perhaps because the entrepreneurs were in too much of a hurry to conduct a proper survey before work began. The many technical problems which were encountered were expensive to rectify. By contrast, the last of the big lake drainage schemes, that of the Schermer, was technically and commercially successful.[44] A map of the Schermer, surveyed and mapped by Pieter Wils and engraved by Salomon Rogiers, was published in 1635.[45]

Despite the success of such schemes, investors found that they could get a higher return on their capital if they invested it abroad. After about 1640 little capital was thus forthcoming for further drainage schemes. There was also relatively little suitable land left. Because of these factors less land was drained annually between 1650 and 1775 than at any time since the Middle Ages.[46]

The Training and Legal Status of Surveyors

Perhaps the Dutch were more aware than others of the importance of codifying rights to land, since the land area of the Netherlands was constantly fluctuating. Inundation by natural floods and storms and by deliberate strategic flooding of the land in war reduced the value of land or destroyed it entirely. Reclamation created new land and hence new title to land but at the same time often damaged others' rights to fish or to gain passage by sea. Because of the need to codify changing land rights, references to surveyors are widespread in the Netherlands from as early as 1300.[47] In the early modern period there was a change in the position of surveyors: they were no longer government officials in the strict sense of the word. They had to demonstrate their competence and be approved by the authority concerned, but they were essentially freelance workers. The method of admitting surveyors evolved gradually, but by the middle of the sixteenth century the *hoogheemraadschap* of Rhineland already had an established procedure. Since there was no central authority in the Netherlands, each authority had different regulations for examining and admitting surveyors. However, in practice there was a great deal of overlap between them. The innovation of the sixteenth century was the emergence of surveyors with a standardized repertoire of skills, including mapmaking. A sworn surveyor normally had to pass tests to show his knowledge and then had to take an oath that he would be honest and impartial in his profession. It has been estimated that in Friesland there was on average one surveyor to every five thousand people between the sixteenth and seventeenth centuries. This extraordinarily high figure shows the importance laid on surveying, but also the fact that surveying was largely a by-employment. Surveyors had a wide variety of other occupations, including those of notary public, solicitor, stonemason, carpenter, schoolmaster, and architect. The nonspecialized nature of surveying, combined with a lack of a central authority to set standards, meant that uniformity and standardization of surveying and mapping were impossible to achieve.[48]

The first formal training for surveyors was instituted in 1600 at the University of Leiden; later it was also taught at the University of Franeker. Most surveyors, however, continued to learn their craft as apprentices to established surveyors, a pattern which changed little until the rise of the engineer in the second half of the eighteenth century.[49] However, seventeenth-century technical advances in the related fields of dike and sluice construction, architecture, town planning, and navigation meant that the knowledge expected of surveyors became both more advanced and broader in scope, and admission as a sworn surveyor was increasingly recognized as a guarantee of surveying competence.

Maps are not necessarily neutral conveyors of information but can be wrong, partial, or indeed deliberately misleading. In land title disputes a surveyor was very often called upon to give an objective account, and the fact that he was a sworn or admitted surveyor gave the commissioner of the map a guarantee that it accurately represented the situation in the field. A map drawn by a sworn surveyor was a legally binding document. In some cases, if a surveyor miscalculated the extent of land which was being sold and thus caused the buyer or seller to lose money, he could be personally liable for the loss.[50]

2.2 State Revenue

Rents and taxes on land were the chief source of income for individual proprietors and official bodies of many types from polder authorities to religious authorities to state bodies. Surveying and mapping were used from a very early date to consolidate rights to, and improve the collection of, rents and dues. A very early example is a picture map of 1350–58 of the tithes of a newly won area of polder land in the districts of Oostburg and IJzendijke near the coast of southern Flanders west of Antwerp. A bishop and an abbot each claimed rights to the tithes, and this map, which shows clearly the plots on which tithes were due, was used in the court case called to settle the dispute.[51]

Mapping was not the norm at this date, however, but it became increasingly important in the sixteenth century under Charles V. As count of Holland, he sought to exploit to the full his rights to rents and taxes on land to pay for his wars and growing bureaucracy. Such exploitation was, of course, controversial, and surveying and mapping proved useful weapons in his efforts. One example of this process is his revision of the *Hondsbossche-contributie* in 1533 (see section 2.1). Another is his tackling of the problem of the *vroonlanden* (domain lands belonging to the counts of Holland) in North Holland, especially north of the town of Alkmaar. The *vroonlanden* consisted of many hundreds of land parcels, large and small, leased out to many different people. It was difficult to exert close control on the income derived from this land, and the position was made more complicated by the fact that many tenants sublet their holdings. Complicated subletting arrangements made the counts suspicious that their tenants were making money at their expense. When Emperor Charles V became count of Holland, he determined to investigate the position and appointed Adriaan Stalpaert van der Wiele as treasurer to the counts in 1523. Stalpaert admitted that he had no accurate knowledge of the extent of the land, only of the income derived from it, which had varied considerably. A commission was set up under Stalpaert to address the problem. Between 1529 and 1531 Stalpaert cataloged all 1,300 land parcels with the help of the monk Pieter Jacobszoon. Together they traveled to the villages concerned, investigated each land plot, and fixed its boundaries. The surveying was carried out by sworn surveyor Maarten Corneliszoon from Rhineland with the help of Pieter Jacobszoon and later Simon Meeuwszoon. Corneliszoon delivered two maps to the Treasury, and a copy was made of part of them in 1532. The original maps show land parcels, but the copy does not, although colored tints indicate different land use (figure 2.7).[52]

The work aroused considerable controversy; Stalpaert received death threats from the villagers, who realized how much they stood to lose if the Treasury gained an accurate picture of the *vroonlanden*. Their suspicions were fully justified. After the survey the lands were newly leased, and the Treasury received far higher sums in rents than it had before. In addition the survey was used in the levying of various taxes due to secular lords and to churches or church officials, and in the levying of charges to pay for the maintenance of dikes, mills, and polder reservoirs.[53]

Another early rental map is that of the Biesbos (Biesbosch) near Geertruidenberg. The Biesbos was surveyed and mapped in 1562 on the instruction of the Treasury in The Hague by Pieter and Jacob Sluyter, important surveyors in Holland between 1540 and 1570.[54] The map is at a scale of 1:12,000 and shows individual landholdings; the accompanying register lists the tenant and the extent and use of each land parcel. The map and register were used as the basis for levying rents.[55]

In the course of the Reformation, which affected the northern Netherlands between about 1570 and 1600, there was debate not only about spiritual matters but also about the practical and administrative form the reformed church should take and about what should happen to church property. In the end, nonmonastic churches kept much of their property for the upkeep of church fabric and the relief of poverty. By contrast, monasteries were dissolved and their property taken over by the state, which disposed of their new lands in a variety of ways.[56] The beneficiaries often commissioned a cadaster of their newly secularized land to aid administration and rent collection. One such cadastral map book is that of the Bailiwick of Naaldwijk made in 1620 by Floris Jacobszoon. The precise reason for the cadaster is not clear, but it seems likely that, since part of the land had been recently acquired through secularization, the new heir, Frederik Hendrik of Orange, later leader of the United Provinces, wanted to gain an accurate picture of what lands he had and what rights he and others had to levy dues on them. The resultant maps include

Fig. 2.7a (left), 2.7b (opposite) Maps of *vroonlanden* near St. Pancras en Oudorp, Langendijck, and Koedijk en Huiswaard begun by Maarten Corneliszoon and Pieter Jacobszoon and completed by Simon Meeuszoon and copied by Cornelius Buys in 1532. The maps are at scales of 1:3,000 to 1:5,000. Source: Algemeen Rijksarchief, The Hague, VTH 2507 and 2508.

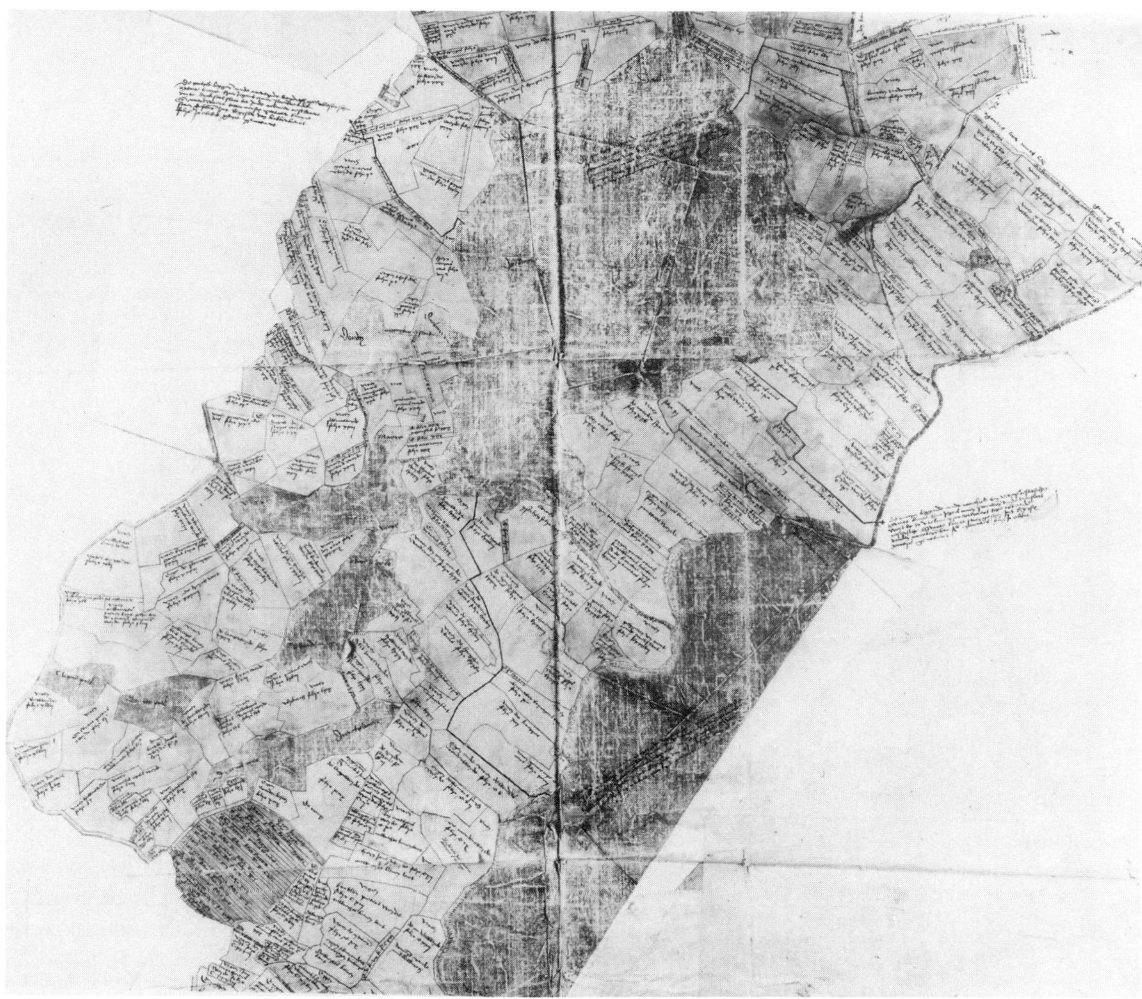

detailed cadastral maps at a scale of 1:6,500 (figure 2.8) and a noncadastral overview map at a scale of 1:17,000. The maps were used in general administration of the estate, for the collection of rents, and for the payment of the *verponding*.[57] Soon after the secularization of the nunnery at Rijnsburg, near Leiden, Jan Pieterszoon Dou(w) was commissioned to draw up very precise plans of the estate lands of the nunnery. A folio of maps was produced between 1626 and 1635, which survives, although the cadastral register which accompanied it is lost (figure 2.9).

Some of the earliest cadastral maps were drawn by polder authorities for the levying of dike taxes (section 2.1). Other taxes, however, were designed to produce general state revenue. Of these the land tax levied in country areas was the most important, although some states like

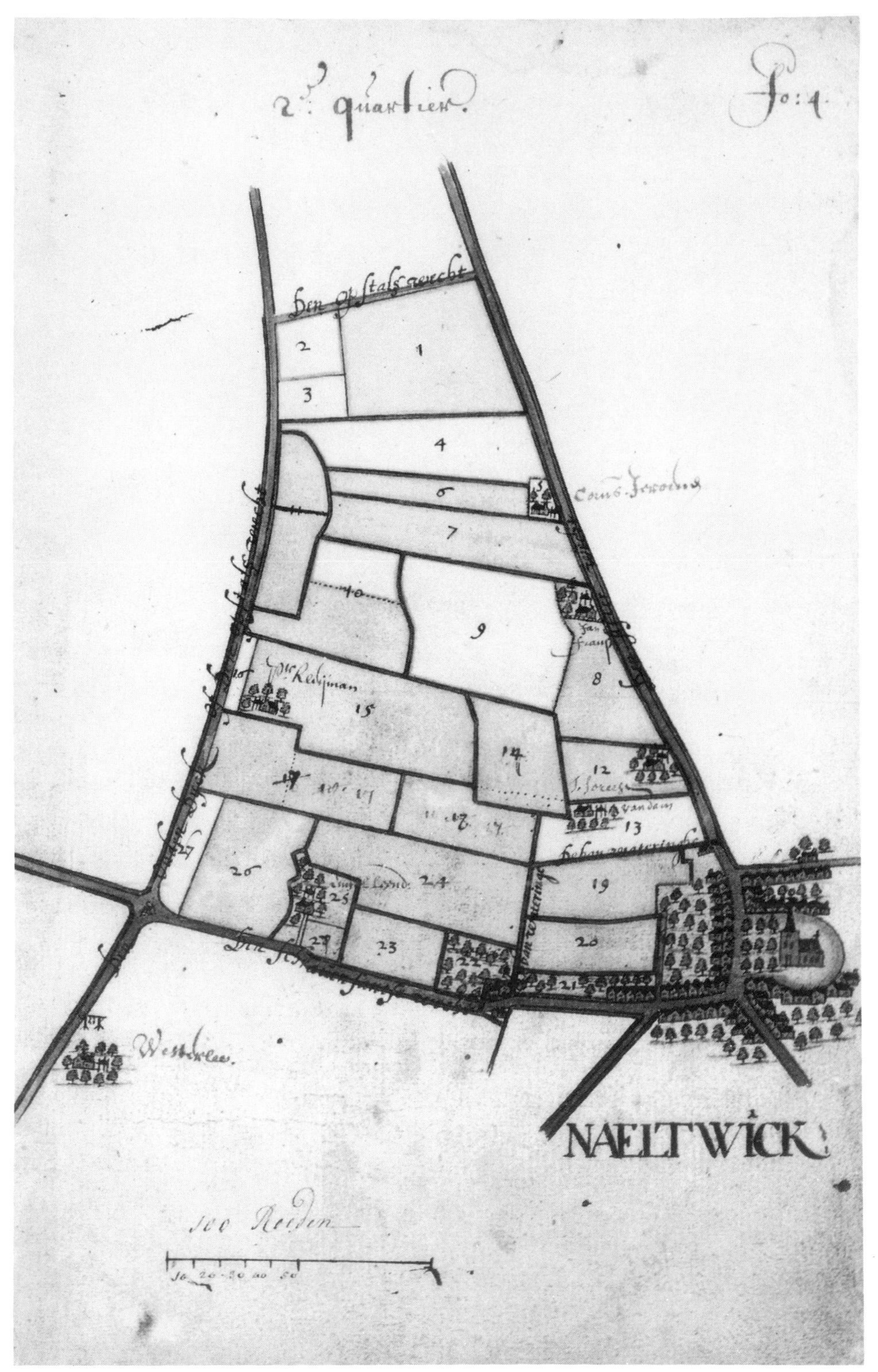

Fig. 2.8 (left) Map 9 from the atlas of the Bailiwick of Naaldwijk (Naeltwick) made in 1620 by Floris Jacobszoon at a scale of 1:6,500. Source: Kok 1985.

Fig. 2.9 (opposite) Map of the lands once belonging to the Abbey of Rhijnsborch (Rijnsburg) in Rhineland by Jan Pieterszoon Dou(w), between 1626 and 1635, with later alterations. Source: Algemeen Rijksarchief, The Hague, VTH M folio 3, "Rhijnsborch and Noortwijck."

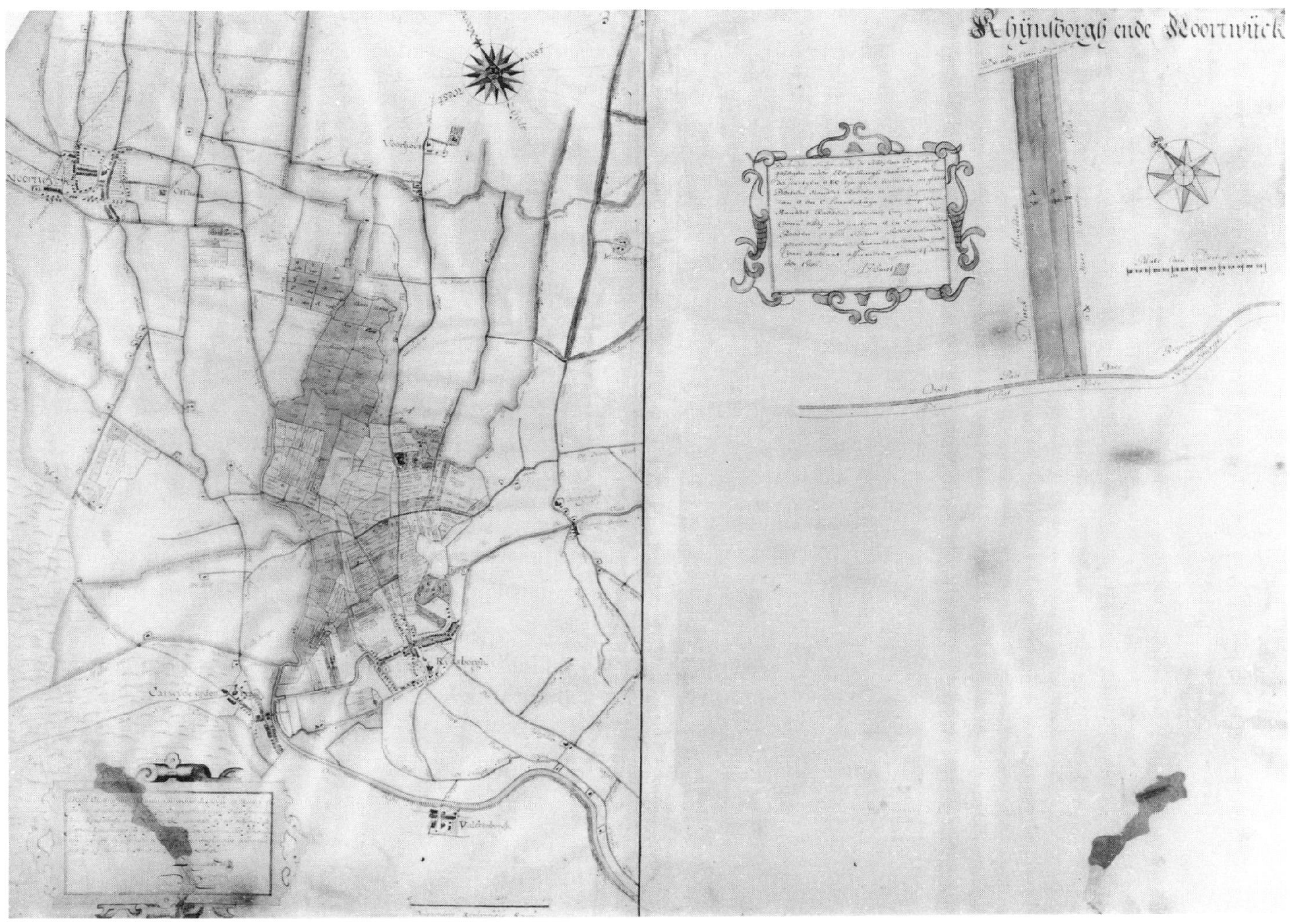

Holland were very highly urbanized and generated much of their wealth from commerce rather than agriculture.[58] Before the *Code Napoléon* was introduced in the Netherlands in 1811, tax on immovable property was administered differently in each local area. Early written cadasters, with such names as *ommelopers, maatboeken, morgenboeken,* and *everingsboeken,* record land and water taxes and were in some isolated cases accompanied by maps.[59] Since there was no uniform land tax, there could be no uniform technique of assessment; in some cases the literature gives an impression of inertia and failure to reform. Neither surveying nor mapping was carried out for the apportionment of some land taxes, such as that levied in the province of Gelderland in the southeast of the country in the mid-seventeenth century. Assessment continued to be based on written information on landownership and tenure and rent and other dues payable on the land.[60] Similarly the land tax in Friesland, the *Floreentaux,* underwent no alteration between its inception in 1511 and the introduction of the French cadaster at the beginning of the nineteenth century, despite riots over it in 1748. In Holland the land tax was known as the *verponding.*[61] It was regulated in 1632, but in the century after 1650 it pressed ever more heavily on the agricultural sector as prices fell while the tax remained fixed at its 1632 level. Demands for reassessments were met finally in 1732, but only houses were reassessed. The land tax remained unchanged until the introduction of the French cadaster (section 2.4).[62]

The impression of universal inertia is, however, misleading. Although comprehensive reform in many areas had to wait for the arrival of the French, surveying, and particularly comprehensive, parcel-by-parcel surveying, was gradually accepted as the best means of apportioning tax liabilities fairly. Survey was initially undertaken to resolve specific local dissatisfaction, but there was a more general movement for reform by the eighteenth century. In 1602, for example, the state of Drenthe, in the northeast of the country, decided on a reform of taxation based on assessment of the quality of the land. Surveying was to happen only in doubtful cases, and as landowners often bribed the surveyors not to survey their land, much escaped measurement. The results of the exercise were so unjust that in 1615 a new and general survey was ordered, which was followed in 1640–41 by a general survey and *verponding.*[63]

Local mapping followed local surveying, although country areas seemed to lag behind the towns. As early as the sixteenth century Leiden had not only a house ledger but also a *kaartboek* (map book) of all properties within the city walls.[64] A recently rediscovered cadastral map is that drawn in 1664–65 of the *heerlijkheid* (manor) of Warmondt (Warmond), near Leiden, by Johan Dou(w) the elder with the help of his son Johan Dou(w) the younger (figure 2.10). It covers the whole of the manor, which contained some 980 land parcels, and was commissioned as a result of a tax dispute. The apportionment of the *verponding* and of a local village tax had become so unfair in the course of time that two local people petitioned the States of Holland for an investigation. Two *verpondings* lists for 1583 and 1632 existed, but the situation had become so complex—not least because some land had recently been reclaimed and was granted temporary exemption from taxation—that a new and thorough investigation with a map was felt to be the most appropriate way to resolve the dispute. The survey took place between 1663 and 1664, and by 1665 the map and a register had been completed and checked by a notary. The original maps are lost, but copies—a wall map of 1667 and a map book of 1669—have survived.[65]

Another example of a mapped survey was that in 1750 of the *heerlijkheid* of Heerlo in Brabant by Adan Blom (Adam Blum). The *boenderboek* (written register) is accompanied by a *kaartboek* and a tax schedule.[66] The Adan family, surveyors to the marquis of Bergen op Zoom for more than one hundred years (1739–1840), drew maps to go with the revised *cijnsregisters* (registers of feudal levies due to the marquis) in the mid and later eighteenth century, as well as maps of new

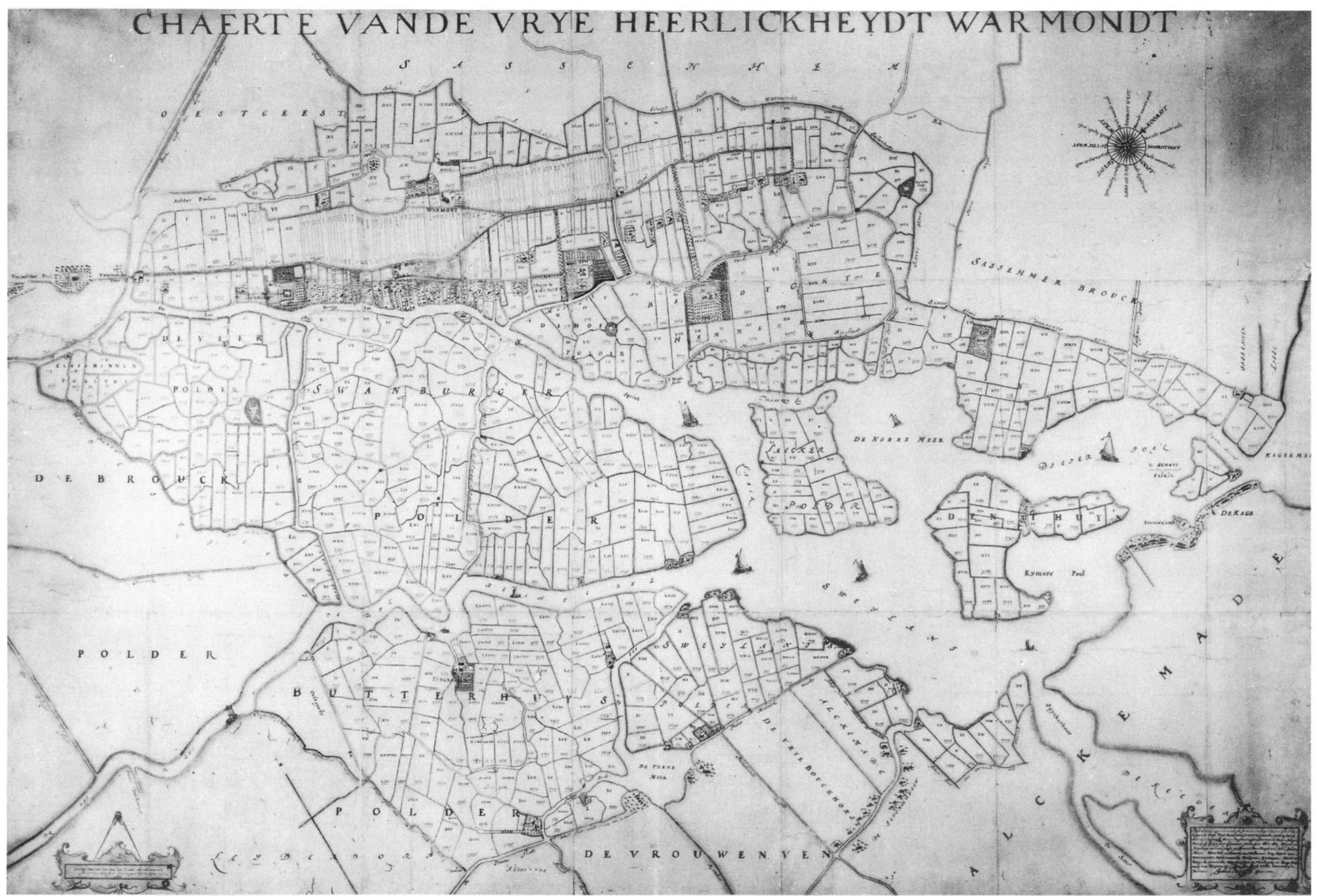

Fig. 2.10 Map of the manor of Warmondt near Leiden for tax apportionment. The original maps of the manor were drawn by Johan Dou(w) the elder with his son Johan Dou(w) the younger in 1664 or 1665 but are now lost. Two seventeenth-century copies survive. Source: van der Steur 1985.

and extended tree plantations on the wastes of the marquisate and numerous maps of dikes, mud flats, and polders. Copies of these very high quality maps were later used for the collection of the *verponding*.[67] Piecemeal local revisions and use of existing maps, however, could not meet the general need for reform, and more general attempts at revision in Brabant and particularly Holland (section 2.3) were made at the end of the eighteenth century.

In the Meijerij, part of the Duchy of Brabant, a new tax on immovable property, the *verponding*, was instituted to replace the old poll tax, or *bede* (*beeden*), after the Peace of Westphalia in

1648. In order to apportion the *verponding* justly, it was decided to carry out a description and assessment of each property. These details were to be recorded in *boomboeken* from which the *verpondings* registers were to be compiled. There was no legal requirement that the *boomboeken* be kept up to date, and no personnel were assigned to the task. It is not surprising, therefore, that the tax burden on individuals and villages came to be unjustly apportioned. The situation became so inequitable that the taxpayers of the four quarters of the Meijerij (Maasland, Cisterwijk, Peelland, and Kempenland) petitioned for reform in 1786. In response, the Council of State set up an investigation on 31 December 1787 under the auspices of the Leen en Tolkamer (Loan and Duty Department) in the town of 's-Hertogenbosch. Caspar van Breugel was in charge of the investigation and began his inquiries with the *verpondings* registers. According to the *regenten* (the local representative of the ruling prince), in some places the registers were fairly accurate, but in others discrepancies were marked. Some areas had had as many as three surveys since 1648, but others had had no systematic survey at all. It was decided that attempts to amend the old registers in a piecemeal fashion would be useless. Accordingly, on 7 October 1791 a new general survey was ordered by the Council of State, and van Breugel began to engage surveyors to survey land plots, make maps showing the plots identified by reference numbers, calculate the area of the plots, and record the information in cadastral registers. The surveyors were to measure only arable land, since houses and meadowland were not subject to the *verponding*. When the survey was complete, the land was assessed. By 1794 all the surveying and assessments were complete, needing only the approval of the Council of State to bring the tax into force. The revolution of 1795 intervened, however, and the tax was not implemented.[68]

Koeman and Hessels state that surveyors were engaged on the Meijerij survey with instructions to map land parcels as well as to compile a land register, and Hessels reports that maps were collected and used by surveyors. Scheffer, however, considers it "highly probable" that land was simply surveyed and that no maps were drawn.[69] The overlap of personnel in the Meijerij, the *Hollandse* cadaster (section 2.3) and the first national cadaster (sections 2.4 and 2.5), and the mention of mapping in the instructions for the Meijerij survey make it seem probable that maps would have been drawn had not the political crisis of 1795 intervened.[70]

2.3 The *Hollandse* Cadaster (1795–1811)

In 1795 the Batavian republic was established, and it was agreed that state revenues, then collected in varying ways in the different provinces, were in need of reform. Isaac Jan Alexander Gogel was put in charge of state finance and on 1 May 1798 organized the pooling of debt from the various provinces to form the national debt. He also planned a new, comprehensive, and uniform taxation. Part of the money was to be raised by a land tax which was to be levied directly by the state and not farmed out. By 1800 a plan for the reform of tax was ready, and on 25 March 1801 a bill was introduced. A tax of ten million guilders on all land was approved. It was to be levied in proportion to the value of rents received from each property during the preceding ten years. Political circumstances in the republic made it impossible to put this law into effect, and not until 14 July 1805 was Gogel able to introduce a *verpondings* law and set up a commission to organize surveying and mapping.[71] Several thousand *verpondings* maps were drawn under the 1805 law.[72] Plans for taxation reform were encouraged by Louis Bonaparte, who came to the throne in 1806 and was inspired by the cadaster begun in France in 1807.[73]

There were disputes about the level of taxation and how it was to be apportioned, but the bill for the imposition of the land tax was finally approved on 20 January 1807.[74] It followed the Dutch system of basing the land tax on the income that the owner-user or owner-landlord derived from the land, rather than on the French system of basing the tax on the value of a property.[75] It was agreed that the entire kingdom would be surveyed and that this survey would be the basis on which land parcels were assessed and valued for taxation. It was estimated that the tax would raise between forty and fifty million guilders, whereas in fact it brought in no more than eleven million guilders. The tax was to come into effect in 1815 based on the average rent gained from the land in the preceding ten years, and a revision was to take place every fifteen years after 1815 based on the rent from the preceding fifteen-year period. All changes in ownership, division of land parcels, and raising of mortgages were to be reported to the cadastral authorities so that the registers could be kept up to date. A High Commission was set up to organize the work under Casper van Breughel, who had been in charge of the surveying for the *verponding* in the Meijerij. Lieutenant Colonel Gillis Johannes le Fèvre de Montigny (born 's-Hertogenbosch) organized the surveying, which was to be done by a regiment of soldiers. Articles 25–31 of the act of 20 January 1807 specified that only qualified surveyors were to be employed and that they were required to carry out their work impartially. Uniformity is stressed throughout the instructions; surveyors were required to use a uniform measure, their equipment was to be checked to see that it conformed to a common standard, and maps were to be drawn to a uniform scale of 1:2,880.

In the papers of the surveying family Adan, Scheffer discovered two model maps and a model survey book issued to surveyors employed on the *Hollandse* cadaster to ensure uniformity of presentation.[76] The model map was drawn up by Le Fèvre de Montigny and includes much topographical detail, reflecting his background as a topographical engineer-surveyor. Van Breughel, however, the commission's director, considered this detail costly and irrelevant to the work in hand. He wanted to show only property lines on the maps and leave other details for the accompanying descriptions. His view prevailed, and the model map was not used. To save time

Fig. 2.11 Map of lands in the second *arrondissement* (district) of Leiden made for the *Hollandse* cadaster. Source: Algemeen Rijksarchief, The Hague, AANW 1932 III.

on the survey it was decided to use as many preexisting maps as possible as a basis for the work and to leave out of the survey wastes, rivers, and seas. A central drawing office was set up in the Dépôt-Generael van Oorlog (War Office), the forerunner of the Topografische Dienst (Topographical Service), and this led to the separation of the operations of land survey and map drawing and to increased standardization both within the cadaster and between cadastral and topographical maps.

In just three years 375 surveyors worked on the cadaster under the supervision of 37 army officers, who acted as supervisors and controllers.[77] Survey work was guided by the manual *Mode d'arpentage pour l'impôt foncier,* which specified that surveyors were to work in teams of eight to ten men under a supervisor. Two different methods of surveying were to be used. Where land was fertile and plots were regular in size and of considerable value, the survey was to be conducted *selon la nature;* that is, external boundaries were to be fixed first, and plot divisions were to be marked on the map subsequently. The resulting map and calculations of area were to be checked with the supervisor and landowners and then submitted to the commission for further examination. Where land was sandy and parcels were irregularly shaped and less valuable, surveying was carried out parcel by parcel. Scheffer suggests that this difference in method might be due to the fact that almost all the fertile, regularly divided, and valuable land in the country was polder land, which was already covered by polder authority maps which showed parcel divisions. Sandy land was not covered by such maps, and so the delineation of plot boundaries was a far more onerous part of the project in these areas.[78]

While the work for the *Hollandse* cadaster was underway, the French annexed the country in 1810. In 64 of the 111 districts work was complete, but the French had scant regard for the *Hollandse* cadaster, as it was not based on triangulation, the scale of the maps was not compatible with their own, and it was not sufficiently comprehensive or uniform. They abandoned the work and began afresh.[79]

The maps and registers from the *Hollandse* cadaster are considered by Koeman to be of relatively little importance as a historical source, although they are some twenty or thirty years older than the Napoleonic cadaster, as they are incomplete and the cadaster was poorly planned and hurriedly executed.[80] Others, however, disagree. Scheffer admits that the *Hollandse* cadaster cannot compare in quality with the subsequent French cadaster and that it in fact never came into force but feels that it has been rather unjustly dismissed, as some potentially if not actually useful surveying, mapping, and tax assessment were carried out (figure 2.11).[81] Muller and Zandvliet point out that the survey was intended as part of a national surveying and mapmaking exercise in which the wastes were to be mapped by army officers, the rivers and coasts by water authorities, and the fields by cadastral surveyors.[82] It was thus one of the first attempts at nationwide mapping by a body of surveyors working with uniform regulations.[83]

2.4 The French Cadaster (1811–13)

The French annexed the Netherlands in 1810. There, as in the rest of their territory, they considered it important to revise the collection of land tax by drawing up a new and accurate mapped cadaster (chapter 6.12). French land law applied to the seven *départements* of the Netherlands from 21 October 1811. Under this law, land taxation was levied on the yield and not on the rental value of immovable property. The maps and registers of the *Hollandse* cadaster were not appropriate to this system, which demanded a comprehensive new survey and valuation. After some early surveying in the present-day province of Limburg in 1807, the project began in 1811, following the official instructions of the French cadaster known as the *Recueil méthodique . . .* , which was published in Dutch in Amsterdam in 1812. For the fixing of assessments immovable property was divided into fifteen classes for houses and five classes for arable land. The taxation system came into operation in 1811 and remained in place until 1976.[84]

2.5 The Netherlands Cadaster (1813 Onward)

Two years after the start of their cadaster the French retreated from the Netherlands. Work on the cadaster stopped, and there followed a period of indecision as to whether it should be continued and, if so, what methods were to be used. The project began again in the southern Netherlands in 1814 and in the northern Netherlands in 1816, but problems concerning such details as the method of surveying and the scale of maps were resolved only by later decrees.[85]

The method of survey and the personnel involved in the project in fact changed very little, despite the political turmoil (figure 2.12). Maps were drawn on sheets measuring 65 cm × 100 cm at a scale of 1:1,250 in areas where there were more than five land parcels per hectare, 1:5,000 in areas where parcels were on average larger than five hectares, and 1:2,500 in all other areas. The maps are island maps of cadastral sectors, which were subdivisions of the *gemeente* (the smallest administrative unit, civil parish). More than seventeen thousand of these maps were drawn over a period of twenty years, and about one hundred surveyors were employed on the project. Unfortunately their field notes have not been preserved. Land parcels are numbered on the maps to correspond with the number scheme of the *oorspronkelijke aanwijzende tafels* (cadastral ledgers), which contain the name, place of residence, and occupation of landowners as well as the extent, land use, quality, and yield of the land parcels. When the surveyor had drawn up the register, it had to be agreed by the local landowners and mayor. Each plan shows all or part of a cadastral section; in addition, *verzamelkaarten,* or summary maps, show whole cadastral *gemeentes* at a scale of 1:5,000, 1:10,000, or 1:20,000. *Gemeente* boundaries were carefully surveyed and fixed in a legally binding court case before surveying of land parcels started in what was the earliest codi-

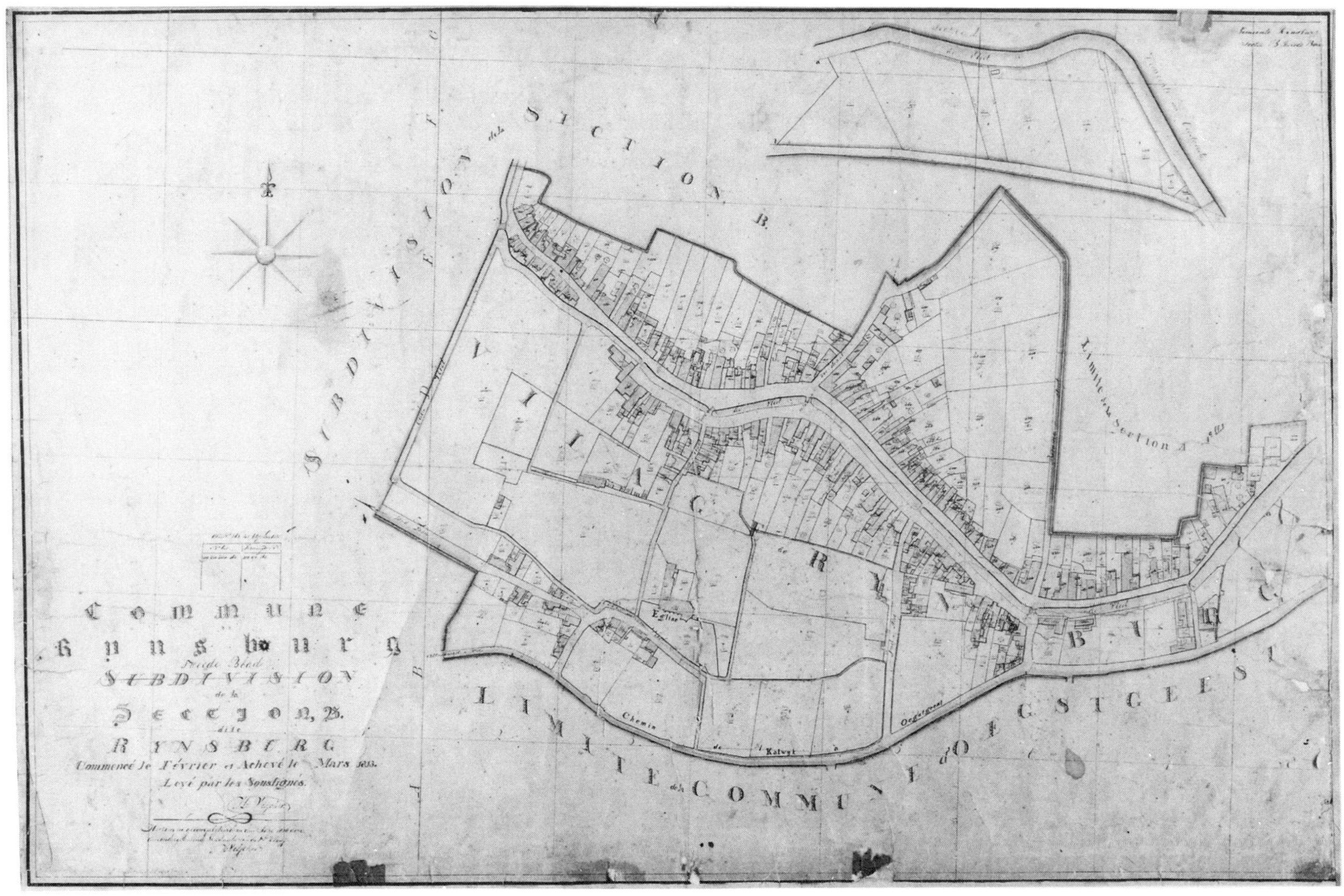

Fig. 2.12 Cadastral map of Rijnsburg compiled for the French administrative authorities in 1813 and revised by the Dutch in 1818. Source: Algemeen Rijksarchief, The Hague, KADOR G B2.

fication of administrative boundaries in the Netherlands. The completed plans, called *minuutplans,* with the written registers and other relevant documents, were delivered to the office of the *ingénieur-vérificateur* (examining surveyor), where the extent of the land parcels was calculated and two copies of the plans were made: these are the *netteplan* and the *gemeenteplan*, which went to the local administrative authorities for their use. The cadaster was organized by *département* under the French administration and later by province.[86]

Surveyors were paid piece rates for their work at a rate of twenty-five *centimes* for each parcel surveyed. It is not surprising, therefore, that the first survey showed more parcels than actually existed. From 1813 to about 1840 the cost of compiling the cadaster was a very heavy burden on precarious state finances, but the government had no option but to persist, since it was dependent on the land tax for all of its direct income. It was hoped that the survey would be finished by 1828, but it was not completed until 1831 in the Netherlands and 1835 in Belgium.[87]

The Duchy of Limburg, now in the far southeast of the Netherlands, formed a special case. When the Belgians broke away from the Kingdom of the United Netherlands in the 1830s, Limburg joined Belgium, although the area around the fortress of Maastricht was effectively under Dutch control. The Netherlands cadaster was introduced in the Maastricht enclave in 1837 and in the rest of the duchy between 1841 and 1844, after Limburg had become to all intents and purposes a province of the Netherlands in 1839, and subject to Netherlands law from 1 January 1842.[88]

Revision of the Cadaster

The Dutch cadaster was to act both as a land register and as the base on which land tax was to be levied.[89] It was realized early on that the cadaster could fulfill neither role unless it was kept up to date, and rules governing its revision were issued in the *Verordening ter instandhouding van het kadaster* of 1832. It was prescribed that when changes were made to the cadaster, additional sheets would be added, but that the original document should not be altered. The original documents thus form a very useful historical source, since they are a complete and unaltered cross section of early nineteenth-century landownership. When changes in landownership occurred, the alteration was noted, and a *hulpkaart* (auxiliary map) of the affected area was drawn. Nevertheless, in the nineteenth century the cadaster deteriorated both because of the lack of a solid geometric basis for surveying and because many changes in landownership, especially where land was inherited rather than sold, were not reported to the cadastral authorities. It was decided that the situation could be improved only by initiating a new cadastral survey controlled by the established national triangulation network. Work began in 1863 in Boxmeer, south of Nijmegen, and in Assen, south of Groningen, but progress was so slow such that by 1950 only 13 percent and by 1978 only 30 percent of cadastral plans had been resurveyed. Fortunately for historians and historical geographers the old cadastral documents were not discarded as resurveying took place.[90]

Use of Cadastral Plans in the Construction of Topographical Maps

The earliest example of cadastral plans being used in the construction of topographical maps was in the mapping of Drenthe by French *ingénieurs-géographes* between 1811 and 1813. Koeman considers these topographical maps, which are at a scale of 1:20,000 and which depict land use, one of the finest cartographic achievements of the French period in the Netherlands. The later military topographical survey, begun in 1815 in the southern Netherlands and extended to the northern Netherlands in 1830, also used cadastral maps as a source. Cadastral maps were also used by civil surveyors working under the auspices of the *ingénieurs-vérificateurs* to compile provincial topographical maps in the middle of the century, in the compilation of the military topographical survey of the Netherlands in 1834–60, and in the compilation of various nineteenth-century water authority maps.[91]

The Southern Netherlands

Early forerunners of state cadasters in Belgium include, as in many other countries, both litigation maps, concerning disputed land boundaries, land rights, or tithes, and private estate maps. Disputes over tithes led in 1639 to the drawing of a tithe map of Bavikhove in western Flanders, northeast of Courtrai (Kourtrijk), by Lowys de Bersacques. A map of lands on which the abbey of Grooten-Bygaerden, northwest of Brussels, had rights to tithe was drawn by J. D. Dekens in 1734–35.[92]

Secular and religious landlords commissioned estate maps from the sixteenth century onward. Early examples of church estate maps are those of 1652–64 of the estates of the abbey of Parc-Heverlee south of Leuven (Louvain) and maps made between 1647 and 1657 of properties belonging to the Jesuits of Brussels.[93] Of greater significance than these in the development of state-sponsored property mapping are the cadastral maps of Duke Charles de Croÿ which date from the late sixteenth to the early seventeenth centuries. These are in one sense estate maps, since they are of the family possessions of the duke, who was engaged in a program of restoration and improvement of his estates and wanted information to carry this out. His estates were so extensive, however, that the registers and the 230 surviving maps cover wide tracts of land in the southern Netherlands and northern France on which the duke had the right to collect *cijnzen* (feudal dues). The maps are thus in another sense forerunners of state taxation cadasters. Charles ordered the making of a cartulary of feudal dues and rents of his lands and entrusted the work to the surveyor Pierre de Bersacques and two of his sons. The Principality of Chimay was surveyed in 1593–98; the Baronies of Aarschot, Bierbeck, and Heverlee in 1597–98; the Barony of Rotselaar in 1600; Beveren-Waas in 1602; the county of Beaumont in 1604; and the seigneuries of Blaton, Wallers, and Avesnes in 1606. The duke sent Bersacques all preexisting registers and maps of the estates, and thus armed, the surveyors went out into the field and, with the assistance of local officials, surveyed and gathered material from which they compiled registers and drew maps. Each plot is numbered on the maps, and the numbers correspond to descriptions in the registers. Relations between the surveyors and the duke were sometimes turbulent, as their surviving correspondence reveals, but the resultant maps are beautiful works of art and also provided a solid basis for the administration of the estates and the collection of dues.[94]

2.6 Early Taxation Cadasters

The Napoleonic cadaster, often referred to in Belgium as the first cadaster, was certainly a revolutionary and pioneering project because it was a comprehensive, scientific, *national* cadaster founded on the political principle that all citizens were equal and hence equally liable for taxation. It was, however, far from being the first cadaster in Belgium, as there were numerous early large-scale cadastral maps which are little known, not least because many of them have been lost or destroyed.[95]

As might have been expected, it was in polder areas of Belgium that the earliest and most carefully surveyed and mapped cadasters are to be found. In these areas, with their very valuable and densely populated land, the first mapped cadasters were produced for levying dike taxes.[96] Polder maps such as these have been discussed in section 2.1 above. It would be wrong, however, to suggest that early cadastral surveying and mapping in the southern Netherlands were confined to polder areas.

Brabant

Surveyors were active in Brabant and Brussels from as early as the thirteenth century. They were called *gesworen paelder* (sworn assessors), and they apportioned taxes according to the possession of land. Between the sixteenth and eighteenth centuries as many as fifty surveyors worked in West Brabant and came increasingly under official scrutiny and control. The first document regulating the activities of surveyors dates from 1451, and revision of the regulations followed in 1657 and 1705.[97]

Land taxation in Brabant can be traced back to 1480 and 1496, and revisions of the tax were made as early as 1525. The revisions failed to stop complaints and appeals about the apportionment of the tax, and in 1666 the States of Brabant decided on a thorough renovation. Despite this intention, the revision does not seem to have taken place uniformly, as some villages are included only with a rough indication of their overall extent, while for others the names of all inhabitants and the extent of their property are carefully recorded. The renovation again proved to be inaccurate, and further revisions followed in the 1680s.

Edicts of 2 January 1680 and 7 August and 10 September 1681 ordered the compilation of a general tax cadaster and assessment in rural areas. Two councillors from Brabant were sent out to each of the main parts of Brabant (Antwerp, Brussels, and Leuven), and commissions of local officials were set up in each quarter, or *meijerij,* to organize the work. The extent of each plot was ascertained by a sworn deposition by its owner, whose evidence was cross-checked with existing survey books and tax lists and with lease documents. Registers of all the land in each *gemeente* recorded details of the capital value of the land and the average income derived from it. There was, however, no surveying or mapping. The results of this cadaster were considered untrustworthy, and on 25 July 1683 the States ordered a general survey of the province, beginning where the population was suspected of having sent in falsified returns. Land was to be divided into classes according to land use and tenure. Both Bigwood and Ockeley state that the intention was to conduct a general survey using sworn surveyors, but Bigwood doubts whether the survey was in fact carried out, as he could find no trace of activity in the archival record. He suggests that instead the authorities continued to rely on depositions made by proprietors and local people. Coppens also doubts whether surveying took place and goes on to stress that, despite the existence of various levels of checks and controls and of dire threats if false evidence was given, falsification was common; the registers were not kept up to date, and the cadaster rapidly degenerated.[98]

There was a further revision of the land tax in Brabant in 1686, and these registers remained in force until the nineteenth-century national cadaster was introduced.[99] There were, however, some local attempts to improve the situation by using surveying and mapping to bring the cadaster up to date. In the parish of Opwijk, northwest of Brussels, information submitted to the local magistrates by landowners and tenants was so inaccurate that in 1718 the authorities decided on a general survey of the parish. The parish received permission for the revision from the Council of Brabant in 1723, and it was agreed that a surveyor would be employed to survey all land plots and draw up a written cadaster and a map.[100] Similarly the sheriffs, mayor, and tax authorities of Uccle commissioned the sworn surveyor Charles Everaert to survey and map the parishes of Uccle, Stalle, and Carloo near Brussels in 1742. The survey covered some 780 plots, which are numbered and identified by owner and extent in written registers and on two maps.[101]

Surveyed and mapped cadasters were drawn at the end of the seventeenth century and beginning of the eighteenth for a large number of villages to the north and northwest of Brussels. An early cadaster of the parish of Grimbergen was compiled in 1696. The *gemeente* of Grimbergen wanted to apportion taxes more fairly and commissioned Peter Meysman to survey the area and draw up cadastral registers and maps. Meysman had already made map books of Beigem and Buggenhout by 1690 and was an experienced surveyor. He died before the work was completed, and the cadaster was finished by Jan van Acoleyen, who later made map books of Londerzeel (1704), Bever (1715), Merchten (1717), and Strombeck (1724). The Grimbergen cadaster consists of 488 written sheets describing the 1,296 land plots and their owners and users and nineteen maps at a scale of 1:2,288.[102]

The Transport of Flanders

The Transport of Flanders was a levy imposed by the king of France from the beginning of the fourteenth century. It was administered by devising quotas for each of the towns and *châtellenies* (local communities). The apportionment of equal quotas, although crude, was originally quite fair, but over time some communities flourished and others declined, particularly in an area where some places benefited from reclamation and others suffered from inundation. Of early revisions, that of 1517 was the most accurate, but even this was so flawed that twenty villages were missed out entirely and not brought into the Transport until an edict of 1550. Further revisions followed, but in spite of these Bigwood, writing at the end of the nineteenth century, considered the Transport the most antiquated of all the southern Netherlands taxes. By contrast, analyses by Coppens and van der Haegen consider the Transport to have been one of the best-organized and most modern cadasters in the southern Netherlands.[103] Van der Haegen traces written registers and maps made in connection with the Transport back to the seventeenth century. The parish of Lichtervelde in West Flanders to the south of Bruges (Brugge) is one example where a land register with a map of 1680 is complemented by written cadasters from 1714 and 1773–74. Outside Flanders land taxation was not so highly organized, but many areas have one or two written cadasters dating from the eighteenth century.[104]

2.7 The Duchies of Luxembourg and Limburg and the Limburg Kempen

In the Duchy of Luxembourg there had been a series of cadasters dating from 1495 and revised up to the beginning of the eighteenth century. These were rather crude counts of hearths, and only that of 1692 included any information about the extent and value of the land belonging to each village. In 1752, after the southern Netherlands had passed to the Austrian Habsburgs, Empress Maria Theresa decided to abandon the practice of counting hearths and to compile a proper cadaster instead. The edict of 24 July 1752 ordered that within three months all inhabitants of the duchy should declare to the mayor of their village their name, estate, profession, their livestock, and the number of houses, mills, or other buildings which they owned. Each landed proprietor had to make a similar declaration of his estate. The Seven Years' War interrupted the project, but work began again in 1766. The work was quite thorough in some ways. Different types of land were to be separately distinguished, and the government sent out printed forms to ensure as uniform a response as possible. There was no attempt at surveying or mapping, however, despite the detail of information called for and the recent precedent of the mapped cadaster completed under Maria Theresa in the Duchy of Milan (chapter 5.4). In fact the work was simplified as it proceeded; original attempts at land valuation were deemed too complex, and a simpler version was instituted. Nevertheless, the returns provide a very comprehensive picture of the Luxembourg rural economy in the period, even if they are primitive by comparison with the Milan survey.[105]

A survey similar to that undertaken by Maria Theresa in Luxembourg was instituted in the Duchy of Limburg, which was also part of the Austrian Netherlands. The survey was initiated by an edict of 15 January 1755, but work got under way only after a further edict of 4 April 1770. The details of the survey are substantially the same as those governing the Luxembourg work. The project was less successful, however, and complaints and appeals against the land tax continued.[106]

These were, however, isolated examples of land tax reform undertaken in the southern Netherlands by the Austrian Habsburgs. In general the Habsburgs were content not to bring in radical tax reform in those areas which contributed a significant amount to state revenues and which had traditionally had a considerable degree of local autonomy and authority in matters such as taxation. Maria Theresa proceeded by making the existing system work more efficiently, by introducing (where possible) agreements with the Estates covering several years so that land tax grants did not need to be renegotiated every year or even every six months, and by concentrating on new forms of raising money—in particular the lottery, bank loans, and customs duties.[107] The success of this strategy reduced the need for more far-reaching and more controversial reforms.

The Limburg Kempen forms a distinctive area within the Prince-Bishopric of Liège (Luik) to the south of the city of Eindhoven and to the north of the city of Liège. It is an area of woods, sandy heaths, meadows, and lakes which before the nineteenth century was only sparsely inhabited. From the mid-sixteenth century to the mid-eighteenth century, the Prince-Bishopric of Liège, although itself neutral, suffered almost continuously from war. The Eighty Years' War (1568–1648), the terrorization of Charles IV of Lorraine (1648–54), the Dutch War (1672–78), the Nine Years' War (1688–97), the War of the Spanish Succession (1702–13), and the Seven Years' War (1756–63) all ravaged the area and increased the need to raise taxes to pay for defense. Some two-thirds of state revenue was raised by property taxation, most of which came from rural areas. As revenue demands increased, great efforts were made to extract the tax in full, with threats of confiscation of property and imprisonment for defaulters. The fundamental problem was that many people were required to pay taxes quite out of proportion to their ability to pay, partly because their farms had suffered the ravages of war, but also because the system of tax assessment was very unfair. The tax was not based on a survey and had not been revised for more than a century, during which time numerous changes had taken place, not least considerable reclamation of land. The situation was becoming so serious that peasants were being forced to surrender their lands and leave the area. A revision of the land tax was made possible by an edict of 12 March 1686. This, however, simply prescribed a revision on the old lines of sworn testimonies given before the authorities, the very system which had failed in the past. It was left to local communities to take matters into their own hands in the course of the eighteenth century and organize surveying and mapping in response to local petitions for reform. Three-quarters of the Limburg Kempen was covered by such surveys during the course of the century, although few maps survive.[108]

2.8 The Nineteenth-Century Cadaster

Belgium was annexed by the French in 1795 and thus became subject to French law, including that governing taxation. As in the northern Netherlands, the French initiated a mapped cadaster following the procedures detailed in the *Recueil méthodique* . . . (chapter 6.12). Work was suspended when the French withdrew but was taken up again after a delay by the new administration. The cadaster was completed in Belgium in 1835 and in Luxembourg in 1843.[109] Since the cadaster in Belgium is so similar to that in the northern Netherlands (section 2.4), it is not discussed in detail here. These cadastral maps are the earliest precise maps for the great majority of Belgian parishes. They are usually at a scale of between 1:1,000 and 1:2,500. In addition, the Belgian National Geographical Institute has a unique series at a reduced scale of 1:20,000 formed by combining larger-scale originals.[110]

2.9 The Cadastral Map in the Netherlands

The remarkable sixteenth- and seventeenth-century development of surveying and cadastral cartography in the northern Netherlands is closely linked to the flowering of Dutch culture and to its mercantile and imperial expansion during this the Dutch golden age.[111] Mapping was most highly developed in polder areas for many reasons. Where land is abundant and of little value, the costs of surveying rapidly exceed its perceived benefits (cf. the Norwegian experience discussed chapter 3.24). In the polder areas of the Netherlands the population was dense and land relatively scarce. Not only was it not abundant, it was not free: considerable amounts of labor, skill, and money had to be invested before it existed at all. The cost of a map was thus minute in comparison with the overall cost of land and with the benefits to be derived from land.

Expenditure did not end with the winning of the land. Constant maintenance of dikes and sluices, locks and windmills required either regular labor service, by which the inhabitants maintained the installations themselves, or regular taxation to pay specialists to carry out maintenance.

As drainage installations became more complex and farmers more specialized, the latter method was favored, and since the taxes were not imposed by some remote despot but by the local polder board, demands for a just apportionment of taxes were normally met. Surveying and mapping were used from an early date in the Netherlands to achieve this just apportionment; *specialist* land surveyors and property mapmakers thus emerged earlier in the Netherlands than in many other countries.

A just apportionment of taxes meant increases in taxes for some as well as reductions for others, and there were also constant efforts to raise more money, to pay for water installations in the case of polder authorities, or to fund growing state expenditure. Mapping was therefore often a contentious and highly political activity.[112]

Agriculture in Holland in the early modern period was the most advanced in Europe, not least because the country was one of the most urbanized and thus required a market-orientated farming sector to feed the urban population. Production for the market meant that land yielded tangible profits and was thus itself a valuable commodity. Where land is not a commodity, rights to it may be customary and not codified. Capitalist agriculture, by contrast, demands a highly developed system of private property to allow individual exploitation of resources and hence demands a system of secure title to land. The development of capitalist agriculture thus encouraged cadastral mapping in the Netherlands.

Perhaps the most remarkable aspect of cadastral mapping in the Netherlands is that there are so many early printed cadastral maps. Small-scale topographical maps were commonly printed in many countries from an early date, but large-scale cadastral maps generally remained as manuscripts housed in state or municipal offices. In the Netherlands, by contrast, the early development of printing and engraving, combined with a need for multiple copies of maps for polder administration or reclamation publicity, led to printing of cadastral maps from the sixteenth century. Furthermore, awareness of and interest in the national territory also meant that cadastral maps were prized as decorative objects and collected by people who had no professional or financial interest in the areas they depicted. Because a market for maps existed outside the circle of those who would actually use them, the cost of production could be spread and the production of maps could be an economic venture in its own right.

All these reasons, together with advances in related fields such as fortification construction and the art of war, combined to ensure that, in the polder areas of the Netherlands, surveying and mapping developed early and rapidly became an indispensable part of public administration. The effects of the development were felt far beyond the boundaries of the Netherlands. Its surveyors and cartographers were active throughout Europe and in the Dutch colonies. The chief Dutch unit of measurement, the Rhineland rod, was used in many parts of Europe. Above all, early Dutch advances in cadastral mapping, as well as in topographical and maritime cartography, were widely known. The printing of cadastral maps in the Netherlands ensured for them a wider audience than any other property maps in Europe.

Acknowledgments

We should like to thank the following for their help: Mr. W. H. C. Blazer of the Provinciale Waterstaat van Noord-Holland, Dr. Marc Hameleers, Mr. Rob van Iterson of the Hoogheemraadschap van Rijnland, Dr. Piet Lombaerde, Mr. Roelof P. Oddens of the University of Utrecht, Dr. Hans Renes, Dr. Erik Swyngedeouw, Mrs. Joke B. M. Terra of the Rijksarchief in Noord-Holland, Mrs. A. D. M. Veldhorst, Dr. Anton Verhoeve, Dr. Kees Zandvliet, and the staff of the Koninklijke Bibliotheek and of the Algemeen Rijksarchief, The Hague.

3

THE NORDIC COUNTRIES

The Nordic area—Sweden, Norway, Finland, Denmark, Iceland, Greenland, and the Faeroe Isles—seems an unlikely place to look for the early development of cadastral mapping (figure 3.1). It is the most climatically extreme area of Europe, and the effects of latitude and altitude combine to ensure that many areas are covered by permanent ice or by snow for many or even most months of the year. Even where present climate is less extreme, the last ice age left features such as eskers, moraines, drumlins, deranged drainage, and outwash plains and removed much of the topsoil so that agriculture is difficult and relatively unproductive. If we accept that cadastral mapping depends at least in part on the value of the land, then the Nordic area seems a relatively unpropitious place to look for its early development.

It might be the case, however, that local scarcity of *good* land was extreme enough to increase its value markedly and prompt its survey and mapping, but in fact land was generally fairly abundant relative to population in the Nordic countries in the early modern period until the marked increase in population of the nineteenth century. Wide areas in the north were not, and in some cases are still not, permanently settled, so that in all countries except Denmark there was a pioneer boundary; if pressure did build up locally, people could migrate within the country.

Except in Denmark and southern Sweden, agriculture was of relatively little importance commercially but was characterized by subsistence production in peasant households. In terms of contribution to the national economy and certainly to exports, agriculture was eclipsed in importance in Sweden by mineral extraction. Elsewhere, activities such as fishing, forestry, and charcoal burning, which were often of critical importance in both the subsistence and the commercial activity of Nordic peasant households, demanded not so much exclusive ownership rights over resources but rather complex rights of usufruct over different resources at different times of the year by different members of the household.

One would expect under these circumstances that the costs of surveying and mapping would

Fig. 3.1 Sweden-Finland: regions, provinces, and places noted in the text. Sources: Droysens 1886; Muir and Philip 1927; Shepherd 1930; Kinder and Hilgermann 1974–78; Darby and Fullard 1978.

be extremely high and that the perceived benefits would be small and thus that very little mapping would be carried out. These expectations are fully justified in some areas: Iceland, northern Norway, and the Faeroes had no cadastral mapping until the nineteenth century—in some places, not until the twentieth century. Sweden-Finland, however, despite its ostensibly unpropitious physical and economic conditions, witnessed some of the most remarkable of all early cadastral cartography. In the sections below both the presence and absence of cadastral cartography are analyzed. Today the Nordic countries are thought of as united and culturally coherent, but for much of the seventeenth and eighteenth centuries, Sweden-Finland and Denmark-Norway were at war as they struggled for supremacy in the Baltic. Sweden-Finland, Norway, and Denmark, with the Faeroe Islands, Greenland, and Iceland, are thus treated in separate sections, after which general conclusions are drawn.

Sweden and Its Empire

3.1 Sweden: From *Stormakt* to Obscurity

It is not fortuitous that the history of Swedish cadastral mapping effectively began in her *stormaktstid* (age of greatness) under Gustav II Adolf (Gustavus Adolphus, 1611–32), one of Sweden's greatest monarchs. He consolidated the internal reforms of Gustav I Vasa and wrested Jämtland, Härjedalen, Skåne, Blekinge, Halland, and Gotland from Danish and Norwegian control and made them integral parts of Sweden. He led Sweden to dominance, albeit temporary, in the Baltic, where Livonia, Karelia, Ingria, Mecklenburg, and Hither Pomerania were added to its existing lands of Finland and Estonia; he also took the Duchy of Bremen and Verden on the North German coast. It is against this background of expansion abroad and consolidation at home that cadastral mapping began, though it continued in less happy circumstances with the loss of most of the Baltic lands other than Finland in the Great Northern War (1700–1721) and Sweden's return to relative obscurity and great poverty.[1] The loss of Finland to Russia in 1809 was partly offset by Sweden's personal union with Norway (1814–1905), but the nineteenth century was characterized by poverty and distress, leading to the emigration of nearly one million people, or one-fifth of the total population, mainly to the New World.[2]

While Sweden's economic distress was shared by many other nations, her cartographic history is unique. According to Sporrong, no other country in the world has so well preserved and systematic a series of large-scale land maps from as early as the seventeenth and eighteenth centuries.[3]

3.2 The Formation of the Swedish Lantmäteriet

The Swedish Lantmäteriet (land survey) was established on 4 April 1628, when Gustav II Adolf instructed Anders Bure (Andreas Bureus) to set up a state organization to survey the entire Swedish kingdom and to investigate the potential for economic development and improvement. Bure was already an active cartographer. In 1603 Karl IX, father of Gustav II Adolf, had instructed him to produce a topographical map of the whole country to replace the existing inaccurate and inadequate maps of Sweden and Scandinavia. The lack of maps was felt acutely at a time of increased political activity, particularly with regard to Sweden's northern border. Bure's map was completed by 1626, and it increased awareness of the importance of maps in defense and the development of the nation's resources.[4] The feeling that a national body was needed to survey and map the nation bore fruit in the establishment of the Lantmäteri.[5]

The surveyors were to produce maps of two kinds, known in Swedish as geographic and geometric. Geographic maps were small-scale topographical maps, and geometric maps were large-scale cadastral maps of individual villages, freeholds, and farms. Geometric surveys for property examination, for the delineation of land boundaries, and for taxation purposes already existed before 1628, but almost without exception the surveyors, known as *revkarlar,* had simply measured land, chiefly cropland, with primitive rods and ropes; they described in written form what they saw but made no maps. Cadastral mapping thus effectively began with the 1628 instruction, when a new breed of surveyors, *lantmätare,* replaced the old *revkarlarna.*[6] The instruction makes clear the king's personal interest in the work and his idea of himself as an improver, pulling Sweden into the modern age through systematic investigation and action. The maps are a "synopsis," a basis for action, not an academic exercise. The instruction reads:

> Whereas it is His Royal Majesty's gracious desire and intent not only to protect his land and realm against the enemy, but also to use every opportunity and means to improve their condition: now in order that His Royal Majesty the better may effect and render the same, His Royal Majesty desires a synopsis in which His Royal Majesty may have set before him the condition of all his provinces and towns, that he may better ascertain and reflect upon by what means and in what manner each of them might be repaired and improved.[7]

Lantmäteriet's task was systematic in that the whole nation was to be mapped, but it was not intended to keep the geometric or geographic maps up to date. Its maps were a reflection of the individual measures, such as boundary settlements, enclosure, and the fixing of taxes, which were effected with the help of maps. In fact there was a steady demand for maps for different purposes, and many areas were regularly mapped from 1628 onward. This regularity, however, was fortuitous and reflected the increasing acceptance of the map as an administrative and judicial aid, rather than governmental desire for a systematic and continuous revision of the nation's maps for its own sake.[8]

3.3 Early Cadastral Mapping: The *Geometriska Jordeböckerna*

In the 1628 instruction Bure was ordered to train surveyors, who were to survey each village with its arable fields, meadows, forests, and waste. They were to suggest improvements and calculate the costs and expected returns on these improvements. Improvements were to be considered not only for farmland but also for rivers and lakes which could be made navigable and forests which could be cleared for cultivation or managed more intensively for timber. Harbors, towns, mines, and fortifications were also to be mapped.[9] There was no existing organization to carry out this daunting task, and those surveyors who could be found were *revkarlar,* who had had no experience in mapping. Bure had himself to train a small group of apprentices, many of whom had studied mathematics, geometry, and astronomy at the University in Uppsala. In 1633 the first six trained surveyors, some of whom had worked with Bure on his great topographical map of 1626, were formally admitted.[10] The training of surveyors remained an entirely internal affair: active surveyors trained their successors as they worked. To prevent surveyors' leaving too much work in the hands of untrained apprentices, it was decreed in 1643 that an *inspektor* must examine each apprentice to determine whether the surveyor needed his help and to test his skill.[11] The surveyors were to be paid by the state.[12]

In the instruction of 1633 the first six surveyors were charged "to go out honestly each year as soon as the ground is bare [of snow], to begin work and diligently to measure the land belonging to each village, both arable and pasture."[13] In areas where the land was cultivated or was good enough to be hedged and tilled, they were to note in the margin of the map whether the soil was black earth, clay, sand, or heath. They were also to give the average hay yield of each meadow, state what wastes, forest, and fishing grounds belonged to each village, and mark clear boundary lines between farms and between villages.

They were finally to calculate each farm's cultivated and uncultivated land. Various units of land measurement were to be used, including the *öresland, örtugaland, penningland,* and *tunnland.*[14] The last—the *tunnland,* or acre, was the most important, although its calculation varied in the early stages of mapping. The 1633 instruction stated that one *tunnland* equaled 13,263 *kvad-*

ratalnar (square ells; 1 ell = 0.5934 m).[15] The 1634 instruction stated that one *tunnland* equaled 14,000 *kvadratalnar* plus meadowland which could bear four loads of hay, although this could vary between three and five loads depending on the fertility of the land.[16] The 1635 instruction stipulated that one *tunnland* equaled 14,000 *kvadratalnar*.[17] This variation makes the interpretation of the *tunnland* of some of the earliest maps rather complicated. In general, however, it seems clear that surveyors did not consider land quality when measuring its extent, and the relationship of 14,000 *kvadratalnar* to the *tunnland* seems to have been fairly uniform.[18]

The 1636 instruction gave surveyors a detailed color scheme in order to standardize presentation of the maps. Cultivated fields were to be colored gray, meadows green, mosses yellow, fences black, lakes light blue, rivers dark blue, boundaries red, forests dark green, and stony slopes white. In addition, a compass was to be drawn at the top of the sheet indicating north.[19] More detailed color systems were prescribed later.[20] Further standardization was achieved by an instruction of 1665 which, mentioning surveyors specifically, decreed that all Swedes were to use the standard linear unit of the Swedish, or Stockholm, *fot* (2 *fot* [feet] = 1 *aln* [ell] = 0.5934 m).[21]

Surveyors used the plane table, which had been designed by Johan Praetorius in Nuremberg in 1590 and reached Sweden in the 1640s.[22] The early maps, including the *skifte* (partition) maps (section 3.4), were drawn up using the so-called graphic method. A *konceptkarta*, or detailed base map, was drawn up in the field immediately after measurement and surveying. This was in contrast to the so-called numerical method, whereby a map would be drawn up at a desk after measurement in the field. The graphic method gave scope for error because measurements were not subsequently checked in detail.[23] This difficulty with the graphic method was realized in the eighteenth century; in 1763 surveyors were instructed to draw in justifying lines to try to correct distortion.[24] The early maps are most reliable in their central parts around the village buildings. Inaccuracies around the edges, however, make it difficult to join up the maps to use them as a base for mapping larger areas.[25]

Because of the varying length of the *aln* in varying parts of Sweden before the 1665 instruction took effect, it is difficult to calculate the scale of the early seventeenth-century maps exactly. A further difficulty is that the paper on which the maps were drawn has shrunk, and there is disagreement as to the extent of the shrinkage.[26] The most widely used scales seem to have been 1:5,000 and 1:3,333. Where landownership was complicated, these scales might be doubled to 1:2,5000 and 1:1,666 respectively. Scales of 1:1,000 and 1:4,000 were also used.[27]

The 1636 instruction ordered that, when each hundred was completely mapped, the maps with their marginal notes were immediately to be numbered and bound in books with an alphabetic index. These are the *geometriska jordeböckerna*, or geometric land books (figures 3.2–3.4).[28] For ease of binding, the map sheets were to be the same size. The presentation, method, and content of the surveys, however, was left largely to the discretion of the surveyors, despite early

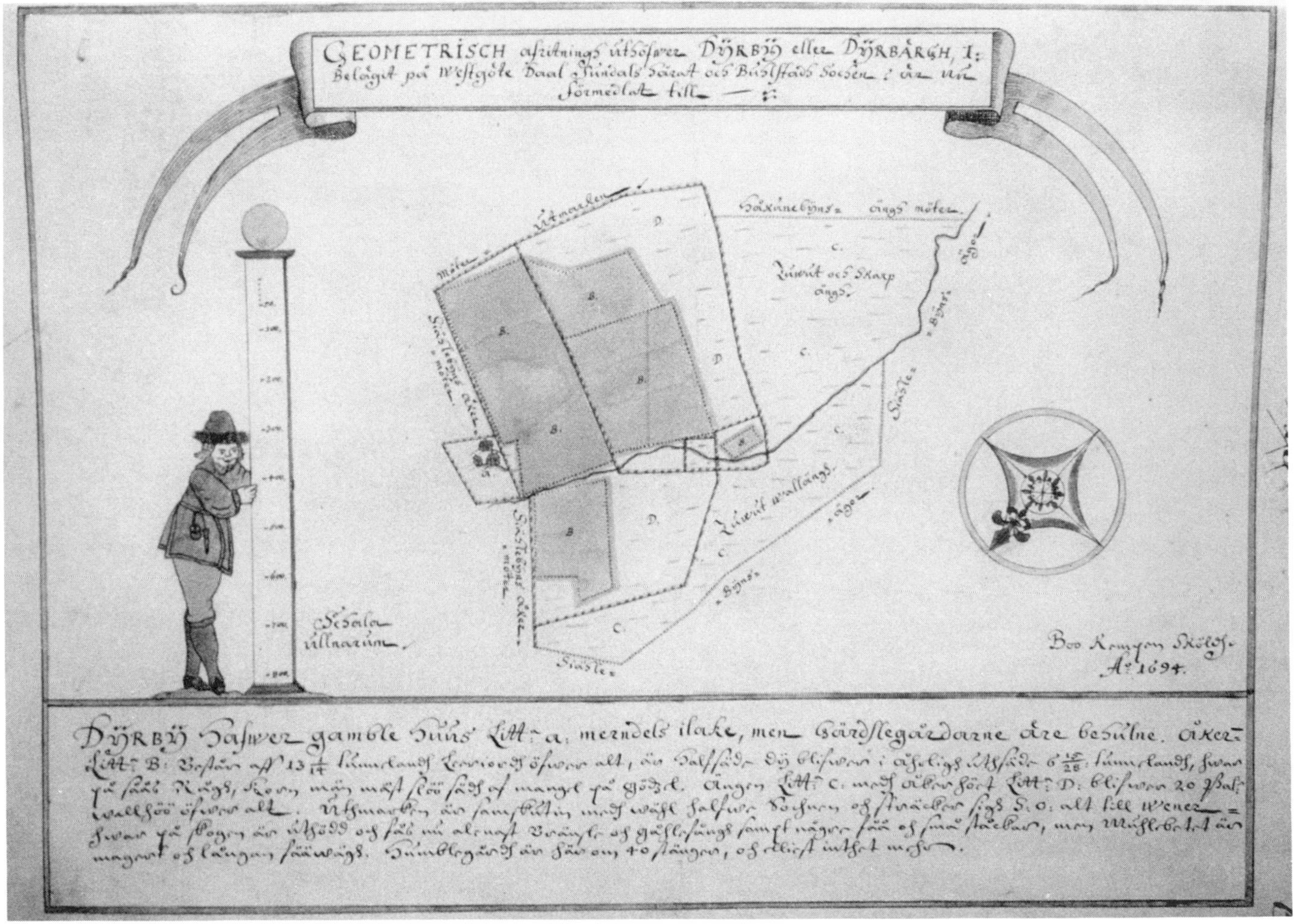

Fig. 3.2 *Geometriska jordebok* map of 1694 from Älvsborg County by Bo Kempe Skjöld. Source: Lantmäteriverket, Gävle.

attempts at standardization. As a result there is considerable variety in the early maps, and many individual surveyors developed personal styles.[29]

The Purpose of the *Geometriska Jordeböckerna*

There has been considerable debate as to why Sweden embarked upon a systematic and comprehensive program of mapping in the first half of the seventeenth century and particularly as to whether or not the *geometriska jordeböckerna* were taxation maps. In 1903 Lönborg suggested that the geometric maps were taxation maps, intended to provide accurate and up-to-date infor-

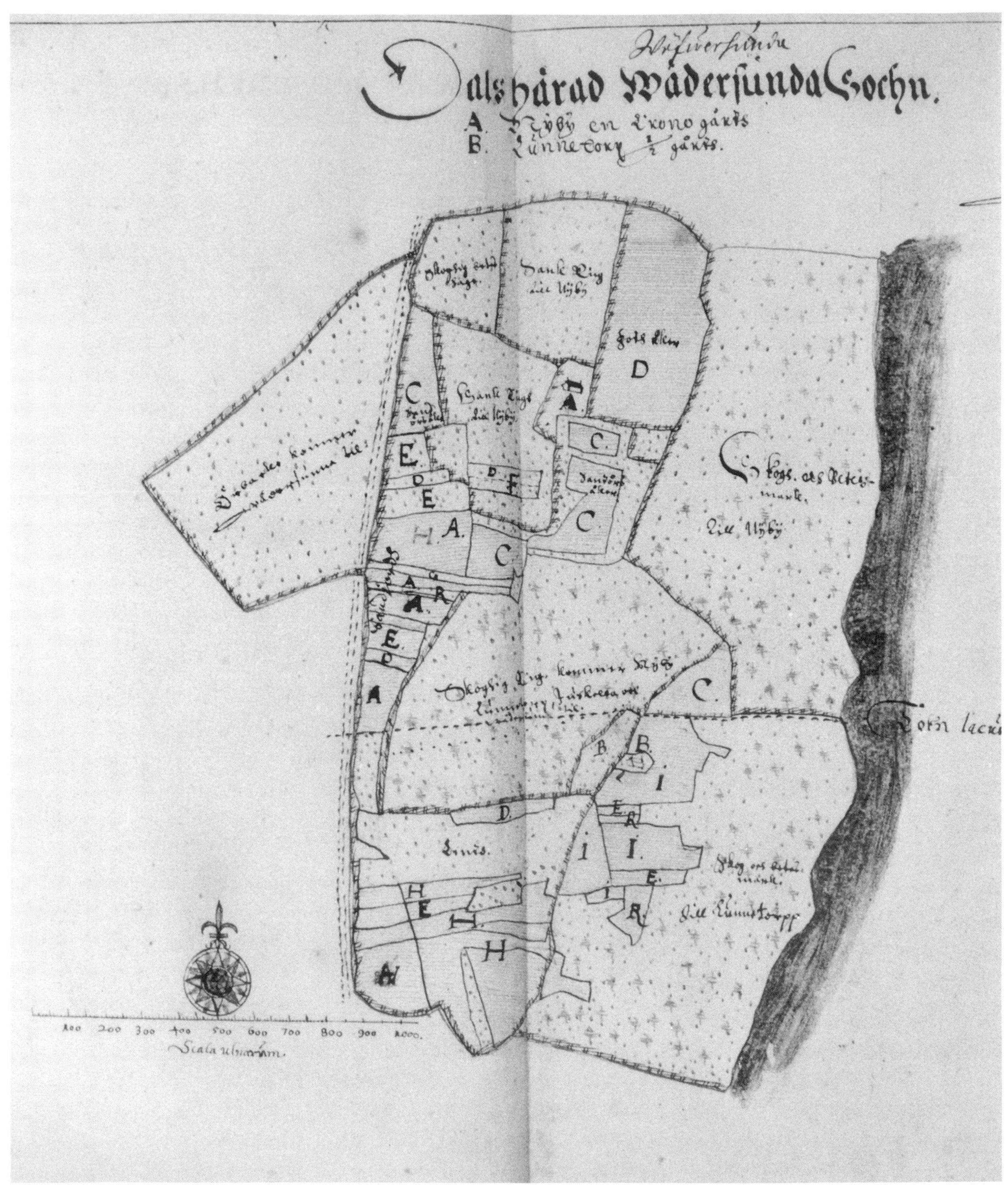

Fig. 3.3 *Geometriska jordebok* map of Väversunda in Dals Hundred in Östergötland by Johann Laurents Erst, 1633–34. Source: Landmäteriverket, Gävle.

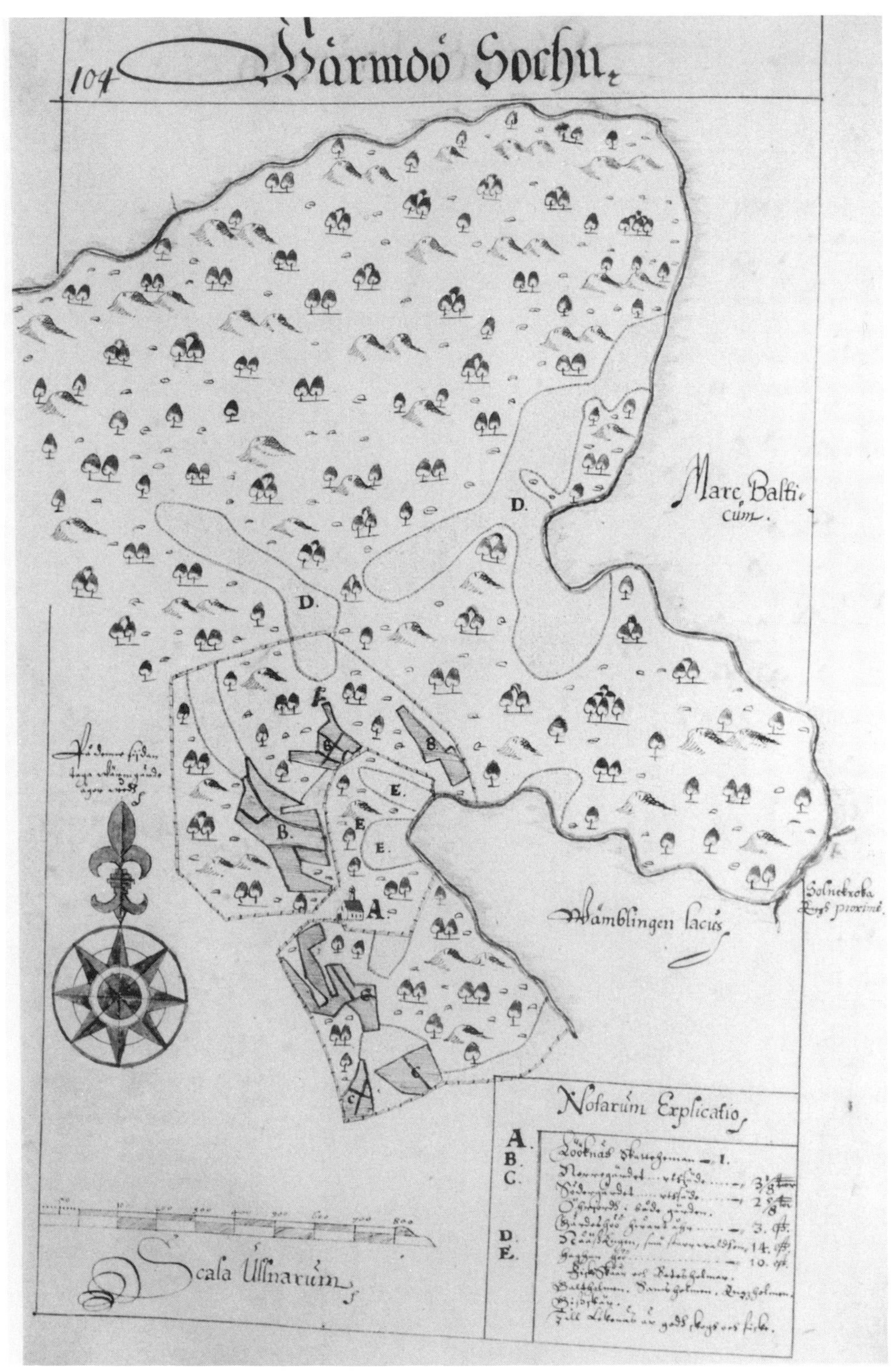

Fig. 3.4 *Geometriska jordebok* map of Värmdö in Stockholm County. Source: Landmäteriverket, Gävle.

mation so that taxes could be efficiently and equitably levied, an interpretation that has been widely accepted.[30] Others, however, have challenged this view. Bydén, for example, suggested in 1919 that the *geometriska jordebok* maps were forerunners of Sweden's current economic maps, which are large-scale maps showing land type and use and which have no link with taxation.[31] Doubts have been cast on the taxation thesis by Hedenstierna, who points out that tax assessments for Stockholm's archipelago were entirely unchanged after surveying and mapping had taken place.[32] There were three types of land in Sweden: *krono* (crown), *skatte* (taxable), and *frälse* (exempt),[33] and the 1634 instruction states that all types of land were to be surveyed.[34] It was thus not only taxable land that was of interest to the king. Helmfrid states that the early maps were an independent undertaking and that there was at most an indirect connection with taxation. The surveyors were called upon by the Riksdag (Parliament) to give information about the land and farming, information which was subsequently used in taxation matters. This was not their main function but rather a recognition of their experience and judgment gained during surveying. In 1634, when there were calls for a general tax revision, no connection with the mapping then in progress was suggested: the revision was part of a long-standing and entirely separate campaign.[35]

The 1628 instruction made clear that the surveyors were both to map the nation's resources and to suggest improvements, which implies that there was official demand for a purely economic map of the nation. Peterson-Berger suggests that Gustav II Adolf saw the maps mainly as an aid in defending "his land and realm against the enemy."[36] There was certainly concern that foreigners should not have access to the early maps, although this restriction applied mainly to geographic maps.[37]

Johnsson suggests, however, that the taxation thesis should not be too lightly dismissed. That the maps were not ultimately used for taxation purposes has no bearing on the intention of those who initiated the project. While the 1634 instruction ordered the survey of all types of land, the 1635 instruction required all surveyors to survey only Crown and taxable land and not exempt land, except where the landowners themselves requested it.[38] It is difficult to see a nonfiscal explanation for this exclusion. In villages with land of several different types and with farms having their land fragmented into tiny plots, it would have been impossible to have excluded exempt land from the survey. There is some evidence to suggest that surveyors included such land simply for practical considerations, explaining the "irregularity" in the text accompanying the map. Further textual evidence and the fact that all survey material had to be submitted to the Exchequer make it difficult to exclude the taxation thesis.[39]

The evidence is inconclusive, although it can be said with certainty that the *geometriska jordebok* maps were never directly used for a revision of tax, whatever their original purpose.[40] The debate over a link with taxation has in a sense obscured the wider political and cultural context in which mapping took place and has somewhat overlooked the personal interest of Gustav II Adolf in cartography. His involvement in campaigns deep inside the German mainland during the Thirty Years' War had shown him how useful maps could be and had brought him in contact with German states where cadastral mapping for the purposes of boundary delimitation and taxation had already begun, albeit rather sporadically (chapter 4). His knowledge of European maps was considerable. As well as founding the Lantmäteri, he also set up the Militäringenjörskår (corps of military engineers) in 1613. He thus laid the foundation for both civil and military cartography in Sweden.[41]

Thorough and systematic collection and presentation of information is characteristic of the Swedes. Swedish ecclesiastical and secular registration, for example, is of astonishingly high quality from the seventeenth century onward, and the *geometriska jordeböckerna* display the same thoroughness and detail.[42] Certainly interest in the gathering and mapping of information about the nation was important militarily and politically in this, Sweden's age of greatness, when very thorough reform of internal administration, contact with continental powers, and the start of state-directed economic development all revealed the need for maps in a country which had none.[43] The production of the *geometriska jordeböckerna* must also be seen as part of the national outburst of creative energy which, Roberts suggests, followed from the widespread acceptance of so-called *storgöticism*, or "megalogothicism." This was the belief that the Swedes were the stem of the Gothic people, the oldest nation in the world with the oldest language, who were now regaining their rightful ascendancy in Europe. The sometimes bizarre claims of *storgöticism* were widely accepted. Gustav II Adolf, who was elevated by *storgöticism* to become the mystic Lion of the North, and his mentor, Johann Bure, Anders Bure's cousin and teacher, were enthusiastic believers. *Storgöticism* gave a confidence and energy to Swedish politics and military campaigns as well as to the arts and sciences.[44] The great project of mapping the nation can be seen as part of this newfound vitality, and its grand conception can perhaps best be gauged from the instruction that the surveying and mapping was to cover not only the then Swedish lands but also areas which could be won for Sweden.[45] This is the vision of a nation fulfilling its historic destiny.

Geographic Coverage of the *Geometriska Jordebok* Maps

By the end of the seventeenth century few parishes in mainland Sweden had not been surveyed and mapped, although the island of Gotland, the southern province of Skåne (which had only recently been gained from Denmark), and much of the unsettled inland part of Lappland remained unmapped. The earliest maps are from Uppland, Södermanland, western Östergötland, Värmland, and Norrland.[46] Johnsson suggests that these findings be treated with some caution.[47] Only cultivated land and grazing land in the vicinity of the village core were mapped. The forests and wastes around the villages were not included, so that, even if every village had been mapped, a large part of the national area would have escaped survey.[48] It is very rare to find every hamlet and farm of a parish covered by the *geometriska jordeböckerna*. Despite these shortcomings, however, the *geometriska jordebok* mapping is unparalleled in Europe at this time, and this achievement is all the more remarkable because of Sweden's sparse population and because surveying was seriously hampered by mountainous terrain and by the land's being covered by snow for much of the year.

The *Geometriska Jordeböckerna* in Their European Context

While the comprehensiveness and quantity of the *geometriska jordebok* maps from the first half of the seventeenth century make them unique, Helmfrid points out that in technical and artistic terms they are unremarkable and certainly not innovative in a European context.[49] The source of most scientific knowledge in Sweden at this time was the University of Uppsala. Surveying was studied and taught there, and its influence on Swedish cartography was considerable. In the early stages of surveying, however, the influence of foreign universities was also important: Anders Bure himself owed his cartographic knowledge to his cousin Johann Bure, who had been educated at Heidelberg.[50] For the *geometriska jordebok* mapping, skilled cartographers had to be brought in from Holland and Germany. There was, however, no Dutch or German precedent for the systematic and comprehensive surveying of seventeenth-century Sweden. The inspiration for the *geometriska jordeböckerna* and the skill to compile them may have come from outside Sweden, but they form the most comprehensive and richest cartographic work in Europe at this time. The Lantmäteri was without doubt the leading mapping body in Europe.[51]

3.4 Lantmäteriet, Property Boundary Maps, and Enclosure

In the early phase of geometric mapping from 1628 onward, very few geographic maps (small-scale topographical maps) were produced. This was rectified by an instruction of 1643 which brought into being a lively period of geographic mapping, at least in part a reflection of the interests of the Lantmäteri's new director Peter (also Pehr, Per, Peder) Menlös, appointed in 1642.[52] As a direct result, few geometric maps were produced in the second half of the century. The number of surveyors was so limited that it was impossible at first to produce geometric and geographic maps simultaneously.

The second phase of geometric mapping, which began around 1700, was marked by ever-closer cooperation between peasant farmers and surveyors. Surveyors were increasingly called in to assist in questions of landownership or boundary disputes or even individual land redistributions before official enclosure.[53] Sporrong considers these early eighteenth-century maps and their accompanying descriptions to be particularly informative for the landscape historian and social historian but admits that they are rather difficult to read. Exceptionally the forest which surrounded almost every Swedish village was mapped as well as crop and meadowland around the village core. The scale of these maps is 1:4,000. Unfortunately they occur only sporadically, as each was drawn up by individual request and not as part of a systematic mapping program.[54]

Boundary maps consist of *gränskartor,* maps of historic and administrative boundaries, and *rågångskartor,* maps of property boundaries. The *rågångskartor* can cover considerable areas and are important cadastral documents for the more remote parts of the country.[55]

From the end of the seventeenth century, demands for agricultural improvements, particularly enclosure, began to affect surveyors' work. The 1725 instruction to surveyors mentioned for the first time their role in the partition of land, and as enclosure progressed in the eighteenth and nineteenth centuries, this work came entirely to dominate the Lantmäteri. Enclosure reached its peak in the 1860s, when it employed up to five hundred surveyors, and it continued to employ many surveyors until the first decades of the twentieth century.[56] The Lantmäteri had originated as a mapping organization but developed more and more into a land-partition body dealing with measures related to the various enclosure movements. Mapping became the chief instrument of its work rather than its purpose, and responsibility for mapping passed to a great extent to other institutions in the nineteenth century.[57]

Improvement and development had from the beginning been included in the surveyors' brief: in the initial instructions to Anders Bure it was clear that the surveyors were not only to describe the existing state of affairs but actively to suggest improvements. The major economic concern of the time was agricultural improvement, and the leading advocate of agricultural reform was Jacob Faggot, one of the first members of the Royal Swedish Academy of Sciences, whose books, lectures, and pamphlets set out his very radical plans for agricultural reform.[58] In 1747 he became the director of the Lantmäteri, which was to be the executive instrument for the *storskifte* (*stor* = "great," *skifte* = "partition"), the first Swedish enclosure. Some early redistribution, which aimed at combining each farmer's strips into one plot, occurred before parliamentary enclosure, notably in the Uppland villages near an estate managed by Faggot himself. The main reform, however, took place from 1757 onward with the less radical aim of reducing to a small number the parcels of land which each family farmed.[59] Maps were an important instrument in codifying the land available and in redistributing it; large areas of Sweden and its provinces were mapped as part of the *storskifte.* The maps are usually at a scale of 1:4,000 and are of great value for the social historian (figure 3.5).[60] They show how the land was distributed after the *skifte,* but there are comments on the previous landholding pattern. In some respects, however, they give an idealized rather than a realistic picture of the landscape, as the surveyors tended to make the fields more regular in the maps than in fact they were.[61]

The rationalization of farming was continued with the *enskifte* (*en* = "one") from 1803 on-

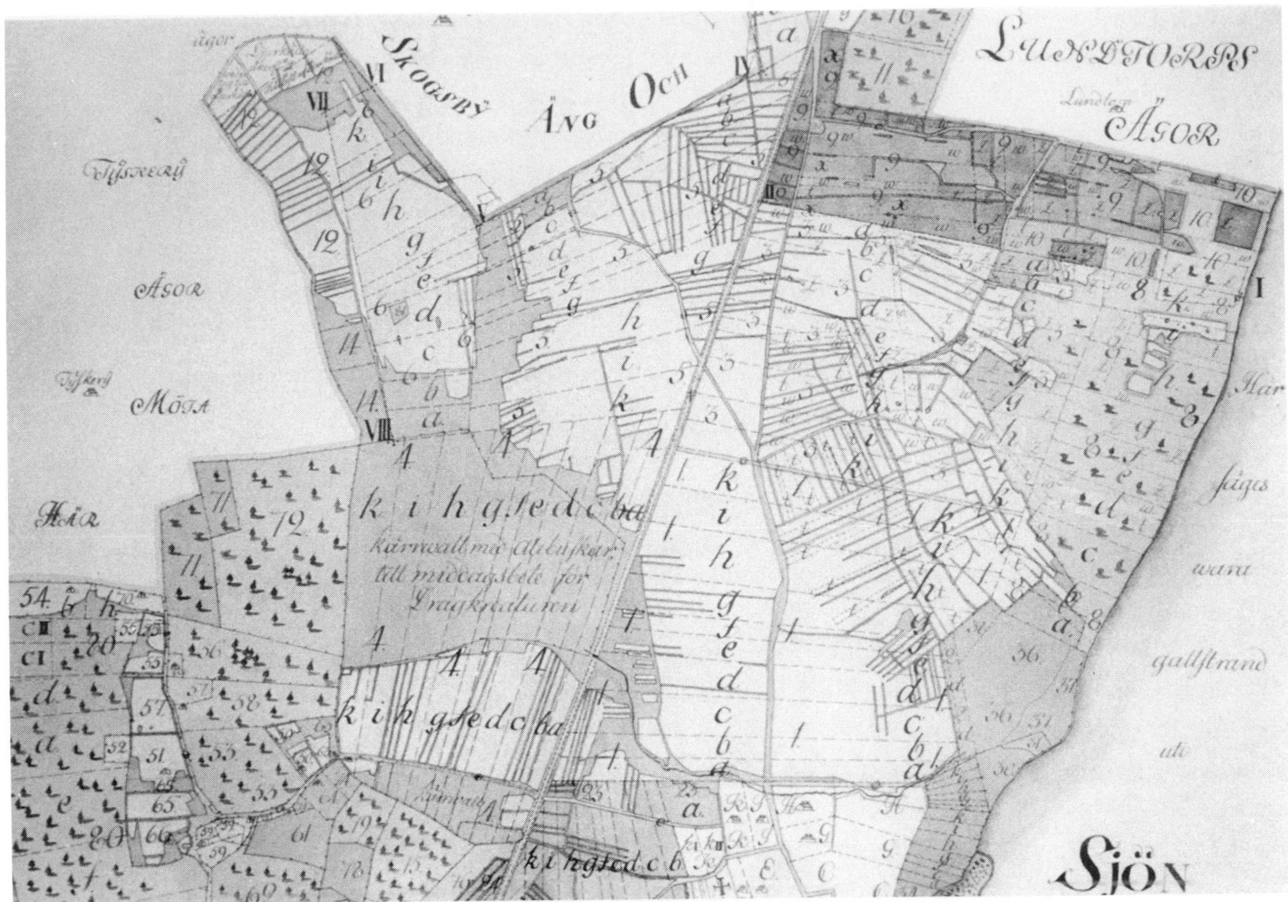

Fig. 3.5 *Storskifte* map of Väversunda in Östergötland, 1765, by Mattias Wallberg. Source: Lantmäteriverket, Gävle.

ward. This took to its logical conclusion the policy behind the *storskifte,* and it led to the creation of single-plot farms and a dispersed settlement pattern as houses moved out from the villages to the new consolidated plots. This proved appropriate only to certain parts of the country—for example, Skåne in the south and other fertile plains—but not to wooded areas or areas of mixed farming. A modified enclosure was carried out with the *laga skiftet* (*laga* = "law") which governed land division from 1827 until 1972 (figure 3.6).[62]

The process of enclosure is clearly documented in the surveyors' maps and reports. The maps were the basis on which the redistribution was effected and were also a legal record of the process. Two copies were made of each *skifte* map, partly as a security against loss by fire. A *publik renovation,* or good copy of the map, was drawn up at the expense of the landowners and depos-

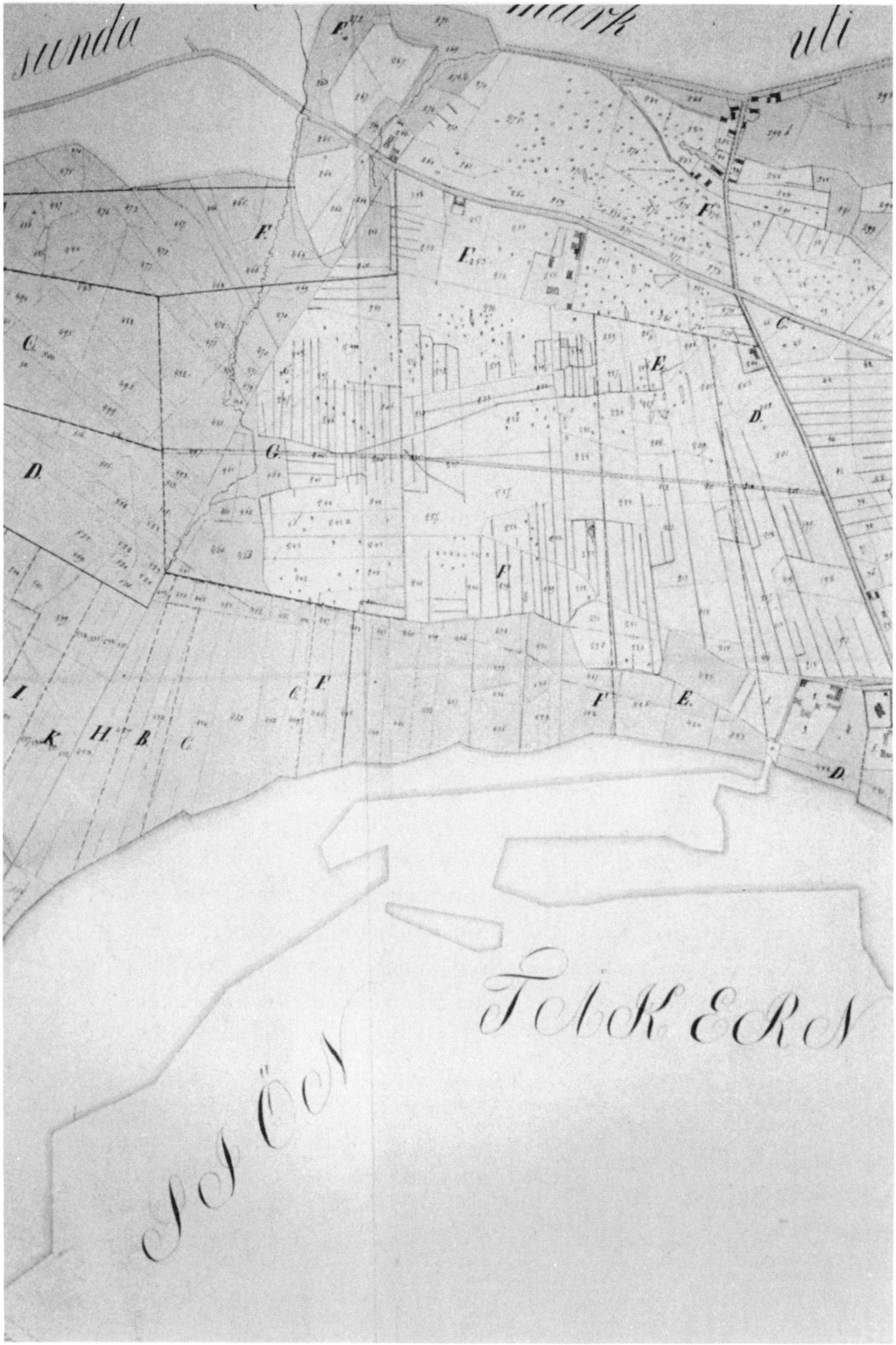

Fig. 3.6 *Laga skifte* map of Väversunda in Östergötland, 1832. Source: Landmäteriverket, Gävle.

ited with the Lantmäteri, so a national collection of maps was made at no cost to the state. The surveyors lodged with the farmers while doing enclosure work and were paid by them. The state through the Lantmäteri paid only the chief surveyor in each county during the period of enclosure work.[63]

Petersson suggests that a surprising proportion of redistribution and enclosure plans went through without acrimony. The surveyor would draw up his proposed redistribution map and would then withdraw while the landowners considered these. He concedes that there were some disagreements but suggests that most were settled amicably on the return of the surveyor.[64] Hoppe takes a less optimistic view and suggests that there were usually at least minor and not infrequently major changes to be made.[65] After agreement was eventually reached, the surveyor presented the maps and documents to the *ägodelningsrätt* (property division court).[66] Land was often apportioned in the new consolidated plots according to the number of strips each farmer had had before. If this was unsatisfactory, the surveyor could measure the width of the strips and take this into account. In the last resort he could use the *Kronans Jordebok* (Crown Land Book), which gave the size of each farm according to its tax assessments.[67]

The *enskifte* and *laga skifte* maps cover almost all of the land in the country, though this did not by now include Finland. They are very well produced and extremely detailed and, like the *storskifte* maps, are normally at a scale of 1:4,000, which became known as the *åkerscala,* or field scale. Details of the yield of the land are given in the accompanying notes. The *laga skifte* maps were directly surveyed and so are of superior quality to the *enskifte* maps, which were often simply copied from *storskifte* maps.[68]

The geographic coverage of the *storskifte, enskifte,* and *laga skifte* maps for southern and central regions has been analyzed by Helmfrid.[69] The *storskifte* (figure 3.7) occurred early in coastal rather than inland areas and especially in the area immediately to the north of Stockholm (in Stockholm, Uppsala, Västmanland, and Gävleborg *län* [counties]), where the personal influence of Faggot was felt; in Värmland, bordering Norway in the west, where, despite the mountainous terrain, partition occurred rapidly, sometimes even spontaneously among the peasants; and round Vättern in the Östergötland plain, which was intensively cultivated. Large parts of Skåne and the area between Lakes Vännern and Siljan remained comparatively little affected by the *storskifte.*

The *enskifte* and the *laga skifte* followed a different course and occurred earliest in Skåne, where the pioneering Scot Rutger Maclean carried through very thorough and radical reform (figure 3.8). They also occurred early on the island of Öland, off the southeastern coast of Sweden, and in parts of Västergötland, where another landowner, P. E. Tham, had considerable influence. Darlana, with its very complicated field system, was again little affected, and neither was central Södermanland (figure 3.9).[70]

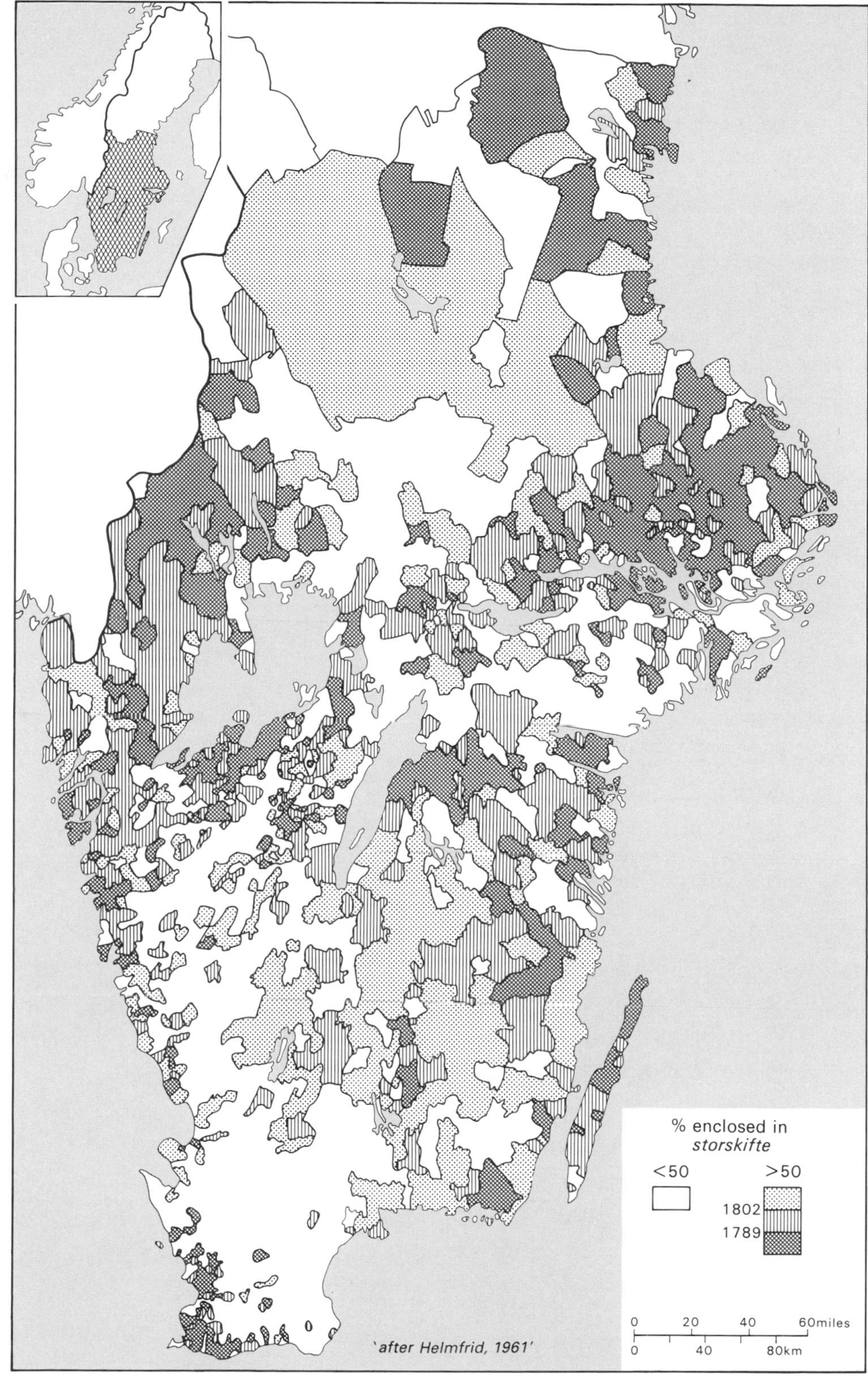

Fig. 3.7 The spread of enclosure under the *storskifte* in Sweden. Source: after Helmfrid 1961 (by kind permission of *Geografiska Annaler*).

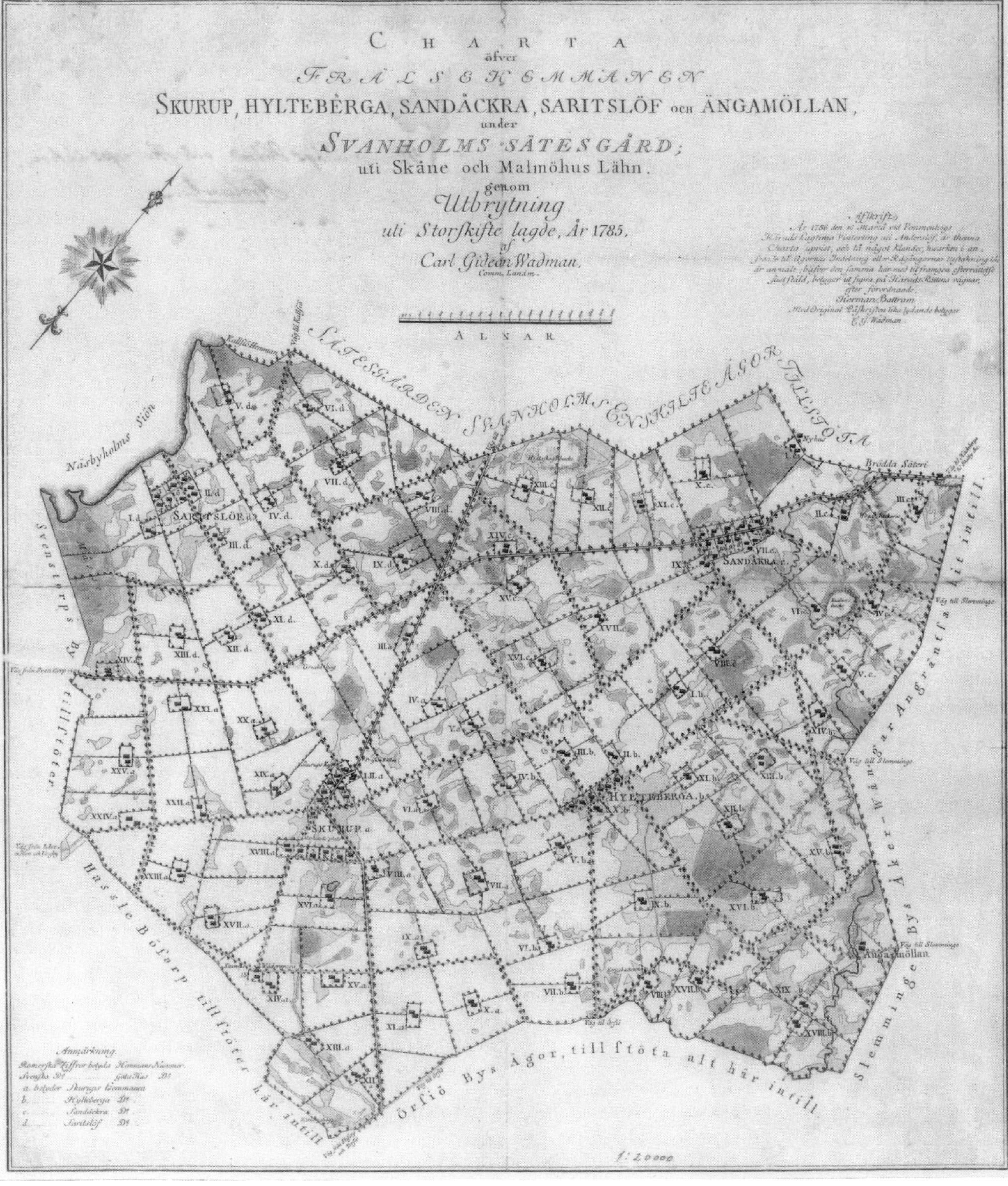
CHARTA
öfver
FRÄLSE HEMMANEN
SKURUP, HYLTEBERGA, SANDÅCKRA, SARITSLÖF och ÄNGAMÖLLAN,
under
SVANHOLMS SÄTESGÅRD;
uti Skåne och Malmöhus Lähn.
genom
Utbrytning
uti Storskifte lagde, År 1785,
af
Carl Gideon Wadman.
Comm. Landm.
ALNAR
SÄTESGÅRDEN SVANHOLMS ENSKILTE ÅGOR TILLSTÖTA
Näsbyholms Siön
Svenstorps Bys tillstöter
Hasle Böserp till stöter här intill
Örsiö Bys Ågor, till stöta alt här intill
Slemminge Bys Åker-Wångar Angrentsa alt intill
SARITSLÖF
SANDÅKRA
HYLTEBERGA
SKURUP
Ängamöllan
Brödda Säteri
Anmärkning.
1:20000

Fig. 3.8 (opposite) Map by Carl Gideon Wadman, scale 1:2,000, of four villages on the Svanholm Estate in Skåne owned by the pioneering agricultural improver Rutger Maclean. Compact farms were created in the *skifte* of 1785, which became a model for later enclosure. Source: Facsimile copy by the Länstyrelsenslantmäterienhet, Malmöhuslän.

Fig. 3.9 (right) The spread of enclosure under the *enskifte* and *laga skifte* in Sweden. Source: after Helmfrid 1961 (by kind permission of *Geografiska Annaler*).

After parliamentary enclosure was complete, surveyors were involved in the process of secondary division of plots (*hemmansklyvning*), which occurred throughout the nineteenth century. This was done with state support around the turn of the century to create means of livelihood for a growing population in a country which was concerned at the loss of population through emigration to North America. *Hemmansklyvning* maps are often of rather poor quality. The landowners wanted to save money, so they simply copied from existing maps.[71]

3.5 Taxation Maps

Whether or not the *geometriska jordebok* maps were property taxation maps has been discussed in section 3.3. There were, however, maps from the second half of the seventeenth century onward whose link with taxation is clear. They were drawn up partly in the normal calculation of *mantal* (taxation units), partly in connection with the new organization of the army, and in the Baltic provinces as well as on the mainland, in connection with the land resumptions and taxation revisions of Karl XI (section 3.7).[72] There was some further tax mapping in the eighteenth century, notably that for the island of Gotland in the 1740s. In the eighteenth and nineteenth centuries there was tax mapping in connection with the taking in of the commons.[73]

A particular group of taxation maps are the so-called *rå-* and *rörskartor.* In the seventeenth century, nobles were exempt from most taxes on their *säterier* (seasonal dairy pastures) and *rå-* and *rörshemman* (boundary farms). Consequently they sought to have as many farms as possible declared boundary farms. This brought them into conflict with Karl XI, who ordered the investigation and mapping of these farms by the Lantmäteri to try to limit tax evasion by the nobles. Many *rå-* and *rörskartor* were drawn up, especially for Uppland, Södermanland, Östergötland, and Skaraborg *län.*[74] Cadastral maps were used by the central government to reveal evidence which the nobility would rather have kept hidden, a phenomenon evident throughout Europe in the eighteenth and nineteenth centuries (section 3.7 and chapters 4 and 5).

A large proportion of Sweden's extensive land area was unsettled, and as the population expanded, new areas of forest and waste were constantly being colonized. This was obviously of concern to the Crown, as newly settled areas were escaping taxation. A decree of 1683 deemed that all land which could not with certainty be ascribed to a village as its common land (*allmänning*) would fall to the Crown. Land was to be assessed to differentiate Crown from private land and to fix tax assessments for the latter. Work to map the commons began with the Swedish forestry commission under the auspices of the Lantmäteri in the last two decades of the seventeenth century mainly in Skaraborg, Älvsborg, Uppsala, and Västmanland *län.* In 1793 and 1805 it was ordered that all remaining commons be mapped. This led to a great flurry of activity in the first decades of the nineteenth century.[75]

This mapping was very similar to the *avvittring* (differentiation of Crown land and land belonging to villages and hamlets), a process which began in the first decades of the eighteenth century with the *äldre* (old) *avittringen*. In the nineteenth century the work gathered pace and was virtually completed.[76] Unlike enclosure work, for which the surveyors were paid by the landowners, *avvittring* work was paid for entirely by the state.[77] The *avvittring* maps are in general rather small scale and not as detailed as other geometric maps for corresponding periods, but they are important, as they are often the only maps of some remoter areas. Their scales of 1:16,000 and 1:20,000 came to be known as the *avvittringsskalor*.[78]

3.6 Military and Technical Influences on Cadastral Mapping

Important advances in Swedish cartography were made in the mapping of the strategically important northern frontier and coastal areas, but the influence of these and other military and political maps on civilian cartographic development was limited, as the maps were rarely published at home and were often classified as secret and remained hidden in archives.[79] Military surveyors commonly used the civil surveyors' maps and descriptions as a source of information. They reduced the scale of cadastral maps and used them to compile small-scale military maps and also civilian geographic or topographical maps. Conversely, Lantmäteriet had access to military maps of certain inner parts of the nation, but not of sensitive border areas. The influence on the cartographers themselves, however, was clearer and more significant. Until the nineteenth century *lantmätare* (civilian surveyors) were commonly pressed into service as *militära fältmätare* (military surveyors) in time of war, just as officers trained as surveyors became *lantmätare* in time of peace, and in the Baltic provinces civilian and military surveyors routinely undertook common tasks (section 3.7).[80]

Despite the success of Swedish cadastral mapping, the well-known surveyors' instruction book by Peder Nilsson Raam *Ortuga deelo-bok* (1670) gives instructions on how to survey land and conduct taxation assessments but makes no mention of mapping.[81] Twenty years later Åke Claesson Rålamb published the first volume of his *Adelig öfning* (1690), which sought to codify the technical practices and instruments used in the Lantmäteri. Ehrensvärd remarks that, with such books and the Lantmäteri's own instructions about the construction of maps, civil surveyors had far greater help and guidance than did their military contemporaries.[82]

3.7 The Duchy of Finland and the Swedish Colonies

Finland

Since the Duchy of Finland was in union with Sweden when mapping began, its cartographic history is very similar to that of Sweden itself until 1809, when Finland became a grand duchy under Russian control. In 1633 the first surveyor, Olaus Gangius, was sent out to Finland to begin geometric surveying.[83] A taxation assessment decree prompted a new phase of mapping from 1749 onward, and the *storskifte* of 1757 onward was based on these maps.[84] *Storskifte* maps are normally on a scale of 1:4,000 or 1:8,000.[85] By 1760 *storskifte* work was occupying 115 surveyors and numerous assistants in Finland. The work was ultimately directed from Stockholm, but Finland had its own chief surveyor.[86] The draft maps for Finland's *storskifte* date very largely from the eighteenth century, although the legal process of land division continued long after this.[87] After coming under Russian control in 1809, Finland kept its own social organization, constitution, and laws. To administer the country a number of new administrative offices were founded, including the National Survey Board (Maanmittaushallitus, or Lantmäteristyrelse) in 1812.[88] Finland was unaffected by the Swedish *enskifte* and *laga skifte,* which happened after Sweden lost Finland to Russia, and so has no maps from these enclosure movements, although it had instead the *nyskifte* (new partition), which affected most of the country apart from the far north.[89]

Surveyors in Finland, as in most other countries, were often involved in settling land disputes (figures 3.10 and 3.11), but in Finland there was also much surveying and cartographic activity associated with the emergence of new land in coastal and lakeside areas. Postglacial uplift in Finland is as much as one meter per century. With the very gently sloping coasts of the Gulf of Bothnia, this uplift exposes considerable areas of land. Changes in drainage due to uplift cause lake levels to fall dramatically and expose new land areas. *Tillandningsskiftet,* or the partition of emergent land, often involved surveyors who were appointed by the villagers to apportion the new land. Where they could avoid doing so, they did not draw maps, thus saving the villagers unnecessary expense. In many instances, however, they did draw maps to help with apportionment, although, unlike *storskifte* maps, these were not legal documents and were never ratified in court.[90] It was difficult to keep tax assessments equitable, as the emergence of land gave new cropland to some but deprived others of their fishing grounds. Cadastral maps were often made in connection with tax adjustments.[91]

Because partible inheritance operated in Finland, plots consolidated under the *storskifte* were often redivided among heirs, a process known as *klyvning.* To try to rectify this, in the last two decades of the nineteenth century surveyors undertook *storskiftesreglering* (reconsolidation of the

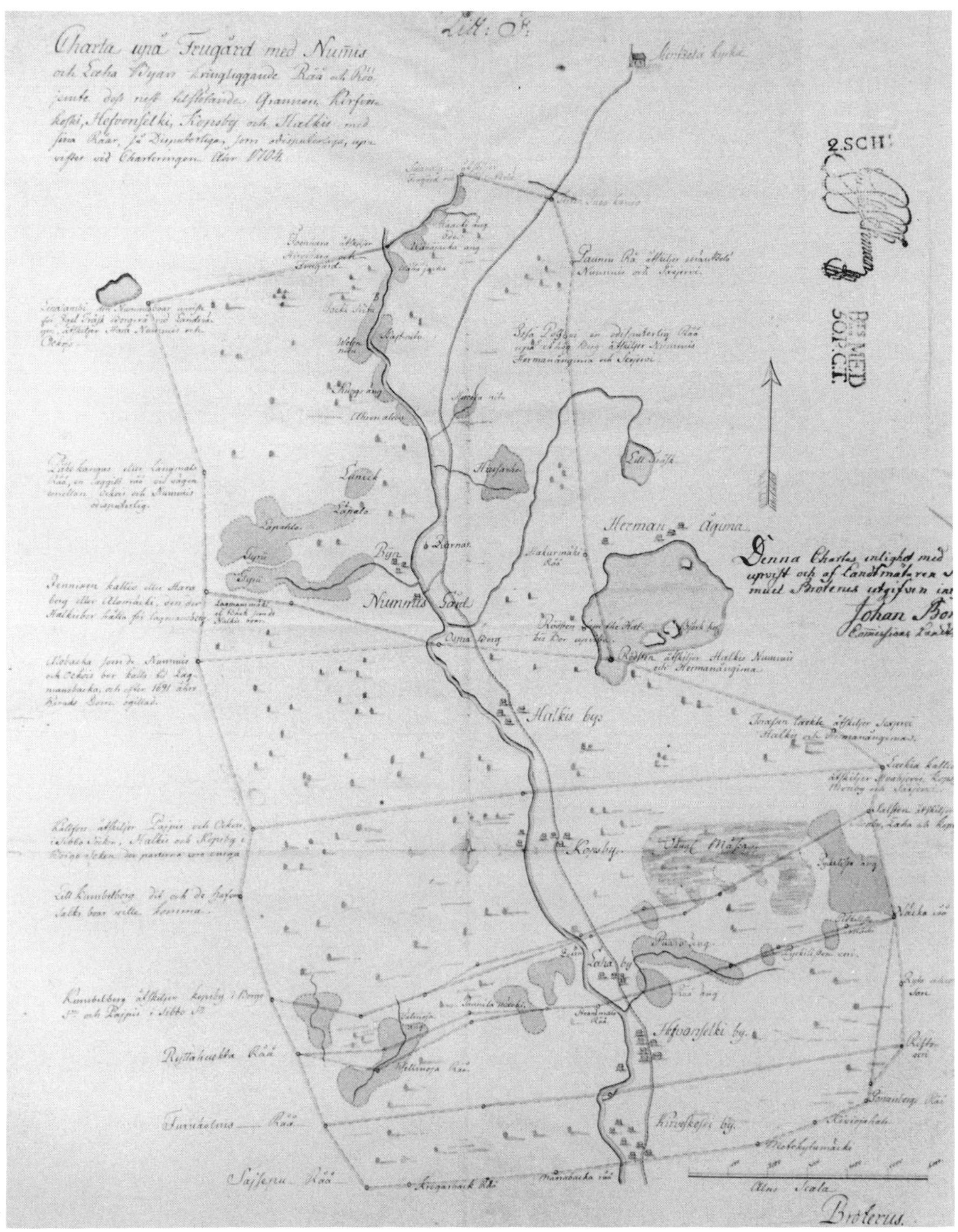

Fig. 3.10 *Rå- och rörskart* (farm boundary map) from Frugård and Numis, Finland, drawn by Samuel Broterus in 1704. Source: Riksarkivet, Stockholm.

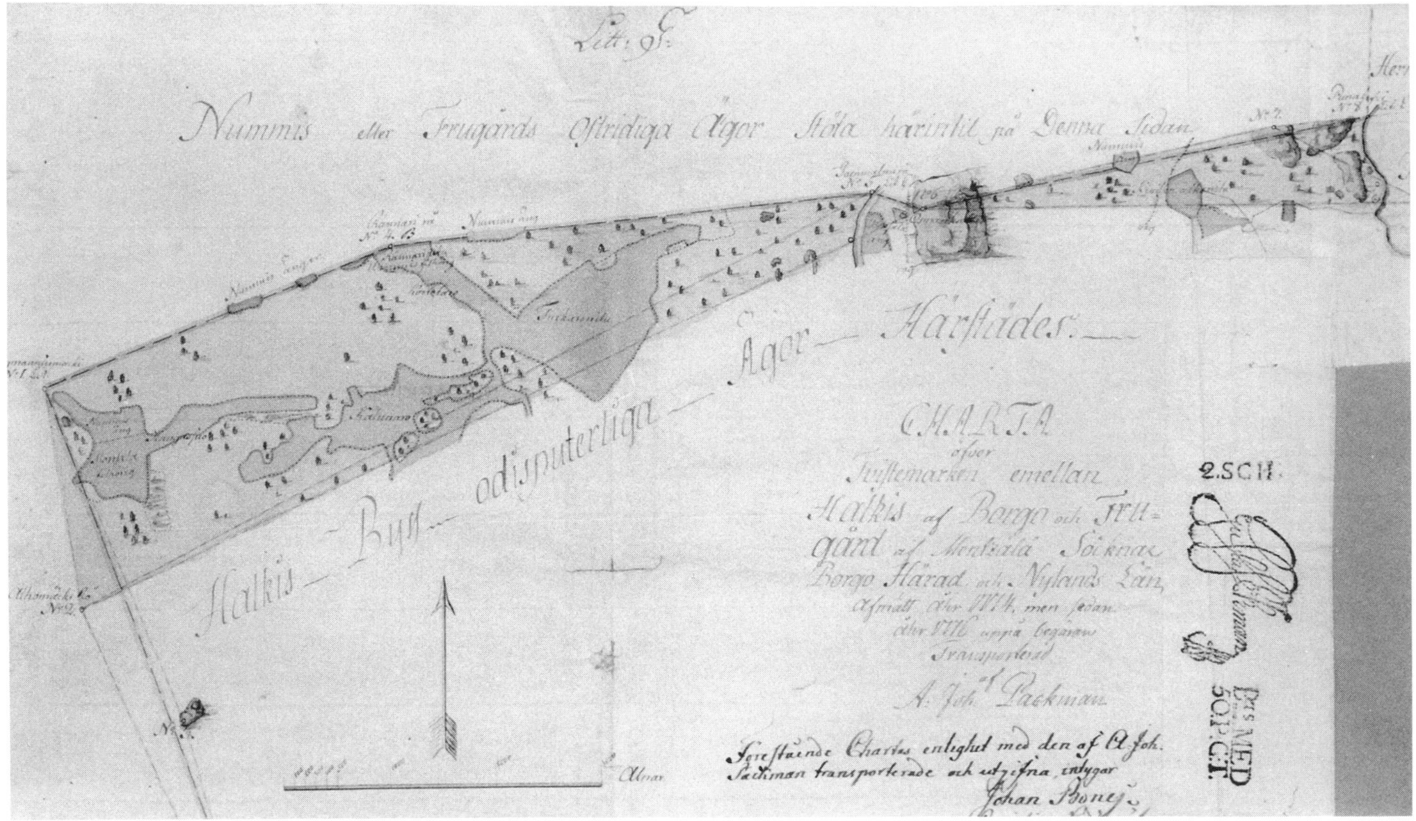

Fig. 3.11 Land dispute map from Nyland County, Finland, 1774–76. Source: Riksarkivet, Stockholm.

plots).[92] Maps from both the division and the subsequent reconsolidation survive, as do maps from the nineteenth-century colonization of crown land.[93]

The Baltic and German Provinces: Cadastral Mapping and Colonization

Swedish rule in Estonia and Livonia (Livland) was not oppressive by comparison with German, Polish, and Russian rule in the Baltic, and the Swedish monarchs made important attempts to improve education and to free the peasantry from their oppressive feudal burdens.[94] Nevertheless, the Swedish Crown and individual noble families were keen to reap the economic benefits of their conquest.[95] In 1699, for example, nearly 50 percent of Swedish Crown revenue came from the provinces, the most important of which was Livonia.[96] Cadastral mapping played an important role, in theory if not in practice, in the economic investigation and exploitation and taxation of the provinces.

Swedish rule in Livonia (southern Estonia, northern Latvia, and later the island of Ösel [Saaremaa]) was established gradually from the end of the sixteenth century and was consolidated in 1629, after which it continued until the treaty which ended the Great Northern War in 1721, when most of Sweden's Baltic possessions passed to Russia.[97] Inequality in taxation was a long-standing and grave problem, and the first revision to the taxation cadaster by the Swedish authorities was carried out as early as 1601. The aim was a thorough revision of the annual land tax, but the work was carried out amid political and military uncertainty. While there is some evidence that surveying was carried out, no maps were drawn, and the survey was based at least in part on preexisting *Wackenbücher.*[98] The *Wackenbücher* (or *wackeböcker*) were books drawn up at the *Wacka,* the annual inspection of the peasants which attempted to establish the number of *unci* held by each peasant.[99] The *uncus* had originally been the area which could be cultivated with one horse and one plow. Its area varied greatly from place to place, especially after revision by different authorities, notably German, Polish, and Russian.[100] By the time of Swedish rule the *uncus* was not an areal but a cadastral unit which took into account the quality of the land, proximity to markets, the capacity of the peasant to render statute labor and pay customary dues, and so on. Farms traditionally had two different *uncus* numbers: the traditional one (*Besitzhaken*) and the one on which they in fact paid tax (*Zahlhaken*).[101] Land revisions, or *uncus* revisions to establish the number of *unci* on which taxes were levied by the Swedish Crown, were held in Swedish Livonia in 1601, 1617–18, 1624, 1627, and 1638. They were not accompanied by surveying, however, and did not put an end to the discrepancies between the number of *unci* farmers held in fact and those they held according to the *wackeböcker.*[102]

Cadastral mapping in Livonia began with the Great Cadaster, which was carried out between 1681 and 1710. The Swedish Diet of 1680 and the Diets which immediately followed it ordered the *reduktion,* or resumption by the Crown of land which had been alienated to the nobility on an ever-increasing scale during the sixteenth and seventeenth centuries.[103] As a result of this alienation of land there was a steady decrease in Crown revenue, until by the 1680s basic military and government expenses could not be met.[104] The resumption was to apply to lands in Ingria, Estonia, and Livonia as well as in Sweden itself. Since Crown lands in Estonia and Livonia had almost entirely been given away by Gustav II Adolf and Kristina, the resumption was particularly dramatic there.[105] The purpose of the cadaster was to fix the rent to be paid on newly resumed Crown land and then to fix the number of *unci* for each estate. On the basis of the *unci* the amount of *Station* and other Crown taxes payable by the peasants was calculated and the *Roßdienst* (military service) was fixed for the nobility.[106] The surveying and mapping for the Great Cadaster were an example of military and civil cooperation, as the thirty *lantmätarna* sent out from Sweden were assisted by thirty junior officers from the Riga garrison. More interestingly, it is an isolated instance of the failure of a mapped survey in Swedish lands. The survey began in 1681 but was opposed by many of the nobles and officials of Livonia, who were mainly ethnic Germans and who were fundamentally opposed to the tax revision and land resumption.[107] They claimed that King Karl XI was acting unlawfully and exceeding his limited powers in Livonia, which was joined to Sweden by a personal union and in which some matters, notably taxation, were formally under the control of the Livonian Diet.[108] They claimed that the young and inexperienced *lantmätarna,* strangers to the country, would be deceived by the peasants, who would try to escape taxation by lying about the extent of the wastes to which they had access. The nobles wanted to revert to the old nonsurveyed system, under which they themselves sent in facts about their estates and under which they were freed from taxation but had to collect taxes in kind from the peasantry. Government officials were also skeptical of the value of mapping, and one even began independently to gather statistics about farms and estates in *jordeböcker.*[109]

The opposition of the nobles was seen to be partly justified when the first series of maps was revealed in 1687. Maps at a scale of 1:10,400 showed accurately the extent of arable land but made no attempt to relate this to the *uncus,* or taxation unit. Adjoining plots which belonged to different farmers had been surveyed and measured together, and no attempt had been made to tackle disputed wastes and commons. The speed with which the surveyors had had to work, combined with their inexperience and the opposition they had met, made the maps inadequate for a revision of taxation and land resumption. After this failure there was a return to the traditional methods of tax revision. Mapping was not abandoned permanently as an instrument of taxation, and the Revisionskommission in 1688 set about resurveying and mapping the country. The second survey produced maps and descriptions of resumed estates and also maps and descriptions of those peasant farms where the so-called *spezielle Einteilung* had taken place. In order to increase revenue from taxation the *spezielle Einteilung,* or enclosure, was to bring together tiny strips of land into consolidated plots, accompanied by the founding of new farmsteads. The fiscal implications of the enclosure were not clear, and in any case it affected only 10 percent of all farms. The work remained unfinished because of the Great Northern War, but the maps which were completed are of a far higher standard than those of the first series. They were used, as planned, to revise taxation. With the first series of maps they provide the most detailed information on the type and quality of land, the rural population, and number of farmers on each farm, as well as on taxation. The maps are of such high quality that they are considered to be one of the greatest achievements of the Swedish administration in the province.[110]

In the northern part of present Estonia, Swedish rule was established in the sixteenth century, extended and consolidated in 1561 and 1595, and ended in 1721.[111] The area formed the Swedish province of Estonia, which bordered the province of Livonia. The Estonian land tax was also revised in 1601, and the work seems to have been carried out more carefully than that in Livonia at the same date.[112] As in Livonia, however, an effective revision of taxation had to wait until the end of the seventeenth century; between 1688 and 1700, thorough surveying and cadastral mapping were used to effect Karl XI's tax reform and resumption of land from the nobility. The resultant maps, like those of Livonia, are of extremely high quality (figures 3.12 and 3.13). More controversially, the tax revision led to a significant increase in taxation in the Swedish period.[113]

Hither Pomerania with Stettin was in Swedish hands partly from 1629 and mainly from 1648. Prussia gained the area in 1721 and 1817. The province had been devastated by the Thirty Years' War and subsequent lesser wars, as well as by pestilence and fire. Much of the once-cultivated land was now unplowed, and many areas were deserted and reverting to scrub. The Swedes could not estimate the tax basis of their colony without a thorough survey to ascertain current ownership and use of land.[114] Mapping in Pomerania began in 1681, but, as in Livonia, the nobles opposed the new surveying methods. They too wanted to continue with the old system, under which they estimated their own land for taxation purposes. They of course often underestimated their land, despite the severe penalties which could be imposed for so doing, and they feared that they would suffer financially under the new system. As in Livonia, the first attempt at mapping failed, this time because of the use of incompetent Pomeranian surveyors. In 1691 Karl XI determined to finish the survey. He sent out eight surveyors from Sweden, and by 1699 the whole of Swedish Pomerania and Swedish Mecklenberg had been mapped. Some 1,800 maps are still extant from the Pomeranian survey, despite their checkered history. They are on a scale of 1:8,000 or 1:16,000.[115] Curschmann points out that the maps are not cadastral maps in the strict sense of the word, in that they distinguish land areas according to soil type, land use, and land cover and do not mark each individual land parcel. As well as the large-scale parish maps, there is a general map of the whole area which was drawn up by Gunno Eurelius, the man in charge of the whole proceeding. The map measures 1.82 m × 3.10 m, meters, notwithstanding its reduced scale of 1:64,000.[116] The accompanying books give detailed information on soil types, land use, livestock numbers, and farmers' names and status.[117] The Swedish maps of Pomerania are unparalleled among contemporary German maps. Ironically enough, although they were designed as part of a tax revision, they were never in fact so used; tax continued to be collected according to the old system, under which the nobles made their own estimates.[118] Interest in the Pomeranian maps has been considerable, but they have been examined as much for the light they can shed on the state of agriculture and colonization in the area as from a point of view of cartographic interest.[119]

In the Baltic provinces Swedish mapping is considered one of the most significant legacies of the colonial period; the Swedish maps of the German provinces are unrivaled by any contempo-

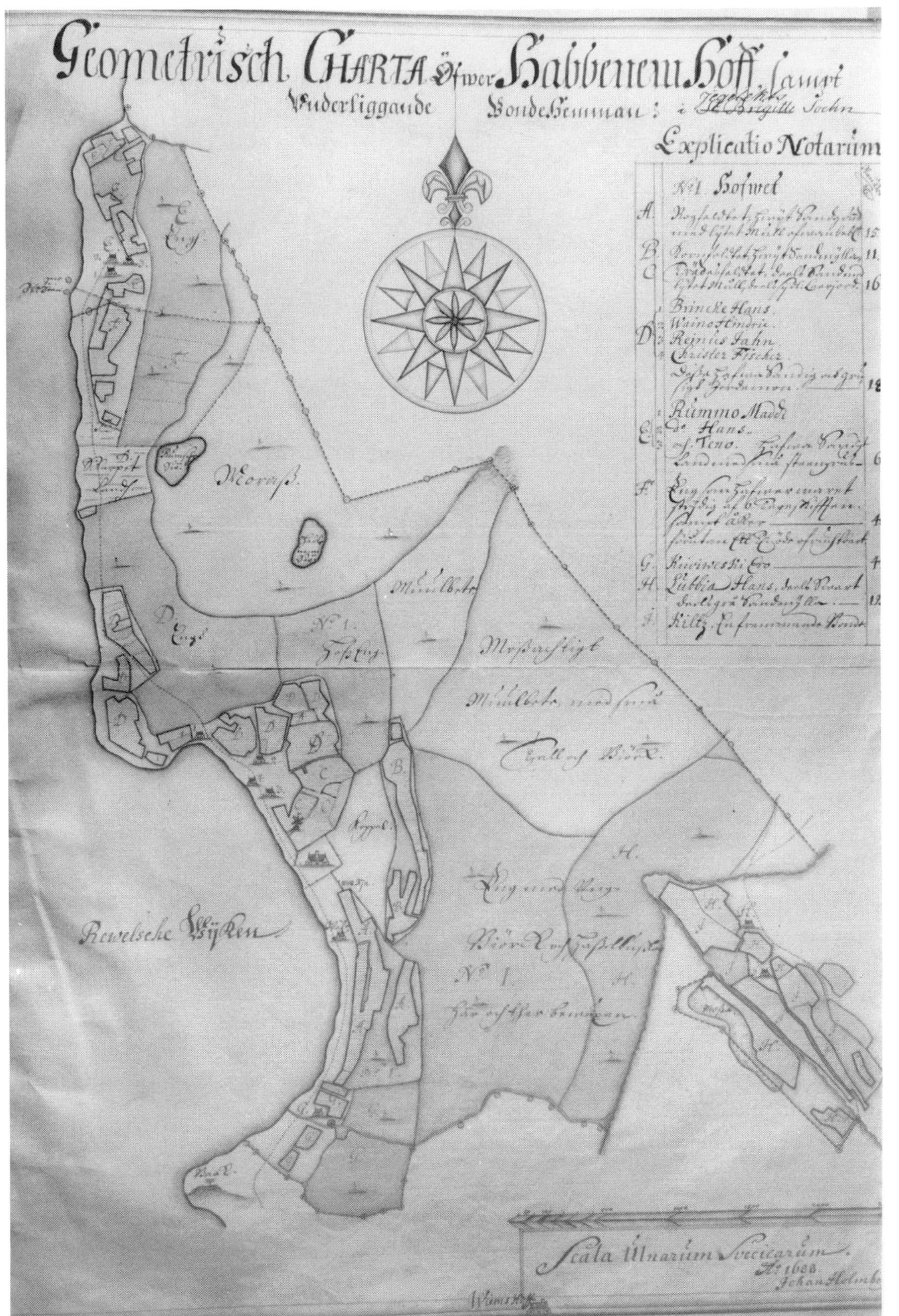

Fig. 3.12 (left) Map of Hasbenem Estate and its peasant farms drawn in 1688 by Johan Holmberg, the principal surveyor on the Great Cadaster in the Swedish province of Estonia. Source: Estonian State Archives, Tartu, I, 2, CIII Map 4.

Fig. 3.13 (opposite) Map of the Kostter Estate and village drawn in 1688 by Johan Holmberg. Source: Estonian State Archives, Tartu, I, 2, CIII Map 6.

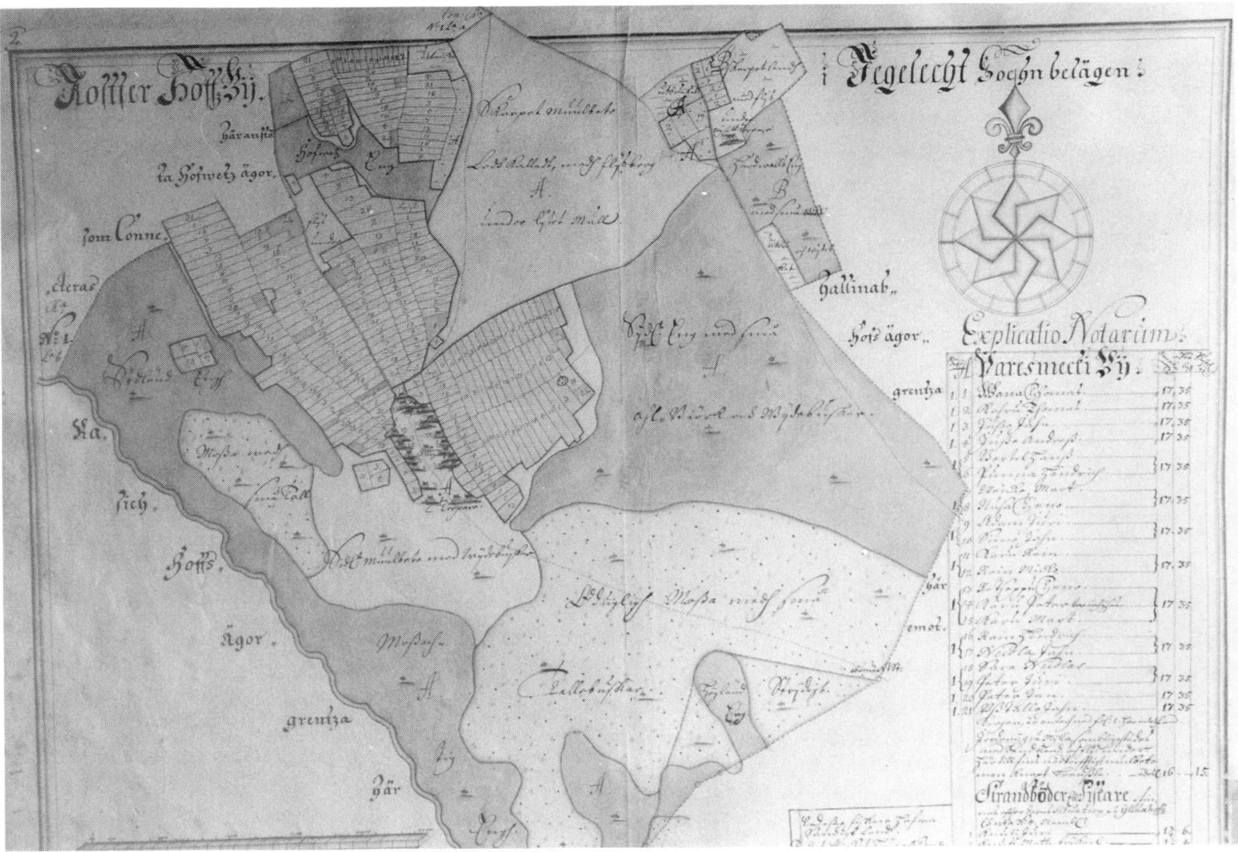

rary indigenous German mapping. When this colonial mapping is combined with the production at home of the *geometriska jordeböckerna,* the extraordinary achievement of Swedish seventeenth-century cadastral cartography can be appreciated: it is without parallel in the Western world.[120]

Denmark and Its Colonies

Denmark began the seventeenth century as the most powerful of the Nordic nations.[121] Norway had been joined in a personal union with Denmark since 1380, and in addition Denmark controlled what is now southern and southwestern Sweden (Blekinge, Halland, Helsingborg, and Skåne), as well as the islands of Gotland and Ösel off the Estonian coast. By the mid-seventeenth century Denmark had lost all of these territories to Sweden, which was rising to prominence in the Baltic. Denmark's union with Norway lasted until 1814, when it was forced to cede the coun-

Fig. 3.14 Denmark: regions, provinces, and places noted in the text. Sources: Droysens 1886; Muir and Philip 1927; Shepherd 1930; Kinder and Hilgermann 1974–78; Darby and Fullard 1978.

try to Sweden. Iceland, Greenland and the Faeroes, which had been under Norwegian rule, remained in personal union with Denmark after it had lost Norway, and they achieved varying degrees of autonomy in the twentieth century.[122] From 1448 the king of Denmark was also the duke of Schleswig (Slesvig) and the count, later duke, of Holstein (Holsten), and these two duchies are considered here. Of the other Danish lands Norway is considered separately, but Iceland, Greenland, and the Faeroe Islands are included in this section (figure 3.14).

Economic surveyors and cartographers in Denmark had five main tasks in rural areas: (1) the 1688 *matrikel* (cadaster), also called Kristian V's *matrikel* or the *landmålingsmatrikel*, for the years 1683 to 1688; (2) the survey of the Antvorskov and Vordingborg *rytterdistrikter* (property appropriated to support the cavalry) in 1768 to 1772; (3) the survey of the Jutland Alheden when this heathland was colonized in the second half of the eighteenth century; (4) *udskiftning* (land reallocation and enclosure) in conjunction with the removal of common farming, which also took place in the second half of the eighteenth century; and (5) the 1844 *matrikel,* preparatory work for which was done between 1806 and 1822. With the exception of the 1688 *matrikel,* all of these were based on a comprehensive series of maps.[123]

3.8 The Danish *Matrikel*: The Land Register

Early Cadasters

Jordebøger, or cadasters of the lands of private individuals or the Crown, are extant from at least the thirteenth century in Denmark. They were commissioned by individuals or secular or church authorities, normally for taxation purposes, and were written rather than mapped cadasters. *Matrikler,* or registers of land for the whole country, exist from the second half of the seventeenth century onward.[124]

An important early *matrikel* was that of Frederik III (1648–70). After the death of Kristian IV (1588–1648), a struggle lasting several months broke out between the heir, Frederik III, and the nobility. As the price of his election as king, Frederik had to guarantee the nobility the economic and political influence which they had gradually acquired without legal foundation. The nobility hoped to enjoy a long period of preeminence, but friction between the king and the nobles and among the nobles themselves combined with defeat in the war with Sweden to precipitate a crisis. The king and burghers united to attack the nobles, particularly over contributions to state finances, then in a parlous state. The council of nobles wanted money raised by an excise; the clergy of the state Lutheran church and the burghers demanded a new tax to be paid by everyone. This would have ended the privileges of the nobles, who claimed exemption from taxation not only for themselves but also for some of their peasants. The determination of the two lower Estates forced the nobility to retreat from its position. The provincial governors, who were nobles and who collected taxes, were brought much more firmly under royal control, and a series of reforms established Frederik III as hereditary sovereign king with a new constitution known as the King's Law. Under the new constitution the king promised to respect the nobles' exemption from existing taxes, so one of his first tasks after the reforms were in place was to introduce a new tax on land to which the nobles would be subject. This move at once respected and made illusory noble privilege. The new land tax was to be fairly apportioned and to bring to the Crown a much higher revenue than that formerly derived from Crown lands.[125]

The tax reforms were designed not only to attack noble exemption but also to try to end the unjust variation in tax assessment in different areas. A revision of 1662 produced a register which contained information on crop type for each farm and on the *landgild* (rent of land) which formed the basis of tax assessment, but this was not based on surveying or mapping. Between 1663 and 1664 in Frederik III's *matrikel* the register for the whole of Denmark, excluding the island of Bornholm, was revised to try to satisfy criticisms of the 1662 *matrikel*, but again the revision was not based on surveying or mapping.[126]

The 1688 *Matrikel*, or Kristian V's *Matrikel*

Frederik III's *matrikel* of 1664 was still inaccurate: rents were not uniform over the country, tenants who paid high rents were also too heavily taxed, and the *matrikel* as a whole bore evidence of having been drawn up in a hurry. In the instruction of 16 April 1680 to the newly established Kammerkollegium, or Rentekammeret (Exchequer), the *rentemester* (chancellor) was ordered to draw up a new cadaster for Denmark. The original plan was for a revision of the old *matrikel*, but it was decided to start afresh. The 1688 *matrikel* thus became the first to be based on a systematic survey rather than a revision of existing cadastral information. Chancellor Henrik von Stöcken initiated the survey, but the real impetus came from the nobleman Knud Thott. He had large possessions in Skåne and became a Swedish subject after Skåne was ceded by Denmark to Sweden in 1660. He played an important role in tax surveys of Skåne between 1670 and 1672. After the war of 1675–79 between Denmark and Sweden, Thott settled in Denmark, where his surveying experience was put to good use. His first task in 1679 was to investigate how the *ryttergods* could best be valued. The valuation was in the end not carried out, but in 1681 Kristian V ordered commissioners and surveyors to prepare a revision of the cadaster of the entire country. Zealand, Fyn, and Falster with their respective islands were surveyed in 1681–82, and Jutland with parts of Schleswig in 1683; Bornholm was not surveyed, and here the *skyld* system (section 3.18) remained in force until 1844. The new cadaster was to be based on comprehensive surveying and mapping under the direction of Jørgen Dinesen Oxendorph, later professor of mathematics at Copenhagen University.[127]

The *matrikel* was to ensure a reasonable assessment of taxation by measuring the extent of arable land in *tønderland* and expressing its productive capacity in *tønder hartkorn.* One *tønde land* was equal to fourteen thousand square Zealand ells, or 1.363 acres. The royal resolution of 8 August 1685 gave the scale of relation of *tønderland* to *tønder hartkorn* so that for first-class land two *tønderland* equaled one *tønde hartkorn* and so on down to sixth-class land, for which sixteen *tønderland* equaled one *tønde hartkorn.*[128] The surveyors classified the land according to its productivity, which depended on soil quality and the proportion of time it would lie fallow.[129] This very complicated system of tax assessment was one of the reasons for the constant grumblings about the inequalities of the taxation system and the frequent and time-consuming local revisions of the *matrikel*.

The 1688 cadaster was to have been based on full surveying and mapping, although it was only ever intended to cover arable land and meadows and not forests and wastes. In the end, because of the shortage of time and skilled men, the meadows were not surveyed, and the cartographic side of the project was abandoned. One map is known to have been drawn by Jørgen Dinesen, who led the work, but this has since been lost.[130] Problems arose because most of the surveyors had had no practical experience of surveying and there were not enough of them, even after more had been appointed during the course of the revision. Several of the surveyors were German. They had served in the Danish army and later joined the project, and they had great difficulty in understanding Danish.[131] The written registers which survive contain details of landholdings, their distribution, the cropping patterns and relationships among land values, soil quality, and parcel size.[132] The great cadaster of 1688 was to form the basis for property taxation for the next 150 years and is the most important source for numerous studies of Danish agriculture, local history, and landscape.[133]

3.9 The Special *Landmåling*: The Special Survey

The 1688 *matrikel* was not kept up to date. By the mid-eighteenth century there was some discussion as to whether there should be a new *matrikel* to take account of enclosure, which had dramatically altered landholding patterns in many parts of the country, and of the many areas which were unsurveyed wastes in the 1680s but had since been taken into cultivation. In 1768 the Exchequer decided on a new cadaster based on surveying and mapping of the whole nation to lead to a new calculation of area and assessment of taxes.[134] The project was to be completed in four years and was to begin in the *rytterdistrikter*. These covered large areas of the country's best land which the strained Exchequer intended to auction off to raise money.[135] The *rytterdistrikter* had already been surveyed between 1720 and 1723, and one map from the earlier survey was reused in this second project.[136] The first survey was in general not detailed enough for the Exchequer's purpose, and new surveying began in the Antvorskov *rytterdistrikt* and later progressed to Vordingborg. In the end the survey never got beyond these areas, together with small parts of the *ryttergods* around Copenhagen. Work ceased entirely following the royal resolution of 3 March 1772 because of lack of money and because the new Guldberg government reacted against reform.[137] Thomas Bugge was in charge of the work and had two *landinspektører* (survey inspectors) and about a dozen *landmålere* (surveyors) under him, but shortage of surveyors was a constant

Fig. 3.15 Detail of the special *landmåling* map of 1768–72 of Vordingborg *ryttergods* (property appropriated to support the cavalry). Source: Matrikelarkivet, Kort- og Matrikelstyrelsen, Copenhagen, Æ 23–10.

difficulty. The surveyors used the most modern equipment, following the example of the Videnskabernes Selskab (Scientific Society), which had surveyed and mapped the country under Thomas Bugge a few years earlier. The 1768–72 survey was known as the special survey to distinguish it from the Videnskabernes Selskab's topographical survey. Surviving from this survey are 139 maps at a scale of 1:4,000 (figures 3.15 and 3.16) and one general map of Antvorskov at a scale of 1:24,000. There are also thirty-seven bundles of documents which specify the ownership of the land, its cultivator, and the tax assessment, together with four journals which contain correspondence and instructions to surveyors. The maps themselves are very finely produced and

Fig. 3.16 The special *landmåling* map of 1769 from Antvorskov *ryttergods*, scale 1:4,000. Source: Balslev and Jensen 1975:map 5. Printed with permission of Matrikelarkivet, Kort- og Matrikelstyrelsen, Copenhagen.

show buildings, place-names, roads, forest, and land use and have fine decorative cartouches and vignettes.[138]

3.10 The Surveying of the Jutland Alheden

Ownership of Alheden heath in Jutland had been vested in the Crown since the Middle Ages, but peasants from the surrounding areas had customary rights of grazing and turvery, and there was steady encroachment on the heath by individuals. From the beginning of the eighteenth century there were attempts to colonize the heath more systematically, and in 1753 a project was begun to

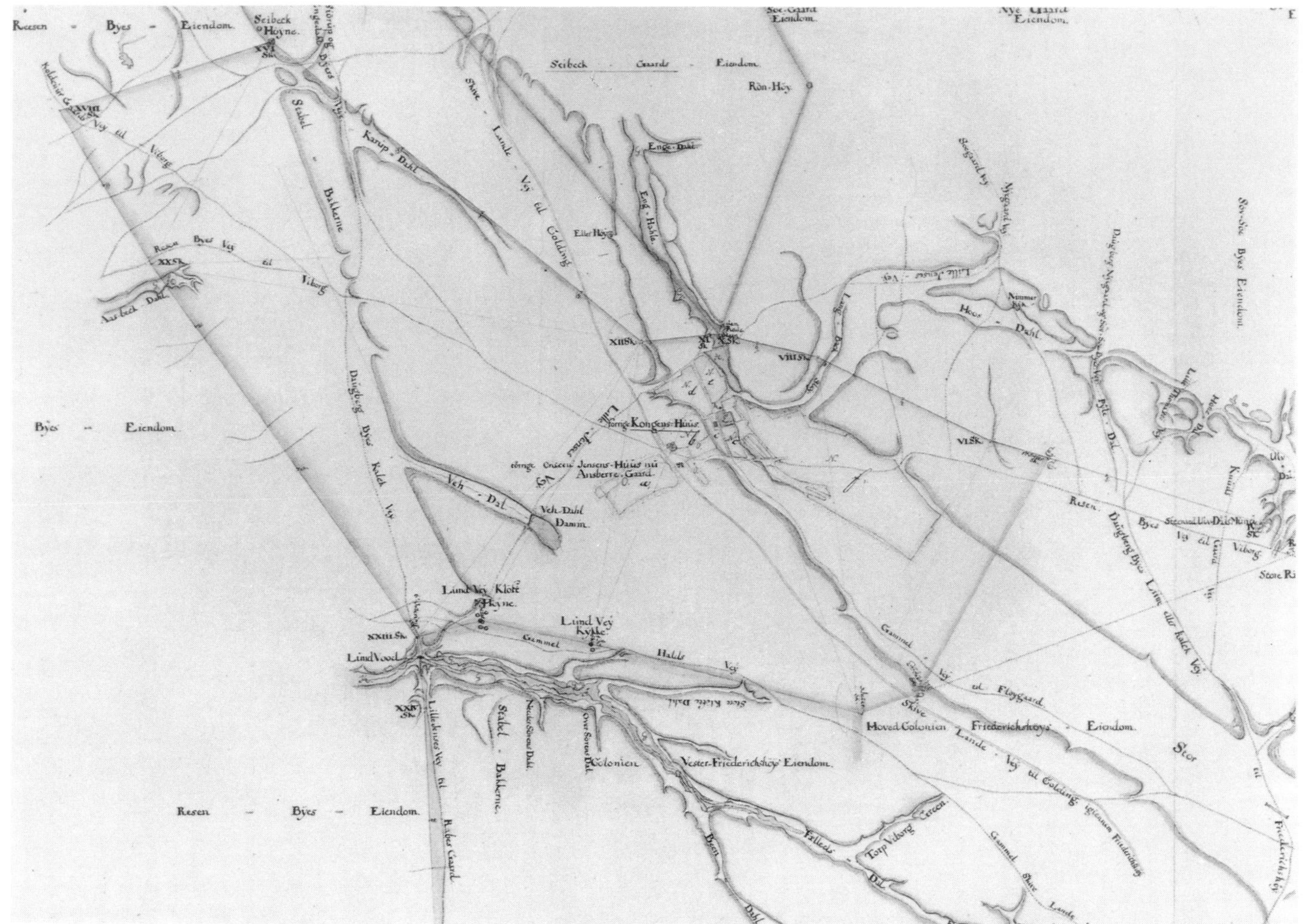

Fig. 3.17 General map at 1:12,000 of the Jutland Alheden, surveyed and mapped in 1770 by Ancker Grolau at scales of 1:4,000 and 1:12,000 as part of a colonization project. Source: Balslev and Jensen 1975:map 4. Printed with permission of Matrikelarkivet, Kort- og Matrikelstyrelsen, Copenhagen.

colonize it by German peasants with official financial support. It was important to differentiate Crown from private land so that the former could be parceled out for colonists. Three survey commissions were set up to survey the land, codify the peasants' customary rights, divide the Crown land into plots, assess these for tax, mark boundaries, and draw maps. There was some suspicion of surveyors because of their connections with taxation, but it was explicitly stated that the purpose of the survey was not to cheat the peasants but to make sure that all got their due.

The work ran into some practical difficulties. The surveyors lacked equipment and were unable to borrow it from civil or military authorities. Eventually the Danish instrument maker Muth was instructed to make new equipment for the surveyors. The Exchequer showed a lack of sympathy with some of the surveyors, notably Lieutenant Ancker Grolau, but he is vindicated by the high quality of his maps at a scale of 1:4,000 (figure 3.17).[139]

3.11 *Udskiftning*: Land Reallocation and Enclosure

Agriculture in Denmark was based on strip farming of open fields with widespread intercommoning, and there was some pressure for a reform of the system from the mid-eighteenth century.[140] The leading publication advocating reform was *Norges og Danmarks Økonomiske Magasin*.[141] The *Magasin* was set up in 1755 on the birthday of King Frederik V (1746–66), and his Danish and Norwegian subjects were invited to send in proposals for economic improvement in the two kingdoms. Publication at the Crown's expense was promised for all useful suggestions, and the results were seen in the eight volumes of the *Magasin* published between 1757 and 1764. Much space was devoted to agricultural reform, which was regarded as the most pressing economic issue of the day. The *stavnsbaand*, which tied the peasant to his lord's estate for as long as he was liable for military service—that is, from his fourteenth to his thirty-sixth year—was condemned in the magazine as serfdom, and *fællesdrift* (cultivation in common) and labor service were criticized. The Crown was hampered in agricultural reform, as it depended on estate owners for its revenue from the land tax and recruits for the army, but steady advances were made.[142] The publishers of the magazine included Erik Pontoppidan, a topographer and cartographer; as in Sweden, a cartographer was thus among the chief advocates of agricultural reform.[143]

There was some early enclosure in Schleswig-Holstein (section 3.15) and, in the Kingdom of Denmark, on the private estates of reformers. An early private enclosure was carried out between 1764 and 1766 on the Bernstorff family estate. Frederick Bernstorff was a famous Physiocrat (chapter 5.7) and reformer.[144] The villages of Gentofte, Ordrup, and Vangede were surveyed by Peter Jacob Wilster, who produced a book containing fifty-one maps, including maps of the whole estate before and after enclosure at a scale of 1:8,000, maps of each village at a scale of 1:4,000 before (figure 3.18) and after (figure 3.19) enclosure, and a special postenclosure map of the whole estate at 1:2,000.[145]

Danish Physiocrats were leading advocates of enclosure and agricultural reform, as they regarded land as the basis of national wealth and power. A leading figure among them was Count Adam Gotlob Moltke, who, like Bernstorff, was German born and who wanted to reform his own estate at Bregentved. He persuaded Frederik V to appoint the first land commission of 26 November 1757, which was to gather information about enclosure from landowners and *amtmænd*

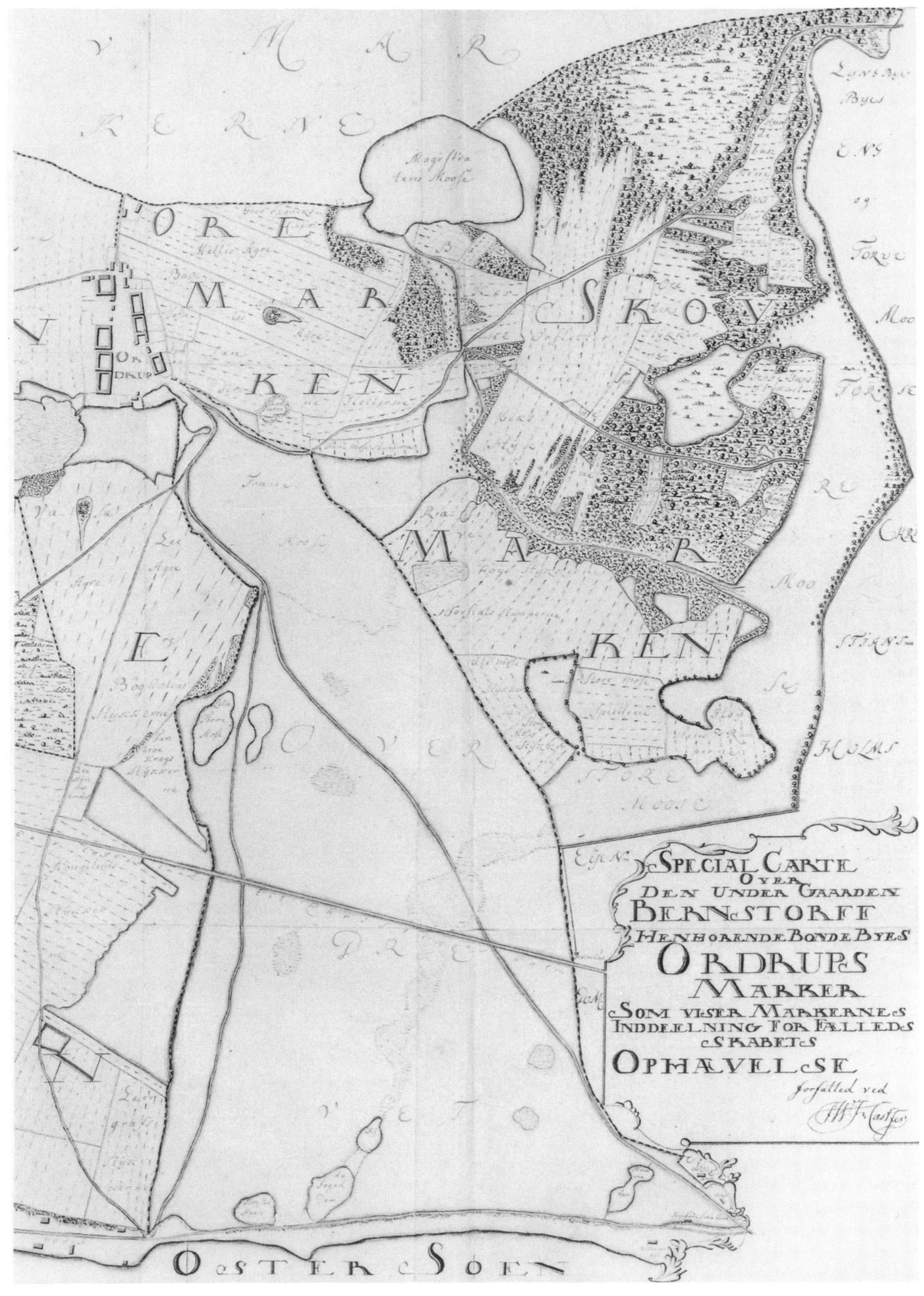

Fig. 3.18 Map of the Bernstorff Estate before enclosure, at a scale of 1:4,000, prepared 1764–66. Source: Balslev and Jensen 1975:map 6. Printed with permission of Matrikelarkivet, Kort- og Matrikelstyrelsen, Copenhagen.

Fig. 3.19 Map of the Bernstorff Estate after enclosure, at a scale of 1:4,000, prepared 1764–66. Source: Balslev and Jensen 1975:map 7. Printed with permission of Matrikelarkivet, Kort- og Matrikelstyrelsen, Copenhagen.

(chief county officials). Many landowners opposed enclosure, which they feared would diminish their control over the peasantry, but the officials generally welcomed it. The work of the commission led to various instructions which tried to encourage the separation of each village's pastureland from that of neighboring villages and which removed legal restrictions on enclosure. These instructions were, however, rather ineffectual, and it was the law of 1781, with its element of compulsion, which laid the foundation for the rapid and comprehensive restructuring of Danish agriculture, with the removal of communal farming and in many places a radical alteration of the settlement pattern.[146] The act stated that an individual landholder could demand the enclosure of all village land. All landholders then had to share the costs of surveying and compulsory mapping.

Enclosure started rapidly after this act and gained momentum from the act of 1792, which gave estate owners the right to pass on the cost of enclosure to their tenants and which prescribed that plots were not to be more than four times as long as they were wide and that there should not be more than 1,500 ells from the farmstead to the furthest field. This contrasted with the 1781 act, in which instructions about how the land was to be reallocated were limited and vague: each farmer was to have a small number of strips which were to have a *bekvem form* (convenient shape). Acts of 1805 and 1858 extended enclosure to forests and turveries.[147]

Enclosure had four stages: surveying, provisional plan of division, taxation of land through the assessment of land value, and final plan of division. In surveying, the *landmåler* fixed the village boundaries, which were previously uncertain, surveyed each individual plot, and drew a provisional map. Although there was some reuse of the *ryttergods* maps from the 1770s, most were drawn specifically for the purpose at the same scale of 1:4,000, which had been used for the *ryttergods* maps.[148]

In 1768 the first *landinspektør* to assist with enclosure on nonestate land was appointed. The law of 1769 said that an *inspektør* was to be appointed if anyone demanded it. In 1772 costs for surveying, taxation, and reallocation were fixed, but in many cases the landowners could not afford this. In 1776, in order to encourage enclosure, it was ruled that a salaried *stiftslandinspektør* was to be appointed in each of the country's seven *stift* (dioceses) to supervise enclosure at no cost to the landowners if they requested it. However, in 1776 there were only thirteen *landinspektører* and about twenty *landmålere* for the whole country, and most of these did not have their own equipment but had to borrow it from the Landmålingsarkiv.[149] The employment of a *landmåler* and a *landinspektør* in enclosure was made compulsory under the 1781 act: costs were to be shared by the villagers, regardless of whether they wanted the enclosure or not.[150]

In the second stage of enclosure the *landinspektør* made a provisional allocation plan which was approved by the agriculture commissioners and lodged with the Exchequer. At the third stage the quality of the land was assessed by two *boniteringsmænd* (tax assessors) under the *land-*

Fig. 3.20 Enclosure map of 1775 of Roved village in Coldinghuus *amt.* This is a fair copy of the enclosure map sent in to the Exchequer. Source: Balslev and Jensen 1975:map 8. Printed with permission of Matrikelarkivet, Kort- og Matrikelstyrelsen, Copenhagen.

inspektør's supervision. The *landinspektør* did the final calculations, and the allotment was carried out. The peasants were obliged to house the *landinspektør* and to take him to his next place of work when he had finished in their village. Surveying was supposed to take place when it would least inconvenience farm work.[151]

Enclosure began in the 1760s, reached its peak in the 1790s, and continued until the 1820s, by which time it was virtually complete throughout the country. Enclosure maps are most numerous for the period 1790 to 1810. The maps by no means cover the whole land area of Denmark, since only villages, as opposed to forests and other wastes or private estates, were involved in the reallocation. The original maps contain a wealth of detail, including the land divisions before and after reallocation, but the copies which were sent to the Exchequer contain only the new land divisions and are far less detailed in general (figure 3.20).[152]

Enclosure was the agricultural reform which depended on the construction of cadastral maps, but parallel with it were reforms governing the peasants themselves. Together the measures dismantled the legal basis of feudalism and allowed capitalist agriculture to flourish. In 1787 tenants were given full protection against the lord of the manor, and in 1788 *stavnsbaand* was formally and finally abolished. Other measures regulated the labor services of tenant farmers, made it easier for them to buy their land, and abolished manorial privileges such as the monopoly in the trade in stall-fed cattle. The combination of enclosure and these other reforms transformed Danish agriculture in the late eighteenth and early nineteenth centuries: arable production was three times higher in 1807 than it had been in 1750, and although landless cottagers remained wretchedly poor and ruthlessly exploited, the position of the independent peasant had improved markedly.[153]

3.12 The *Matrikel* of 1844

By 1800 about one-sixth of the total land area of the country was exempt from taxation. This cast an unfair burden on those who were not exempt, and the variation in their taxation was itself very considerable. A revision of the 1688 *matrikel* was set in motion by an order of 1802. On 16 May 1804 the king approved the Exchequer's suggestion that the revision be based on surveying and mapping, and the *Matrikulsinstruktion* of 6 June 1806 began the revision which led to the *matrikel* of 1844.[154] There was consensus that a new cadaster should be based on maps. To save money it was decided to use enclosure and *rytterdistrikt* maps to draw up the new *matrikel*. Work began under the direction of Christian Rothe and continued, despite the interruption of the Napoleonic Wars of 1807–14, during which many surveyors who were army officers were recalled to their regiments. The Exchequer required that the enclosure map for each village be sent in to be copied by army officers. The instruction of 1806 declared that no enclosure map could be used in the *matrikel* until it had been tested in the field and any mistakes corrected. If the map was inaccurate by more than 0.5 percent, it was rejected. Further checks ensured that adjacent maps joined up satisfactorily, and then they were brought up to date with the addition of boundary changes made after enclosure. Areas with no maps or for which the maps had been rejected were surveyed afresh, but not always thoroughly. The new maps were tested in the same way as the old; if they were found to be inaccurate, a new full survey was carried out at the expense of the surveyor concerned. The Danish islands were in general relatively well covered by existing maps, while the Jutland peninsula was poorly covered, especially in the west and north. The precise number of new maps is not known, as they are very difficult to distinguish from the enclosure maps. In total there are about 8,500 maps which differ greatly in size and format. They are normally at a scale of 1:4,000 but may be at 1:2,000 or 1:8,000. They show, among other things, property boundaries, public and private roads, common gravel pits, clay pits, buildings, ditches, forest, heath, moor and watering places, and, for each plot, its land registry number and the tax payable on it (figure 3.21).[155]

Surveying and mapping were carried out from 1806 to 1822, the productivity of the land was assessed between 1822 and 1826, and taxation assessment then began. Since it had been decided to continue with the system of *hartkorn,* tax assessors had to decide how many square ells of *bonitiert* land (cf. actual areal ells) should equal one *tønde hartkorn.*[156] *Bonitering* was measuring of the productive quality of the land. In the 1844 *matrikel* this was measured on a scale of 0, for the poorest land, to 24, for the best (cf. section 3.8).

Protokoller or *matrikler* (written cadasters) were drawn up for each parish. They record precise details of the ownership of land, while the maps are intended as a summary and overview. The 1844 *protokoller* give information on the owner of the land and its cultivator or user, its old taxation assessment in *hartkorn,* its geometric area, its new *hartkorn,* and a description of the plot. The public roads round the plot are described, and the owner's total landholding of combined plots given. Today 1,630 of these *protokoller* from the 1844 *matrikel* are extant, together with *koncepter* (drafts) of some of them.[157] There are also work journals of the *matrikel* work (it was a statutory obligation to keep such journals), which give valuable information on such things as plowing methods, forest resources, and roads. There are also *hartkornsekstrakter,* which are summaries of cadastral information for each hundred.[158]

The perennial problem with the cadaster was that of keeping it up to date. An instruction of 1806 required this for the first time, and a further instruction of 1810 required that when a plot was divided, a map of the relevant land was to be sent to the Exchequer. The changes on the map were to be noted, and the map returned to the landowner. From 1819 onward, maps were to be drawn in duplicate, one for the landowner and one for the Exchequer. There continued to be difficulties, however, in keeping the maps current: railway lines, for example, were never added

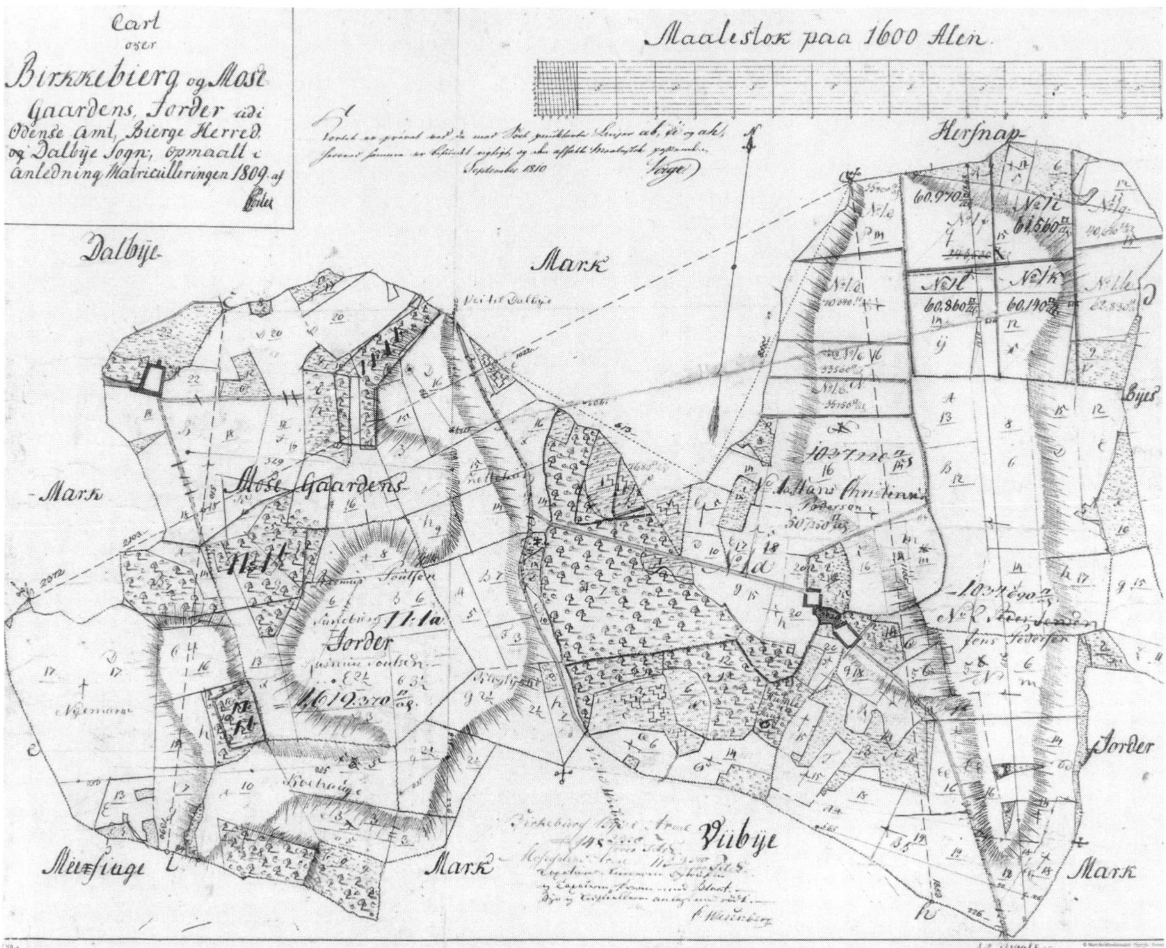

Fig. 3.21 *Matrikel* map, 1809, of Birkkebierg and Mose in Odense *amt,* which was in use from 1844 to 1852. Alterations were made in red on the original as property relations changed. Source: Balslev and Jensen 1975:map 9. Printed with permission of Matrikelarkivet, Kort- og Matrikelstyrelsen, Copenhagen.

to the *matrikel* maps, and where there were frequent changes, it was difficult to keep the maps legible. *Tillægsberegninger* (supplementary documents) were used from 1810 to 1857 to update the *protokoller.* After 1857 less-detailed *forandringsprotokoller* (records of alterations) were used to keep the register current.[159]

The 1844 *matrikel* has remained in force to this day for all areas other than market towns, which were removed from the general *matrikel* in 1809.[160] From about 1870 an annual sum of up to twenty thousand *kroner* was set aside for surveying to bring the *matrikel* maps up to date. With this money the most densely populated rural districts were surveyed, and maps drawn at a scale of 1:800. Triangulation was prescribed, and maps were to be checked against other maps and in the field. Toward the end of the nineteenth century there was increasing demand for a cadaster which was unconnected with taxation; such a cadaster was created following the recommendations of the 1903 Commission of the Department of Agriculture.[161] The maps used by the Matrikeldirektorat today are, in the main, updated versions of maps surveyed in the early nineteenth century for the 1844 *matrikel,* although some areas were resurveyed from the end of the nineteenth century and new maps drawn at a scale of 1:800 or, later, 1:2,000.[162]

3.13 Mapping, the *Matrikel*, and Ownership

One of the most interesting aspects of the 1844 *matrikel* is the role of the map in changing the concept of ownership. Ownership in Denmark has not been fully explored in either the historical or the legal literature. The word *ejerlav* first passed into the juridical terminology at the end of the nineteenth century, when it was defined, if it was defined at all, as the entity registered in the *matrikel* of 1844. Originally *ejerlav* was defined not territorially but with regard to a group of people who had rights of use of particular areas of land, typically arable, meadow, waste, and forest. In seventeenth-century *matrikel* documents it was the village with its distinctive name, the community of villagers, which was the *ejerlav.* Through time the concept of *ejerlav* attached itself to the land rather than the people.

From the Middle Ages there were laws demanding the physical fencing of one village's land from the next, but in fact intercommoning was frequent. Enclosure brought a considerable regulation of village boundaries, which had often previously crossed parish boundaries, and of parish boundaries, which had previously often cut across hundred boundaries. Enclosure was also a significant step away from the notion that property resided in people rather than land. The use of enclosure maps in the 1844 *matrikel* meant that the area of the enclosure map became equated with that of the *ejerlav.* Enclosure normally took place in each *dyrkningsfællesskab* (unit of common farming), which could cover many villages. In the 1844 *matrikel* and in subsequent general usage, it was the property covered by one enclosure map, not the people of one village with their assorted use rights, which was the *ejerlav.*[163] Cadastral maps often record changes in the concept of ownership, but their role in changing such concepts is rarely so apparent.

3.14 Technical Considerations and Training

Some of the most obvious aspects of Danish cadastral cartography are the widening availability of instruction books and the increasingly systematic training. The influence of cartographers such as Thomas Bugge and John Johnsen in the late eighteenth and early nineteenth centuries helped to standardize the conventions used by cadastral cartographers.[164] The flamboyant and individual decoration of some early *rytterdistrikt* and enclosure maps gave way to simpler styles. Lack of time, money, and personnel also encouraged the trend toward simplicity and uniformity.[165]

Until the 1780s surveyors were trained in a wide variety of ways. The Danish Landkadetakademi trained some military surveyors; others received their training abroad. Some received fairly formal instruction, and others learned their craft in the field from experienced surveyors. Still others were self-taught. This variety of training, or lack of it, gave rise to great variety in the earlier maps. In 1781 and 1782, respectively, examinations for aspiring *landmålere* and *landinspektører* were set up. For the next eighty years *landinspektører* and *landmålere* had to pass examinations in mathematics, surveying, and land reallocation before they could be appointed to public bodies. In the 1850s the examinations were combined, and the *landmåler* in effect disappeared. Since 1858 the Royal College of Veterinary Medicine and Agriculture has provided practical and theoretical surveying instruction for *landinspektører.*[166]

3.15 Sønderjylland: The Duchies of Schleswig, Holstein, and Lauenburg

The Duchy of Holstein, at the southern end of the Jutland peninsula, and, to the south of it, the Duchy of Lauenburg were part of the Holy Roman Empire. The Duchy of Schleswig, which adjoined the Kingdom of Denmark to the north, was not. The House of Oldenburg supplied the dukes of Schleswig, the counts (after 1474 the dukes) of Holstein, and, from 1448, the kings of Denmark. Schleswig and Holstein were thus joined in personal union to Denmark, and in 1460 the two duchies were declared joined in *Realunion,* “forever indivisible.” Lauenburg passed to the Danish Crown in 1815 at the Congress of Vienna. In spite of the *Realunion* the Danes attempted to incorporate Schleswig into Denmark in 1848, but this was opposed by the Germans, who insisted on the indivisibility of Schleswig-Holstein. Uprisings in Schleswig-Holstein and military intervention by Prussia led in 1864 to the ceding by Denmark of Schleswig-Holstein and Lauenburg to Prussia. In 1864 Denmark also gave up certain royal enclosures of land which lay chiefly between Ribe (Ripen) and Tønder (Tondern). As a compensation Denmark received the eight parishes of Dalby, Hejls, Stenderup, Sønder-Bjert, Taps, Vejstrup, Vonsild, and Ødis, which all lie south of Kolding, and the island of Ærø (Ärö) in Schleswig. In 1918 at the Treaty of Versailles northern Schleswig was ceded to Denmark. In 1920 the *ämter* (counties) of Haderslevhus (Hadersleben), Sønderborg, Toftlund, Tønder, and Åbenrå (Apenrade) in North Schleswig were reunited with Denmark following a plebiscite in which the majority of the population voted to join Denmark rather than remain part of Germany (figure 3.14 above).[167]

There are still some administrative peculiarities of Sønderjylland: it has a separate, decentralized cadastral organization which follows the Prussian Katasterämter, although this organization is under overall control of the Danish national Matrikeldirektorat (Cadastral Board).[168] A comprehensive catalog and description of the mapped surveys of the duchies up to 1864 exists; the following section is a selective account of the course of mapping in the area.[169]

The first cadastral survey with maps in Sønderjylland is that of Johannes Mejer, who surveyed the villages in Åbenrå *amt* in northern Schleswig between 1639 and 1641. Mejer was later to become royal mathematician to Kristian IV and to make important early topographical maps of the country.[170] The Åbenrå maps are some of his earliest and consist of one map of the whole *amt,* one of each of the two hundreds in the *amt,* nine parish maps, and fifty-one maps of the villages with information about landholdings, property boundaries, roads, wastes, and meadows. The village maps are at a scale of 1:12,500, the parish maps are at 1:25,000, and the hundred and county maps are at 1:50,000. All maps show their construction lines. Mejer used a unit of length which was based on a rod of eighteen feet, but it is not wholly clear which foot he used. The maps together with a *jordebog* (terrier) form the book *Erdt Buch des Ambts Apenrade,* which was drawn up on the order of *Amtsforvalter* (county administrator) Joachim Danckwerth in an effort to make the taxation of the area more equitable.[171]

In the early eighteenth century Tyrstrup hundred in Haderslevhus County together with adjacent small parts of Tønder *amt* in northern Schleswig had two surveys, possibly connected with the work of a commission which was to make a complete cadaster of Haderslevhus *amt.* Between 1714 and 1715 the surveyor Peter Petersen surveyed seven parishes in the area using surveying methods of the 1688 *matrikel,* namely, measuring the length and breadth of individual plots but making no maps. He was followed between 1716 and 1718 by Samuel Gries, who surveyed the eight parishes which became Danish after 1864. The survey was to provide the basis for a more equitable taxation policy. It nevertheless proved unpopular, and there were many requests for exemption from the survey. These were not accepted, however, and surveying and mapping went ahead. Work was broken off suddenly in 1718, so the maps which survive are only in draft form. They are at a scale of 1:4,000.[172]

A survey of the island of Ærø was carried out in 1734 by First Lieutenant Christian Ludolf Pape, who surveyed church land on the island and produced both written cadasters and colorful maps at various scales between 1:1,000 and 1:2,500. The maps have an eastern orientation and include bird's-eye views of houses and trees. Church land is clearly marked. He used the Zealand ell, as the introduction of the Rhineland ell in Denmark in 1683 had not applied to Ærø. The cadaster contains some rather gross errors—for example, some land areas are doubled because of the calculation method used. The reasons behind the survey are unclear; it was perhaps an attempt by the bishop on economic grounds to have church land clearly demarcated, or perhaps an effort to prevent or settle land disputes with the peasantry. Indications of such disputes are found in the written cadasters.[173]

Enclosure began earlier in the duchies than in the Kingdom of Denmark. At the end of the seventeenth century enclosure began both spontaneously on the initiative of independent peasants and through the action of large landowners, and it increased rapidly in the eighteenth and through the nineteenth centuries. Early enclosure was privately organized and piecemeal. The permission of the *amt* was needed before it was carried out, and thus it very often involved legal documents including maps. The most important phase of enclosure in the duchies followed the instructions of 1766 and 1770, which predated but were less radical than the Enclosure Act of 1781, which initiated enclosure in the rest of Denmark. The instructions served to speed up the process where it had already started and to bring it to those areas where it had not. The authorities now approved of enclosure and other economic reforms designed to intensify the use of church and secular lands, especially common land. Instead of being seen as expensive and superfluous, surveying and mapping were increasingly regarded as an integral part of the process to protect the state and individual land proprietors from potential financial claims on their lands, to increase the efficiency of taxation, and to settle land disputes.[174] The following are some of the more important mapped surveys carried out in connection with the enclosure and other reform of agricultural land in the duchies.

The mid-eighteenth century brought the famous survey by Pierre Joseph Duplat the elder between 1743 and 1750. In order to clarify the boundaries between various types of forest and meadow, he surveyed much of *amt* Schwarzenbek, other isolated villages, and the *Vogtei* (Bailiwick) of Schönberg with Franzdorf. By 1752 Duplat had completed twenty-four maps, mostly at a scale of 1:3,600, together with a general map of part of the area. These maps were part of a more general survey with the aim of enclosing land to allow more intensive farming. Later surveyors involved in the project included Emil Frederich Roth, Ernst Eberhard Braun, and Pierre Joseph Duplat the younger.[175]

Enclosure was the reason behind the outstanding survey from 1767 to 1799 under Johann Bruyn, which produced maps of 252 *Gemarkungen* (parishes) in the Duchy of Schleswig, of 168 *Gemarkungen* in the royal part of the Duchy of Holstein, and of part of the island of Fyn in the Kingdom of Denmark (figure 3.22). The maps are at a scale of 1:1,800 and often show the boundaries before and after enclosure, as well as some topographical features.[176]

The Slesvig-Holstenske Landkommission, for which Bruyn undertook his work, also ordered the survey of the whole of the county of Haderslevhus. This was carried out between 1789 and 1797, not by Bruyn himself, since he was already too busy, but by the surveyors Klaus Matzen, Frederich Feddersen, and Hans Lund. They were to survey and supervise the enclosure of all Haderslevhus villages. They worked systematically, parish by parish, and produced excel-

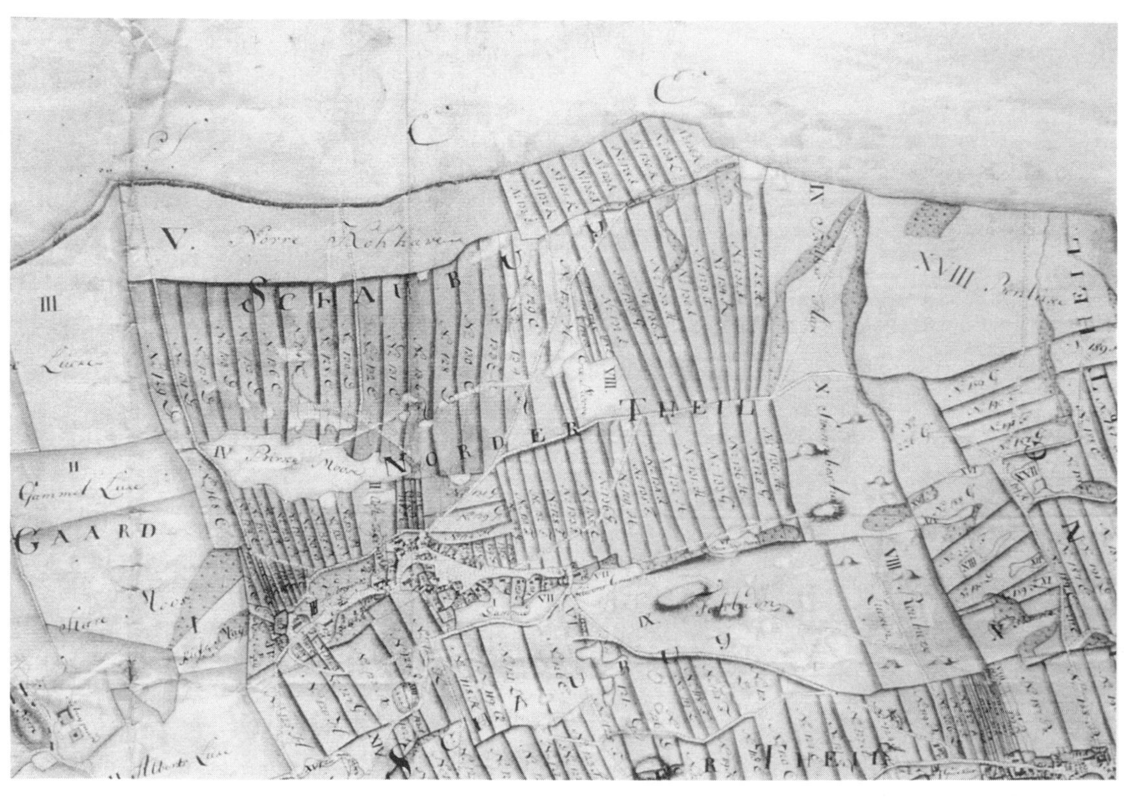

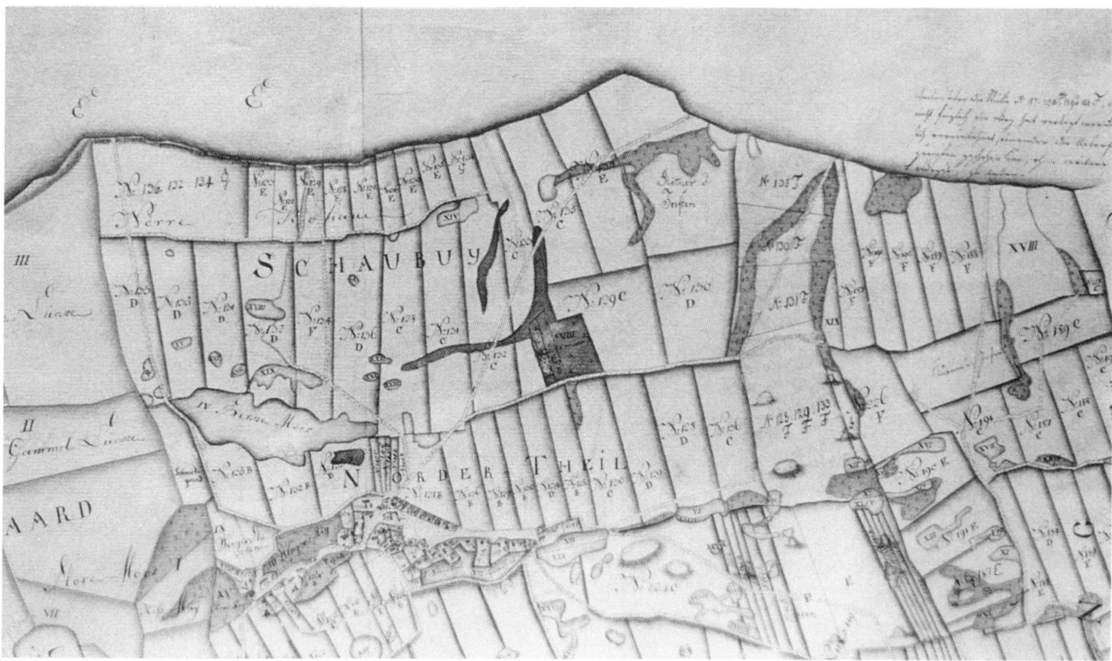

Fig. 3.22 Maps by Johann Bruyn of Seebuygaard (Skovby) Ærø, at a scale of 1:1,800: (*a*) 1772, before *Verkoppelung* (enclosure); (*b*) 1776, after *Verkoppelung*. Source: Rigsarkivet, Copenhagen, Kielerafleveringen II-3 and II-4.

lent maps. Of the surveyors Feddersen had trained and qualified in Copenhagen, and Lund had trained under Bruyn.[177]

Most of the maps which survive from the duchies are enclosure maps, but the areas were also subject to land registration for taxation and other purposes. The seventeenth-century Danish *matrikler* did not extend to south Jutland, where the basis for taxation was the *plovtall* (plow tax). Attempts by the Danish authorities to revise land taxation in the duchies in the early nineteenth century were unsuccessful. The plow tax remained the basis for taxation until 1868, after Schleswig had passed into Prussian hands, when a cadaster based on the Prussian land tax was introduced. This revision involved the surveying of the great majority of the duchy using polygonal nets linked to the triangulation of the district. The Prussian system works on the differentiation of individual land parcels which have different land uses; only after this has been done are all the plots belonging to one owner collected. Prussian maps therefore show only parcel numbers and not *matrikel* numbers, which show the ownership of the land, and each map is accompanied by two protocols: the *Flurbuch,* or *parcelbog,* lists the parcels of land, and the *Mutterrolle,* or *matrikelbog* or *artikelbog,* gives details of ownership. This parcel system is currently being phased out. The taxation assessments of the land are given in a *Reinertrag.* The maps to accompany these documents are at scales of 1:1,000, 1:2,000, and 1:5,000.[178]

Ærø and the eight parishes which came into Danish control after 1864 were registered at that time, and the data included in the 1844 *matrikel.*[179] On 12 June 1865 a cadaster of those parts of the Duchy of Schleswig which had remained Danish was ordered. There was detailed new surveying using the polygonal surveying method with occasional triangulation and plane-table surveying. Discrepancies were not to exceed 0.5 percent after shrinkage of the paper had been taken into account.[180]

3.16 Iceland, Greenland, and the Faeroe Islands

The earliest record of settlement and land ownership in Iceland is the *Landnámabók,* which records the names, land claims, and farmsteads of each of the four hundred settlers who came to Iceland between 870 and 930. Settlers and settlements on the island are described, but the aim is historical, topographical, and genealogical. It is by no means a cadastral document, although land claims have been reconstructed from it.[181] In Iceland the landholding pattern was different from that in much of Europe, as the manorial system did not exist. There were, however, large landowners who held land in various parts of the country. Mapmaking effectively began in Iceland in the nineteenth century, and for this period many maps exist. They relate to boundaries between farms; disputes over the division of coastline between farms, which was important, as farmers depended on their rights to catch fish, seals, and seabirds and to collect driftwood; and to other specific matters such as the wishes of farmers to use a church technically not their parish church but far more easily accessible or to be joined in a combined parish where the population was very sparse.[182] Systematic investigation of the maps has begun, and they have been shown to be of considerable interest for the light they shed on the Icelandic way of life. From a purely technical point of view, however, they are primitive, as they were usually drawn by people who had little if any formal knowledge of surveying or mapmaking.[183]

The drawing of cadastral maps was first mooted in the Faeroe Islands in 1896 when the Ministry of Justice expressed an interest in a large-scale map of the islands to show property boundaries in cultivated areas. It was suggested that this be combined with the topographical mapping which was then being carried out in the Faeroes. The plan came to nothing at the time, but in 1899 there was some cadastral mapping when Tveraa Inmark and Sand Inmark were surveyed and mapped at a scale of 1:5,000. The maps give a striking picture of the complex property relations of the Faeroese villages, which had developed a very high degree of subdivision and fragmentation of holdings. In the 1920s and 1930s the Matrikulsvæsen (Cadastral Board) ordered the production of cadastral maps at scales of 1:2,000 and, in densely populated areas, 1:500.[184] The Faeroes achieved a measure of autonomy from Denmark in 1948, and cadastral responsibility passed to the Matrikulstóvan (Cadastral Board) in Tórshavn.[185] Greenland has no cadastral machinery, although Greenlanders are encouraged to register property voluntarily. Rights of property for both owners and lessees remain to be codified.[186]

3.17 The Cadastral Map in Denmark

The significance of the Danish contribution to cartography is not found in technical innovation or in any specifically Danish social or economic policy which demanded cadastral maps for its realization. Danish surveyors used units of measurement, techniques, and equipment developed abroad. Individual reformers were often foreign born, as in the case of Bernstorff, or guided by foreign ideas, as in the case of the Physiocrats, whose intellectual origins were French. The Danish contribution to cadastral cartography probably owes most to the establishment in 1660 of the absolute monarchy. This enabled the introduction of a series of reforms, notably of taxation and agriculture, which both consolidated and reflected the power of the absolute monarch. The Danish kings used precise cadastral surveying and later mapping to consolidate state revenue and hence their independence from aristocratic power. Their ability to institute sweeping agricultural reforms, of which enclosure was a prominent example, rested on their absolute power and completed the subjugation of the nobility by removing feudal and manorial rights. Danish history shows clearly the link between cadastral cartography and the end of feudalism and the development of capitalism.

Norway

Norway's political history is closely bound to that of the other Nordic powers. It was in personal union with Denmark in 1380, ceded to Sweden in the Peace of Kiel in 1814 in the course of the Napoleonic Wars (figure 3.23), and achieved full independence only in 1905.[187] The unions with Denmark and Sweden were personal unions of the monarch, and under them Norway remained culturally, and certainly cartographically, distinct.

In Norway's cadastral and cartographic history the government led the way, and private maps were of little importance. Norwegian mapmaking is noteworthy, not for its technical innovation, but for the doggedness with which projects were carried out in inhospitable terrain. There were notable cartographic projects in connection with land reallocation, the resolution of property disputes, and the settling of some of the huge tracts of common land in the country. Even more interesting, however, are the occasions on which mapping was *not* carried out. Examination of the lack of maps in Norway draws attention to the general preconditions and prerequisites for cadastral mapping.

3.18 The *Landskyld* System, the *Matrikkel*, and Cadastral Mapping

The oldest surviving cadastral records, apart from isolated pre-Reformation documents, are the *skattemanntall* (taxation registers) from 1514, 1519, and 1520–22. The *skattemanntall* and *lensjordbøker* (county land books, registers of Crown land) were the forerunners of the *matrikkel,* the register of properties and their ownership. The need for a tax reform to raise state income and to share the burden of taxation more equitably was long standing and became acute in the 1640s. In 1647 investigations were begun to find out the size of each property, the form of tenure, and the length of lease. The result of this investigation was known as the *skattmatrikkel* (taxation cadaster), and it replaced the earlier *skattemanntall.*[188] A further revision of the *matrikkel* was carried out in the 1660s when a Landkommisjon was set up to assess each property for tax as accurately as possible. This *matrikkelen* was completed between 1665 and 1670 and remained largely unaltered until the nineteenth century.[189] Preliminary work in 1723 resulted in a new *matrikkel* only in the counties of Hordaland and Sogn og Fjordane.[190] From 1803 the *jordavgift* (land duty) formed a supplement to, and means of correction for, the *skattematrikkel.* In 1838 a revision which had taken twenty years to prepare was completed, and in 1863 a new and important revision of the *matrikkel* began and was accompanied by a detailed account of each farm, commune by commune. Although the detailed accounts are not published, the cadastral lists were published by commune for the whole country in the years following 1886, when the new *matrikkelen* came into force. In

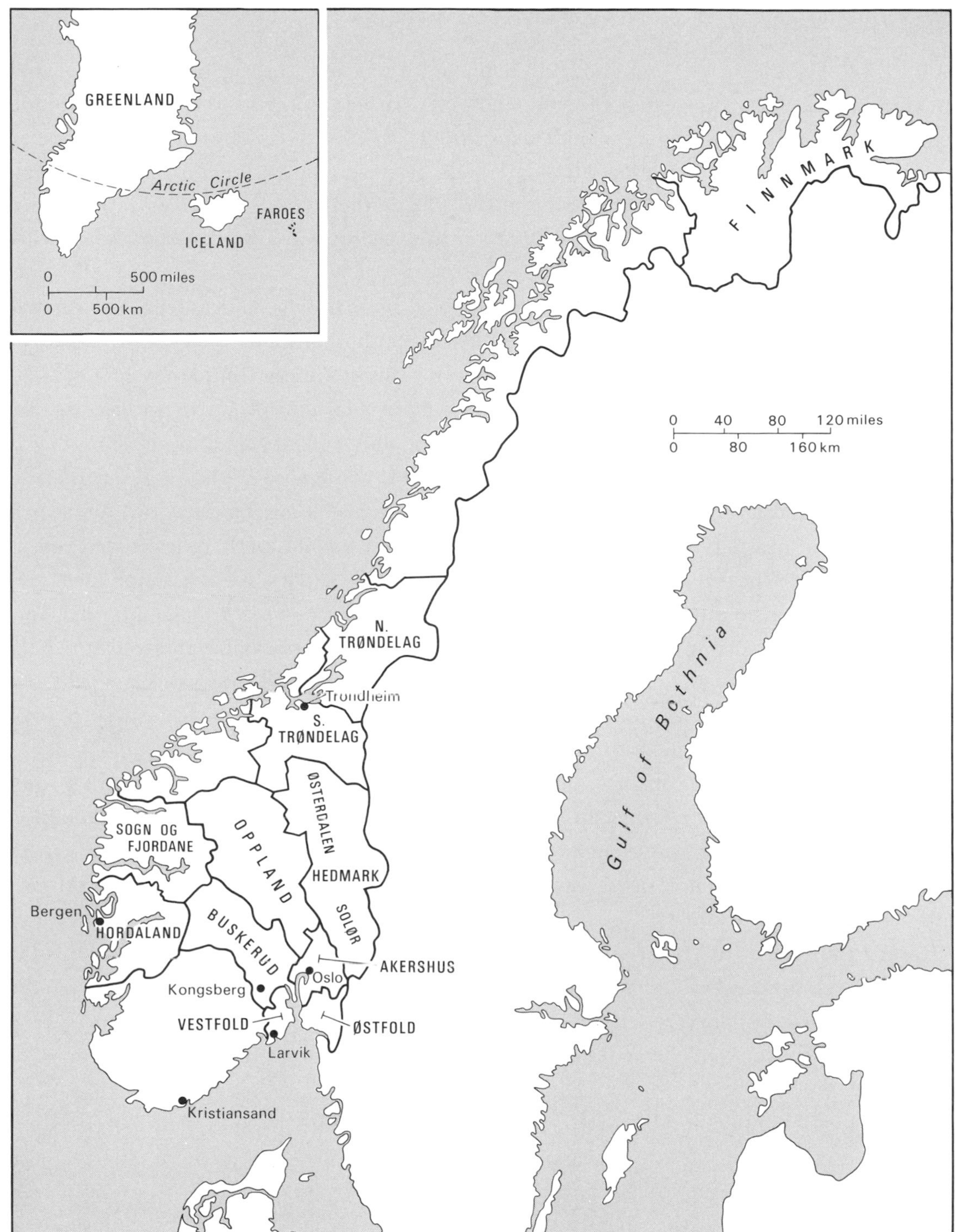

Fig. 3.23 Norway: regions, provinces, and places noted in the text. Sources: Droysens 1886; Muir and Philip 1927; Shepherd 1930; Kinder and Hilgermann 1974–78; Darby and Fullard 1978.

1904 the 1886 *matrikkelen* was brought up to date, and it was kept current from this date until about the 1950s. In the nineteenth century, land taxes payable to the state and the commune were largely replaced by income and wealth taxes, and the *matrikkel* became a register of property rather than an instrument of taxation. Finnmark in the far north was excluded from the *matrikkel* until the eighteenth century, when the state claimed all land there. After 1775 land in Finnmark was apportioned to new settlers, and the allocation was recorded in a special *matrikkel* for that county.[191]

None of these cadastral registers was accompanied by mapping, and indeed Norway has never had a national cadastral map which shows legally binding property boundaries as part of the statutory land registration system.[192] The reason for this lies in the complex system of *landskyld. Landskyld,* or *jordskyld,* originated as the yearly rent of a farm in proportion to its size and value, payable by lessees to the owners of the land. It became a fixed sum and, until 1838, was expressed in kind, commonly in hides, fish, or grain. *Skyld,* a measure of the value of the land rather than the land itself, became the object of possession and lease. Property owners owned *skyld* in a *gård,* and peasants leased and farmed *skyld* in a *gård.* A *gård* is perhaps translated best as "cadastral farm," the historic unit of settlement, which usually originated as a single farm and which was often later subdivided into two or more *gårdsbruk* or *bruksenheter* (working farms). Various forms of joint ownership of land were extremely common in Norway, but normally the joint owners could not point to any part of the farm which was their own. It was value, not physical land, which was divided and in which the owners had a share. The owner of the largest share of the *skyld* on any farm or the one with the highest social status, if the shares were equal, enjoyed the *bygselrådighet,* the right to organize the tenancy of the farm and the payment of entry fines and annual rent to the owner. Ownership or lease of *skyld* conferred rights of access to common resources such as the *utmark* (wastes) in direct proportion to the amount of *skyld* owned or leased. The corollary of these common rights was the obligation to pay tax, which was levied in proportion to the amount of *skyld* held.[193]

Toward the end of the nineteenth century *skyld* lost most of its functions as the basis for taxation, and taxation and valuation disappeared from the cadaster completely in 1980 when the GAB (*gård,* address, building) system replaced the *matrikkel* system. At that point the *skyld* system was not replaced by another system of valuation, as property tax was by then such a small part of state revenue.[194]

Mapping costs money, and the perceived benefits of mapping must outweigh the perceived costs before it is accepted as a necessary part of the cadastral system in a country. The special form of ownership and property law governing land in Norway was such that those most concerned with land—the state, the landowners, and the peasants—had little to gain from cadastral maps. Taxes due to the state were levied on *skyld,* not land, so the state had nothing to gain from the production of a cadastral map which recorded the physical extent of properties. Similarly landowners owned *skyld,* not land, and had little to gain from maps. It was important to them to mark out *gård* boundaries, but *bruk* (working farm) boundaries did not matter to them. Peasants, however, owned or leased *skyld* but farmed land and were thus interested both in *bruk* boundaries within an individual *gård* and in boundaries between *gårder.* Even peasants, however, had relatively little interest in maps, since their interests were best served by the erection and maintenance of physical boundary markers in the field. In special cases, such as disputes or reallocation, recourse to a map was necessary, but in general the costs of cadastral mapping outweighed its benefits for the peasants.[195]

Until 1980 property division was called *skylddeling* (division of *skyld*) in both official and common parlance. It was only when property owners became farm owners that property division or amalgamation took a physical form. This happened in connection with the transition to peasant freeholding, when in the course of the eighteenth and nineteenth centuries peasants came to own the land they farmed. The users of the land, rather than the owners of *skyld,* became property owners. A significant step in this process was the law of 1764 which decreed that *skyld* division must coincide with physical division; after this date farmers came increasingly to hold land in fee simple rather than to lease *skyld.* In brief, then, it can be said that Norway's land tenure system was one of division of value from the Middle Ages to 1764. From 1764 to 1980 there was division of value and physical division, and from 1980 there has been physical division only.[196]

Although with the transition from division of value to division of land it became both practicable and of benefit to the interested parties to draw maps, there was no systematic *matrikkel* mapping. The first *matrikkel* maps were drawn around 1890. They are in themselves both beautiful and interesting, but they cover only a few areas. Plans to link enclosure maps to the *matrikkel* came to nothing (section 3.23). A contributory factor was that the *matrikkel*'s tax function was greatly reduced at the end of the nineteenth century. In towns where there was a demand for maps from the authorities, obligatory cadastral surveying and mapping were introduced. However, this mapping was confined to the towns and did not spread to rural areas. In any case, large areas of the country—Finnmark, the towns, and mountains in Nordland and Troms—were not integrated into the land register system until 1980, so a nationwide cadastral map has become a possibility only in the last decade.[197]

3.19 Forest Maps

In the eighteenth century Norway's forests were of great value to the Danish-Norwegian state, both commercially and for the navy. In order to manage them more effectively the Skog- og Sagkommisjon (Forestry and Timber Commission) was set up in 1725 to investigate the forests of

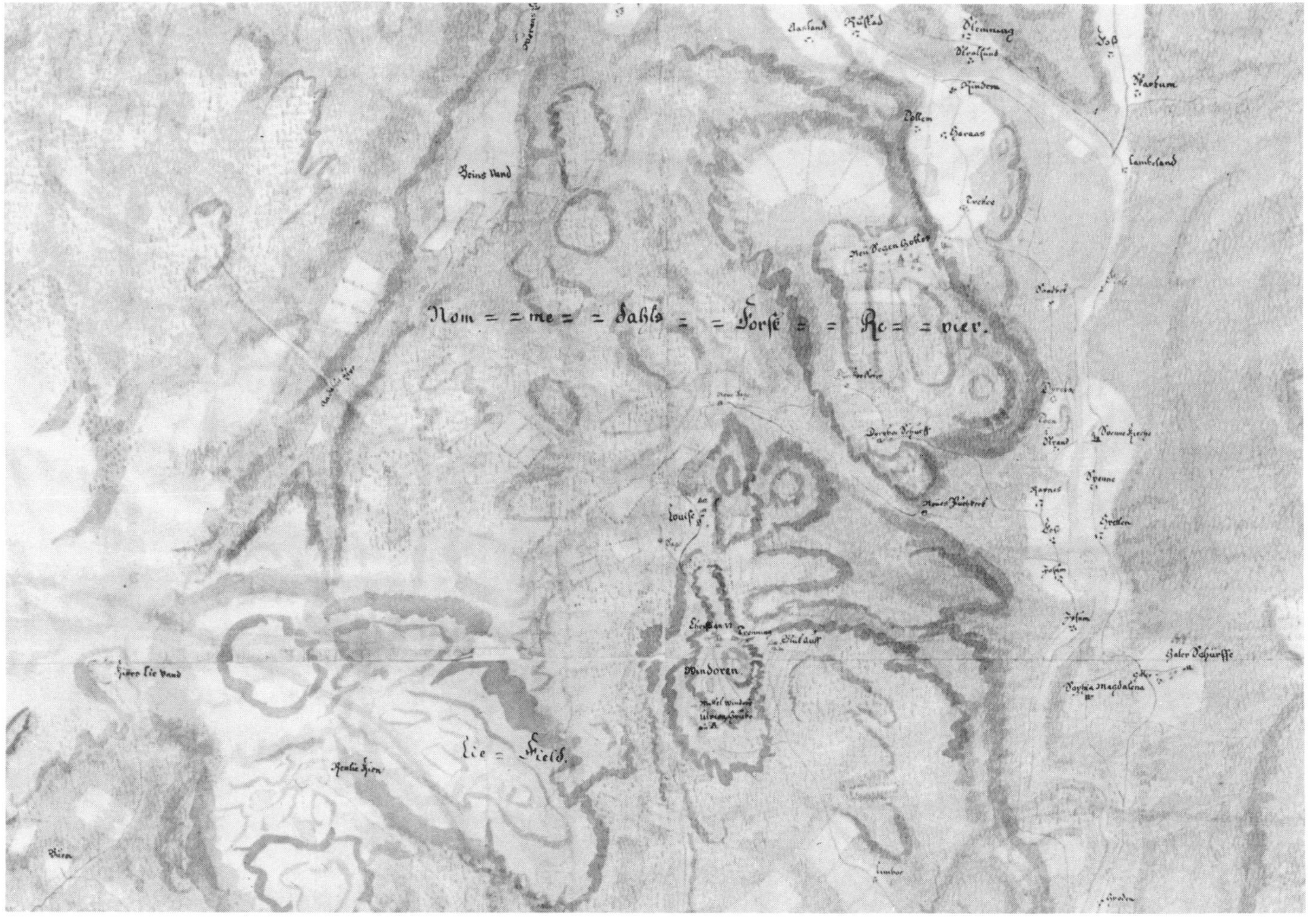

Fig. 3.24 Map of the forest area around Kongsberg in 1746 by F. P. von Langen at a scale of 1:34,000. Source: Norges Geografiske Oppmåling, Hønefoss (now Statenskartverket).

Trondheim, Bergen, and Kristiansand dioceses. In 1730 Kristian VI (1730–46) came to the Danish-Norwegian throne. He took a personal interest in forestry and realized that his Norwegian forests were underexploited compared with those of his German kinsmen. In 1737 he thus set up the Forstkommisjon (Forestry Commission) under the German forester brothers Johann Georg and Franz Philip von Langen. They began work surveying the area around Kristiania (Oslo), using German surveyors, as Norway lacked trained men. In 1739 the work was expanded when the Forestry Commission became the Generalforstamt (General Forestry Board), which was based at Kongsberg and was also led by the von Langen brothers. It consisted of foresters and representatives of the mining and civil administrations. They were required to promote forestry and hunting, to compile maps and descriptions of the forests, and to control their exploitation. One of the board's principal tasks was cartographic; mapping was seen as an indispensable basis for economic development. The maps were to be drawn by foresters with assistance from *fogdene* (sheriffs) in some areas, and they were to be supplemented by topographical-statistical descriptions, giving information about agriculture, forestry, sawmills, mining, frontiers, and waterways, and including "special descriptions" of farms in the form of a *Hofmatrikel* (property register of farms). There were about thirty foresters active in southern Norway; Trøndelag and northern Norway were in effect not covered by the Forestry Board.

Much work was done in the seven years of the Forestry Board's existence, but it was abolished in 1746, partly because of the death of Kristian VI, who had a personal interest in the work, and partly because lack of money and rivalry with other branches of the civil service had hampered work. Many of the German surveyors went home, although the Forestry Board continued to exist sporadically in various guises after 1746.[198] The von Langen brothers are renowned in Germany for making forest maps which were more reliable and which covered greater areas than previous maps, and also for the techniques of hill shading and coloring which give a clear indication of terrain. Their maps were formed by joining together many separately drawn *koncepter* (sketch maps).[199] The maps produced by the Forestry Board were not published, but in 1943 thirty-four *koncepter* and fair copies were known to be extant.[200] Others have since been dated to this period, and there is a suggestion that some may have been given up to Sweden in the nineteenth century.[201] The best known is of the area round Kongsberg (figure 3.24). It shows small rectangular areas which were the result of the Forestry Boards's order to divide wooded properties into small strips to aid systematic felling and management. This was a new departure for Norway introduced by Johann Georg von Langen here and in the Weser district of Germany.[202] Despite the initial schemes for the production of maps combined with property registers of farms, the surviving maps are essentially estate management maps rather than property registration maps.

3.20 Property Dispute Maps and the *Landmålingskonduktører*

Maps had been important in Norway from at least the end of the seventeenth century in the resolution of property disputes (figure 3.25). A royal order of 1719 empowered "skilled and impartial" surveyors to arbitrate in and decide property disputes. This order put such surveying and mapping on a more precise legal footing, although in fact surveying under the 1719 order did not begin until the mid-eighteenth century, as no machinery had been set up to carry it out.

Fig. 3.25 This map of Raundalen's *allmenning* in Voss *fogderi,* constructed in 1684 for a court case involving Claus Henriksen Miltzow, who had unlawfully claimed royal land, is one of the earliest known Norwegian maps. Source: Riksarkivet, Oslo, RK 226.

The first *generalkonduktør* (surveyor-general), Christopher Hammer, was appointed in 1752 in Akerhus district. He was appointed quite fortuitously as a compensation for his having been denied another post. He appealed to be created *generalkonduktør,* pointing out that personnel were needed if the 1719 order was to be carried out properly and citing the work of the Swedish Lantmäteriet, which had a surveyor in every district, as an example of what might be done. Instructions for Hammer were drawn up based on those issued to Danish *stadskonduktører* (town surveyors), but this aroused opposition from rural and urban magistrates who feared that their positions would be undermined and that the employment of a *konduktør* would be too expensive.

By the compromise reached, the *konduktør* was empowered to draw up a map for anyone who needed one in return for reasonable payment. Hammer's appointment was followed by similar ones in the dioceses of Kristiansand and Bergen in 1774 and in the diocese of Trondheim in 1776. Like Hammer, these men were appointed because of their own initiative in applying for the posts, rather than as a result of a general recruitment policy. In Kristiansand Michael Rosing, a product of Kristiania Mathematiske Skole (later the Krigsskole, or Military Academy; section 3.24) and an army officer, was appointed; in Bergen the appointment went to Werner Hosewinchel Christie, a skilled mathematician who had already begun to survey on his own initiative in the Bergen area to resolve disputes among impoverished fishermen who were being forced to turn to agriculture for a living; and in Trondheim Niels Dorph Gunnerus was appointed, as he could combine the post of surveyor with his existing official position at no extra cost to the Crown.

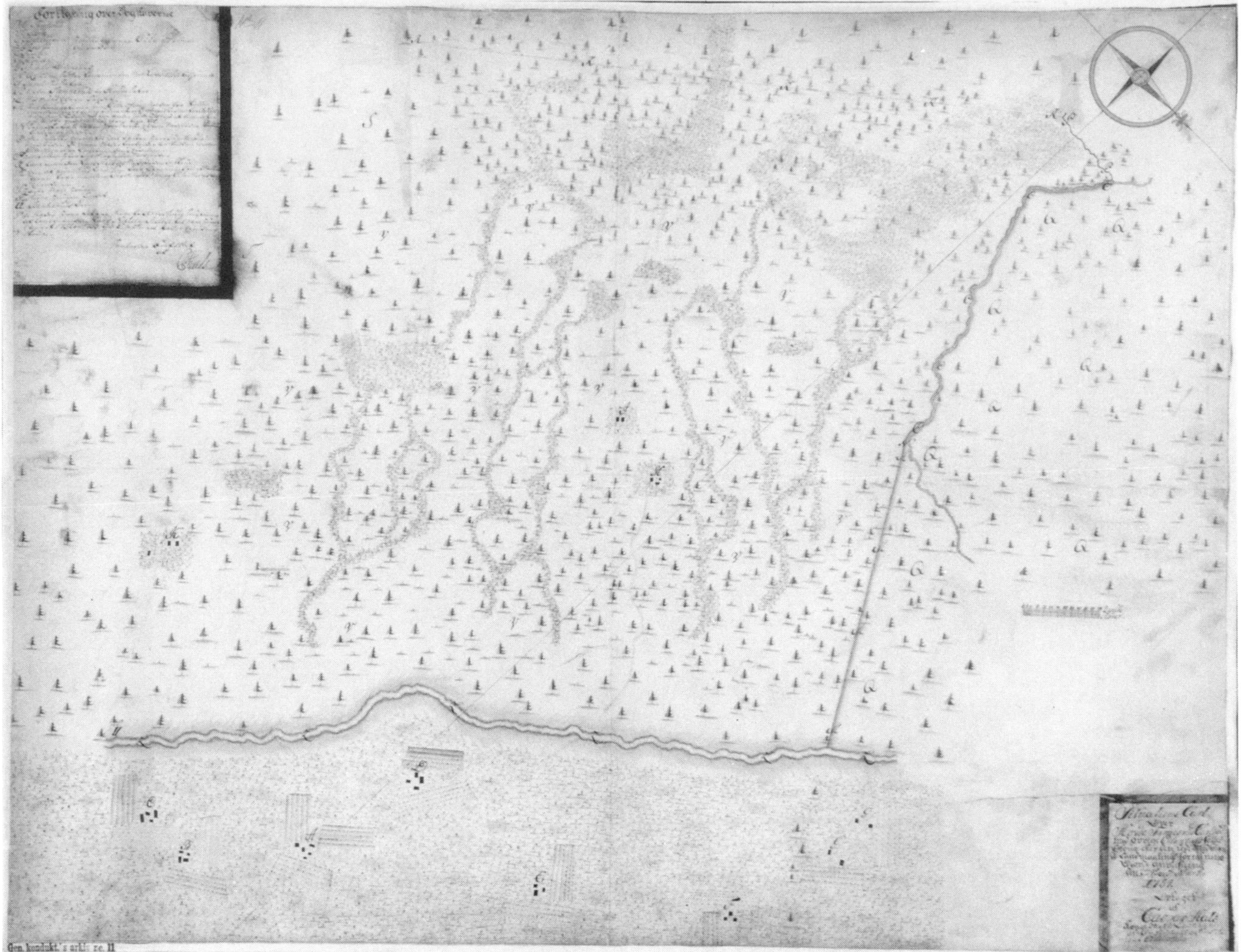

Fig. 3.26 Map used to settle a land dispute in Totens *fogderi,* Kristians *amt,* by Caspar Hals, 1754. Source: Riksarkivet, Oslo, Gen. kond. 11.

When *konduktørene* were called in to sort out property disputes, they drew maps which showed existing boundaries between properties and the proposed new boundaries. The maps were signed by the *konduktør* and the *lagrettemenn* (jury men) with their seals. Disputes could be between individuals or between villages over common land, and costs were borne by the contesting parties. Surveyors were required to make an original and a copy of each map (figures 3.26 and 3.27) and had the right to see all other earlier maps which might be relevant or useful.

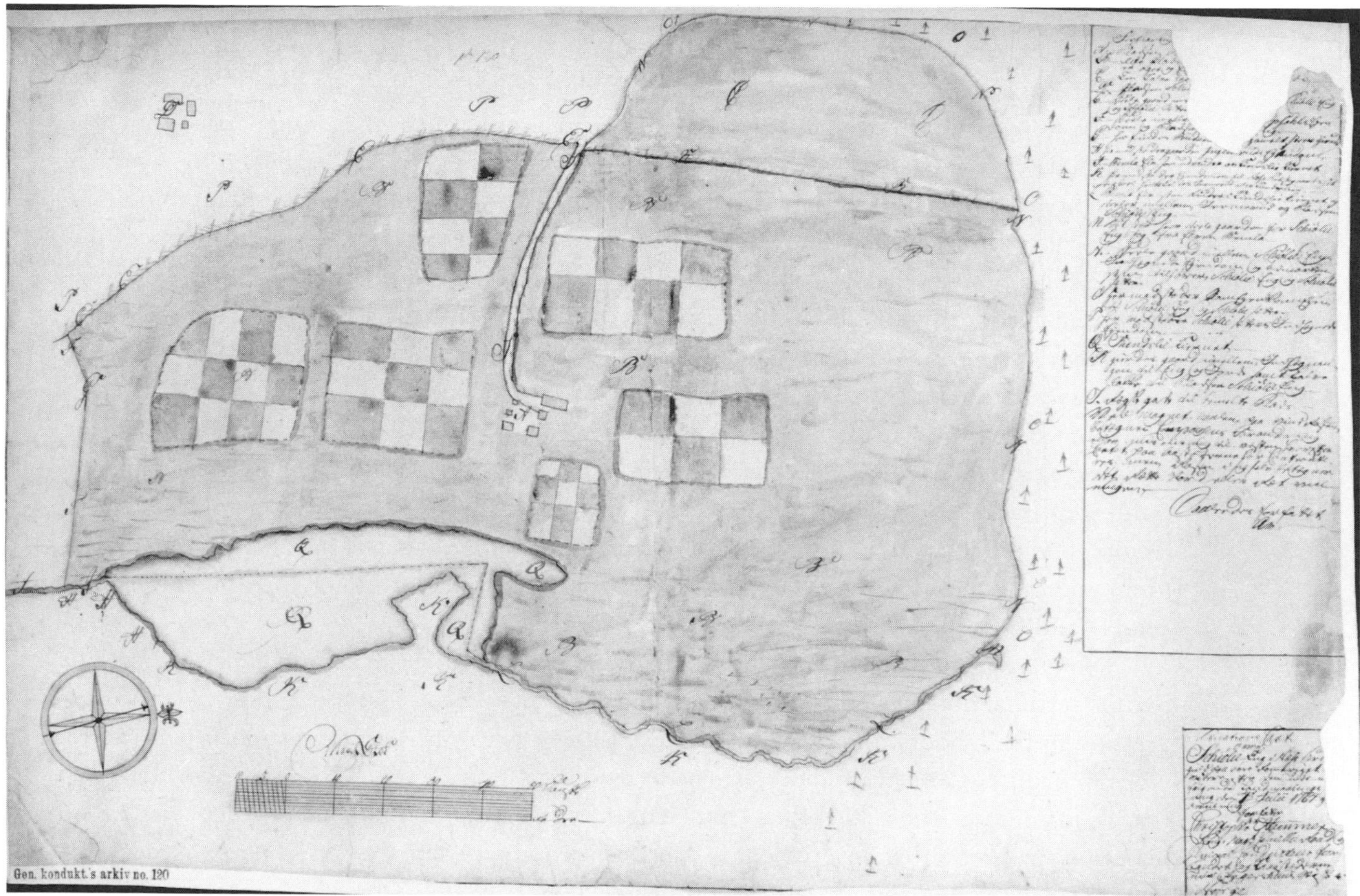

Fig. 3.27 Map used to settle a land dispute in Øvre Romerike, Akerhus *amt*, by Christopher Hammer, 1764. Source: Riksarkivet, Oslo, Gen. kond. 120.

Landowners had the right of appeal against the *konduktør*'s decision. Hammer had fifty officials, mostly army officers, working under him in his district. Most of these officials had their own small area in which they were active. To ensure their impartiality, *underkonduktører* had to swear loyalty and rectitude before they could take office. Surveyors were hampered by the harsh climate and could often work in the field for only four months of the year.[203]

From 1750 onward there was concern over the cost of the maps in land disputes. Costs could exceed not only the value of the disputed land but even that of the entire village, and powerful men could take over areas of land, knowing that their poor neighbors could not afford to appeal.

Efforts were made to rectify this, including new instructions in 1781 that maps were not required in straightforward cases of boundary demarcation. If a dispute went to court, a map was still required, but it could be drawn up by a layman and just checked by the *konduktør*. The costs of the mapping were to be shared among the contesting parties or borne by the losing party. These new instructions were introduced first in the diocese of Bergen in 1781 and extended to cover the diocese of Trondheim in 1787 and that of Kristiansand in 1799. The modification was important for these areas in which agriculture was a secondary occupation and fishing the mainstay of the people. Where forestry or agriculture were the main livelihoods, however, as in Akerhus diocese, maps remained a prerequisite in demarcation. In the former areas a further saving was to be effected by less-accurate surveying; boundaries were not invariably to be measured out but could be plotted with reference to local landmarks.

In 1784–85 the *konduktør* in the diocese of Bergen was moved to another office and not replaced, and in 1801 Christopher Hammer resigned. This pattern continued in the nineteenth century, as the responsibilities of *konduktørene* were gradually taken over by army officers in the country and other city officials in the towns. The system of diocesan *landmålingskonduktører* came to an end, although a government resolution of 1816 gave authority for the appointment of county land surveyors.[204]

The Norwegian experience is interesting as an example where mapping depended greatly on the personal initiative of individuals angling for appointment, and where there was a retrogression in the amount and quality of surveying done simply because it was too expensive. In Norway mapping was made more expensive by difficult terrain, harsh climate, and huge distances; in addition, the value of the land being surveyed was often rather low. There was a tendency on the part of the government to romanticize and overstate the contribution of independent peasants producing from their own land, but in reality agriculture was often only a by-employment of fishermen and foresters or their wives and children, and where land was not scarce, mapping could be an unnecessary luxury.[205] The system of *landmålingskonduktører* seems to be unique in the history of cadastral mapping. Maps were often commissioned by individuals or institutions to prosecute or defend a legal action involving land boundaries, and government certification of "sworn" surveyors is known from the Netherlands (chapter 2.1) and Germany (chapter 4.1), but the provision by the Norwegian state of a body of surveyors to carry out such work is without precedent.

3.21 Enclosure

As in Sweden, the main enclosure movement in Norway was government directed and included maps as an integral part. Enclosure followed the normal pattern of the consolidation of fragmentary plots, accompanied in many cases by the moving out of farmsteads, although the severe

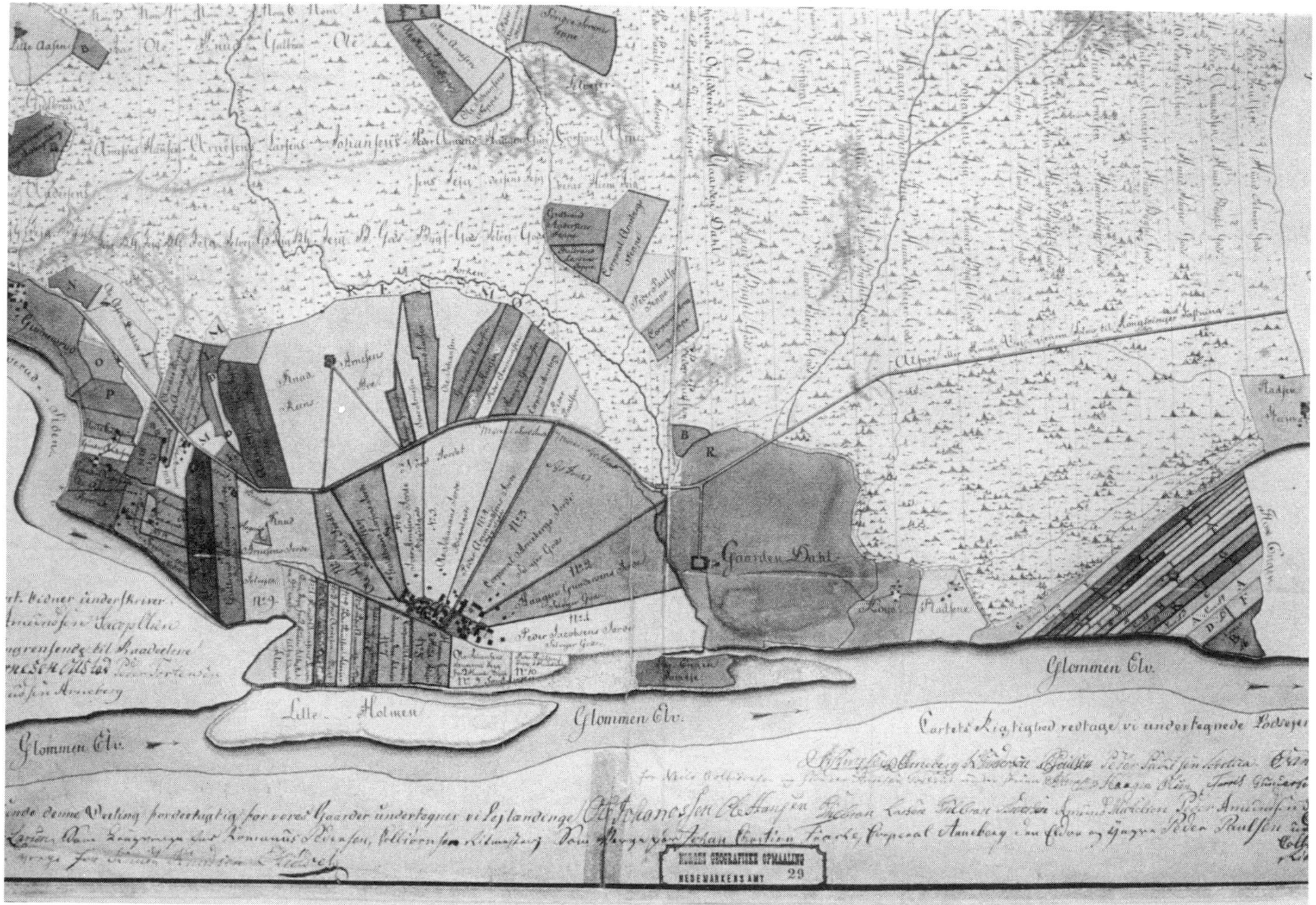

Fig. 3.28 Three extracts from the *jordskiftekart* (enclosure map) of Sorkanæs, Hedermarks *amt,* by Hans Lemmich Juell, 1803–4, at a scale of 1:8,000: (*a*, above) before enclosure (note small strips of land); (*b*, page 110) after enclosure, central section; and (*c*, page 111) after enclosure, bottom right-hand section. Note that only land near the fjord is affected by enclosure; steep land on mountain slopes is not reallocated. Source: Statens Kartverket, Hønefoss, NGO A v 29 Sorkanæs, Hedermark.

climate and terrain often modified ideal patterns.[206] Land consolidation and enclosure first appeared in Denmark-Norway in Sønderjylland in the seventeenth century, but in Norway itself private enclosure was limited, and effective enclosure generally involved government action of some sort.[207]

The Danish-Norwegian Rentekammer in Copenhagen began the work for a Norwegian parliamentary enclosure in the 1790s, and by 1803 they had a comprehensive collection of reports

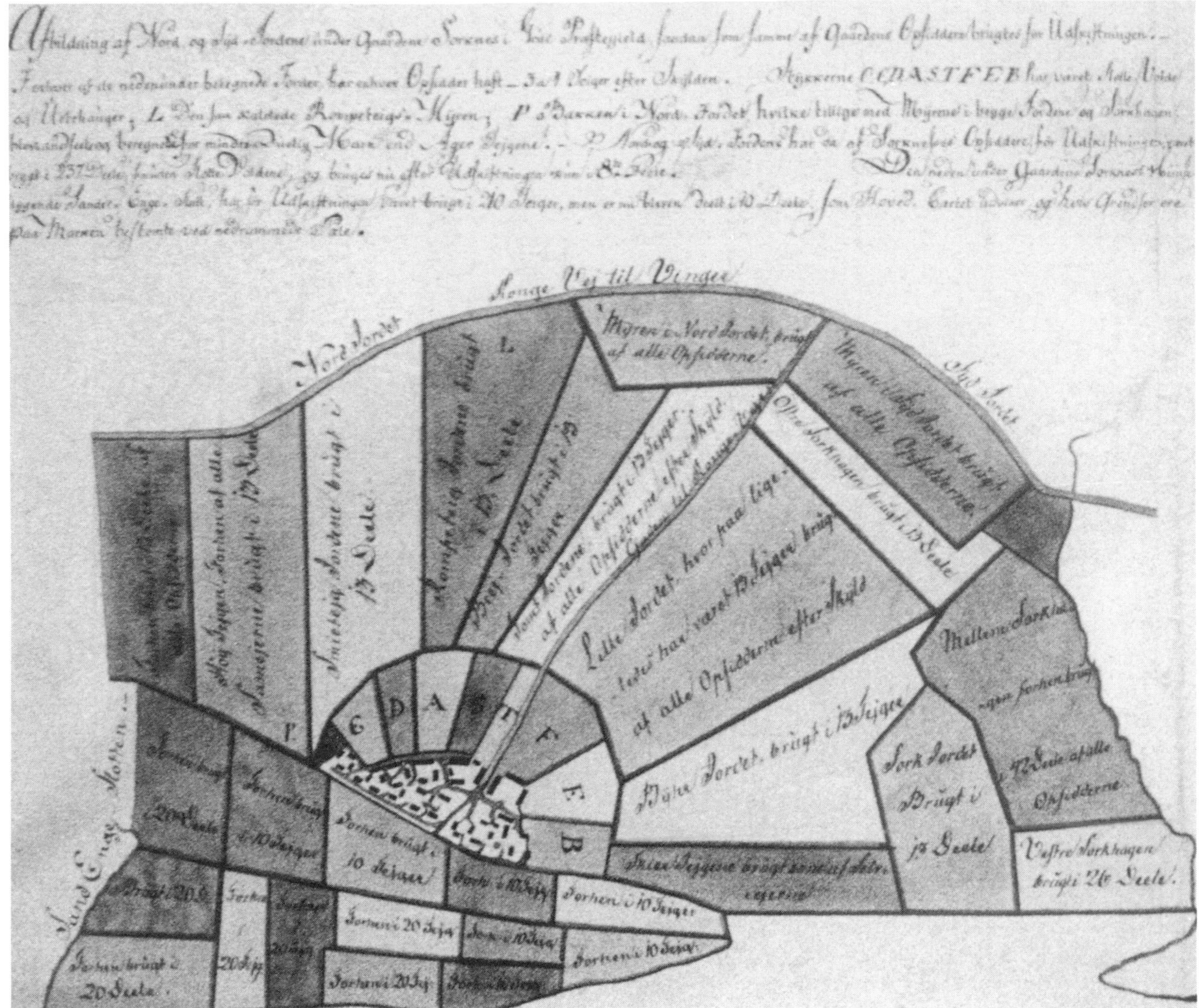

on the agricultural situation throughout the country. This was intended to be the basis for enclosure legislation, but it is doubtful to what extent it was actually used in drafting the first land reallocation act of 17 August 1821.[208] The 1821 enclosure act resulted in only minor changes, as the tax penalties for those not reallocating their land proved insufficient.[209] The changes were normally confirmed in law court protocols; maps were not required, although they were sometimes drawn. These documents and the Exchequer's reports of 1803 form a useful supplement to the descriptions and maps which were made for enclosures from 1857 onward.[210] On 12 October

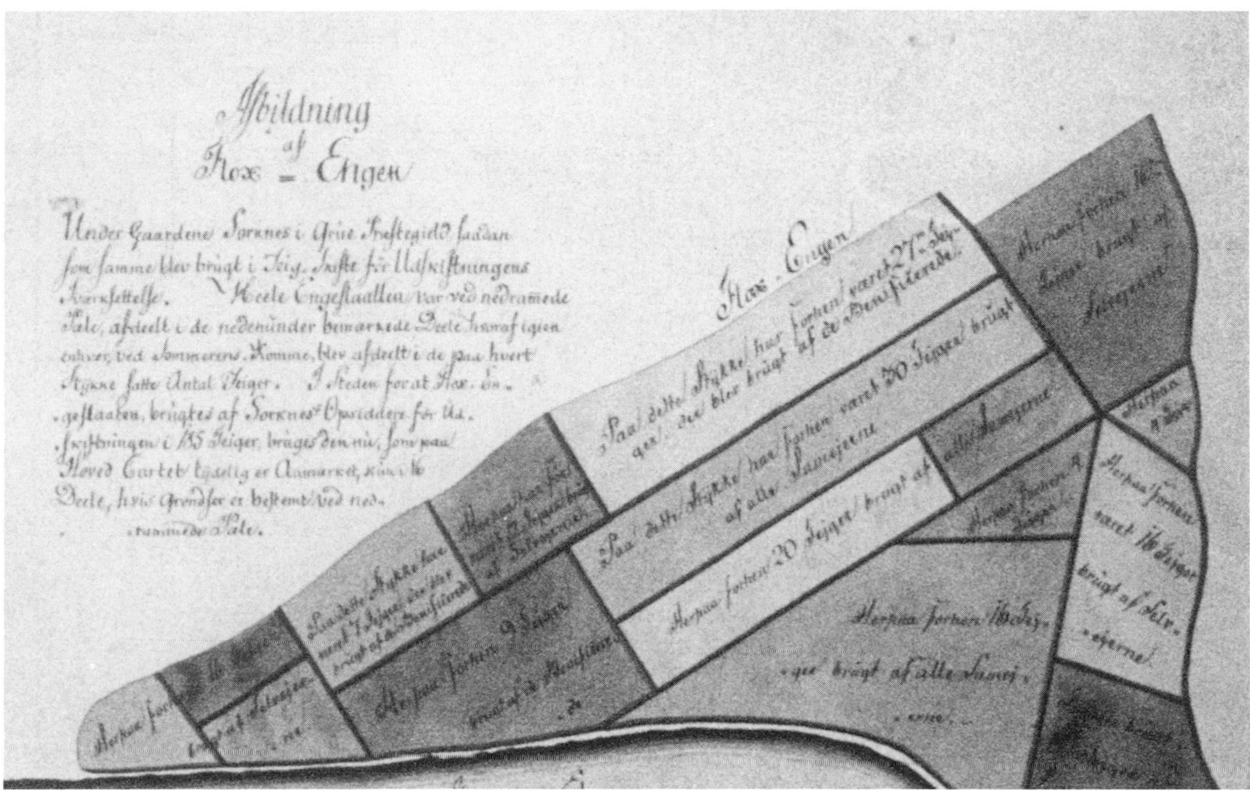

1857 a second enclosure act was passed. It came into force in 1859 and, with later amendments, formed the real foundation for enclosure. To carry out its task the Udskiftningsvæsen, later Jordskifteverk (Enclosure Board), was brought into being.[211]

It was the establishment of machinery to effect enclosure under the 1857 act which started the rapid spread of enclosures.[212] The act was also the first which prescribed that fields be surveyed and mapped and that the maps be carefully preserved. There had been some sporadic enclosure mapping before, but the maps had not been standardized, and many were subsequently lost. The 1857 law ordered that maps be delivered to the Enclosure Board (figures 3.28–3.30) and a copy be kept in the *gård* by one of the farmers. Most of these village copies have unfortunately been lost.[213] Maps show the old boundaries in stipple and new boundaries in continuous lines. Ownership of both old and consolidated plots is marked with letters which correspond to individual farmers. Mills, smithies, and other buildings are marked, and later maps show moors, meadow, and forest, as well as roads and sheep and cattle walks. The maps were accompanied by *jordskifteprotokoller* (enclosure awards). In the northern fishery districts the enclo-

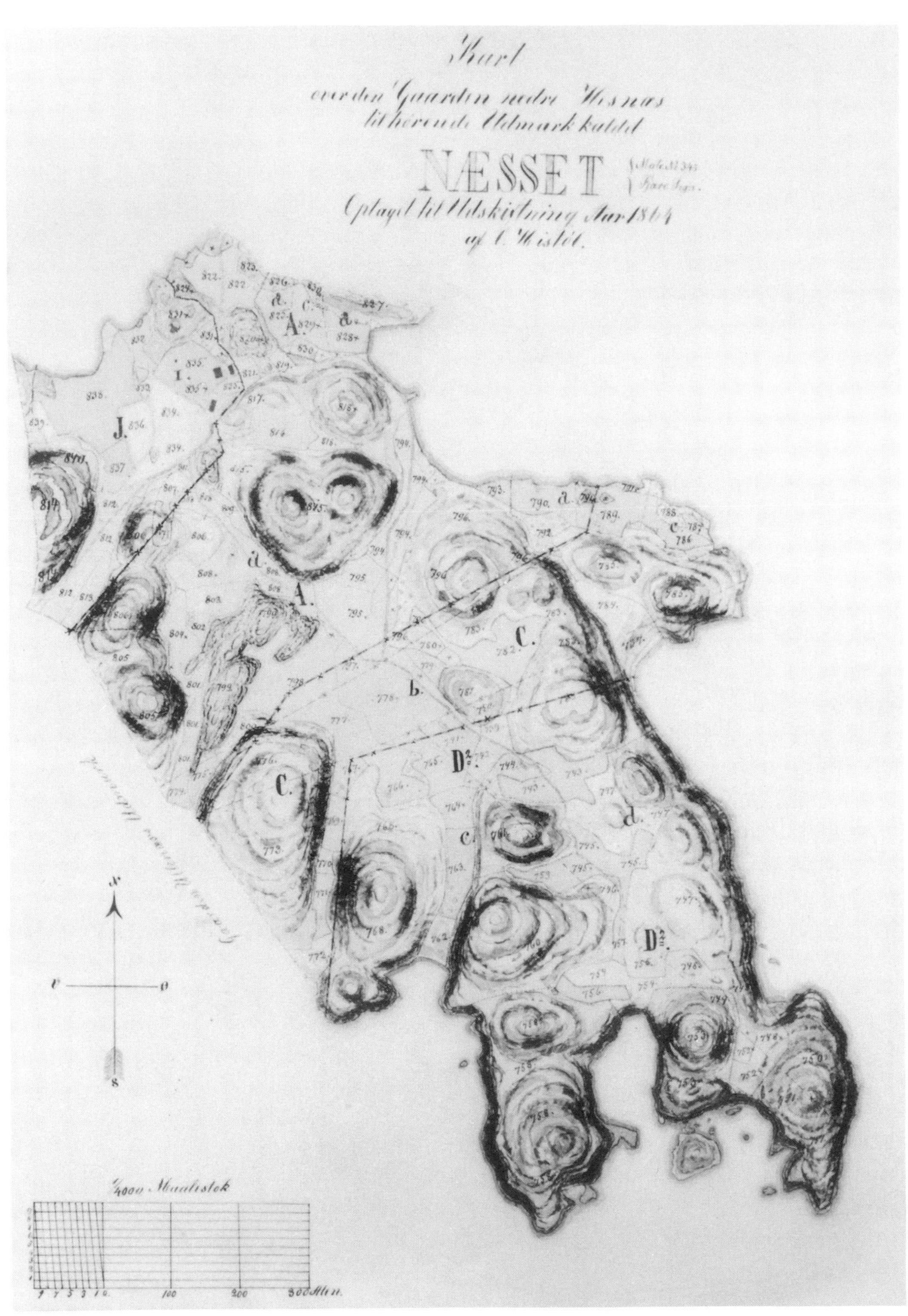

Fig. 3.29 Næsset in Fjære Parish after *udskiftning* (enclosure) in 1864. The map is at a scale of 1:4,000. Rather unusually, relief is represented by hill shading. Source: Jordskifteverkets Kart- og dataavdeling, Ås, VIII 24.

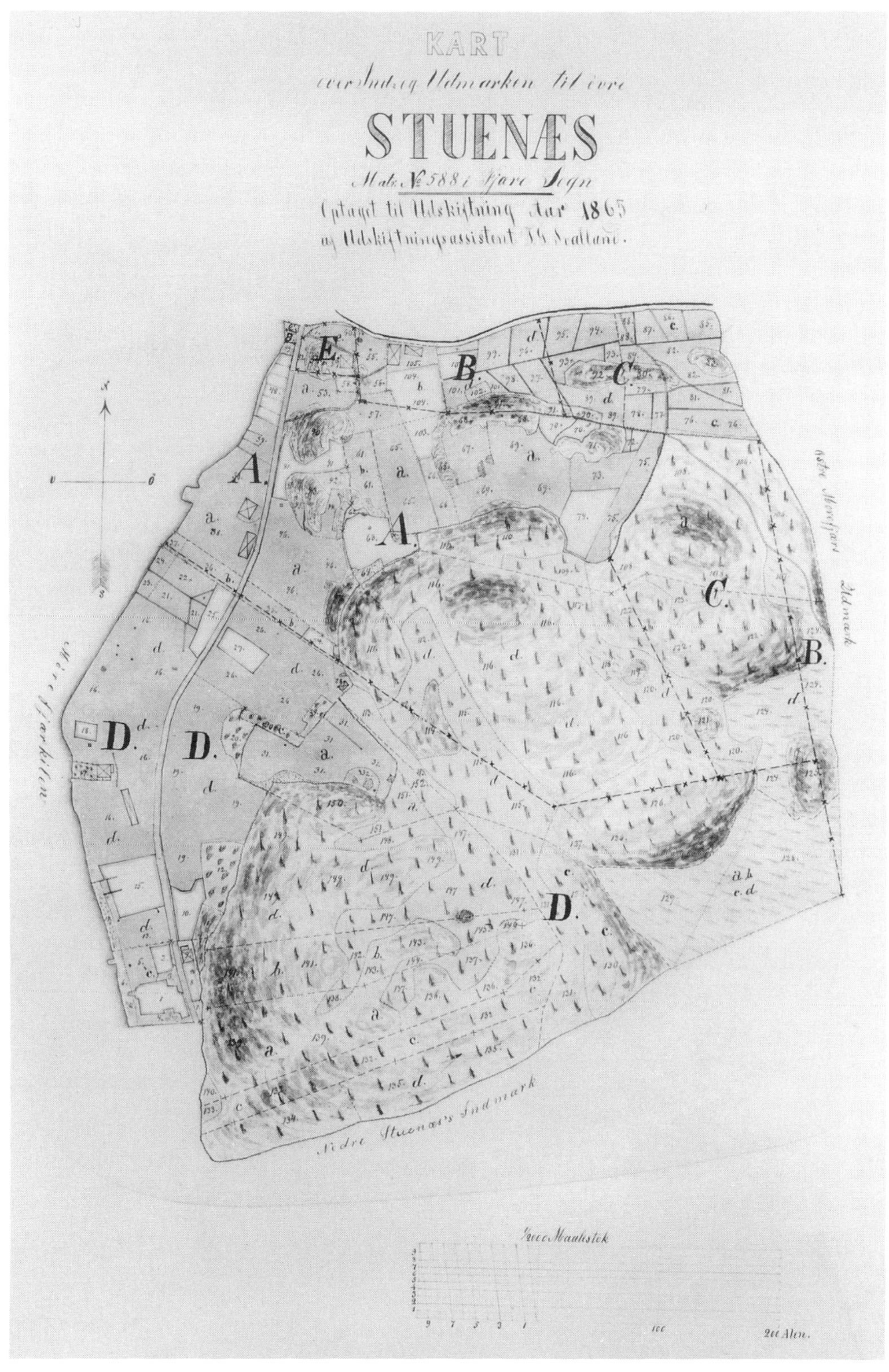

Fig. 3.30 Upper Stuenæs in Fjære Parish after *udskiftning* in 1865. The map is at a scale of 1:2,000. Relief is indicated by hill shading. Source: Jordskifteverkets Kart- og dataavdeling, Ås, VIII 25.

sure maps often do not show preenclosure property divisions, quite simply because they did not exist. The fields were worked entirely communally by women while the men were out fishing. Only on the coast itself, where access to land was important for landing and mooring, were land divisions fixed. This common ownership persisted almost to the twentieth century in remote areas.[214]

Little is known of the reasons behind the geographic spread of parliamentary enclosure in Norway, but it seems likely that it was closely related to the extent to which the money economy had penetrated agriculture in the region. The widely accepted official interpretation of the need for, and benefits of, enclosure as argued by Grendahl is now being reconsidered.[215] A farm system geared to produce subsistence for farming families has less need of enclosure than one geared to produce for the market, even if the land is very highly fragmented. In many of the remoter areas of Norway land reallocation began late and is still going on. By contrast, in the east of the country, especially in the open lowlands of southeastern Norway and Trøndelag, even before enclosure farms formed individual units with houses and grounds fairly clearly separated from neighboring farms.[216] Frimannslund finds evidence of a more highly stratified class society in these eastern districts. In the eastern areas the *landskyld* system had first been superseded by the freehold system from the later seventeenth century, as this permitted more intensive capitalist exploitation in these areas.[217]

3.22 Maps of Crown and State Land

Allmenning, or Crown land and land of rather vague status in the north of the country which in historic time has not been subject to the law of private property, accounts for vast tracts of Norwegian land, especially in the north. Traditionally, local people had customary use rights over it, but the Crown had a type of property right over it. In the seventeenth, eighteenth, and the early part of the nineteenth centuries considerable areas of *allmenning* were either sold or parceled out to those with customary rights over it. Various commissions were appointed to supervise this work, define the extent of the *allmenning,* divide it, and draw boundaries which were legally binding. From the 1750s there was a systematic program of mapmaking (figure 3.31).

The state also controlled much land taken over from the church after the Reformation. This included forest tracts and *enkeseter* (land used to maintain priests' widows). Some of this land was disposed of between 1793 and 1848, partly to military and civil personnel and partly to buyers. Mapping of this land had begun on a small scale by church departments in the 1850s, and it was continued in a more systematic way between 1867 and 1874 under the auspices of Norges Geografiske Oppmaaling (Norwegian Topographical Survey Board). The maps are on a scale of 1:4,000.[218]

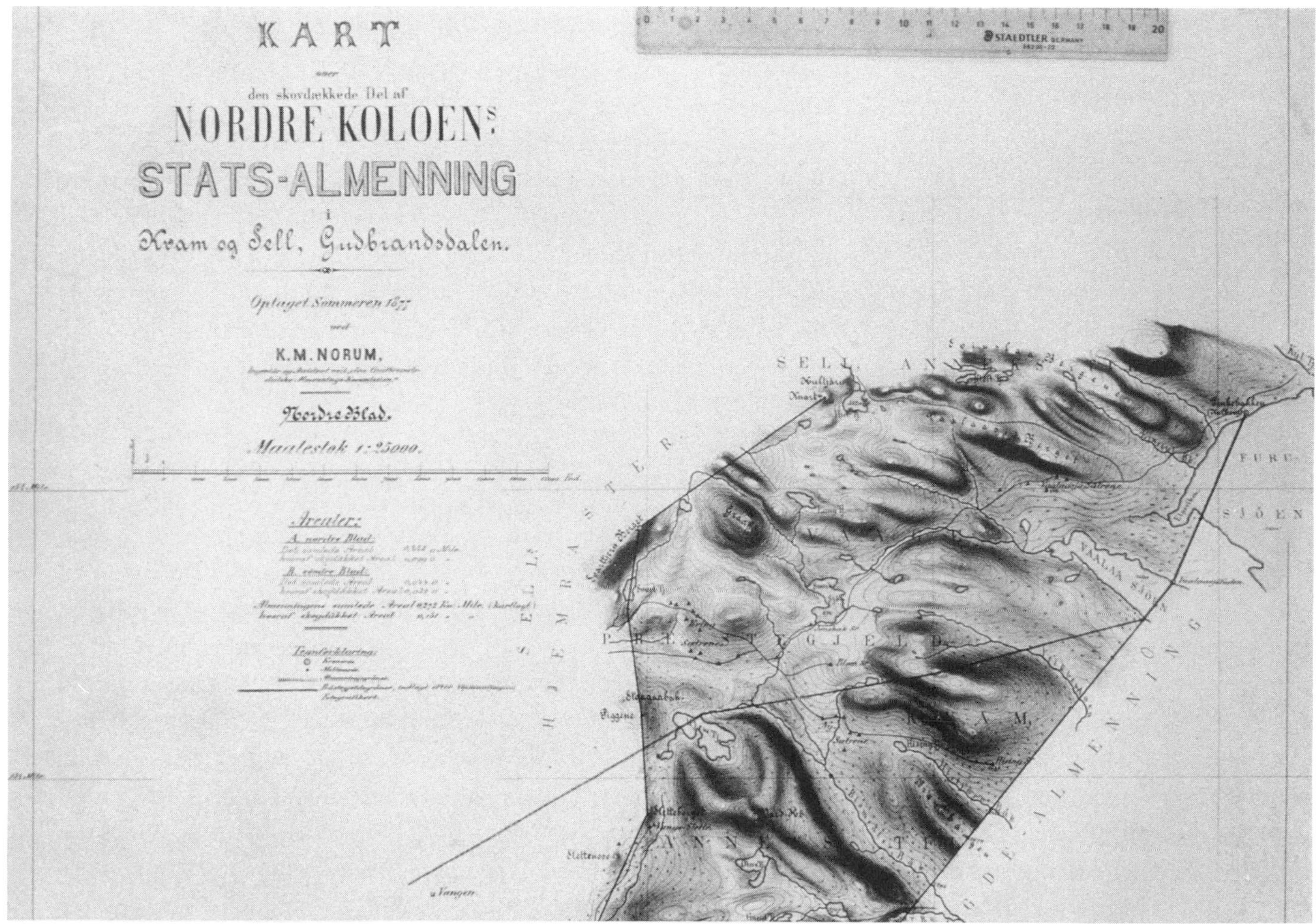

Fig. 3.31 *Statsalmenning* map of Nordre Koloen, Oppland, 1877, at a scale of 1:25,000. Source: Jordskifteverkets Kart- og dataavdeling, Ås, 0517/2–1*, sheet 1.

3.23 Statens Kartverk (National Survey Board)

Statens Kartverk was founded in 1773 and was initially concerned entirely with detailed topographical mapping.[219] From 1804 there were discussions about the need to produce large-scale land-use maps, and a royal resolution of 20 March 1805 ordered a combined military and land-use mapping. Surveyors were to map all property boundaries; measure all fields, woods, and meadows; and classify and list all land types and grazing areas. Each property was to be described and measured in square ells. All this was in preparation for a new *matrikkel* in conjunction with enclosure. The project was troubled from the start, was suspended during the political crisis of 1813–14, when Sweden annexed Norway, and was abandoned completely in 1814. By this time, however, 138 square Norwegian miles had been surveyed (1 mi = 18,000 ells) in Solør, Odal, parts of Østerdalen around Røros, and Trøndelag. These areas are covered by maps at a scale of 1:10,000 and by detailed *Oeconomiske Tabeller,* or descriptions.

At the same time in Vestfold (In Larvik, [Laurvik] County), mapping had started, following the Danish precedent. One hundred maps on a scale of 1:4,000 were produced. They show areas of land in *tønner* (acres) and give information on households, sown areas, and hay yield. These economic maps are of mixed quality. Army officers were responsible for the mapping, so progress was affected by military crises; many maps are obviously unfinished, although many others were completed. After 1814 this economic mapping was not continued, largely for the reason encountered before in Norway—namely, lack of money.[220]

3.24 The Cadastral Map in Norway

Like the Danes and the Swedes, the Norwegians made no major technical innovation in land mapping; indeed the earliest maps of Norway were drawn by Germans, Danes, and Dutchmen in the absence of skilled native surveyors. This pattern was broken to some extent in 1750 with the foundation of Den Matematiske Skole, whose methods gradually became accepted practice within the country in response to demands from the military for more accurate maps and from economic developments.[221] Despite this advance, Norway was not innovative technically. What is most interesting about Norwegian cadastral map history is that it shows more clearly than in many other countries the link between capitalist exploitation of the land and demand for cadastral maps. The operation of the precapitalist *skyld* system of landholding was inimical to cadastral mapping, and in the far north, where land was held in common, it would have been impossible until very recently to construct a cadaster. The pressure to change to holding land in fee simple to aid its exploitation by individuals brought into being cadastral maps, which were at once the reflection of and the tool for capitalist resource exploitation.[222]

3.25 The Cadastral Map in the Nordic Countries

The Nordic experience shows the link between cadastral mapping and the presence of a powerful monarch. The connection is most evident in Denmark, where the absolute monarchy was established in 1660 and the hitherto considerable economic privileges and political powers of the nobility were severely curtailed. This paved the way for comprehensive survey and subsequent mapping, which had the direct aim of raising state revenue and the indirect effect of consolidating the power of the Crown over that of the nobles. The link between mapping and the strength of the monarchy may also be seen in Sweden under Gustav II Adolf. He was not an absolute monarch, as for much of his reign large parts of the country were under the effective control of the royal dukes and especially the queen mother. Parts of Finland were outside the control of the nominal ruler, and powerful individuals there acted as they pleased.[223] With the death of these rivals or their subjugation to royal power, Gustav strengthened his position as monarch. However, he rewarded the nobility for their part in his military campaigns by granting them large tracts of land, which increased their power vis-à-vis his own. Although not an absolute ruler, Gustav was undoubtedly a strong one who elevated Sweden to the status of European power and gained in stature himself as a result. His glory and that of the Swedish Empire were reflected in the strikingly ambitious project to map Sweden and all lands which might one day become Swedish.

The consolidation of the power of the Crown markedly affected the progress of agricultural reform in the Nordic countries. In parts of Germany and Austria, where the power of individual nobles was still entrenched, the most important agricultural reform was the abolition of serfdom. For such a reform it was important to know the extent of the area over which a lord had feudal rights, but it was essentially a nonspatial reform: the precise delineation of boundaries on maps was of little importance. In the Nordic countries the peasants had traditionally been far less oppressed, and in many areas were free or fairly free, although there had been an increase in feudal burdens in parts of Denmark in the early modern period. In the Nordic countries the abolition of the labor service and other impositions was important, but enclosure of the common fields and the transfer of the commons to private ownership was a far more important part of agricultural reform. For enclosure the provision of an accurate map was of greater relevance.

Colonization was also a stimulus to mapping. Sweden wanted an inventory of its new possessions in the Baltic and a basis on which to administer them efficiently and exploit them effectively, whether by granting land to Swedes as a reward for loyal service or by extracting state taxes. Mapping in the Baltic colonies was not an unqualified success, not least because the monarch's power was not absolute. Livonia, for example, was joined in a personal union with Sweden, but the nobility retained many of their feudal privileges, notably over taxation, and the Swedish monarchs were held to be exceeding their constitutional powers and infringing those of the indigenous nobility when they attempted to impose a taxation cadaster. The cadastral mapping itself may be regarded as a success, but certainly the measures which the maps were intended to realize were not. There was also some cadastral surveying and mapping in Denmark's small but economically important American possession of St. Croix. The maps were used in the administration of the sugar-plantation island.[224]

If the link between cadastral maps and royal power is clear, so is the connection with capitalism. The Nordic area still contains areas where private property rights have never been fully established, although normally the Crown has taken over such land. Where private property rights are not established, the development of cadastral mapping is problematic if not impossible. The Norwegian system of *skyld* is a good example of this. *Skyld* is a measure of the value of the land rather than of the land itself and was the object of possession and lease. Joint ownership was common in Norway, but joint owners could not point to any part of the *gård* which they owned, since what they owned was a share of the value of the land not a share of the land per se. To have mapped ownership of *skyld* would have been impossible, just as today mapping ownership of shares in a company is impossible. In the eighteenth and nineteenth centuries the users of a particular piece of land, rather than the owners of *skyld,* became property owners. Only after this transition to the holding of land in fee simple was cadastral mapping possible.

Research from Norway suggests that there was very limited pressure for enclosure in areas where the role of agriculture was to provide subsistence for the peasant household, members of which might also be involved in commercial fishing and/or forestry. Enclosure appeared first and spread most rapidly where agriculture produced goods for market exchange and where desire to increase profit led to the need for absolute individual ownership of land. Recent analysis of enclosure maps has shown that in the remoter parts of northern Norway, there were no preenclosure property divisions. The fields seem to have been worked entirely communally by women while the men were out fishing. Only on the coast itself, where access to land was important for landing catches and mooring, were land divisions fixed. Common ownership of land in northern Norway persisted until just before the twentieth century and made it impossible to construct a meaningful cadaster.

In Denmark a significant change in the concept of ownership occurred as a direct result of the use of maps, although linked to a much deeper process of the establishment of absolute property rights. In the seventeenth century *ejerlav* was defined with regard to the villagers of a named village in which they had complex use rights to arable land, meadow, wastes, and forest. By the nineteenth century *ejerlav* was defined territorially as the land which was represented on an enclosure map.

Surveying and mapping also brought about the introduction of units of measurement appropriate to the capitalist system. At one level there was a need for standardization of existing measurements, but more interestingly, there was a change in types of measurement. Early Swedish instructions decreed that one *tunnland* should equal fourteen thousand square ells of arable land plus meadowland which could bear between three and five loads of hay. It was a measure dependent on the quality of land, not just its area. The definition was rapidly simplified such that one *tunnland* equaled fourteen thousand square ells of land in the interests of efficiency and uniformity. In Livonia one of the major technical difficulties of the cadaster was that the Swedish surveyors did not appreciate that the *uncus* was a cadastral and not an areal measurement of land. It took into account the area of the land, but also the quality of the land, its proximity to the market, the capacity of the peasant to render statute labor and pay customary dues, and so on. The complexity of such early units of measurement, their incompatibility with other measures, and their variation made them unsuitable for cadastral mapping and with capitalism, for which uniform and simple systems are needed.

In Norway maps were recognized in the eighteenth century as a clear and efficient tool for solving landownership disputes, and their use was for a time compulsory. The maps were indeed efficient but not impartially so. The costs of drawing a map could exceed not only the value of the disputed land but that of the entire village. Powerful individuals could take over land in the knowledge that their poorer neighbors could not afford to contest the legality of their actions.

In the Nordic countries the links between cadastral mapping and the presence of a powerful monarch and, more particularly, between mapping and the development of capitalism are too strong to ignore. The most striking developments in Nordic mapping are, first, the compilation in seventeenth-century Sweden-Finland of the *geometriska jordeböckerna,* initiated by Gustav II Adolf at the height of his reign, a reflection of the nation's glory and a symbol of its power; and second, the mapping of the Swedish Baltic colonies to consolidate imperial power there. Equally striking, though, is the complete absence of mapping in much of Norway, where precapitalist forms of land tenure were inimical to its development. Cadastral mapping is not exclusively linked to capitalist developments or dependent on it, as the existence of Roman cadastral maps show; rather, those in whose interest it was to consolidate capitalist developments rediscovered the cadastral map and used it to further their own ends and thus the ends of capitalism itself.

Acknowledgments

We should like to thank the following for their help with the Swedish, Danish, and Norwegian sections of this chapter:

Sweden: Dr. Ulla Ehrensvärd, Dr. Göran Hoppe, Dr. Michael Jones, Prof. William Mead, Prof. Ulf Sporrong, Dr. Leo Tiik, Dr. Riho Mallo and the staff of Lantmäterietsforskningsarkivet, Riksarkivet, Kongelige Biblioteket, Krigsarkivet, Stockholms Länsstyrelsenslantmäterienhet, the Departments of Geography of the Universities of Stockholm and Tartu, the Library of the University of Tartu, and the Estonian State Archive in Tartu for their generous help. Some material in the Swedish section was previously published in *Geographical Journal* 156 (1990): 62–69.

Denmark: Dr. Karl-Erik Frandsen, Prof. Viggo Hansen, Indrid and Knud Hjerting, Archivist Hans Ejner Jensen, Dr. Robert Newcomb, Dr. Haraldur Sigurdsson, and the staff of Matrikelarkivet, Københavns Universitetsbibliotek, Rigsarkivet, and Kongelige Biblioteket.

Norway: Mr. Francis Herbert, Dr. Michael Jones, Dr. Stewart Oakley, Prof. Hans Sevetdal, Mr. Ulf Hanssen and the staff of Statens Kartverk, Arkivist Alfhild Nakken and the staff of Riksarkivet, and Mr. Arne Johannessen and Mr. Paul Østavik of Jordskifteverkets Kartarkiv.

4

GERMANY

It is impossible to define "Germany" unexceptionably.[1] For convenience and clarity this chapter describes the development of cadastral cartography in the area which in 1871 formed the second German *Reich* (figures 4.1 and 4.2). The *Reich* was a confederation of states dominated by Prussia, which accounted for over 60 percent of its area and population and whose king was also hereditary emperor. This second empire was the successor to, although by no means coextensive with, the Holy Roman Empire, or Holy Roman Empire of the German Nation, as it came to be called. The Holy Roman Empire was a loose confederation of a large number of increasingly autonomous territories under an elected emperor whose office brought with it no land or effective source of revenue as of right. Real power lay with the territorial princes and was often based on a number of different legal titles to scattered possessions. However, particularly after the end of the Thirty Years' War in 1648, the more important dynasties strove to imitate the great nation-states and establish a strong central government in their lands. The Holy Roman Empire became increasingly anachronistic and finally collapsed in 1806 after the illegal formation under a Napoleonic protectorate of the Confederation of the Rhine, to which sixteen southern and western German states ultimately belonged, and after the Emperor Francis II under pressure from Napoleon had laid down the imperial crown. In the turbulent Napoleonic period much German territory was either occupied by or dependent on France. After the defeat of Napoleon the German Confederation was formed with thirty-nine members in 1815, and this was followed by various customs unions from 1834 and finally in 1871 the second *Reich*.[2] The area of the *Reich* of 1871 includes land now in the Republic of Germany, as well as areas now in the Netherlands (part of Cleve), Denmark (North Schleswig), and Poland (most of Prussian territory east of the rivers Oder and Neisse), France (Alsace-Lorraine [Elsaß-Lothringen]), and Lithuania (part of East Prussia). Parts of the 1871 *Reich* had previously been subjected to cadastral surveying and mapping by other ruling powers: thus, for example, the Swedes mapped Hither Pomerania (Vorpommern) and Mecklenberg (chapter 3.7), the Danes mapped parts of Schleswig-Holstein (chapter

Fig. 4.1 The Holy Roman Empire in 1648, the German *Reich* in 1871, and modern political boundaries. Sources: Droysens 1886; Muir and Philip 1927; Shepherd 1930; Kinder and Hilgermann 1974–78; Darby and Fullard 1978.

3.15), and the French mapped Westphalia (Westfalen), the Rhineland (Rheinland), and Alsace-Lorraine (chapter 6.9).

Just as it is impossible to define "Germany" satisfactorily, so it is impossible to give a complete account of German cadastral mapping. Vast numbers of maps were produced by very many different authorities, sometimes as part of well-known surveys, but often as isolated instances which escaped documentation. To compound the problem, many maps and their supporting docu-

Fig. 4.2 The German Empir and the extent of Prussia in 1871. Sources: same as figu 4.1.

ments and catalogs have been lost without trace, not least in the Second World War. What follows here is first an account of influences on the development of cadastral maps: these are listed in chronological order, but it is not intended to suggest that all were felt sequentially in all parts of Germany. Many eighteenth-century maps were technically and stylistically wholly within sixteenth- and seventeenth-century cartographic traditions, and it was not axiomatic even in the early nineteenth century that a cadaster be mapped or even thoroughly surveyed, as the experiences of Bavaria in 1801 and Hanover in the 1820s show (section 4.7).[3] Second, various types of map are illustrated with examples to show notable early developments and typical as well as exceptional nineteenth-century maps. Finally, in the appendix there is a structured guide to the very extensive literature on German cadastral maps.

4.1 Forerunners of State Cadastal Maps

Because of the political complexities of Germany and the absence of a central state authority, it is often difficult and perhaps contentious to divide private from public mapping. Where a lord was also a territorial prince, individual mapping of estate boundaries could at the same time be state mapping of political boundaries, while inventory maps of private estates could simultaneously be maps of state land resources. Similarly public and private revenues are not easily differentiated, since much state revenue was derived from rents from the prince's lands (section 4.5). However, a gradual transition from individual to state mapping can be seen, together with increasingly clear demarcation of the two, so it is useful to describe some of the private forerunners of the later state mapping.

Land Title Maps

Perhaps in Germany more than in other more-centralized states there was scope for dispute over title to land. Peasants and princes were exercised about questions of sovereignty, possession, ownership, use rights, and jurisdiction. Boundaries were zones rather than lines: some areas were claimed by more than one authority or person, and some were jointly owned or administered, while other land was not initially claimed by anyone. The process by which each area was ascribed to one and only one owner and ruler was slow and often contentious.[4]

Two types of action were taken to establish boundaries. The first was the surveying and marking out of boundaries in the field with stones; the second, which was sometimes needed before such stone setting could take place, was litigation to establish land title. From the sixteenth century the practice began of using picture maps in legal disputes about possession and ownership of land and the relation of peasants to estates and of landholders to the state. These picture maps form the oldest part of most German cartographic collections. The conventional view of picture maps is that they are in fact landscape paintings, often drawn by a professional artist or a "sworn painter," such as the one specially appointed by the court at Speyer in the sixteenth century to stop the contestants' producing too partial a picture. Often the disputed area is well represented, but the surrounding landscape is more of a decorative border around the "map." The picture maps are not photographic representations of reality but compilations of salient features in the dispute, not all of which could in fact be seen from one spot. Scherzer suggests that most were not surveyed, although he concedes that on some of them the representation of each parcel of land is so accurate that a survey probably had taken place.[5] Aymans considers, however, that those who criticize picture maps as inaccurate have misunderstood their role. He contends that comparison between court papers and the disputed land itself generally shows the land to have been very carefully surveyed. The picture maps themselves, however, were never intended as accurate graphic representations. Rather, they complemented the accurate survey data as an aid in picturing the situation of the land.[6] In either event, picture maps were important in that they established very early the value of graphic evidence in court cases. The move from a landscape painting to a surveyed map picture and then to a surveyed map was not straightforward, and costs increased at each stage. When the change happened therefore depended on the perceived costs and benefits in each individual case. Grees describes a map of Obersteinach in Baden-Württemberg which is on the border between pictorial representation and abstract delineation. While the map itself is accurately portrayed and delineated, it is combined with a bird's-eye view and has an elaborate border which, Grees suggests, has very little to do with the content of the map and was designed more to show the cartographer-painter as a cultured person and a gifted artist.[7]

Many sixteenth-century picture maps and later surveyed maps were used in litigation about individual property plots. Such maps are usually at a very large scale and cover only those landholdings involved in the dispute. There are many thousands of these maps in German archives, but they are often poorly cataloged and separated from other documents relating to the case.[8] Until about 1800 it was the responsibility of each individual landholder, rather than an official land register office, to have the deeds to his property drawn up and preserved. Many maps were drawn not for court cases but simply to complement deeds proving land title when land was bought or sold, rented or inherited.[9]

While most land title maps are of individual property plots, many others show the boundaries of large estates or large landholdings or forest possessions. These were often drawn to resolve disputed rights, particularly hunting rights, but also rights to graze animals or gather wood.[10]

One such dispute occurred in 1563 between Herr v. d. Malsburg and the villagers of Niederelsungen in Hessen. The villagers claimed the right to collect wood on land which was previously part of the Carthusian monastery of Niederelsungen. Herr v. d. Malsburg had acquired the

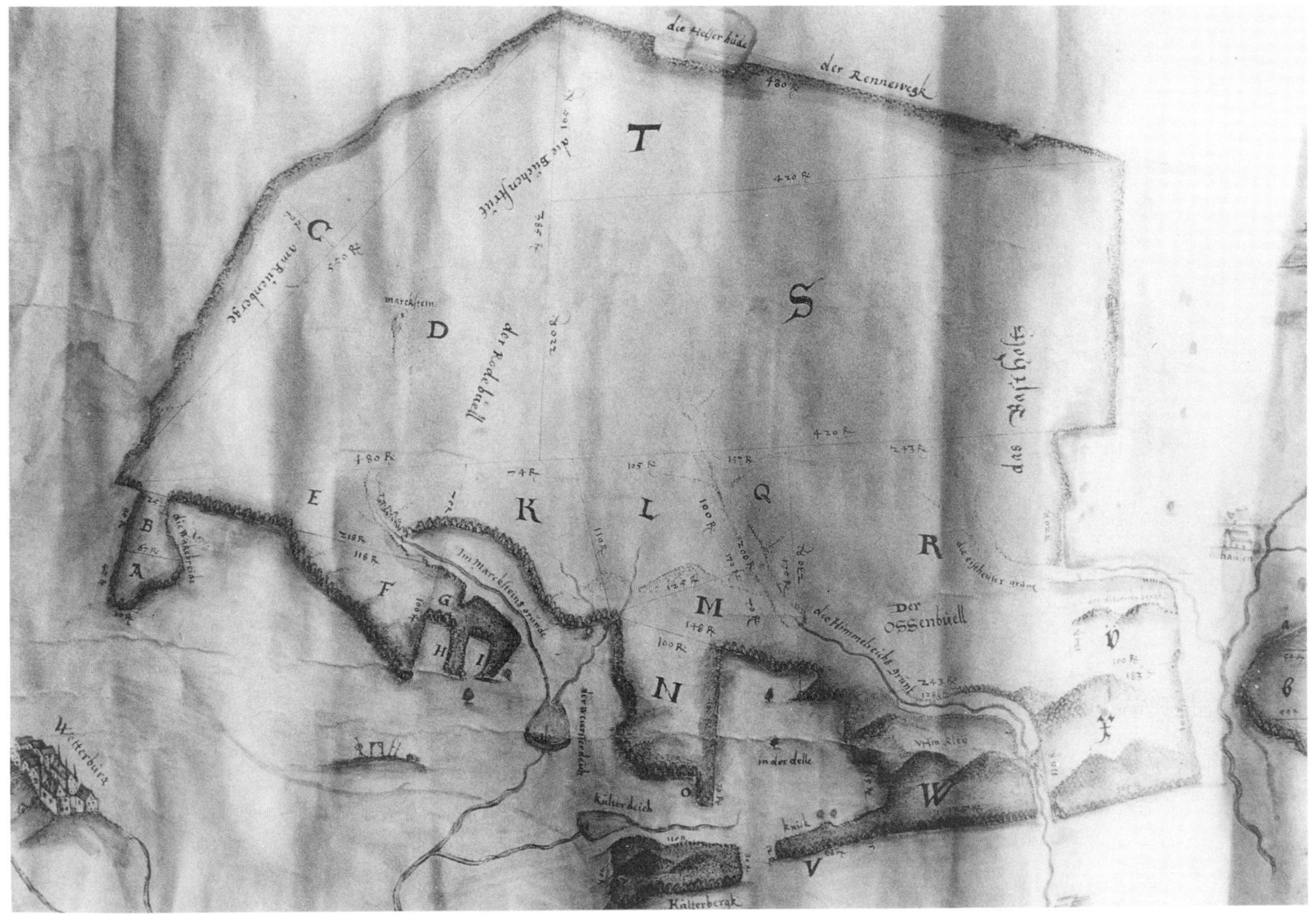

Fig. 4.3 Map of woods near Külte, Hessen, c. 1585, by Joist Moers. This map is among the earliest examples of property mapping in Hessen. Source: Hessisches Staatsarchiv Marburg, Karten P II Nr. 10. 469.

land on the dissolution of the monastery and wanted exclusive use of it. A map was used in the resolution of the dispute.[11] Another important early map drawn in a land dispute is that made by Joist Moers in 1584 or 1587. It is of Volkmarsen in Hessen and represents disputed woodland at a scale of 1:25,000.[12] Further woodland boundary maps are those which Moers drew in 1585 of Külte, also in Hessen (figure 4.3).[13]

In Germany maps of territorial borders occupy an intermediate position between cadastral maps of landholdings and topographical maps of regions because many politically independent units were extremely small, and political units and estates were often coterminous. Territorial rulers often held many noncontiguous pieces of territory, and territorial enclaves and exclaves were common. Political boundary maps can cover very small areas at very large scales, where the area in dispute or the total territory was small, or, at the extreme, they can be virtually indistinguishable from topographical maps. As with other types of land-title maps, the earliest examples of territorial boundary maps are picture maps, but there was a gradual move to more accurate and abstract cartographic representation.[14]

The difficulty of differentiating such boundary maps from topographical maps is illustrated by the example of the boundary map of the Electorate of Saxony (Sachsen) begun in the 1570s and finished in 1635. Elector August I (1553–86) conceived the project, whose purpose was to describe the boundaries of his territory and particularly to show the forest areas over which he had hunting rights. The first mapping was started by G. Öder the younger in the 1570s. He produced manuscript maps and a series of beautifully bound printed books containing maps of forest boundaries, roads, and paths. Fifty-six of these books survive. After the death of August I his heir, Elector Christian I, continued the work. The surveyors, under Matthias Öder, produced thousands of maps covering twenty thousand square kilometers in an area which corresponds broadly to the Saxony of 1815–1945. There is a set of original maps by Matthias Öder at a scale of 1:13,333.3 and a smaller set of copies by Balthasar Zimmermann based on Öder's surveying at a scale of 1:53,333.3. Although often called topographical maps, they are in fact boundary maps.[15] Most administrative boundaries are shown with great precision, and all the elector's vineyards, fishponds, forests, and especially his hunting grounds are portrayed. The immediate border areas show details of agricultural land use, soil type, and vegetation. The survey was by plane table without triangulation control but is of a quality unrivaled by contemporary surveyors and not equaled in Saxony until the nineteenth century. However, the maps had no influence on cartography in the region because they remained hidden in an archive until rediscovered in 1830.[16] Whether these maps are considered as boundary maps, topographical maps, private forest and hunting maps, or state inventory maps is largely a matter of personal choice; they illustrate well the difficulty of classifying early maps in Germany.

Disputes over the ownership, tenure, and use of property were not confined to land itself: they extended to water and to resources on the land, with the right to hunt game over land the most contentious. Disputes over water, especially the right to divert water to power a mill or to collect tolls on traffic using waterways, were often as bitter as those over land, and again maps were often used to argue a case in court. A very early example of such a picture map is that of 1528 showing the Alster-Beste Canal, which links Hamburg and Lübeck. The building of the canal caused disputes over the right to both territory and water. [17] A map of 1552 of three mills near the town of Fulda was used in a legal dispute between the town of Fulda and the owner of one of the mills. He had recently dug a new channel and storage pond to ensure that he always had enough water to power his mill, but the town disputed his right to divert the river, claiming it infringed their rights to the water.[18]

Maps were important not only in establishing existing property boundaries but also when such boundaries changed. There could be many reasons for change, other than the individual buying and selling of plots. Enclosure is an obvious example (section 4.4). Other large-scale projects, such as the building of a new road, railway, or canal, changed property rights over wide areas, and maps were regularly used to codify such changes.[19] Land reclamation in low-lying areas adjacent to the Netherlands and in other areas in the east (section 4.3) and changing river courses could affect rights to both land and water, and property disputes which arose from such changes were often resolved with the aid of maps. Before its canalization, the course of the Rhine was constantly changing. Land on the riverbanks and islands in the river appeared or disappeared, and ditch systems had to be altered. Large-scale maps were regularly used to record changes in the river course and resultant property changes. It was not only individual property rights which could be changed by natural processes. The course of the river Schwalm in Hessen altered considerably over time, and a map of 1583 records a change which entailed a revision of the territorial boundary between Hessen-Kassel and Hessen-Marburg.[20]

Maps were thus used from the sixteenth century onward to establish title to property. The sheer numbers of picture maps and of later, more abstract maps prove both how pressing was the need to establish land title in Germany and how useful maps were seen to be in this process.

Inventories of Private Estates and Forests

Throughout Renaissance Europe large-scale maps came to supplement medieval written cadasters as inventories of property. In Germany such instruments of administration and control were particularly important for both secular landowners and religious orders and monasteries, who often found it difficult to keep control of their widely scattered land possessions. Landowners needed inventories of their land and of the rights to demand labor service or payments in cash or kind vested in it.[21]

In 1713 Johann Philipp, bishop of Würzburg, called in the cartographer Mattäus Beck(er) to "renovate" his estates in Swabia (Schwaben) which had been laid waste in the Thirty Years' War. Becker measured the land plots precisely, ascertained their use and possession, and recorded the information on large-scale maps and very detailed registers.[22] The Hansa towns of Lübeck and Hamburg each had farmlands attached to them. Surveyor Simon Schneider mapped several areas around Lübeck between 1651 and 1669, and at the very beginning of the eighteenth century surveyors Schmid and Schumacher mapped more of the town lands. A mapped survey to determine the extent of the town lands of Hamburg was proposed as early as the 1620s, but it is doubtful that this actually took place.[23] Such private inventories are difficult to distinguish from early state taxation cadasters. The important "renovation" commissioned by the authorities of the Imperial City of Schwäbisch-Hall in Swabia at the end of the seventeenth century was in a sense a combination of the two. It was in effect an inventory of lands under the jurisdiction of the city but is described in section 4.6 as an early state taxation cadaster, as it covered land owned by more than one person.

Surveyed and mapped inventories were often compiled when estates changed hands. In the early eighteenth century the city of Leipzig acquired the estate of Pfaffendorf, bought some lands adjoining the estate, and took over the agricultural management of the land. The importance of precise statistical information was realized by the town council, which had been carrying out surveys, with and without maps, of its farmlands, meadows, and forests since the late seventeenth century. The estate of Pfaffendorf was investigated thoroughly by order of the council, and their surveyors produced an atlas containing seventy-eight very fine maps, as well as information on such matters as manuring and sowing practices, harvest totals, and hay yields on the estate.[24]

Religious houses often ordered mapped surveys of their lands to improve estate management. A map was made in 1723 of lands at Marienburg which belonged to the Hildesheim Cathedral Chapter. The chapter wanted to fix new ground rents and to this end made a new and accurate survey with a map, which survives, and a register, which has been lost. Within Marienburg the map carefully depicts land use and boundary markings, but the surrounding villages are shown only schematically.[25] In 1770 the fragmentation of land, legal uncertainties over ownership and possession, and a difficulty in collecting revenues led the nunnery of Heinsberg in the lower Rhine area to order a thorough investigation of all the land it owned or over which it had rights to fish or to collect tithes. Large-scale maps were drawn, and descriptions compiled. The maps were later used by the French when they sold off the nuns' lands during their occupation of the area.[26] In 1779 the Chapter of Trier Cathedral ordered an inventory of the Kührer Estate in the Rhineland palatinate. The land was surveyed and mapped at a scale of 1:4,000 by J. P. Dilbecker, and the precise area and use of each land parcel, together with all the dues and services required of its owner, are recorded in the map and in the register on the border of the map (figure 4.4).[27]

In the first half of the seventeenth century an inventory which included maps was taken of all *Vorwerke* (central farmsteads on manorial estates) belonging to the *Grafschaft* (county) of Oldenburg.[28] One map drawn before 1644 of Wurp, an area attached to the new farmstead of Jade in East Friesland, shows an area divided into thirteen plots enclosed by drainage dikes covering altogether an area of more than one hundred hectares. The map is simple with a colorful cartouche and a picture of the farmstead.[29]

In much of early modern Germany potentially valuable forests were greatly underused, and

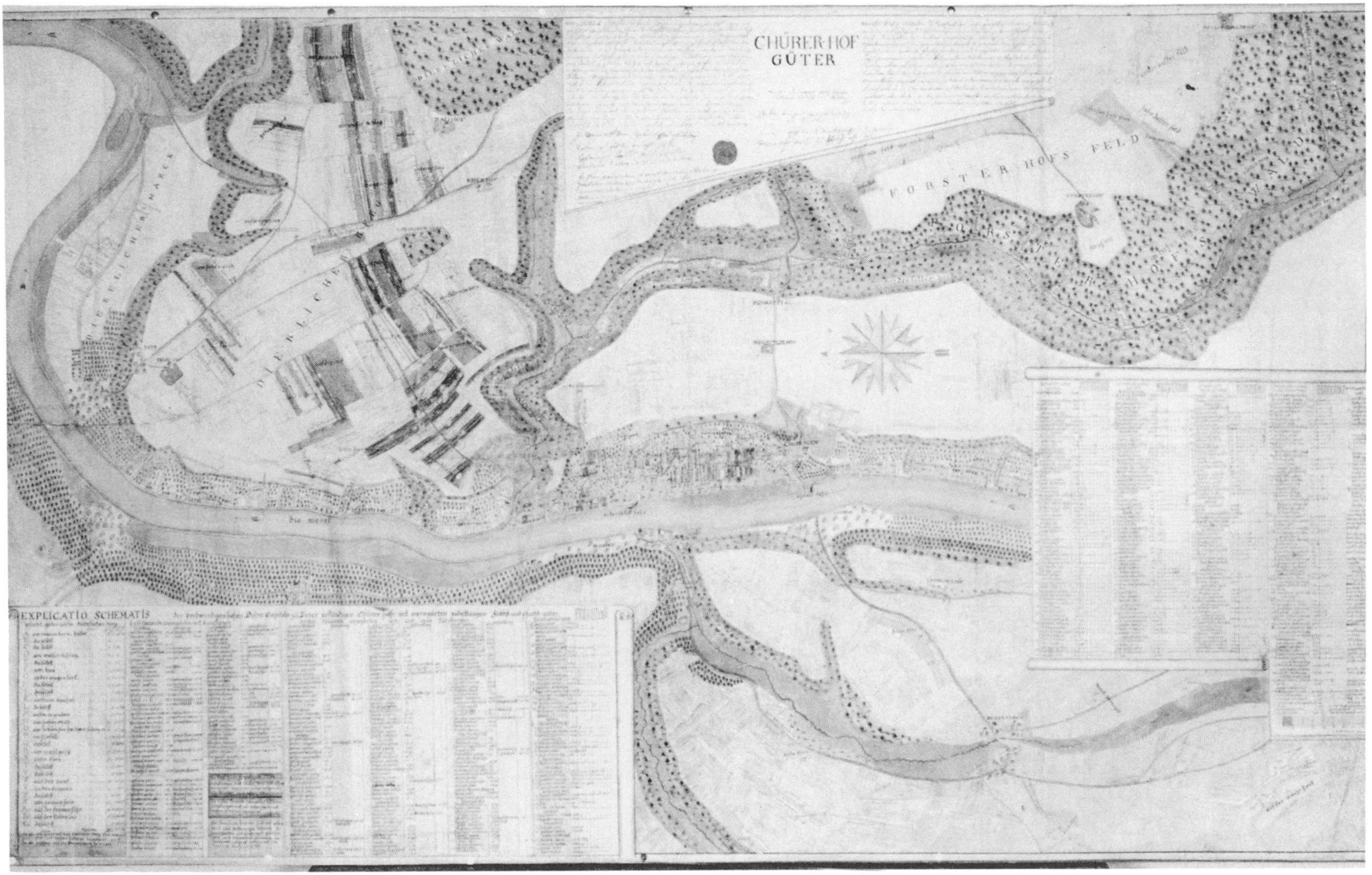

Fig. 4.4 Map at 1:4,000 drawn in 1779 of the Kührer Estate. The estate belonged to the cathedral chapter at Trier, which wanted a complete inventory of the estate. Source: Landeshauptarchiv, Koblenz. Best. 702 Nr. 235.

surveying and mapping were often undertaken in an effort to manage such forests more efficiently. At the Reformation many church lands, including considerable amounts of forest, passed into the hands of secular authorities and landowners. In Swabia in the eighteenth century there was a recognition that little was known about or had been done with the erstwhile church forests, and many of them were surveyed and mapped to aid their more intensive exploitation.[30]

By the later eighteenth century, forest management was becoming more systematic and scientific,and forest owners took a keener interest in them as they realized their value. One such estate owner was Hacke, the owner of the estate of Trippstadt in the Palatinate. He was particularly interested in forestry and commissioned the architect Siegmund Haeckher to map his estate.

In 1767 Haeckher produced a map which had several purposes: it was a boundary map of his patron's land, a wall map to adorn his new country seat, but perhaps most important it was a detailed forest map. Haeckher's map shows carefully the extent of his patron's forest and also the predominant tree species. Oak, pine, birch, and aspen trees are each distinguished by colored shading.[31]

4.2 Maps in the Management of State Forests

The private land title and estate maps described above are important in their own right, but they also led to a greater awareness of the map as a tool for consolidation or extension of power. Litigation maps had helped assert rights to property, and inventory maps had proved useful in the administration and exploitation of private estates. It was only a short step to the wider use of maps as general instruments of state policy and control and in the more systematic management and exploitation of state lands and particularly state forests.

Forest maps form a large part of many cadastral map collections. Forest management maps are generally outnumbered by those which show forest boundaries, which often remained unclear until as late as the eighteenth century, and those which show hunting rights.[32] There was, however, a gradual increase in the number of forest maps conceived as inventories and used for more systematic and scientific forest management, and state mapping came to dominate over private forest mapping.

An important early example of state rather than private forest mapping in Germany is that of forests of the Duchy of Württemberg carried out in 1680 for Duke Friedrich Karl by Andreas Kieser, with the assistance of Johann Niclasz Wittich and Johann Jacob Dobler. Together they produced 280 maps at a scale of 1:8,250 which cover the seven forest districts of Kirchheim, Böblingen, Stromberg, Reichenberg, Leonberg, Tübingen with Bebenhausen, and Schorndorf, about four thousand square kilometers of land. Woods belonging to the Free Imperial Cities within the area were excluded, as the cities expressed no interest in joining the survey. Unfortunately Kieser could not finish the work before the French army invaded in 1688, and he died soon afterward. Kieser based his work on an earlier survey by Wilhelm Schickhardt, but he enlarged the surveyed area significantly. His maps show the boundaries of the forest very accurately, but their internal configuration and that of the nonforested areas are shown only schematically. Artistically the maps are unremarkable, but their reliability and accuracy represented a noteworthy advance over previous maps. In a sense the maps are transitional between the picture maps of the previous century and the abstract representations of later centuries. Settlements are represented pictorially, and some are stylized portrayals copied from other maps. By contrast, the woods, which were the focus of interest, are very accurately portrayed: coniferous and deciduous woodland are distinguished, and Kieser had copperplates made of symbols for woods from which he printed off sheets of symbols which he cut out, pasted on to his maps, and colored. Together with the maps, Kieser produced comprehensive cadastral registers which contain calculations of the area of forest and information about its owners. At the time Kieser's work was heavily criticized, and one of his assistants was sent out to resurvey the area before the survey was finally abandoned. There was further difficulty with proposals for the production of small-scale topographical maps from Kieser's originals, and it was not until 1693 that the maps were combined at a reduced scale of 1:80,000 by M. Johann Mayer.[33]

Brandenburg-Prussia is most noted for its state forest mapping, not least because its royal forests were so extensive, particularly in the eastern provinces, and because surveying and mapping were considered important in the management of the forests from an early date. From the mid-seventeenth century onward, *Waldmesser* (forest surveyors) are known to have been in the regular employ of forest owners. The first mention of state forest maps is from a forest instruction of 1704 which ordered that maps be drawn of heaths and woods to show their extent and to differentiate the various species of tree. Despite such instructions, however, there was much less concern for the forests themselves than for the game within them and the hunting rights over them. An *Oberjägermeister* (head master of the hunt) remained in charge of the woods, and forest management per se was fairly rudimentary. Improvements were made following an instruction of 1764 that all the forests of Brandenburg-Prussia be surveyed and mapped and the quality of the trees noted, but the project ran into difficulty because of lack of both money and trained surveyors. Considerable progress was made in the Kurmark (Electorate of the Mark of Brandenburg), where fifty-one forests had been surveyed by army officers by 1767; in the eastern provinces of Pomerania, Lithuania, and East Prussia, however, little was accomplished. A special team of surveyors was sent in 1769 to Lithuania to try to fix the disputed forest boundaries in the area. A much more important step was taken in 1770 with the setting up of the Forestry Department, which was directly responsible to the king. The department immediately issued a plan for the improvement of forest management and demanded copies of maps of forests from the Kriegs und Domänenkammer (regional bodies with responsibility for the administration of war and state lands). The Kurmark was able to comply with this demand at once because of its earlier successful surveying, and some other areas also sent in maps.

It became clear that the areas west of the river Elbe had good maps of their small and shrinking woods, while the areas east of the Elbe, which were rich in forests, were almost entirely unsurveyed and unmapped. New instructions in 1771 and 1772 ordered the surveying and mapping of all forests to show the number and size of the *Reviere, Blöcke*, and *Schläge* (divisions of the forest made to improve management by making felling more systematic and for control of game), but again work was hampered by lack of money and personnel. The maps were often too poor to be useful: the areas for felling, for example, were often marked out near to the roadside, but not deep in the forests. Some peasant forest was marked in as royal forest, which made calculations of the extent of royal forest very inaccurate. Such mistakes were not rectified at the time, as the surveyors' work was not checked in the field. The maps were often huge and unwieldy and were described at the time as *blos eine Tapeten- und Dekorations-Malerey* (just showy wallpaper). To try to overcome these faults the Forestry Department sent surveyors model maps in 1780 and in 1782 ordered that work be checked. If it did not come up to standard, the surveyor was not to be paid, and his license to practice for public authorities could be revoked. In 1783 six hundred copies of map symbols and conventions were printed and issued to those surveyors in Forestry Department service. After about 1780, maps started to come in regularly. Trainee surveyors were set to make copies of them at a reduced scale, and they were very carefully stored and cataloged. The maps range in scale from 1:800 to 1:105,000 and cover all of Brandenburg-Prussia except Silesia (Schlesien). Silesia had its own forestry administration, but its history is similar to that of the rest of Prussia, with a new phase of forest management based on accurate surveying starting in the 1770s.[34]

In 1750 the Forestry Department of Karlsruhe proposed a survey to fix forest boundaries which had become unclear during the many wars of the sixteenth and seventeenth centuries. The Forestry Department aimed to mark forest boundaries on the ground with stones and to record the boundaries in accurate maps. One extant map from this survey dates from 1756–57 and is of forests in part of the Hardtwald, north of Karlsruhe. They had been in the possession of the Margravate of Baden-Durlach since the Middle Ages, but the local peasants still had customary rights to wood and grazing in the forest. It is not simply a boundary map, as deciduous and coniferous trees are differentiated, and the age of various stands of trees is noted.[35]

Forest maps are thus an important category of early state maps. Initially concentration was on delineation of boundaries and codification of the rights of the monarch, the nobility, and the peasantry to ownership or use of the forest. Exploitation of the forests was necessarily predicated on establishing rights over them, and so in a sense all these forest maps are management maps. There was however a trend toward the drawing of what might today be more normally considered as management maps. Increasing production of timber for the market encouraged states to use maps to establish inventories of forests, both of their extent and of the type of timber they contained, and to aid profitable exploitation by systematic felling.

4.3 Internal Colonization and Land Reclamation

Within Germany there was considerable scope for the establishment of new settlements. Important colonizations took place from the twelfth and thirteenth centuries onward in the vast and comparatively underpopulated areas east of the river Elbe, but there were also underused heaths, mosses, forests, and moors throughout Germany, and in some areas land could be created by making polders or draining marshy ground. Some settlers went to planned colonies, the houses and land plots of which were often laid out with the help of maps.[36] One such orderly colony was that of the remnants of the Asperden heath south of the town of Cleve in northwestern Germany. In 1775 it was planned to colonize the heath with twenty-eight colonists from the Palatinate. The surveyor H. van Heyss surveyed the land, parceled it out, and mapped it. The resultant maps show the regular rectangular plots set out for the colonists (figure 4.5).[37]

The Reformation brought changes in landownership and displaced groups of refugees, and these changes were sometimes a stimulus to colonization schemes. In Protestant areas the Reformation entailed the dissolution of religious houses and secularization of church land. Landgrave Wilhelm IV of Hessen acquired the lands of a Cistercian monastery which had been dissolved in 1527. He wanted to use the lands to his own glory, and in 1580 commissioned a map of the land which was used to organize the foundation and settlement of the new village of Wilhelmshausen.[38] When the Edict of Nantes was revoked in 1685, French Huguenot refugees were welcomed by many Protestant German princes, who recognized their agricultural and manufacturing skills. One group of refugees fled to Hessen, and a map of 1696 shows the new settlement of Carlsdorf in Hessen which was laid out for them. The model village, the name of which again celebrates the secular ruler, with its surrounding agricultural land was carefully planned by the architect Paul du Ry and is represented on the map at a scale of 1:2,500.[39]

Polder mapping has been discussed in detail in the Netherlands chapter (chapter 2.1), but polder construction was also important in the low-lying coastal areas of East Friesland which adjoin the Netherlands to the west of the river Ems. Cadastral maps were used to lay out the regular fields on newly won areas. The earliest known map of East Friesland is that of 1628 by the famous surveyor Johan Sems. Fens in the same areas were also reclaimed, colonized, and settled from the 1630s onward. The colonization was organized sometimes by a large-scale entrepreneur, sometimes by fen cooperatives, and sometimes by the state. A map of 1704 by J. Tönnies of the second oldest fen colony in East Friesland, founded in 1637, depicts the drainage canals and the colonists' regular plots.[40]

In Brandenburg maps played an important role in the improvement of the *Bruch- und Luchgebiete*, low-lying marshy areas round the Oder, Warthe, Netze, Rhin, and Dosse rivers, which drain northward into the Baltic. In the mid to late eighteenth century Frederick the Great promoted schemes to canalize rivers, manage dikes, and increase the cultivated area by organizing settlements of new colonists. The aim was to revive the area after the destruction of the Seven Years' War. In Neumark maps on a scale of 1:5,000 were used in mid-eighteenth-century schemes to make cultivable the areas round the Oder. By 1763 over one thousand smallholders and their families—in all, more than six thousand people—had been settled on the fertile soil reclaimed

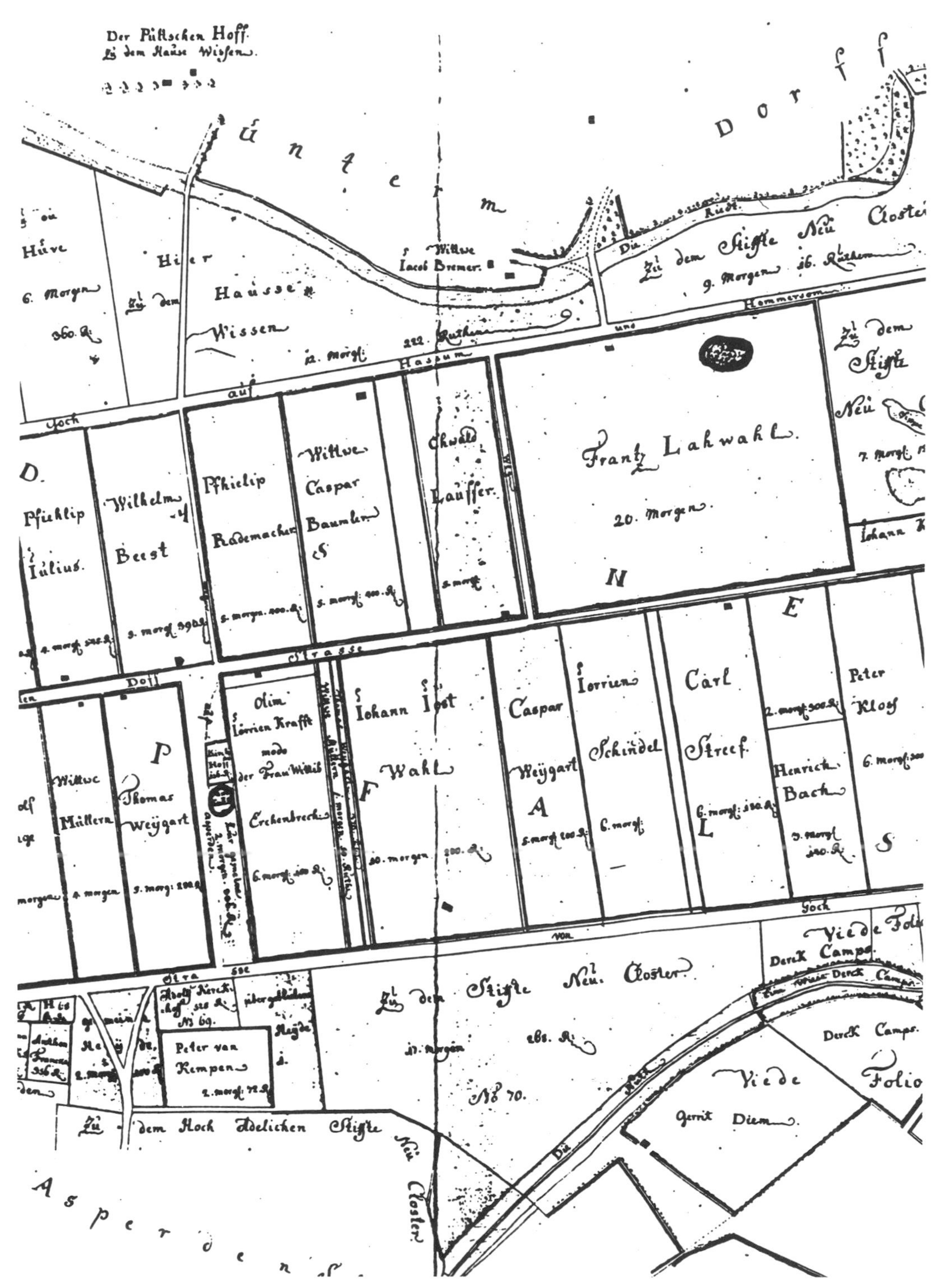

Fig. 4.5 Map of Asperden heath by H. van Heyss, drawn in 1775 in preparation for the settling of twenty-five colonist families on the heath. Source: Archiv Gaesdonk, Grafenthaler Akten.

from the Oder marshes. Lieutenant Colonel Petri, who was involved with the Oder reclamation project, went on to projects in the valleys of the Netze and Warthe between 1766 and 1776, where more than seven thousand people settled in the first nine years of the work and where the population soon rose to about fifteen thousand. Maps played an integral part in the planning and execution of these successful schemes.[41]

4.4 Agricultural Reform in Germany in the Eighteenth and Nineteenth Centuries

Agricultural reform in this period in Germany as elsewhere had two physical aspects. *Teilung,* later often called *Separation,* involved the division of common lands into privately owned plots and the ending of common use-rights; *Verkopplung* involved the amalgamation of strips of land into enclosed consolidated blocks. Neither aspect necessarily involved the abandonment of the prevailing two- or three-field system of the breakup of the village, and in the early stages of reform each farmer might still have a large number of plots. Later consolidation, or *Flurbereinigung,* was a far more radical process, which continues today. However, the main object of agricultural reform in Germany in the eighteenth and particularly the nineteenth centuries was not one of physical rearrangement of landholdings and hence not one in which mapping could play a significant role. The overwhelming preoccupation was with the emancipation of the peasantry, the removal of the legal basis of personal servitude. Emancipation was sometimes, but not invariably, accompanied by a physical rearrangement of landholdings. To understand how the two processes were linked, it is necessary to sketch the variety of landholding types in Germany during the eighteenth and nineteenth centuries.

Land Tenure and Its Reform

In the great majority of cases the landlord was a ruling prince or nobleman, although townspeople, the Roman Catholic church, and the Free Towns also owned estates. There was a wide range of relations between peasant and landlord, but a broad division may be drawn along the line of the river Elbe. To the east was an area of *Gutsherrschaft* (manorial system of estate agriculture), and to the west was an area of *Grundherrschaft* (feudal lordship with peasant agriculture). Recent research has shown the east-west division along the Elbe to have been oversimplified, but in a brief discussion such as this the twofold division remains useful.[42]

In the east, land was under the direct personal control of the lord of the manor—the *Rittergutsbesitzer* or *Junker.* Since the sixteenth century the *Junker* had been increasing the size of their domains by *Bauernlegung* (suppression of the peasantry) and colonization of the wastes. In the eighteenth century the *Junker* had often let out their land, but by the nineteenth century they normally farmed it themselves. The estate remained the standard unit for administration (*Gutsherrschaft*) as well as production (*Gutswirtschaft*). The peasants were not strictly speaking *Leibeigene* (serfs) but were *Gutsuntertanen* (subject to manorial jurisdiction) in criminal and civil cases. They had to meet heavy demands for labor, sometimes up to six days a week of *Spanndienst* (the provision of a team of draft animals and their drivers), but correspondingly had to make only light payments in money or kind.[43]

To the west of the Elbe (northwestern and southwestern Germany, Bavaria, Saxony, and Thuringia), by contrast, the *Gerichtsherr* (legal lord) and the *Grundherr* (seigneurial lord) were often different people, and the nineteenth century witnessed an increasing centralization of seigneurial rights in the hands of central government. Under the system of *Grundherrschaft* or *Rentengrundherrschaft* serfs were liable to dues in kind and rents to their lords, but the lords themselves had at most only small estates, and labor service often amounted to only a few days a year. Feudal service had been commuted for money payments, and peasants often held their land on a hereditary basis and could dispose of it freely. The exception to this pattern was Bavaria, where conditions were more similar to those east of the Elbe.[44]

There were local variations within each of these two blocs according to whether the peasants had legal property rights in their holdings, to the degree of penetration by market forces and commodity production, and to locally specific factors such as the existence of proto-industry or a vigorous land market following the dissolution of church property.[45]

As well as these different legal bases for agriculture there were varying physical arrangements of fields and settlements in the countryside. In some areas consolidated landholdings were the norm. Westphalia, for example, was characterized by a loose settlement pattern of widely spread villages with farm plots which were already consolidated; here there was no need for enclosure of the arable, although the commons remained to be divided. In some Alpine areas plots were already consolidated. In areas of recent colonization such as along the North Sea and Baltic coasts and in some forest areas, especially in the east, land was set out from the beginning in compact holdings and peasants were free of communal restrictions on cultivation and rotation. There were also enterprises run on capitalist lines in Rhineland vineyard areas. These examples of consolidated holding were, however, the exception to the rule; the majority of the land in both east and west was characterized by central village settlements surrounded by open fields divided into strips.[46]

From the late eighteenth century agricultural reforms took place with very different results east and west of the river Elbe. In Prussia and most other German states east of the Elbe, as in most of continental Europe, peasants were freed from their feudal bonds and burdens but had to compensate the old feudal lords. Emancipation was combined with enclosure and in many cases involved the sweeping away of ancient villages.[47] Peasants on the royal demesnes of Prussia were emancipated in the years following 1798 in both Westphalia and in the eastern provinces. Some landlords in eastern Prussia followed the king's example, but most other peasants had to wait until the reforms of the early nineteenth century. On 9 October 1803 emancipation was extended to all peasants in all Prussian provinces, and the statute of 14 February 1808 encouraged the consolidation and enclosure of peasants' strips.[48] Decrees of 1807 and 1808 made the peasants on private estates, like those on royal demesnes earlier, the direct subjects of the king; their landlords were no longer intermediate authorities. In this sense the nonroyal peasants had been emancipated, but they had not been relieved, like the domain peasants, of the services with which their land was burdened. Under decrees of 14 September 1811 and 29 May 1816, peasants who had the right to pass on their land had to cede one-third of it to the lord to become freeholders of the rest; peasants whose holdings were not heritable ceded half of their land to become freeholders of the remainder. In many cases peasants were evicted from land which they had recently acquired, with the result that they faced the depression of the 1820s seriously impoverished. Many became landless laborers. The nobility, in contrast, kept their freedom from the land tax and other privileges until 1861, and many discovered that by working their estates as capitalist enterprises free from any obligation to the erstwhile peasants, they were far better off than under the old system.[49]

In the east there had been some separation of the demesne from the open fields and from the common pasture in the eighteenth century. As emancipation progressed, so did *Separation:* the lords consolidated the pieces of land that they received as peasants bought out their holdings, and they separated their share of the commons from the peasants' joint share. The more substantial peasants also separated their land, moved out of the village, and built new houses on enclosed plots which they could use more intensively. The process proceeded slowly, but by 1850 most lords had compact holdings.[50]

Until 1872 consolidation of holdings was in a sense incidental to the main object of emancipation, although some enclosure took place. Legislation of 1872 began the direct promotion of enclosure and allowed the rearrangement of all or part of the common-field system if a simple majority of the landowners demanded it. If enclosure was not explicitly promoted by emancipation, it was certainly encouraged by the coincidental demand for food for export and for the expanding home market of urban-industrial and landless agrarian workers. Owners of consolidated and extensive estates in the east who were able to draw on an abundant pool of wage labor were best able to exploit these new markets, and this acted as a growing spur to enclosure in the nineteenth century.[51]

The process of agricultural reform west of the Elbe is poorly documented by contrast to that in the east, and because political authority in the west was far more fragmentary, it is more difficult to generalize about the nature and speed of reform.[52] However, it can be said that in the west enclosure did not usually accompany peasant emancipation, and there was generally a sharp gap between the consolidation of strips in the open fields, which happened in the nineteenth century, and the division of common lands, which often took place in the eighteenth, for example in Westphalia, especially in Osnabrück. There were very few instances of peasants in the west being cleared off their land as a result of emancipation, and as a result very large capitalist estates like those east of the Elbe did not develop. There were, however, exceptions. In Schleswig-Holstein, for example, emancipation and enclosure were carried out together in the eighteenth century. Northwestern Germany in general formed an advanced agricultural area.[53]

In central southeastern and especially southwestern Germany, as in the rest of western Germany, the consolidation of strips did not usually occur in the period of emancipation but gained impetus only toward the end of the nineteenth century or even later or became arrested halfway in the form of the rationalization of holdings within an open-field system. The division of the commons proceeded more rapidly in these areas. In Baden, for example, serfdom was abolished very early, and the peasants' legal position was strengthened as the *Code Napoléon* was introduced in its entirety. The first law facilitating the rearrangement of fields followed much later in 1856, and it was not very successful.

As in Baden, in Bavaria the question of emancipation was separated from that of enclosure. In Bavaria emancipation began in 1779 on Crown land, but the main reforms came with the laws of 1826 and particularly 1848. There were some early campaigns to reduce the size of holdings, since small holders used their land more intensively. Subdivision was first legalized and then actively promoted by acts of 1762 and 1772. The secularization of 1803, which involved the dissolution of monastic foundations and the confiscation of ecclesiastical estates, affected a large proportion of the Bavarian peasantry, who passed under the direct control of the state and in the majority of cases strengthened their claim to the land they farmed. Land subdivision gathered pace, which mainly benefited smallholders. By the mid-nineteenth century the traditional pattern of cultivation still predominated, and what land reallocation there had been was mainly the result of subdivision and reflected peasant instability and the extension of the subsistence farming sector. The pace of change was too slow for the authorities, and the laws of 1858, 1886, and 1899 attempted to speed up land redistribution by introducing the element of compulsion. By the 1880s the pace of land redistribution had increased dramatically but continued to strengthen the position of smallholders.[54]

In the west in general, peasants gained more than they lost, in time becoming owners of by far the greater part of the soil. Before 1840–50, pressures for further reform, including redistribution and enclosure of land, were limited in western districts, with the exception of Schleswig-Holstein. The peasants in the west were often small farmers who could not afford to take the risks inherent in innovation: they were in any case not always bound to follow common crop rotations, and they were primarily aiming to provide subsistence for their families, not a surplus for the market. The breakdown of the common-field system was protracted as the peasants gradually bought off of their burdens, but while in 1840 the three-field system predominated, by the 1870s small but consolidated holdings were the norm.[55]

By the end of the nineteenth century, then, Germany remained a land of peasants on small holdings in the south and west and of large estates in the north and east.[56] Physical reform was limited, and the country remained full of scattered holdings in the twentieth century, though in most cases before 1875 the system of common fields had been abolished. It was only in Brunswick that the old system completely disappeared.[57]

Agricultural Reform Maps

From the description above it will be appreciated that it would be meaningless to generalize about the use of maps in agricultural reform in Germany; it is indeed striking that the literature on the process of reform is entirely divorced from the literature on land maps. What follows here is a tentative attempt to link land reform and land mapping in some parts of Germany, particularly northwestern Germany, which, as has been suggested, was an area of advanced agriculture.[58]

Brunswick (Braunschweig) saw some of the most radical agricultural reform in Germany. Between 1746 and 1784 the Braunschweig *Generallandesvermessung* (general survey) took place, prompted partly by the need for financial reform and partly to give an opportunity for joining land strips into *Koppelweiden* (enclosed pastures used for alternate husbandry). This did not involve the ending of the three-field system or the breakup of the village but provided the opportunity to reduce fragmentation of cultivated land, divide common grazing land, improve roads and ditches, and demarcate boundaries clearly. Work started in 1746 under a commission whose task it was to survey all villages and their fields, meadows, and forests, together with all demesne land and separately to survey the state forests. Two procedures were adopted. The first was the *speziell* (special) survey, in which the existing division of land was surveyed and mapped. The second was the *allgemein* (general) survey, in which the surveyor investigated the existing situation through available written cadastral material and then, with the help of local villagers sworn in for the purpose, went on to allocate consolidated land plots to the farmers according to the amount of land they had held before and to survey and map the new landscape. Consolidation was unnecessary in some areas, such as the northern heath lands, and would have been impracticable in others, such as areas of very rugged terrain; these areas were covered by the *speziell* survey. The maps, then, may show either a regulated or an unchanged landscape, and since the surveying was not done in any systematic order, maps of neighboring villages may be many years apart in date (figures 4.6 and 4.7).

Enclosure in Brunswick was not dependent on the prior agreement of the villagers, but the

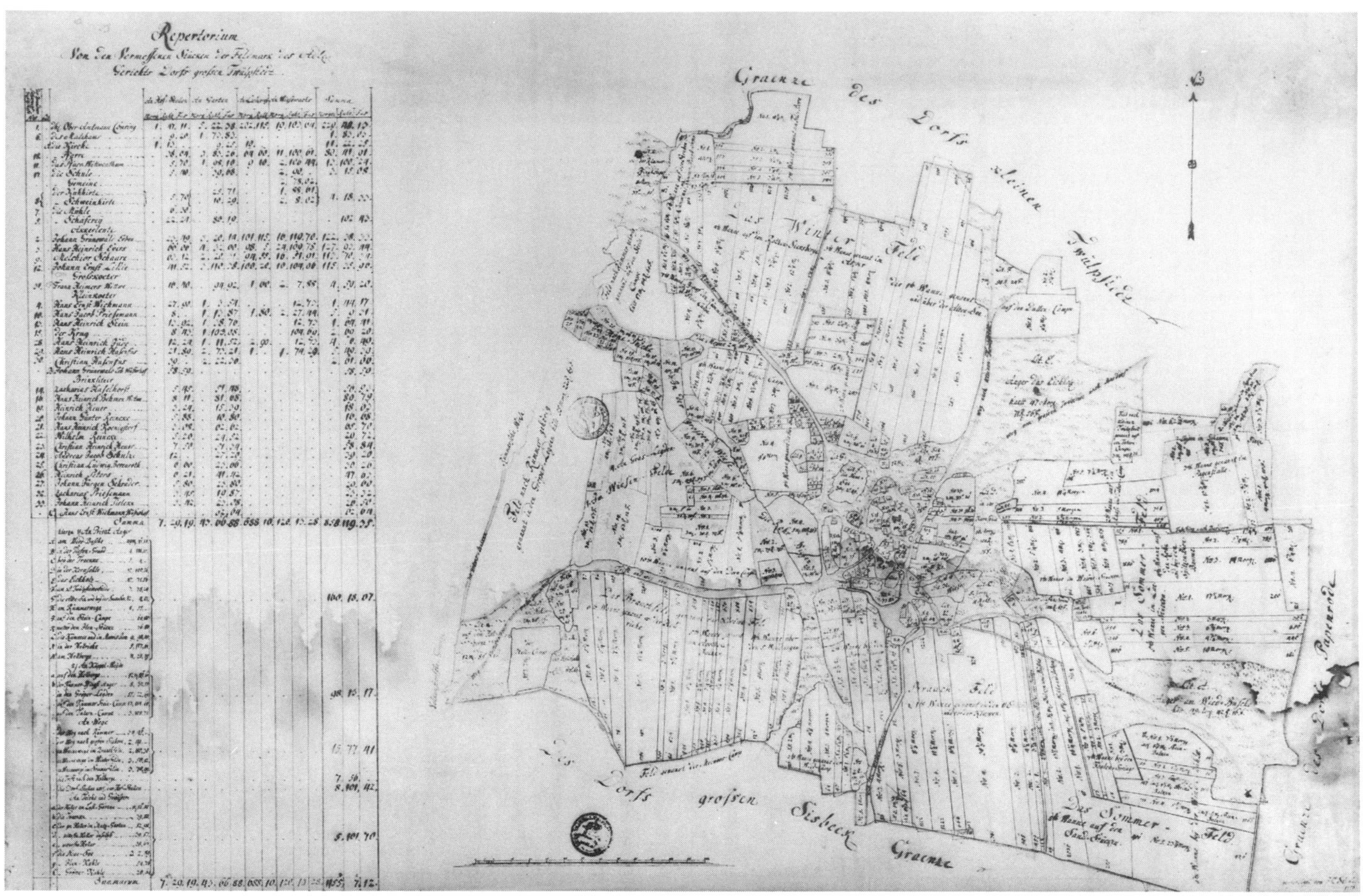

Fig. 4.6 Map of Groß Twülpstedt drawn in 1758 as part of the Braunschweigische Landesvermessung. The scale is 1:4,000. Some consolidation of holdings took place in the survey, and the map shows the situation after consolidation. Source: Niedersächsisches Staatsarchiv in Bückeburg, Wolfenbüttel, k3311.

process generally was carried through without dispute, not least because allocation had to take place in the presence of interested parties, and no one was to gain or lose land by the process. Field sketches were produced by the plane-table method at a scale of 1:2,000, and written descriptions were made in the field. Fair copies at a scale of 1:4,000 were made soon afterward and sent in to the survey commission. The work was finished in 1784, by which time the whole of the then Duchy of Brunswick, including the exclaves of Thedinghausen and Calvörde but with the exception of the Principality of Blankenburg, had been mapped. Only the village areas with their fields were surveyed and mapped. Surveyors excluded forests, mountains, and other unsettled

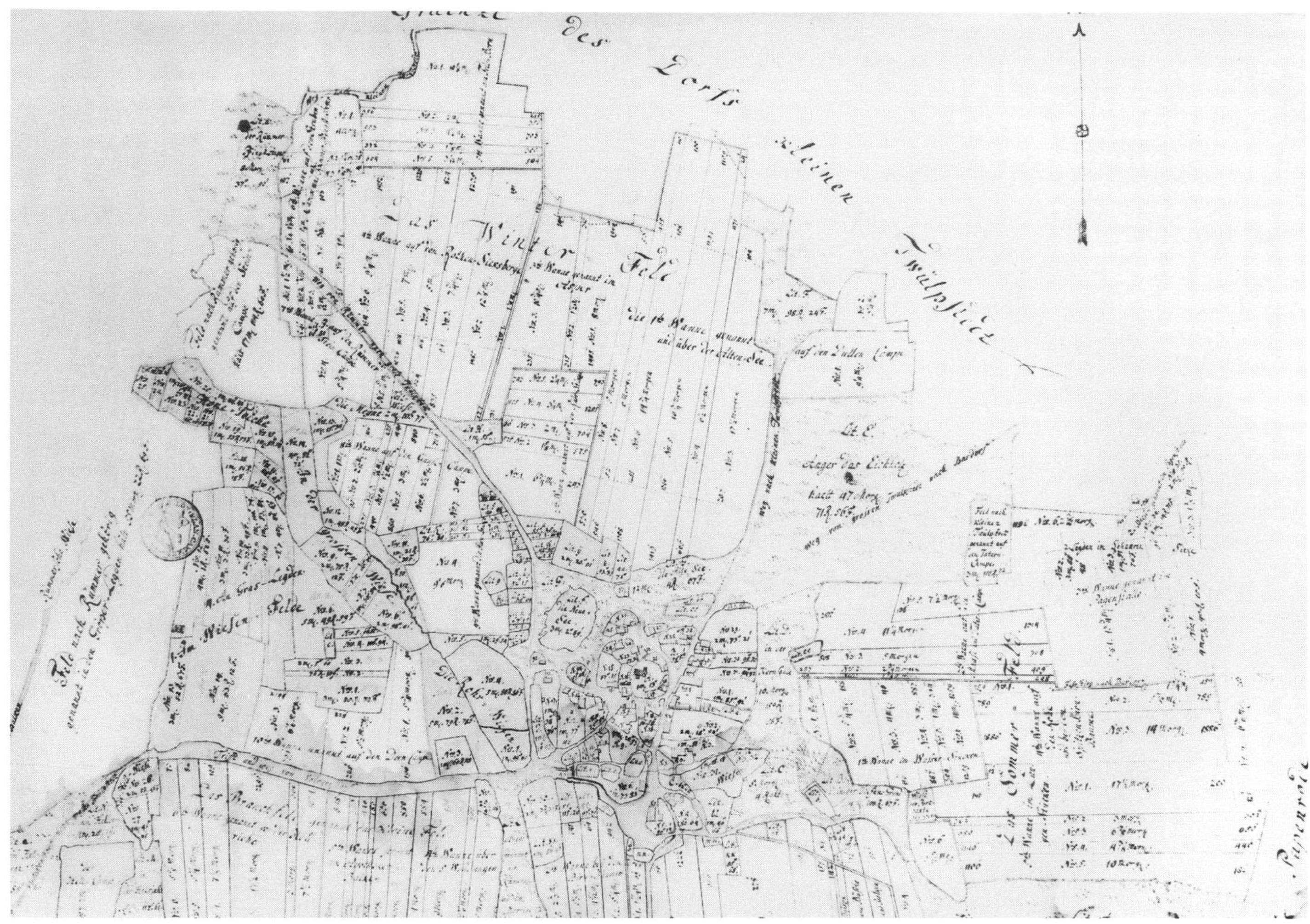

Fig. 4.7 Detail of map of Groß Twülpstedt in figure 4.6.

areas not in private or common ownership. They produced 424 maps, drawn at a scale of 1:4,000. These do not represent relief, as had originally been planned, and some were so poorly drawn that they had to be redone. Generally, though, the surveying was successful in promoting agricultural and fiscal reform. The maps show the settlement pattern before the later *Separationen*, or more radical agricultural reforms, which were in turn based upon the general survey maps.[59] The Brunswick reforms were some of the most radical and comprehensive physical reforms of landholdings in Germany, and mapping and surveying played a key role in their execution.

A survey similar to the Brunswick general survey occurred in Nassau in the second half of

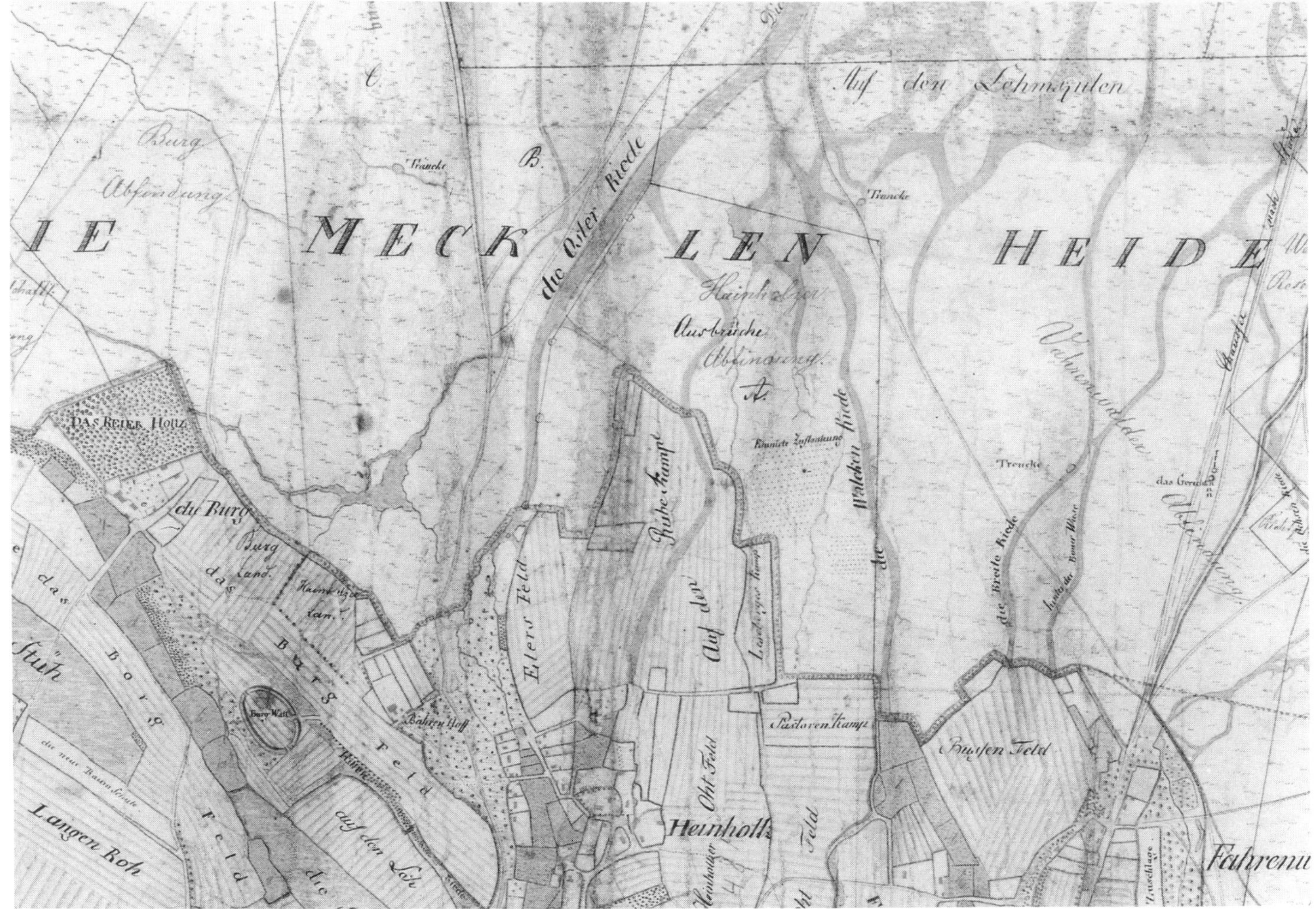

Fig. 4.8 Map of the area around the city of Hanover in 1775 at a scale of 1:5,760, showing the division of commons belonging to several villages. Source: Amt für Agrarstruktur, Hanover.

the eighteenth century. It was prompted by the need for a reform of taxation, but in some areas it encompassed land consolidation as well. In Nassau-Usingen the existing holdings were simply mapped and registered, but in Nassau-Orianien after 1784 land strips were consolidated into more compact holdings, and the resultant landscape was mapped.[60]

A notable promoter of agricultural improvements was George III of England, who also ruled Hanover. He helped to found the agricultural society in Celle in 1764. This society was active in agricultural reform, and it and the British Crown were instrumental in putting enclosure in

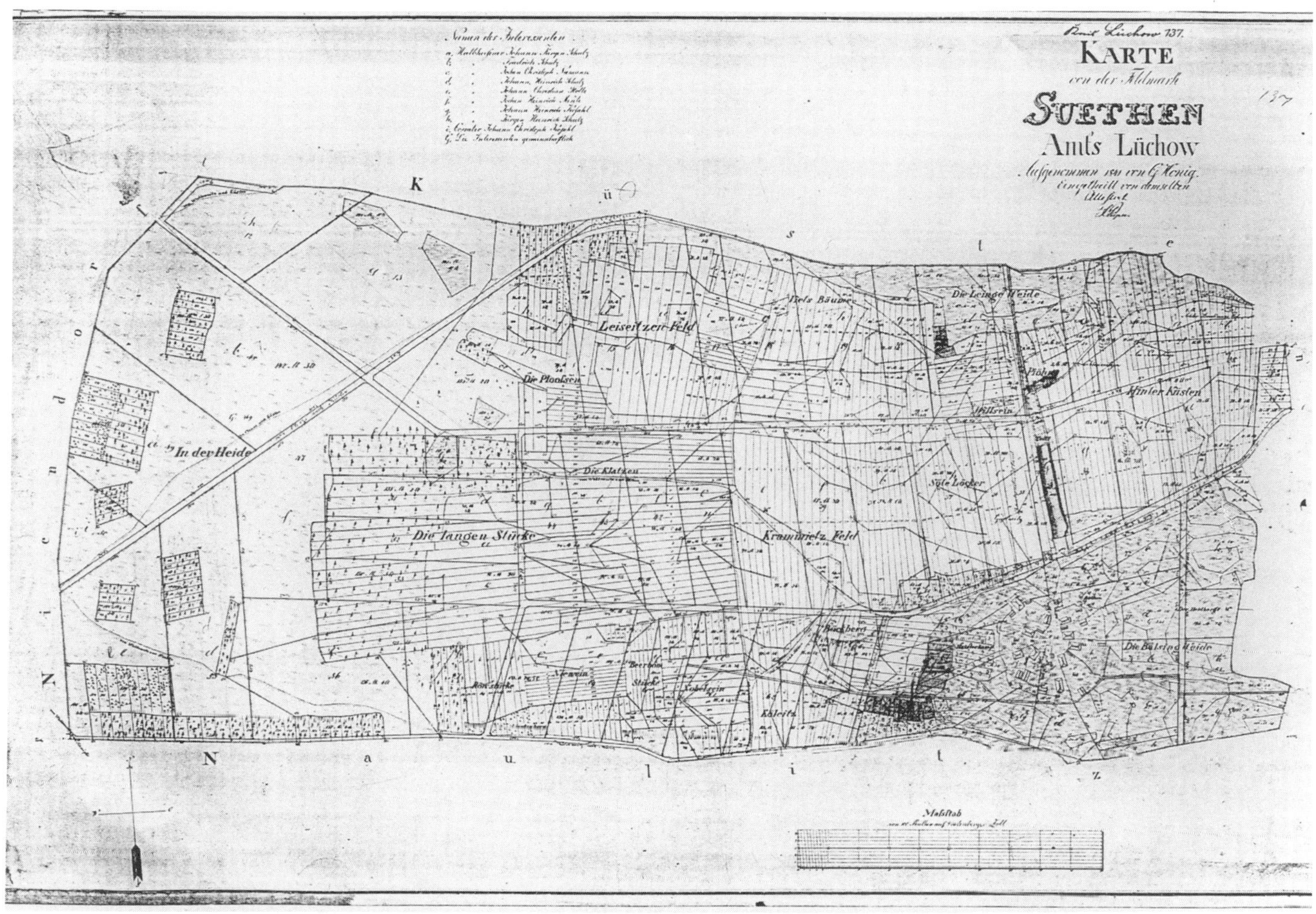

Fig. 4.9 Enclosure map of the village of Suethen in the southeast of the Kingdom of Hanover in 1841. The strips indicate land division before enclosure. The superimposed lines (in red on the original) are the boundaries of the new, large fields. Source: Amt für Agrarstruktur, Hanover.

Hanover on a more precise legal footing in the 1830s. Enclosure of the many extensive heaths and commons had been taking place in almost every part of the province of Hanover from the 1770s, despite the lack of a comprehensive legal basis for it, and there are fine maps from the 1770s which show areas both before and after enclosure (figure 4.8).[61] Reform in nineteenth-century Hanover proceeded on Prussian lines: the lords took part of the commons in absolute ownership, and the peasants retained the rest for their joint use (figure 4.9). Rearrangement of the fields did not start until 1842 and was modified in 1856. Peasant clearances were prevented through protection measures enacted by the state as part of the emancipation.[62]

As has been mentioned, northwestern Germany witnessed early enclosure of the commons. An important early law governing the division of common lands was that of 1802, which applied to Lüneburg and which was stimulated by agricultural reforms in neighboring Denmark. Whereas division and enclosure of the commons had previously been possible only if all parties agreed, from 1802 the assent of only half the people affected was needed. The 1802 act became a model for almost all other areas of Lower Saxony. Further legislation increased the element of compulsion, and surveying and mapping were important instruments by which the reform was effected. From the 1830s precise instructions were issued to surveyors about the equipment they were to use, how various boundaries were to be designated, how disputes were to be resolved, and how they were to carry out reform of taxation consequent on division. They also specified that maps were to be produced at a scale of 1:3,200 and 1:2,133. A general order of 1842, with various amendments passed up to 1856, brought together existing legislation and gave a further impetus to the division of the commons. It set up a commission, normally led by and employing trained surveyors, to carry out division. Between 1835 and 1892, a total of 386 *Separationen* took place in about the same number of parishes.[63]

In 1804 an order was issued for the division of the commons in the Duchy of Oldenburg. Bailiwick maps at a scale of 1:20,000 produced from 1781 onward as part of taxation reform were used as a basis for the division. Other larger-scale maps at 1:3,771 and, after 1844, 1:3,000 were drawn especially for the purpose.[64]

In the Rhineland there was some early, private consolidation of land in which larger landlords increased the size of their land plots by buying or exchanging strips. From 1786 to 1790 the government of the Electorate of Cologne attempted land reorganization on larger estates with the main purpose of improving road communications. The spread of French revolutionary ideas had concrete results in that the French abolished serfdom without compensation to the lords in the Rhineland and the Kingdom of Westphalia, although it was later reestablished in Westphalia.[65] As part of their reforms the French abolished *Flurzwang*. The reform gave peasants the theoretical right not to follow a field system and to do what they liked with their own plots, but in practice most plots were inaccessible by road or track, and the field system stayed in operation to avoid farmers' trampling over other people's land to reach their own. In 1886 the Rheinische Generalkommission (Rhenish General Commission) was set up with powers to reform agriculture where at least half the landowners who together owned half the land area wanted reform. In fact there was little significant reduction in the number of strips which each farmer owned, but the reform ensured that each strip was accessible by road so that each farmer could determine what he did with his own land.[66]

In the Duchy of Holstein-Gottorp, which formed enclaves within the Duchies of Schleswig and Holstein, there were long-standing desires to increase state income and to improve agriculture. In 1745 it was decided to carry out a systematic survey, but no attempt was made by the ducal exchequer to realize the project until 1753. Between 1753 and 1757 Lieutenant Bernard von Wörger surveyed the *Ämter* (counties) of Cismar and Oldenburg. The maps produced were designed to aid the enclosure and consolidation of fields.[67] Between 1763 and 1787 two further important surveys took place in Holstein-Gottorp. Privy Councillors E. J. von Westphalen and Caspar von Saldern initiated surveys as the basis for agricultural and tax reforms. The first survey covered the *Ämter* of Holstein-Gottorp and produced a total of 149 maps between 1763 and 1766. The second survey covered Norderdithmarschen on the North Sea coast between 1769 and 1777, from which fifty-two maps and accompanying documents survive. The surveys were extremely comprehensive: each field was surveyed, and its land use and ownership before enclosure are carefully noted. The maps are of a high technical standard and, together with written descriptions, provide an excellent picture of eighteenth-century preenclosure agriculture.[68]

Schleswig-Holstein witnessed early emancipation and enclosure, following the lead of neighboring Denmark, whose king ruled much of Schleswig and Holstein (chapter 3.11 and 3.15). Peasant protection was tied to emancipation, and the peasantry emerged strengthened by the reform. Enclosure and tax revisions prompted the surveying and mapping of many villages in the Duchies of Schleswig, Holstein, Holstein-Gotthorp, and Lauenburg in the eighteenth and nineteenth centuries. The maps, which are now poorly documented, were left in the keeping of the farmers in newly enclosed villages. More is known about maps from large landed estates which were mapped in connection with ownership changes or, more often, enclosure and agricultural restructuring on the estate. By the later eighteenth century the Hansa towns of Lübeck and Hamburg each had regular salaried surveyors on their staff who carried out surveys of the towns' lands in connection with enclosure.[69]

In conclusion, the main agricultural reform in Germany attacked the legal basis of feudalism. Surveying and mapping played no direct part in legal reform, although written *Urbaria,* which codified the rights and obligations of inhabitants of each seigneurial jurisdiction, were widely used. The juridical emphasis of reform is mirrored by modern historical studies which also stress legal aspects of reform.[70] Other reforms were, however, both discussed and effected. Improved crop rotations and animal husbandry were introduced, and these often depended at least in part on physical changes in landholding which again were often, although not invariably, dependent on legal changes in feudal relations. The north, and especially the northwest of Germany, stands out as an area of advanced agriculture and as the area where the most prominent state mapping in connection with reform took place.

4.5 Taxation in Germany

Germany in the seventeenth and eighteenth centuries was a patchwork of states, many of them tiny, ruled by more or less autocratic hereditary princes. Until the nineteenth century, revenues from the personal estates of the rulers formed the greater part of state income; the rest was composed of income from various currency rights, customs duties, and taxation. Similarly in the expenditure of the princes, there was no clear dividing line between personal expenses, such as the cost of his court and hunting, and what we would understand by state expenditure, such as the costs of foreign policy, state bureaucracy, and the police. In these circumstances there could be no precise dividing line between private and state mapping. If a prince ordered a survey to increase rents from his personal estates, he was at the same time aiming to increase state revenues. Estate and forest maps of ruling princes are discussed in section 4.1; here we deal with state mapping for the purposes of taxation and include surveys which extended over more than one person's property and which aimed to increase the revenue of the collecting authority, whether this was the church, a prince, or, as in the case of dike tax maps, a local authority with a specific responsibility.

Interest in tax reform increased greatly in the late seventeenth and eighteenth centuries in the aftermath of the Thirty Years' War. The war had brought financial dislocation to much of Germany, since the many armies supported themselves by levying contributions from the population of the areas they were occupying at the time. After the war German rulers began to keep standing armies, the costs of which rapidly exceeded their personal income. There was also a steady increase in the complexity and hence cost of government in almost all states, and these trends led to the reorganization of state finances. As the need for money increased, so the states became increasingly interested in accurately surveyed and mapped cadasters as instruments for increasing tax revenues.[71]

State tax revenues came from many sources, the most important of which were land taxation, excise duties, tax levied on manufacturing (*Gewerbesteuer,* or *Gewerb- und Nahrungssteuer*), and poll taxes. These were clearly state levies, conceptually separate from the services and dues of peasants to their landlord and legal lord. The land tax was by far the most important tax, chiefly because the German economy remained dominated by the agricultural sector until well into the nineteenth century. In Bavaria and Brunswick in the early nineteenth century, land tax accounted for 68 percent and 88 percent respectively of direct tax receipts.[72]

By the late eighteenth century the need for reform of the land tax was becoming increasingly apparent. In Bavaria and Hessen, for example, regulations governing the land tax had been laid down in the sixteenth century, and the tax had become incomplete, unequal, and inefficient. Reform, though pressing, was difficult. An accurate survey was expensive: in Bavaria it was estimated that it would take 48,822 man-days to survey the country in the nineteenth century, and the actual costs of surveying regularly exceeded estimates.[73] However, the main reason for delay in reforming tax was, as in Austria and France (chapters 5 and 6 respectively), the opposition of the nobility. The land tax was levied on those least able to resist: the peasants bore the burden, while nobles' estates and some ecclesiastical estates were exempt. The nobles naturally feared that a survey would reveal the abuse of their privileges, which then might be curtailed or even abolished. A centrally organized taxation cadaster would effectively remove from the nobility their control over local repartition (allotment of a share of the locality's tax burden to individuals) and collection of taxation and might also be used to attack their theoretical control over the granting of tax. Thus one of the most powerful political forces in Germany was opposed to reform; in many areas reform had to wait for the changed political climate of the nineteenth century when reassessment was combined with efforts to end tax privileges. These privileges were commuted, for example, in Hessen in 1806 and in Saxony-Weimar in 1817.[74]

The following review of German taxation maps include a brief discussion of maps drawn in connection with special taxes (tithes and taxes for the maintenance of dikes), a description of taxation cadasters from the late seventeenth and eighteenth centuries, and finally a discussion of the nineteenth-century taxation cadasters, which were a major part of the reform of state revenues throughout Germany.

Tithe Maps

Tithe maps form a small but interesting part of many map collections in Germany.[75] An important early tithe map is that of 1625 by Wilhelm Dilich (figure 4.10), one of Hessen's most celebrated early cartographers. His tithe map of Niederzwehren is at a scale of about 1:6,500, and its legend shows the name and extent of all cultivated plots which were tithable. The structure of the three-field system is identified on the map by colored shading. A map of 1750 of Hattenheim in Hessen shows the lands on which the archbishop of Mainz and various monasteries and other religious foundations were entitled to collect tithes.[76]

In the late Middle Ages the area of cultivation and settlement shrank in Germany, as elsewhere in Europe, and much land on which tithes were due was deserted. Some of this was later recolonized, but often the colonists escaped paying tithes. One recolonized village was that of Steers in Lower Saxony. The religious house which had the right to the tithes surveyed and mapped the area in 1764 to help collect them.[77]

Originally tithes were due to the church, but by the seventeenth and eighteenth centuries they had been appropriated in many places by secular lords. One such secular authority was the elector of Hanover. In 1744 the Hanoverian treasury ordered a complete survey, mapping, and description of areas liable for tithes so that they could be collected more efficiently. The maps

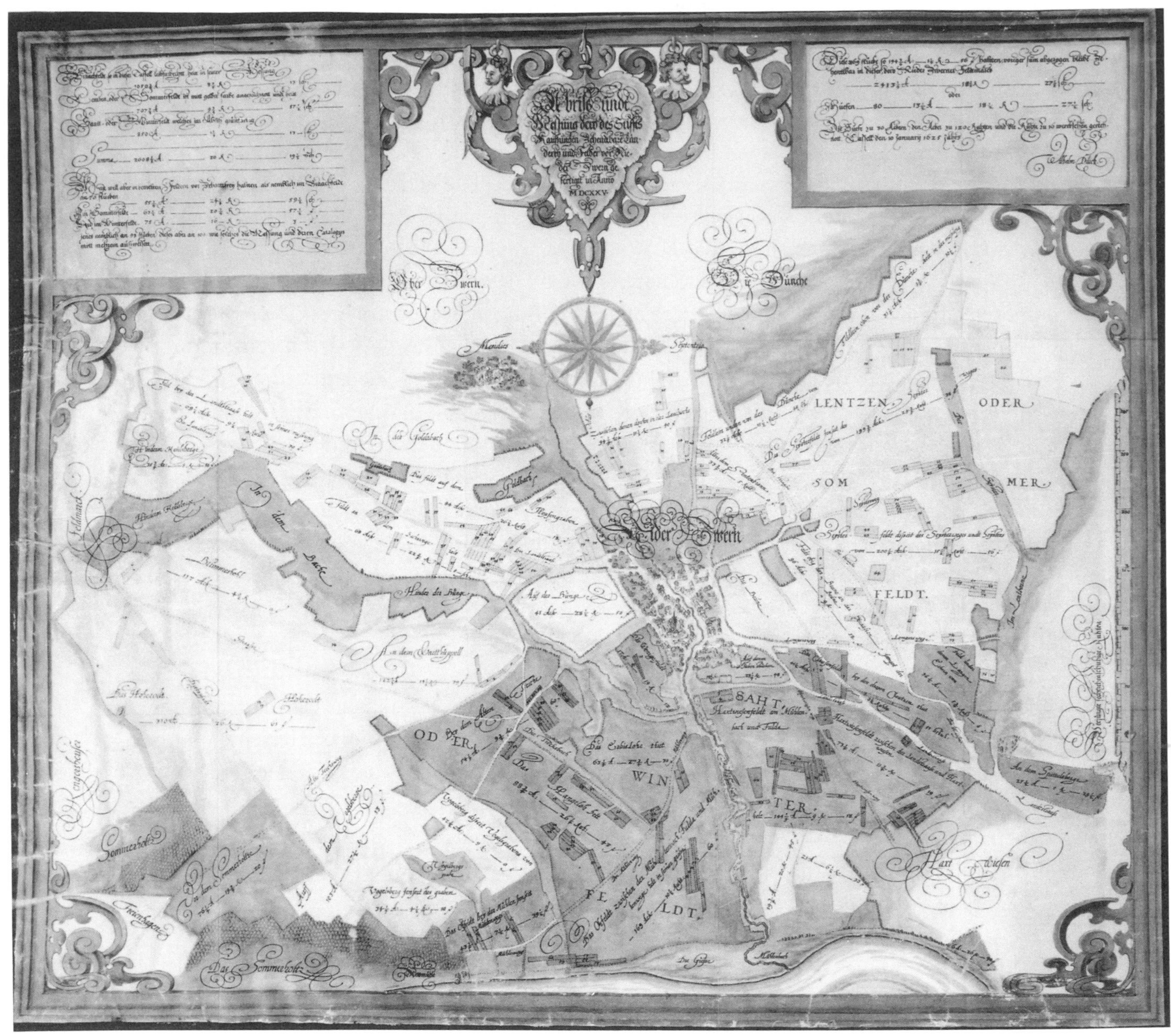

Fig. 4.10 Tithe map of Niederzwehren compiled by Wilhelm Dilich, 1625, at a scale c. 1:6,500. Source: Stiftsarchiv Kaufungen Best., 340 Karte R III 1.

show tithable lands accurately, but areas such as moors and forests, which were not subject to tithes, are shown schematically.[78] Most tithes were probably collected without the aid of maps, but tithes formed a considerable part of many institutions' or rulers' revenue, and they found it worthwhile to commission maps to resolve disputes over tithes and to clarify or revise tithe liability. Tithe mapping thus increased awareness of the map as an instrument of fiscal control and administration.

Dike Taxes

The role of maps in the levying of dike taxes is dealt with at length in the Netherlands chapter (chapter 2.1), but the regulation and efficient collection of such taxes were also important in the adjoining low-lying areas of Germany. In coastal areas of Friesland the responsibility for keeping up dikes lay with the owners of the plots of land adjoining the dike. This system was managed very carefully from an early date, as dike maintenance was both very expensive and vital to the survival of the community. *Deichkataster, Deichregister,* or *Deichrollen,* which are often accompanied by maps from the mid-eighteenth century onward, show the ownership and extent of each plot of land on the dike bank and the amount of labor or money to be spent on dike upkeep.[79]

As with tithe collection, dike areas were often surveyed and mapped where tax liabilities had become unclear. An early case of this is the survey of 1669 to 1673 of an area east of Emden in East Friesland where a reassessment of dike tax liability had become necessary. Most seventeenth-century surveys consist only of written registers, but this Friesland survey includes twenty large-scale maps and one smaller-scale general map which are bound into the register. The villages are represented pictorially, but the point of interest—the dikes and abutting land plots—are very carefully surveyed and depicted.[80]

Surveying and mapping were expensive, and the tithe and dike maps described above reflect circumstances in which the collecting authority considered the scale of nonpayment sufficient to warrant the drawing of a map. There was, however, still a long way to go before tax authorities considered maps an indispensable part of more general tax collection.

4.6 Taxation Cadasters of the Seventeenth and Eighteenth Centuries

Southwestern Germany

Cadastral mapping for taxation purposes began early in the *Landgrafschaft* (Landgraviate) of Hessen-Kassel. There was a clear need for reform, since the Thirty Years' War had disrupted the economy of the area, and the failure systematically to renew and correct the information on which the *Kontribution* (land tax) was based had resulted in the imposition of unfair and often completely arbitrary burdens. From the 1630s attempts had been made to improve matters, but all efforts had foundered mainly for the familiar reason of conflict between the ruling *Landgraf* (landgrave) and the nobility. Comprehensive surveying to reform taxation began in 1680 as part of the *Steuerregulativ* (regulation of tax), which for the first time brought in a unified taxation system for the whole of the landgraviate. In 1699 a new tax office, the Steuerstube, was set up by Landgrave Karl to provide a lasting solution to taxation injustices. Although it set about its task with energy, it found the problem of fixing capital values of properties in a period of rapid economic change intractable. The Steuerstube sent out surveyors and issued regulations to ensure that they used uniform linear and areal measurements, but little apart from this was regulated, and the maps which the surveyors sent in showed great individuality of style.[81] The surveying and mapping failed to provide a lasting solution to the tax problems, and dissatisfaction continued after the death of Landgrave Karl in 1730.

In 1735 his successor, Landgrave Friedrich I, decided to survey the noble lands in his territory to reform land taxation. The central authorities sent out very detailed forms to be completed and returned by noble property owners. No maps were to be drawn, although Friedrich I was also king of Sweden and presumably aware of Swedish cartographic advances (chapter 3). In 1736 a special commission for the reform of land tax on nonnoble lands was set up. The order initiating the commission spoke explicitly of the oppression of the peasants which had arisen from past confusion and from the "unchristian" tendency of the nobility to shift their tax burden onto their peasants. The commission sent out forms which were to be completed within four months. Work on the cadaster was interrupted by the Seven Years' War, which itself made the need for financial reform even more pressing. A new order of 1764 encouraged surveying as part of the cadaster, and the work continued until 1791, when the landgrave broke off the project as too expensive.[82] The example of Hessen shows the familiar problem of conflict between ruler and nobility over land taxation and shows also that the role of mapping in taxation cadasters was far from decisive. Several methods were tried in Hessen, all of which foundered in the face of opposition.

In the nearby *Fürstbistum* (Prince-Bishopric) of Fulda systematic surveying began in 1718. Work proceeded far more rapidly than in Hessen-Kassel, and twenty years later most of this admittedly rather small territory had been surveyed and mapped (figure 4.11). The maps are far more uniform than those of Hessen-Kassel, and surveyors obviously worked to strict instructions governing the use of symbols and color washes.[83] In the Landgraviate of Hessen-Darmstadt a cadastral revision began in 1700, when instructions were issued for the surveying and mapping of the territory. It is not clear whether the intention was changed and maps were never produced, or whether the original scheme was carried out and the maps subsequently lost. Whatever the

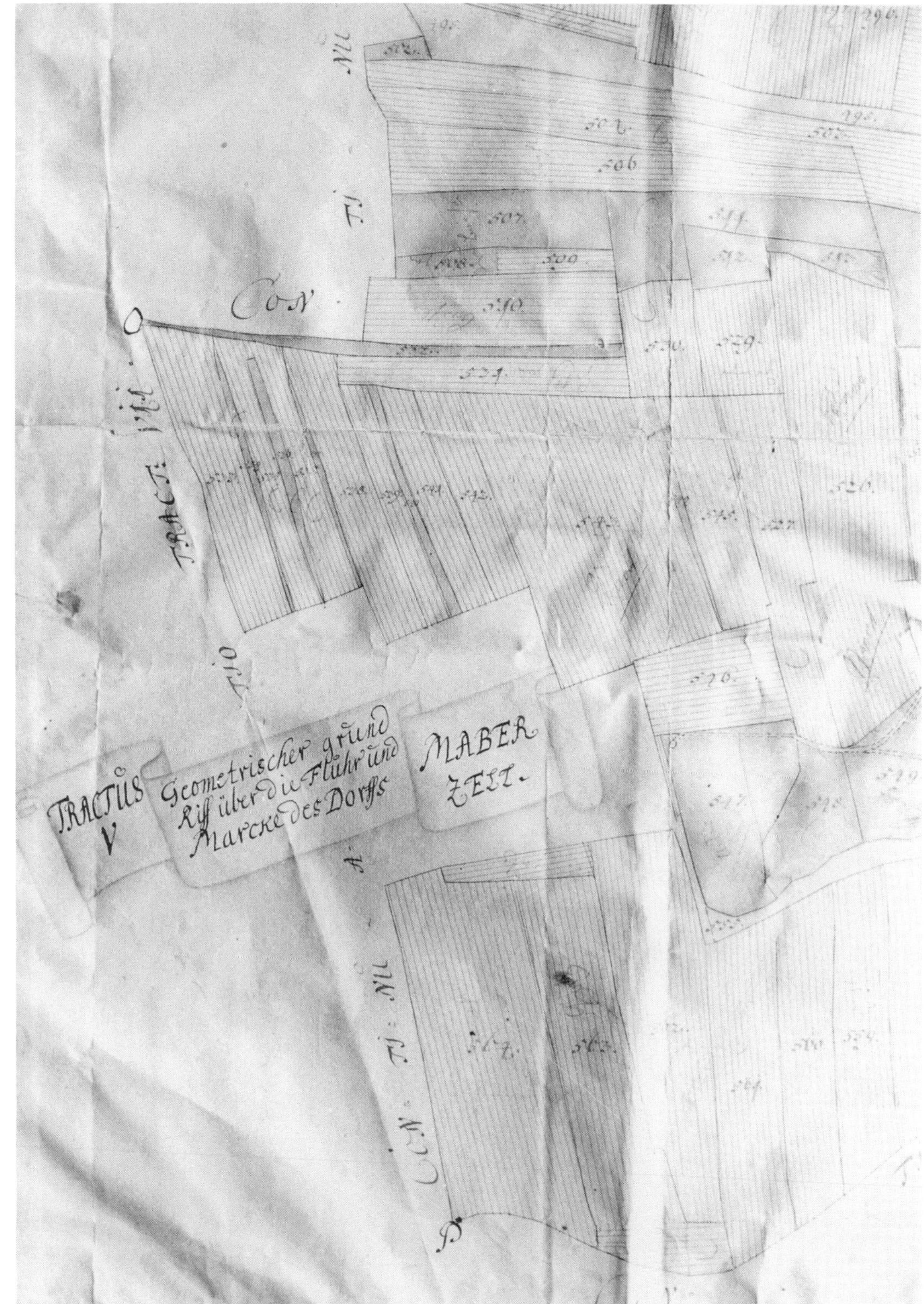

Fig. 4.11 Cadastral map of Maberzell, Hessen, 1722; it is a typical example of the Fulda cadastral maps of the eighteenth century. Source: Hessisches Staatsarchiv Marburg, Karten P II Nr. 13.568/5.

reason, there are no important extant cadastral maps of Hessen-Darmstadt before those of the 1824 cadaster (section 4.7).[84]

In Swabia an important early mapped survey for taxation purposes was that commissioned by the authorities of the Imperial City of Schwäbisch-Hall to put city finances on a sounder and more equitable footing. The city had on its permanent staff men skilled in compiling written cadastral descriptions and assessments, but there was no competent surveyor. The town council decided on some trial surveying in 1696, using existing personnel. One rather poorly drawn and inaccurate plan survives from this trial survey, which proved to the council that a qualified and experienced surveyor was needed. The council consulted the Imperial Cities of Ulm, Rothenburg, and Nuremberg (Nürnberg), which had recently carried out mapped cadasters of their lands and in 1697 started work on a written cadaster to cover secular and religious lands and to describe in meticulous detail their extent, land use, condition, and legal status. The lands were to be surveyed but not mapped at this stage. The *Renovatoren* (surveyors and assessors) went about their work assiduously, sending in their results to the Renovaturstube (tax revision office).

In 1699 Daniel Meyer (Mayer) from Basel joined the project. He had already had considerable experience as a cadastral cartographer in other Imperial Cities such as Nördlingen and possibly Augsburg and came from a long-established Swiss mapmaking family. Meyer was called to Schwäbisch-Hall in the first instance to survey and map the town itself, but he arrived in May just when the townspeoples' gardens were coming into flower. To minimize the damage which would inevitably accompany the survey, it was decided that he should begin work in the rural areas. It was agreed that he should start with the fallow fields, then, when the harvest was safely in, go on to the arable areas and finally return to map the town. The Schwäbisch-Hall authorities seem to have shown remarkable sensitivity in this: surveys in other areas often ran into trouble because surveyors and their assistants damaged crops and aroused the opposition of the peasants. Meyer was sworn in in May 1699 and was specifically charged with mapping as well as surveying. Meyer's skill as a surveyor and cartographer was unquestioned, but he was constantly in trouble with the authorities, as he was unreliable, quarrelsome, and in debt. The last was the most pressing problem and caused him to interrupt work for the Schwäbisch-Hall council to work for neighboring authorities to raise extra money and finally to abscond, leaving the work in Schwäbisch-Hall unfinished and taking some survey material with him. He was brought back by the authorities, who wanted the work finished, and he did complete most of it, although the town plan, begun in 1710, was not completed before his death in July of that year. Although the town council spent much time fending off Meyer's creditors and trying to make him complete his work, Meyer left them outstanding maps of most of the city's lands. He drew or partly drew thirty-eight island maps on which land parcels and land use are delineated in such detail that even different tree species are distinguished. The maps are scientifically accurate and show great artistic flair in their cartouches and marginal drawings. Plans to continue the work after the death of Meyer came to nothing, but his survey, although comparatively little known, remains one of the high points of early cadastral cartography in the region.[85]

In Baden-Durlach the first mention of maps for cadastral purposes was in 1751, when it was decided to map the Bailiwick of Rötteln. Each soke was to be surveyed, mapped, and assessed for taxation. Work was begun under the French surveyor Fressons, who surveyed land parcel by parcel and numbered each plot on maps which he sent in to the authorities. In 1756 Fressons suggested that instead of surveying each parcel he should just survey fields according to land use (*par masses de cultures;* cf. chapter 6.8). This would have been satisfactory for the subsidiary aim of the survey, that of producing base maps for a small-scale topographical map, but as the Treasury pointed out, it was useless for the desired revision of taxation. Eventually Fressons was dismissed from the work after this difference of opinion and more general criticisms about the standard of his work.[86]

Northwestern Germany

One of the most important of all early surveyed and mapped cadasters was that carried out in Cleve (Kleve, Kleef) in the 1730s. The Duchy of Cleve, in what is now part of northwestern Germany and the adjoining areas in the Netherlands, was joined by a personal union with the Electorate of Brandenburg and, through Brandenburg, came eventually to Prussia. In Cleve the Prussian king faced a population ruined by the Thirty Years' War, royal demesnes leased out and largely exempt from taxation, and state revenue which did not even cover the interest on public debts. State revenues comprised the *Kontribution,* paid by the country districts, and the *Accise* (excise), raised in the towns. The *Kontribution* was a tax introduced in 1653 with the agreement of the Estates under pressure from the great elector of Brandenburg. It was in effect a permanent tax; in granting it, the Estates lost their main function, that of controlling taxation. This tax was levied on the basis of a seventeenth-century cadaster which had never been more than a rough estimate. Areas which had been brought into cultivation since the seventeenth century escaped taxation, and dramatic changes in the course of the Rhine inundated some areas and made others newly available for cultivation. Injustices in the amount of taxation paid became more obvious and pressing as the total tax burden increased. As early as 1632 the elector of Brandenburg had tried unsuccessfully to reform taxation in Cleve by planning a thorough cadaster which was to be revised annually. Between 1658 and 1666 the *Matrikel* (written cadaster) was revised, but there was only a limited redivision of existing taxes. The revision was not based on a new survey, and no maps were drawn. There were several further unsuccessful attempts in the seventeenth century to reform tax thoroughly. In 1705 some surveying did take place, but it was not completed.[87]

Despairing of official action, some landowners even ordered their own survey and tried to get their tax altered in accordance with the results.

A turning point came in 1730, when the king appointed Friedrich Wilhelm von Borcke to administer all the Prussian territories west of the river Weser. Borcke was interested and experienced in problems of taxation and in 1731 secured the agreement of the king to a new, thorough, mapped cadaster in Cleve. There then followed the first complete eighteenth-century cadaster in any Prussian territory. Other territories had partial taxation reforms accompanied by isolated surveys, but Cleve needed a complete survey, as land ownership was so fragmented. This need had indeed long been recognized, but others had thought the cost too high. Ketter suggests that Borcke and others close to the king were influenced by their knowledge of the seventeenth-century mapped cadaster of Hither Pomerania conducted by the Swedish authorities (chapter 3.7). Hither Pomerania had passed to Brandenburg-Prussia in 1720 and remained the only Prussian territory whose cadaster was based on general survey and mapping.[88]

Work began in Cleve using Prussian army officers, and by 1735 most of the land had been surveyed. Progress, however, was hardly smooth. There were two main problems: the first was the opposition of local authorities in Cleve to the whole project, and the second was its rapidly escalating cost, which made the king's own support equivocal. Local opposition came from the *Ritterschaft* (knights) and the towns. They maintained that the existing taxation system could not be altered without arbitrarily injuring some parties and benefiting others and questioned the power of the king to order taxation reform without their consent. They were seeking in fact to protect their own interests, as they feared that a complete survey would greatly reduce their exemption from the land tax. They had also seen a considerable increase in their powers during the Thirty Years' War, when, isolated from their nominal rulers, they had had to negotiate with occupying armies. After the war they had retained the right to determine how tax was to be levied, although they could not fix the total sum required, and they were reluctant to relinquish this power. The peasantry petitioned in favor of the survey, as it was on them that the burden of taxation fell increasingly heavily. The peasants in each area had collectively to pay a fixed sum in tax. With depopulation during and since the Thirty Years' War, the sum payable per capita had increased greatly, and many impoverished peasants had been forced to leave their land. Those who remained had to pay even more to make up the unchanged total sum. The Regierung (originally the governing authority, after 1723 the Supreme Court) of Cleve, like the nobility, was against the survey, while the Kriegs- und Domänenkammer, the organ of the Generaldirektorium which had responsibility for Crown rents, the management of Crown Lands, and the collection of taxes, was in favor of it.[89]

The king was in favor of the survey but was concerned that its costs were all too apparent, while its benefits appeared negligible. The cost of the survey was estimated initially between 7,000 and 8,000 *Reichstaler,* and this seems unaccountably low, especially as surveying in areas around the Rhine was difficult and therefore expensive. By 1735 it was obvious that the costs would be at least 19,000 *Reichstaler,* and in 1736 Borcke estimated that the work could be completed only at a cost of 23,000 *Reichstaler.* The Kriegs- und Domänenkammer, which was in charge of the survey, remained convinced of its benefits but was alarmed at the costs and particularly at how these were to be apportioned. Eventually Borcke, who was the main spirit behind the survey, fell from power; though work on the cadaster continued, its purpose had changed. Borcke saw a just apportionment of tax based on a survey as an instrument for the promotion of greater prosperity and the recolonization of lands deserted in the Thirty Years' War: his successors had no such vision and saw the reform as a purely fiscal measure. Frederick II, who succeeded his father, Frederick-William, in 1740, had little interest in Cleve but turned his attention to Silesia, which he had newly won and to which he transferred his best administrators, thus bringing to an end the survey in Cleve.[90]

The surveyors in Cleve were left to survey and map largely as they saw fit. They were helped in the field by trustworthy locals who showed them where property boundaries ran and told them who held the land. The maps were intended to be at a scale of 1:2,045 and are in fact remarkably close to this. The maps measure 66 cm × 105 cm, and they show each land parcel, which is numbered and colored according to land use (figure 4.12).[91] The Prussian officers working on the survey were joined by civil surveyors who were in general less skilled and more steeped in tradition. Buildings, for example, were represented by the officers as red, carefully surveyed house plots, while the civil surveyors drew pictures of houses.[92] Opinions differ as to the technique employed by the surveyors: Ketter suggests that the maps were not based on triangulation, but Aymans cites a letter of 1731 from the director of the Engineering Corps to Borcke which states that the latest accurate triangulation methods are to be used in preference to the old ways. Aymans also suggests that the maps are too accurate at their edges not to have been based on triangulation. This and the astonishing accuracy of the maps bears out Ketter's suspicion that, if more were known about the maps, they would make the later French mapping of the Rhineland look a great deal less revolutionary than it is normally held to be.[93]

The Cleve maps are accompanied by registers which are ordered by map sheet and parcel number and contain the name of the owner of each plot, his place of residence, the size of the plot, and a further description of the nature of the plot. The maps are bound in large volumes ordered by town, *Amt* (county), or *Herrlichkeit* (estate).[94] The 1731–36 cadaster did not set a precedent for surveys in Cleve. It was the high point but also the last example of a tradition of cadastral surveying which Schulte traces back to the early sixteenth century. The next significant sur-

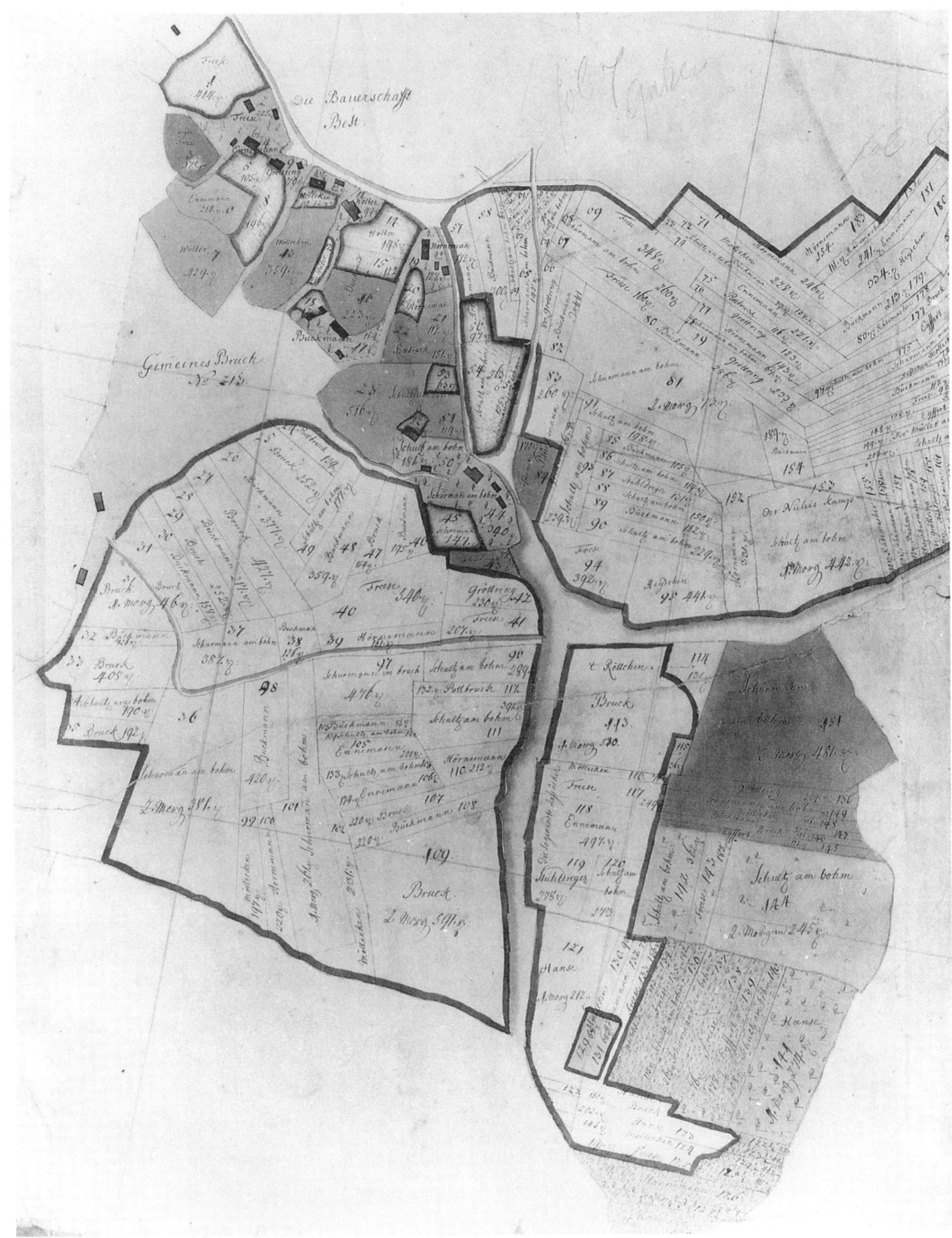

Fig. 4.12 Map of Gahlen-Bühl drawn in 1738 as part of the Cleve cadaster at a scale of c. 1:2,000. Source: Nordrhein-Westfälisches Hauptstaatsarchiv, Düsseldorf, Karte VIIb Nr. 40I.

veys in the area were those of the occupying French troops and the Prussians in the nineteenth century. Theirs were completely different surveying traditions which owed nothing to local Cleve roots.[95]

The Cleve cadaster shows the classic problem of a ruling prince attempting to reform land taxation and meeting the opposition of the local nobility, who realized how much they would lose if a comprehensive cadaster revealed their abuse of tax privileges. It also shows that, though a mapped cadaster might ultimately be an important instrument of power and control for the ruler, the costs of such a cadaster were heavy and immediately obvious and could jeopardize its completion. As in Hessen, in Cleve the ruler abandoned the project because he could no longer afford it.

Another systematic survey in northwestern Germany with the purpose of providing a basis for taxation began in 1743 in Schaumburg-Lippe. Surveying in the two southern districts rapidly came to a halt, but the sworn surveyor Johann Casper Giesler managed virtually to complete surveying in the two northern districts. Giesler's maps are particularly beautiful. Each parish was covered by an island map called a special map at a scale of 1:4,800 and a cadastral register. There were also smaller-scale district maps at a scale of 1:24,000.[96]

Land taxation had long been a vexed issue in the Prince-Bishopric of Osnabrück, but there was reluctance to go to the expense and trouble of surveying and mapping, even though all parties conceded that the old system, based on outdated estimates, was unrepresentative and unjust. The first cadaster to include surveying in Osnabrück was that of 1664, but a surveyor was to be called upon only in disputed cases and was not required to make maps. In 1694 and 1712 the idea of a comprehensive survey was mooted, but nothing came of it, as it was considered too expensive. In 1717 Ernst August ordered a comprehensive survey, and work continued from 1718 to 1723. The initial intention of drawing maps was not realized for lack of time and money. For reasons which are not clear, this surveyed cadaster was never used to revise tax assessments, which continued to be based on the 1667 estimates.[97]

It was against this background that in 1784 a survey of the entire Prince-Bishopric of Osnabrück was ordered. It was to cover heathland and the *Freigüter* (nobles' estates which were exempt from taxation). The taxation of the heath lay at the heart of the problem, as old surveys did not extend to areas of common which had been newly taken into cultivation. This meant that the tax burden was very unfairly distributed. There is some doubt as to the intended place of maps in the survey. One memorandum stated that "maps would not be required"; another report expressed the desire that the maps should be so accurate that they could be used not only for taxation but also for the division of the commons, work on turveries and canals, the clarification of property boundaries, and military and civil communications. In the end, maps did form an integral part of the cadaster. Surveying began in 1784 but encountered opposition from the peasants, who resented the surveyors' damaging their crops and woodland. Reports of peasants using violence against the surveyors were matched by threats of punishment if the trouble continued but also by instructions that surveyors try to minimize damage. Further opposition to the survey came from the nobles who resisted the inclusion in the survey of their land which was exempt from taxation. Their opposition persisted, and in the end some of their land escaped surveying. This and other land which was not taxed was surveyed for the first time during the Napoleonic period.

Despite this opposition 476 island maps were produced between 1784 and 1790 at a scale of 1:3,840. The project was led by Johann Wilhelm du Plat, whose brother, General Georg Joshua du Plat, had led the famous topographical survey of the Electorate of Hanover from 1764 to 1786 (section 4.7). There was at first a shortage of surveyors, but in 1785 Georg Joshua du Plat sent his brother officers from the Hanoverian survey, which was just coming to an end. Not surprisingly, the Osnabrück and Hanoverian maps are very similar in appearance. Registers for the Osnabrück survey identify the holders of the numbered land plots, the extent of the land, and its use. Because the maps were not based on triangulation, because some surveyors were inexperienced, and because some peasants who opposed the survey had deliberately interfered with the work, the Hanoverian army found them too inaccurate at their edges to use them as the basis for their maps in 1841. The maps are, however, both beautiful and informative, showing such details as vegetation, settlement, and even tree type in forest areas (figure 4.13). Associated tax assessment continued until 1808, when it was stopped by the French, who were then in authority and who considered the project too expensive and incompatible with their own projected cadastral reforms.[98]

Lower Saxony

The survey in the Duchy of Oldenburg began in 1781 and was one of the first in Germany after the Cleve cadaster to be based on triangulation. Georg Christian von Oeder, who led the survey, had visited Denmark and was impressed by its agricultural reforms and cartographic advances. He wanted to produce maps to be used in the first instance to make taxation fairer, but also to promote colonization of moor and heath, the division of the commons, and use by the police and other authorities. He appointed the Dane Caspar Wessel as technical director, and he constructed a triangulation net which joined up with that of Denmark. It was astronomically orientated and very accurate. Using it, maps were drawn at scales of 1:4,000 (*Spezialkarten,* or special maps) and 1:20,000 (*Vogteikarten,* or bailiwick maps). After von Oeder's death in 1791 the idea of producing maps at 1:4,000 was abandoned on grounds of cost, although some had already been produced at this large scale. The survey continued for eight years and produced twenty-eight bailiwick maps, from which a general map was produced at 1:160,000. The maps are not strictly cadastral: individual land strips are shown accurately, but there is no register to identify their owners.[99]

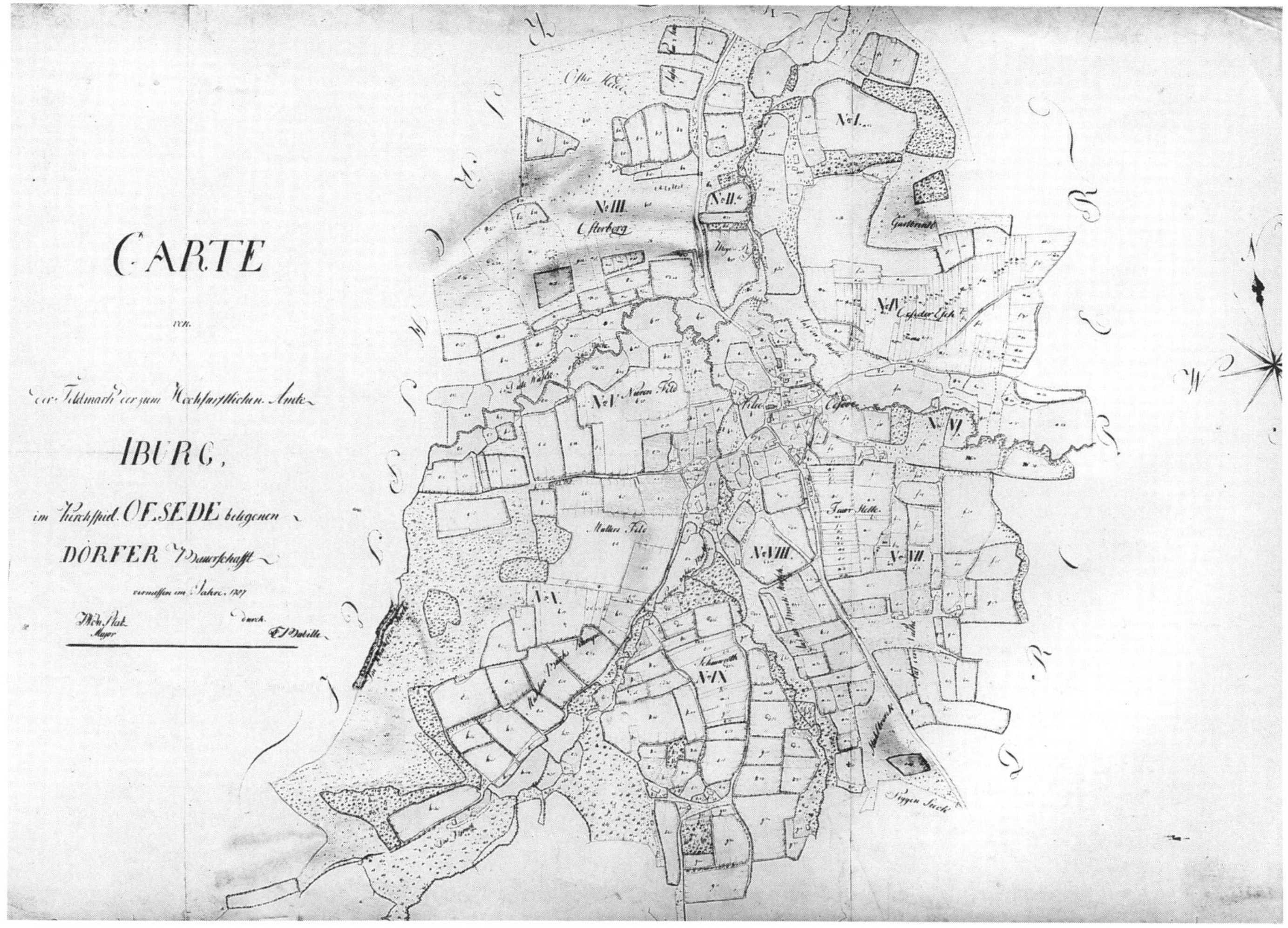

Fig. 4.13 Map of the village of Oesede drawn in 1787 as part of the Osnabrücker Landesvermessung. The scale is 1:3,840. The survey was of taxable land within the Principality of Osnabrück. Source: Niedersächsisches Staatsarchiv in Osnabrück, k100 N. 1H v 3a.

Following the personal union of the Electorate of Hanover and the Duchy of Lauenburg, the Exchequer in Hanover in 1709 issued an instruction (now lost) that surveying take place in Lauenburg as part of a revision of taxation. It was hoped to make the tax burden more equitable and to take into account the new forest clearances. The survey was carried out by Engineer-Lieutenant Georg David Michaelson from Hanover. Between 1709 and 1713 he surveyed the district of Neuhaus, and between 1722 and 1724 he surveyed Lauenburg. It is not clear why he did not go on to survey the districts of Schwarzenbeck and Ratzeburg.[100]

These were all important early mapped cadasters, but the notion that a cadaster should be surveyed, let alone mapped, was not yet universally established. In 1726, for example, a cadastral revision in the Principality of Halberstadt was based not on surveying but on legal evidence such as title deeds, deeds of sale, leases, and notaries' testimonies.[101] Prussia, where Frederick-William had put through the Cleve cadaster in the face of considerable difficulty and where cadastral surveys were so energetically prosecuted in the nineteenth century, did not feel compelled to survey or map its newly won territory of Silesia. Frederick II contented himself with a revision of the Austrian cadaster of 1724. In 1748, with the help of written inquiries as to the situation of the land, its ownership, and other details, the Austrian cadaster was revised.[102]

There was no universal recognition in eighteenth-century Germany that land tax reform should be based on a mapped cadaster—indeed, there was concerted opposition to the idea from groups, especially the nobility, with a vested interest in preserving the status quo. With justification, they feared that the accurate and comprehensive knowledge which a mapped cadaster would put into the hands of its commissioners would be used to limit their evasion of tax. A cadaster organized from the central authority would also undermine their traditional control over the collection and repartition of taxation. With the exception of the Cleve cadaster, large-scale centrally organized mapped cadasters had to wait until the nineteenth century.

4.7 Taxation Cadasters of the Nineteenth Century

The post-Napoleonic political settlement in Germany resulted in territorial expansion for many states, notably Prussia and Bavaria, which inherited systems of land taxation which were often outmoded and in any case nonuniform. In order to promote uniform and efficient land taxation throughout their territories, cadasters were carried out in many states. Advances in military surveying during the Napoleonic Wars had raised the standard of civil surveying, which was mostly done wholly or in part by military engineers. Maps were based on triangulation and were more accurate, although perhaps less aesthetically pleasing, than earlier maps.[103]

Bavaria

One of the earliest nineteenth-century cadastral revisions was carried out by the Topographisches Büro (Department of Topography) in Bavaria. The department was founded by Napoleon in 1801 and took responsibility for both cadastral and topographical surveys. One of its most important tasks was to reform the land tax, the structure of which had been laid down in successive agreements between the elector and the Estates in the sixteenth century, particularly in the preface to the Tax Instruction of 1507–8. In spite of revisions in the eighteenth century, by the nineteenth century taxation assessments had become incomplete and unequal, and even diligent and honest officials could not locate the owners of some strips of land.[104] The case for reform was made more pressing because 70 percent of state income was raised through land taxes (compared with 1.7 percent today), and also because the Napoleonic Wars had left Bavarian state finances in a parlous state, while treaties of 1801, 1803, and 1805 had endowed the Kingdom of Bavaria with greatly enlarged territories in which 114 different systems of land tax were in operation.

The Department of Topography began plans for a cadastral revision immediately after its foundation, but it was not clear how the tax reforms were to be effected. A first attempt followed the French method, in which surveyors measured and calculated the extent of blocks of land which shared the same land use, such as arable, pasture, or wood (*par masses de cultures;* chapter 6.8). Sworn local assessors then estimated the extent of the land parcels within the land-use blocks. There were early doubts as to the reliability of this method. When checks revealed that the assessors' estimates had errors of up to 525 percent, the method was abandoned. It was acknowledged that a properly surveyed and mapped cadaster was needed, but it was also accepted that this would take a long time. From 1808, therefore, attempts were made to rectify quickly the worst injustices of the existing systems. Landowners completed and sent in tax returns, but many were so seriously underestimated that this, like the method of surveying *par masses de cultures,* was abandoned. Simultaneously, however, surveying was begun based on accurate triangulation carried out by soldiers with civilian assistants.[105] An early attempt was made to follow the so-called parallel method of the Danish coastal survey under the geographer Thomas Bugge (chapter 3.9 and 3.14). This proved to be expensive, very destructive of crops, and not very accurate. The Bavarian peasants were highly suspicious of the surveyors who trampled over their fields, and it was reported that the surveyors had to use firearms to defend themselves. The Danish method was abandoned, and surveyors went back to the old plane-table method of drawing maps directly in the field but with a triangulation net to control results. Surveyors used French and German equipment, the latter from Munich, whose Optical Institute was a noted center for the manufacture of surveying equipment. Surveyors were assisted by posts which landowners were obliged to erect on the boundaries of their property.

By the time the survey was finished in 1867, twenty-two thousand maps had been produced at a scale of 1:5,000 as well as a small number of others at larger scales (figure 4.14). Alois Senenfelder, who had discovered lithography at the end of the eighteenth century, won the right to reproduce the maps using his new technique, just as he was doing with the Austrian Franciscan cadaster (chapter 5.9). Lithography was much cheaper than copperplate printing and enabled the Bavarian cadastral maps to be used for a number of administrative and legal purposes in fulfillment of the original aim of producing a body of data to be used for purposes beyond the purely fiscal (figure 4.15).[106] Lee states that the tax assessment which accompanied the mapping and surveying probably resulted in a reduction of the tax burden on smallholders but suggests

Fig. 4.14 Manuscript original map of Scheinfeld and environs from the Bavarian cadaster, 1828, at a scale of 1:2,500. Source: Bayerisches Landesvermessungsamt, Münich. Nr. 10153/91.

Fig. 4.15 Lithographed map of Scheinfeld and environs, 1828, at a scale of 1:2,500. Note that the field numbers which appear on the original map are not shown on the lithographed maps, although all property boundaries are. Source: Bayerisches Landesvermessungsamt, Munich.

that this might well have happened anyway, as there was less need for state income now that the country was at peace. He comments also that the assessments were in fact remarkably similar before and after the nineteenth-century cadaster.[107]

Prussia

The Kingdom of Prussia became a particularly important advocate of mapped cadasters in Germany in the nineteenth century. After 1815 the Prussians regained control of their western provinces from the French and acquired new land, so that they ruled large parts of the Rhineland and Westphalia as well as their traditional lands in the east. One of their most important tasks in the west was to unify the various legal, administrative, and taxation systems which operated in the provinces. There were, for example, nine different taxation systems in Westphalia alone. In some places the farmers estimated their own property and set the tax; in others there was a cadastral register in which each piece of land was noted and its worth assessed; in yet others the amount that each parish had to pay was centrally determined, and the parish authorities divided the sum among the inhabitants. The first systematic cadastral mapping in the Rhineland provinces followed the Napoleonic law of 1808. Work started immediately on the left bank of the Rhine, but on the right bank only those areas which were joined to the Kingdom of Holland (North and West Münsterland) were surveyed. When the Prussians reassumed control of the Rhineland, they made use of the survey material produced by the French and continued the survey using the *Receuil méthodique*, which governed French cadastral surveying (chapter 6.10). A cadastral commission was set up to survey land parcels, draw up cadastral registers, and assess taxes. Surveying in the two western provinces took fifteen years, from 1820 to 1835. The two western provinces thus had a cadaster over forty years before the eastern Prussian provinces. Maps at scales of 1:2,500, 1:5,000, 1:10,000, 1:20,000, and 1:30,000 were produced, together with cadastral registers which give details of all landholders and their parcels and taxation assessments. Information for the cadaster was gathered in the field by surveying and from local people who gave information about land ownership.[108]

This cadaster remained in force until a revision in the 1860s, which was necessary because economic conditions had changed radically in many parts where commons had been divided and railways and new roads constructed. It was also felt to be unsatisfactory that the eastern provinces still had no cadaster and that there were still more than twenty different taxation systems in Prussia as a whole. Work started in 1861, and by 1865 surveying and assessment were complete in both the eastern and the western provinces, and the new tax assessments were in force. Work was carried out by a general commission under the Prussian finance minister. In just four years the whole of Prussia had been surveyed, while it had taken fifteen years to survey only the two western provinces in the 1820s. This was largely because in the 1820s the surveyors had to start from the beginning, whereas in the 1860s there were many existing maps and cadastral registers which were used as the basis for the new cadaster.[109]

As Prussian territory expanded, so more areas were subject to rigorous land tax revision. Cadastral revision, using existing maps where possible and involving new surveying where necessary, was undertaken throughout those areas of Lower Saxony which were annexed by the Prussians in 1866. Prussian cadastral surveys took place in Hamburg in 1847 to 1871, in the *Fürstentum* (Principality) of Lübeck from 1855, in Schleswig-Holstein from 1868 to 1878, in Lauenburg from 1876 to 1880 and in 1890, and in the Free City of Lübeck and its surrounding lands after 1876. In all of these surveyed and mapped taxation cadasters, large-scale maps, often at a scale of 1:2,000, and detailed written registers (*Mutterrollen* and *Flurbüchen*) were produced.[110]

Lower Saxony

In 1817 the authorities in the Grand Principality of Oldenburg decided on a new cadaster, but it was not until 1833 that work began in earnest under the leadership of Freiherr von Schrenck, who had played a leading part in the Westphalian cadaster. The survey was based on triangulation, and most of the surveyors were brought in from the Rhineland and Westphalia, where surveying work was just coming to an end. Three thousand maps were drawn mostly at scales of 1:2,000 or 1:3,000. Surveying lasted until 1850, and it was 1866 before the new tax assessments had been fixed, so the whole operation took fifty years to complete.[111]

An interesting exception to the nineteenth-century trend toward basing land tax on a mapped survey was that of Hanover. The rulers of Hanover were unmoved by continental advances in surveying and in 1817 ordered a tax revision which was not to be mapped or even based on a survey. Instead, landowners were required to submit details of their holdings. Predictably enough, many falsified their returns, despite dire threats from the authorities, and only about a quarter of the total taxable land was recorded. Realizing that this traditional method of gathering tax information was useless, the elector decided in 1823 to carry out what was described as a *geometrische Überschlagung* (geometrical estimate), similar to that which had been tried but abandoned in Bavaria in the early years of the century. This was a very rough survey *par masses de cultures* in which the extent of blocks with the same land use was measured and that of land parcels estimated (chapter 6.8). In Hanover it was carried out by unqualified surveyors who made calculations but drew no maps. The results were far from accurate, but the method was quick and cheap: within three years, in 1836, the revised tax assessments were being collected, compared with a delay of nearly fifty years in nearby Oldenburg, which had had an accurate survey based on triangulation and mapping. The Hanoverian "estimate" remained, however, an isolated instance; every other German cadaster was based on the French-Prussian example of detailed, full, and mapped surveys.[112] It is interesting that the electors of Hanover were at the forefront of topographical mapping, with their famous *Kurhannoversche* survey of 1764–84, while at the same time they had no interest in cadastral mapping. Because of the dynastic links between England and Hanover, the absence of cadastral mapping in Hanover is perhaps explained by the fact that the land tax in England was collected without mapping, surveying, or even a proper written cadaster, but on the basis of lists of properties and notional income derived therefrom (chapter 7.5). Since this succeeded in England in raising large amounts of money, the perceived need for an expensive mapped cadaster in Hanover might well have been reduced.

In 1866 the Kingdom of Hanover became the Prussian province of Hanover, and in 1867 it came under Prussian land tax laws. The Prussians wanted a new cadaster as soon as possible, but only 60 percent of the province was covered by maps of a standard high enough to form the basis for new cadastral maps. The Prussians dismissed the *geometrische Überschlagung* as entirely useless, but many enclosure maps and some forest maps were used, sometimes with additions to bring them up to date. Surveyors brought in from as far away as Russia and Switzerland newly surveyed 41 percent of the area. They used triangulation to produce island maps of an extremely high standard at a scale of 1:2,000. Surveying and assessment of taxes were finished by 1875, and the revised tax was levied for the first time in 1876.[113]

Saxony

In 1811 it was finally decided to put into effect a decision made in 1711 by the Saxon Estates that a survey be carried out and the land tax revised. A trial survey was begun in 1819, and maps at a scale of 1:2,730 were produced. Next there was some experimental surveying which sought to establish the technical basis for the survey and which produced 110 sheets of maps at scales of 1:4,800 in the countryside and 1:2,400 in the towns. When it was decided that the most advanced surveying techniques were too expensive, surveyors fell back on the plane-table method without a trigonometric basis, although, as a check, field boundaries were measured on one survey and parcel boundaries on another, and then each was tested against the other. The field boundary maps were drawn at a scale of 1:4,800, and the parcel boundary maps at 1:2,730. In addition, general maps were drawn at a scale of 1:9,600. The survey lasted from 1835 to 1841. Nearly two million parcels of land were surveyed, mapped, and entered in cadastral registers.[114]

Similar land-tax reforms were put into operation on the basis of accurate surveys in 1824 in Hessen (the *Definitivum*), in Brunswick in 1849, in Saxony-Weimar in 1817–19, in Württemberg in 1827, and in Saxony-Meiningen in 1831.[115] Thus by the later nineteenth century in most areas of Germany, the land tax was collected on the basis of a surveyed and mapped cadaster. Prussia stood out among the various states for its comprehensive and scientifically accurate mapping, but the general principle had been established throughout Germany. Although the urban and industrial population was growing rapidly in Germany, the rural population was still in the majority, and land tax levied on rural property still formed the bulk of state income. The mapped cadaster was established as an important part of state fiscal policy and had begun to assume a wider role as increasing accuracy and ease of reproduction led to the maps' wider use in general administration.

4.8 The Cadastral Map in Germany

The history of cadastral mapping in the German lands is so varied and the German tradition of writing detailed regional and especially local accounts of mapping so strong that to date no attempt at a general history of German cadastral mapping has been made. The final section of this chapter, while not pretending to be a comprehensive history, outlines some of the influences on German cadastral mapping.

Political and Economic Influences

One clear influence on the course of mapping is the impetus provided by war. Any substantial war brought severe economic stress as armies foraged locally for supplies and often destroyed crops, livestock, and settlements as they retreated to prevent their use by enemies. If this continued for any length of time, it could bring considerable economic dislocation, which exacerbated demographic problems caused by the fighting itself. Germany's central position and its political and religious fragmentation meant that it suffered more than most from the effects of war. The Thirty Years' War in particular brought unprecedented devastation and depopulation, which in turn made the taxation system in many parts of the country outdated, unfair, and completely arbitrary. Reform of taxation in many areas, such as Cleve and Baden, though it did not happen until the eighteenth century, had its roots in the economic and demographic dislocation caused by the war. The Seven Years' War brought similar, though less severe, economic dislocation and made the need for tax reform even more pressing. The wars fought to end French domination in the nineteenth century precipitated financial crises in parts of Germany, such as Bavaria, and gave a further impetus to tax reform.[116]

As well as making thorough reform of taxation necessary, wars also profoundly affected the training of surveyors and recognition of the value of maps. During the Seven Years' War in particular, armies realized the advantage brought by reliable maps, which became more generally accepted as a result of the interest of the military.[117] The increased military activity and the practice of training military engineers following French precedents spread knowledge of cartography and trained new surveyors in the late eighteenth and nineteenth centuries.

Enlightenment monarchs needed precise, detailed statistics as the basis for administrative reforms and mercantilist state direction of the economy, especially where resources such as forests were under the direct control of the ruler. This gave a further impetus to the map as an accurate medium for such information. There was a gradual move away from drawing individual maps for specific purposes to regarding maps as a necessary basis for a variety of administrative tasks. With the development of the territorial state, rulers wanted to get as full a picture as possible of the extent and condition of their territories and found the map a useful instrument in this task.

One of the earliest instances of this was in Baden-Durlach in 1773, when the cartographer Jakob Friedrich Schmauss began the integration of topographical with cadastral mapping. The aims were uniformity of standard and style and the construction of cadastral maps which could later be used as a basis for a topographical map. Schmauss started the large-scale mapping of the Oberland of Baden-Durlach, and in 1788 work was extended to the Unterland. The work was not completed, but it is a notable early attempt to produce a unified cartographic service and, in the case of the 1780s work, to provide a body of maps available for future reference rather than for any specific, immediate project.[118]

Maps were not only useful administrative tools but were also symbols of power. Early maps often juxtapose a portrait of the prince and the map of his territory and are emblazoned with his coat of arms. The maps were thus a reflection of the ruler's power as well as a means of extending that power.[119]

The political fragmentation of the German lands was both a hindrance and a help to the development of mapping. It gave impetus to mapping in the sense that rulers needed to know where boundaries ran and often established them by using maps in legal disputes. Political fragmentation, however, could also lead to frequent changes of ruler, as land passed by inheritance or marriage from one ruler to the next, and to disputes between various layers of government, as, for example, in Cleve, where disputes between local authorities and the Prussian authorities in Berlin made the construction of the 1730s cadaster wearisome, expensive, and prolonged. It was also impossible for a comprehensive program of mapping to be carried out in Germany as it could be in the unified and centralized states of Sweden, France, or England.

Technical and organizational advances in other fields, especially the art of war and mathematics, but also in engineering, agriculture, and architecture, affected cadastral cartography. Cadastral mapping itself benefited from technological advances in printing. The maps of the nineteenth-century Bavarian cadaster, for example, were among the first products of lithography, which made the maps far more useful to the Bavarian administration. Influenced by military advances, cadastral cartographers changed from representing individual plots on *Einzelkarten* or *Inselkarten* (island maps) to undertaking *Kartenwerk,* projects to map the whole ground area. This in itself demanded technical change to ensure that the edges of the maps joined satisfactorily.[120]

The Costs and Benefits of Mapping

Whether surveying and/or mapping occurred in any one area depended on the cost-benefit ratio as perceived by those initiating the mapping. In Cleve the king was all too aware of the costs of the 1730 cadaster and kept asking what its benefits were. In the early eighteenth-century mapped survey of the village of Obersteinach, progress was very slow because of arguments about its costs and benefits. The situation was particularly difficult, as there were three main landowners in the village; it proved difficult to persuade one of them, the Teutonic Knights, of the merits of the detailed survey. In the 1823 "geometrical estimate" in Hanover the Hanoverian authorities deliberately decided against mapping. Instead, they tried the cheapest possible means of gathering information and, when this failed, tried the "estimate," which made no great claims to accuracy but was certainly quick, cheap, and effective. In Bavaria in the early nineteenth century it was only after two abortive nonsurveyed attempts at cadastral revision that a mapped survey was started, and even then a cheap method was preferred to the most accurate one.[121]

It is difficult to generalize about which groups in society were for and which against surveying and mapping: this depended on whether they stood to lose or gain by the particular measure being carried out with the aid of maps. The nobles, much of whose land was exempt from tax, were in the main opposed to surveys, as they were liable to suffer financially if an accurate picture of the disposition of land was known. They might also have to contribute to the considerable costs of surveying. Thus in Cleve and Osnabrück the nobles were vociferous in their campaigns against the Prussian cadaster and the du Plat survey respectively. Peasants, who bore the brunt of the land tax, were divided in their attitude. In the long term they might well have benefited from tax revision, and in Cleve they petitioned regularly in favor of a cadastral revision. In the short term, however, they saw surveyors' trampling over their crops and often reacted violently against them. In Osnabrück the peasants, like the nobles, opposed the du Plat survey, even though many of them suffered from the injustices of the prevailing taxation system. There had to be special orders to protect the surveyors from angry peasants. In Bavaria in the early nineteenth century the peasants were suspicious of the surveyors, who reportedly had to resort to pistols to defend themselves. In Obersteinach the peasants opposed the "renovation" of taxation, as they were suspicious of the project. Disagreement and hostility also surrounded the surveying and mapping undertaken for agricultural reforms, and the opposition could come from the peasantry or the nobility, depending on particular circumstances in each locality. Some of this reported suspicion, however, may well be the product of the author's own assumption that surveying was "progressive" and hence "naturally" opposed by the "backward" peasantry. Given this hostility and given the disagreement among the ruling princes about the benefits of surveying and mapping, it is hardly surprising that the acceptance of the mapped cadaster in the German territories was neither unanimous nor uniform.[122]

Cadastral Influences from Abroad

Because of Germany's central location in continental Europe, it was natural that reciprocal movements of people, ideas, and techniques would occur between Germany and other countries.[123] Of influences on cartography the most widely acknowledged and discussed is that of the French in the late eighteenth and early nineteenth centuries. It is undoubtedly true that the French example of a scientific, uniform, and centrally organized mapped taxation cadaster (chapter 6.9–6.11) gave an impetus to such cadasters in Germany. French influence was directly imposed in many areas of Germany. At its height Napoleon's hegemony extended over much of northern Germany, and several dependent states of the Confederation of the Rhine, notably Westphalia, Baden, Frankfurt, and Berg, introduced the *Code Napoléon*. The French occupying or controlling forces carried out surveying for military purposes and tax reform in various parts of Germany, especially in the Rhineland.[124] Even after the defeat of Napoleon, French methods were widely emulated or directly adopted, especially by Prussia and by some southern German states. Württemberg had a survey on French lines between 1818 and 1849, and Baden followed in 1853. In each area the cadaster followed French precedent and was based on triangulation. When the Prussians regained control of their Rhineland and Westphalian provinces from the French in 1815, they immediately realized the value of the French surveying. They used French maps in their administration and continued the surveying started by the French, using the same methods.[125]

The reputation of French surveyors was high, and their skill widely acknowledged. The first cadaster of the town of Korbach and its fields and forests was carried out from 1749 to 1756 by Johan Carl Daniel von Römer. He had been born in Strasbourg in Alsace, and his contemporaries remarked of his work that it was so good that he must have been trained in France.[126] However, French surveyors did not always live up to expectations. At exactly the same time as von Römer's work was being praised as an example of French excellence, the French surveyor Fressons was dismissed from the survey of Baden-Durlach for unsatisfactory work.[127] French influence similarly proved a mixed blessing for the nineteenth-century Bavarian cadaster. The first post-revolutionary French method of surveying blocks of land with the same land use (*par masses de cultures*) and estimating the extent of parcels of land within each block was abandoned as quickly in Bavaria as in France for reasons of its inaccuracy (chapter 6.8). The surveyors on the project, however, continued to use the technologically advanced French surveying equipment.[128] It seems that French influence has been overstated, and it is now being realized that French survey methods were not necessarily the best of their time and, more important, that a considerable amount of cadastral surveying occurred in Germany before the classic French period of the late eighteenth and early nineteenth centuries and quite independent of French influence. Perhaps the best example of this is the Cleve cadaster. In 1929 Ketter voiced the opinion that, if enough were known about the Cleve mapping, it would make the later French Rhineland cadaster seem considerably less revolutionary than it was then held to be. In the intervening sixty years precisely that has happened: Aymans's work in Cleve and the work of many other historians of German cartography have proved the existence of scientifically advanced and administratively useful cadastral maps which predate and owe nothing to the influence of France.

It remains true, however, that Germany was not uniformly recognized for its cadastral advances; indeed, some princes were astonishingly resistant to the ideas of cadastral mapping which came from other countries. It might have been expected that Swedish influence would have been widely felt in Germany, as the Swedes carried out early mapping in their German provinces, often using German surveyors (chapter 3.7). This was indeed the case in Cleve, where Ketter suggests that the Prussian king and his advisers wanted to emulate the seventeenth-century Swedish survey.[129] However, the Swedish example was not always fully appreciated. Brandenburg-Prussia acquired Hither Pomerania from Sweden in 1720 and by 1724 had received the Swedish cadastral registers and maps of 1694. The king, however, rather missed the point of the survey when in 1750 he ordered that the material be translated into German and that the units of measurement be converted to Prussian ones. The Treasury pointed out that the maps were already outdated and that a new survey was needed, but nothing came of this.[130] There were personal as well as territorial links with Sweden. Landgrave Friedrich I of Hessen-Kassel was also king of Sweden and presumably knew of the important work of the Lantmäteri (chapter 3.4), but he was equivocal about the use of maps in his revision of land tax in Baden in the mid-eighteenth century.

If the influence of Swedish cartography was unexpectedly slight, that of the Netherlands was clear. The Netherlands were without doubt the most technically advanced surveying nation in the first half of the seventeenth century, and Dutch influence was felt strongly in neighboring Germany. German surveyors such as the famous Wilhelm Dilich were trained in the Netherlands, and seventeenth-century Dutch surveying handbooks were translated into German very soon after publication and were used widely. Further influence can be seen in the similarity of some northern German units of measurement to Dutch ones. The influence of the Dutch is clear in the low-lying lands of northern Germany both in the construction of the polders and dikes and in the mapping of the areas.[131]

Danish influence was important in the northwest of the country. The technically advanced cadastral survey in the Duchy of Oldenburg which began in 1781 owed much to the Danes. The technical side of the project was led by a Dane who based the survey on a triangulation net which joined up with the Danish net. Danish agricultural reforms (chapter 3.11 and 3.15), in which mapping played an important role, gave an impetus to similar reforms in neighboring areas of Germany in which reform took place very early by German standards.[132]

The influence of Britain was more direct, since Hanover was in personal union with Britain between 1714 and 1837, but also it was more ambivalent. George III of England ordered the topographical survey of Hanover in 1764–86, which in turn influenced cadastral mapping in Osnabrück, and he also promoted agricultural reforms, having seen their effect in England. However, English influence could also be negative, as in the tax reforms of 1817, when the Crown deliberately decided not to produce maps or to carry out an accurate survey.[133]

Just as Germany profited from the experience of other countries, so it in turn influenced the course of surveying and mapping in neighboring countries. German surveyors were active in the country's immediate neighbors, such as the Netherlands, where some Germans were involved in polder mapping (chapter 2.1), and Denmark, where Germans were involved in the *matrikel* of 1688 (chapter 3.8). German surveyors also worked in countries further afield which had ambitious surveying projects but lacked the skilled men to carry them out. They were active in the *geometriska jordebok* mapping in Sweden (chapter 3.3) and in Norwegian forest-mapping projects (chapter 3.19). German surveyors had no special advantages in terms of technique or training. They were in general familiar with trigonometric but not astronomical techniques, and they almost always learned their trade in the field as apprentices to established surveyors.[134] German surveyors often had considerable experience of working both free-lance and as employees of forestry departments or other official bodies and were used to traveling long distances in search of work. Civil surveyors who joined the army often ended up in other countries and took part in surveying projects there. This could lead to difficulties for the host country, however, as the Danish authorities found when they employed German ex-soldiers who spoke no Danish on their survey of 1688 (chapter 3.8).

It is impossible to generalize about the course of German cadastral mapping. Almost every stimulus and obstacle to mapping can be found in its history. Perhaps the single most important aspect of this investigation is that it brings a very fragmented and often inaccessible body of literature together for the first time. Germany's cadastral map history was not only varied but rich. The fragmentation of the literature, caused in part by the widely scattered archive holdings and the strong tradition of writing regional and local histories, has allowed the impression to reign, at least among foreign scholars, that German cadastral history was eclipsed by that of its neighbors. When the literature is brought together, however, the rich indigenous mapping tradition of Germany is plain to see.

Acknowledgments

We should like to thank the following for their help: Prof. Gerhard Aymans, Mr. Peter Barber, Dr. Dietrich Denecke and his family, Prof. Herman Grees, Dr. Susanne Hübschle, Prof. Helmut Jäger, Dr. Helga Kallenbach, Dr. Lachmann, Prof. H-J. Nitz, Dr. Werner Vogel, Magister Ulrike Weihl; also the Staatsbibliothek Preussischer Kulturbesitz, Berlin; Geheimes Staatsarchiv Berlin; Niedersächsische Landesbibliothek, Hanover; the libraries of the Universities of Bonn, Marburg, and Göttingen; Hessische Staatsarchiv Marburg.

Fig. 5.1 The Austrian Habsburg Lands c. 1648 and the Habsburg-Lorraine lands (with the Principality of Piedmont) c. 1860. Source: Droysens 1886; Muir and Philip 1927; Shepherd 1930; Kinder and Hilgermann 1974–78; Darby and Fullard 1978.

5

THE AUSTRIAN HABSBURG LANDS, WITH THE PRINCIPALITY OF PIEDMONT

The territories ruled by the Austrian Habsburgs until 1740 and then by the Habsburg-Lorraine (Habsburg-Lothringen) family comprised a varying number of kingdoms, duchies, and other lands united in the person of the monarch.[1] Despite consistent attempts to weld these territories into a more coherent whole, stimulated by the example of neighboring Prussia, the Austrian Habsburg territories were never transformed into a modern unified nation state, and thus the titles "emperor" and "empress" used by the Austrian Habsburgs should not be misinterpreted. Charles V (1516–56), Ferdinand I (1556–64), Leopold I (1658–1705), Charles VI (1711–40), Joseph II (1780–90), Leopold II (1790–92) and Francis II (1792–1835) were titled "emperor" by virtue of their position as head of the Holy Roman Empire of the German Nation. Maria Theresa (1740–80) was titled "empress" as the wife of Holy Roman Emperor Francis I (1745–65). Francis II was the last Holy Roman Emperor. In 1804 he assumed the title of Francis I, emperor of Austria, and in 1806 the Holy Roman Empire disintegrated.

The heartland of the Austrian Habsburgs was largely in the area of present-day Austria, together with some small German territories (figure 5.1). This core consisted chiefly of the Archduchy of Austria, the Duchies of Styria (Steiermark), Carinthia (Kärnten), and Carniola (Krain), the *Grafschaften* (counties) of Tyrol (Tirol) and Gorizia (Görz), and the Margravate of Istria (Istrien). All these were part of the Holy Roman Empire. In 1526 there was a major expansion of Austrian Habsburg territory when, in fulfillment of an earlier treaty, the future Ferdinand I won the claim to the united crowns of Hungary (Ungarn) and Bohemia (Böhmen), as well as the Duchy of Silesia (Schlesien), the Margravate of Moravia (Mähren), and the Kingdom of Croatia (Kroatien). Of these, Hungary and Croatia lay outside the Holy Roman Empire. The claim to the crowns was subsequently fulfilled, and at the end of the Thirty Years' War the Austrian Habsburgs were the main power in central Europe. In the next two centuries they extended their territory in three main directions: to the east against the Turkish Ottoman Empire, to the south into the Italian peninsula, and in the far west into the Netherlands.

Effective Habsburg dominance in the east was established gradually as the boundary of the Ottoman Empire was pushed back. Turkish Hungary was recaptured between 1683 and 1689, and Transylvania (Siebenbürgen) was added in 1690–91. In 1687 at the Diet of Pressburg the Estates accepted that the Hungarian Crown would in the future be hereditary in the male line of the House of Habsburg. Habsburg rule in Hungary and the Principality of Translyvania was confirmed by the Turks at the Treaty of Carlowitz in 1699. The Kingdom of Slavonia came under Austrian Habsburg rule in 1699; the Banat of Temesvàr followed in 1718, the Kingdom of Galicia and Lodomeria (Galizien und Lodomerien) in 1772, and the Duchy of Bukovina (Bukowina) in 1775. Serbia (Serbien) and Little Wallachia (Kleine Walachei) were held between 1718 and 1739, and West Galicia was held briefly from 1795 to 1809.

In the Italian peninsula Austrian Habsburg lands included small but rich northern territories and larger but poorer southern lands. Under the peace treaties which ended the War of the Spanish Succession (1701–13/14), the Austrian Habsburgs gained the Kingdoms of Sardinia and Naples and the Duchies of Milan and Mantua. The Duchy of Parma, the Grand Duchy of Tuscany, and the Kingdom of Sicily were won later, but most of the southern Italian territories had been relinquished by the mid-eighteenth century. Austrian Habsburg presence in the north of the country was greatly strengthened by the Treaty of Vienna in 1815 and continued until Italy emerged as a nation-state in the 1860s.

In the west the Austrian Habsburgs gained the Spanish Netherlands, including the Duchy of Luxembourg, in 1714. Control of these was effectively lost in 1794, and they were formally relinquished in 1797 at the Treaty of Campo Formio.

In short, steady Habsburg expansion eastward was combined with acquisition and subsequent loss of many southern and western lands and of those territories within Germany which were conquered by Prussia. The chief of these was Silesia, the loss of most of which was confirmed at the Peace of Breslau in 1742, which marked the end of the First Silesian War (1740–42). The Austrian Habsburgs thus ruled over a group of territories which differed widely in culture, economy, political organization, and, as is discussed below, cartographic history.

5.1 Early Mapping: Estate and Boundary Maps

Estate mapping by the nobility, which dates from the seventeenth century at the earliest, generally predated that of the state in Austrian Habsburg lands, as elsewhere in Europe. Most private maps remain in private archives and are unresearched. Exceptions are, for example, the maps of Clemens Beuttler (1623–82), Georg Matthaeus Vischer (1628–96), Franz Anton Knittel (1671–1744), and Johann Jakob (Giovanni Giacomo) Marinoni (1676–1755).[2]

Marinoni, the chief instigator of the surveying and mapping of the early eighteenth-century

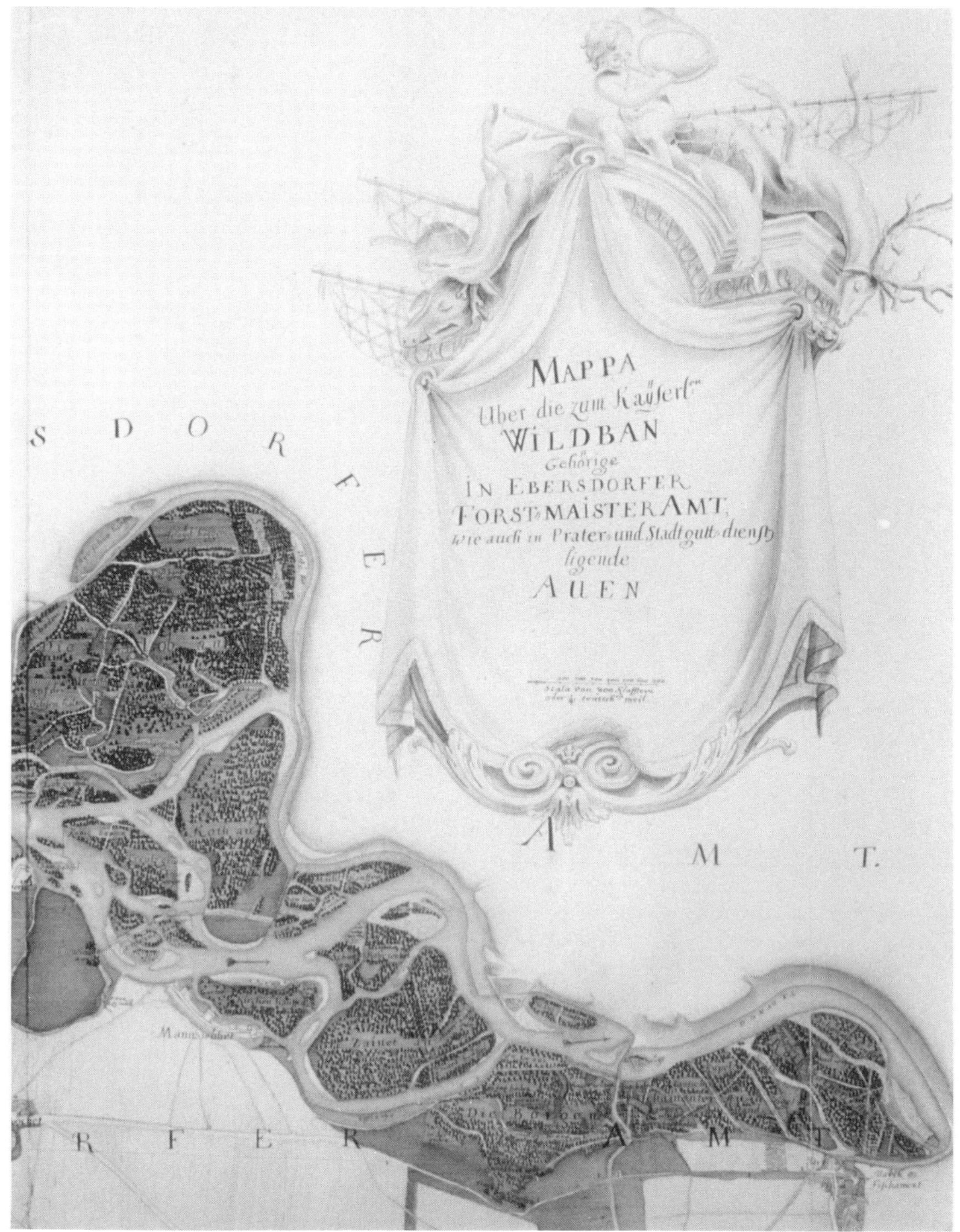

Fig. 5.2 An extract from Emperor Charles VI's personal copy of a map of the areas around Vienna in which he had the right to hunt. It was drawn in 1726 by Johann Jakob Marinoni at a scale of 1:41,000. Source: Österreichische Nationalbibliothek, Vienna, KI 98480.

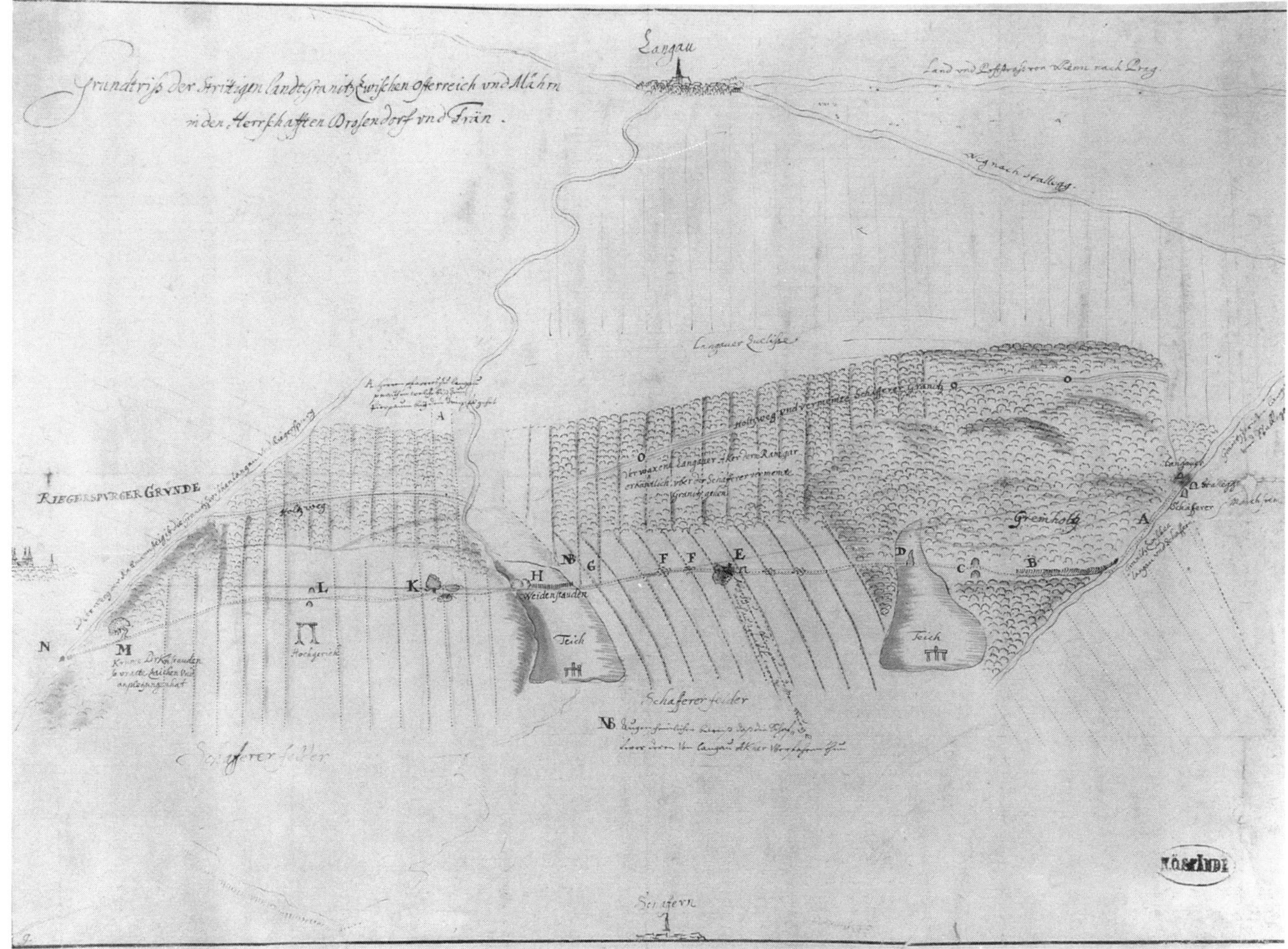

Fig. 5.3 Map of the disputed border between Austria and Moravia in the *Herrschaften* (manors) of Drosendorf and Frain by Georg Matthaeus Vischer, 1674. Source: Niederösterreichische Landesbibliothek, Vienna, B III 140.

Milan cadaster (section 5.4), was simultaneously active as a cartographer for private patrons. Notable examples of his work are the maps of the estates of Count von Hardegg and the hunting atlas drawn for Emperor Charles VI (figure 5.2).[3] The estate maps were drawn between 1715 and 1727 and consist of an atlas containing twenty-three maps and at least three other separate maps, mostly at a scale of 1:10,800.[4] It was Count von Hardegg, as the emperor's *Oberstjägermeister* (chief master of the hunt), who commissioned Marinoni to produce an atlas of the imperial hunting grounds in the 1720s. The maps are contained in two volumes and show the imperial *Wildbann*, or *Jagdreviere* (hunting grounds), which stretched along the river Danube from Vienna.[5] Two cop-

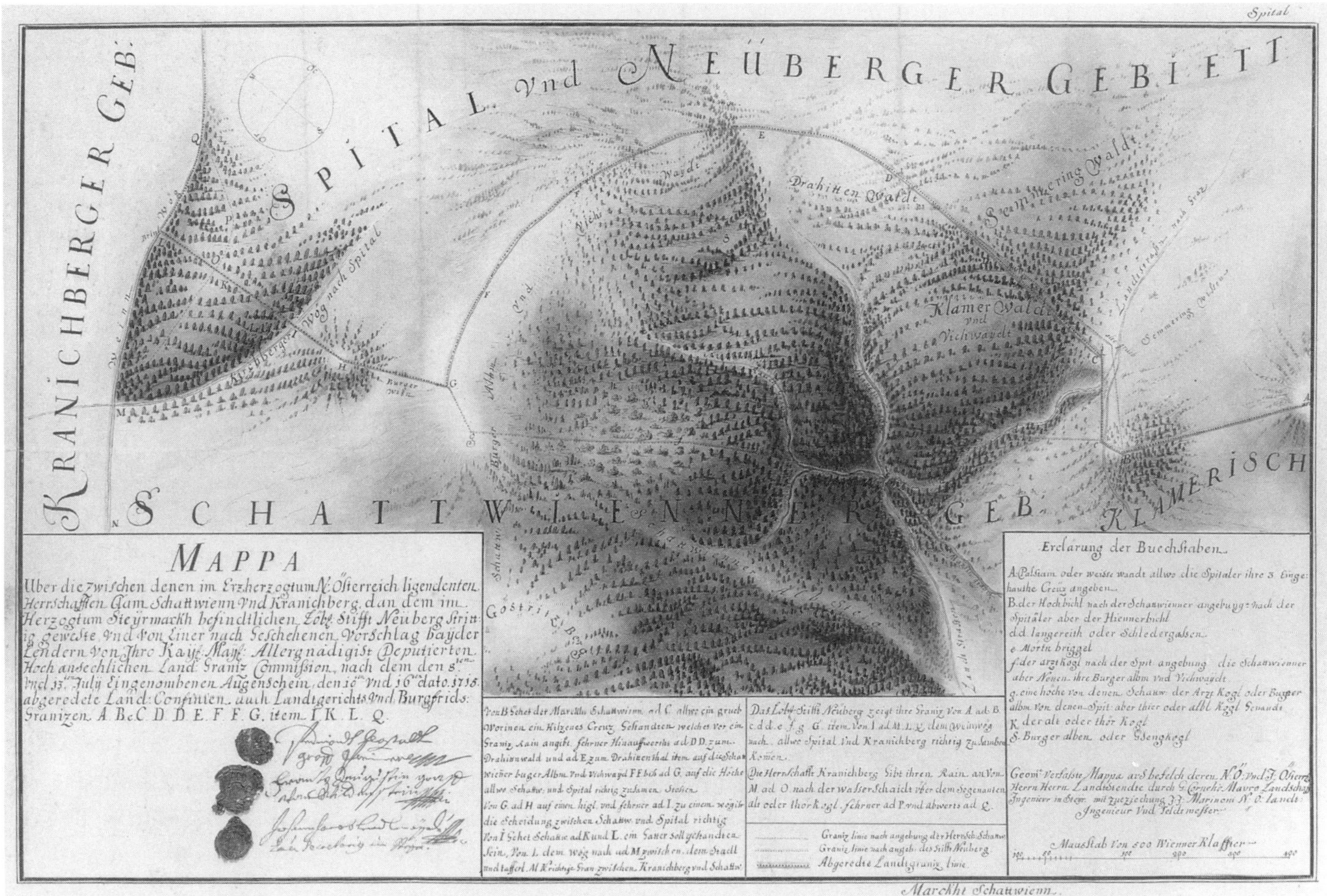

Fig. 5.4 Map of the boundary between the *Herrschaften* Schattwienn and Kranichberg in Lower Austria and the *Stift* Neuberg in Styria at 1:7,600 by Johann Jacob Marinoni, 1714. Source: Niederösterreichische Landesbibliothek, Vienna, B IV 154.

ies of each volume were made. Those for the personal use of the emperor are bound in green velvet with gold and silver mountings and have a title page portraying Charles VI hunting. The maps themselves have elegant cartouches. The other was a more utilitarian set, designed for everyday use in the administration of the hunting grounds. Each set contains thirty maps, again mostly at a scale of 1:10,800.[6]

Private mapping cannot be said to have inspired state mapping in the Austrian Habsburg lands. The earliest private mapping does predate the first state mapping, but in general the two proceeded in parallel and often involved the same surveyors, notably Marinoni. In the mid and late eighteenth century, the nobility used material produced by nonmapped state taxation cadasters to compile their own maps and atlases (cf. sections 5.7 and 5.8).

The Habsburg state undertook civil surveying mainly as part of its taxation reforms, but boundary disputes also gave rise to state surveys, notable early examples of which are by Georg Matthaeus Vischer in the 1670s and by Marinoni between 1711 and 1718. The borders in question were between Lower Austria and Moravia (figure 5.3) and Lower Austria and Styria (figure 5.4) and were simultaneously borders between estates.[7] The Estates set up boundary commissions to survey and map the disputed boundaries so that limits to jurisdiction and tax collection could be firmly established. One such is a map by engineer G. Cornelius Maurus and Marinoni drawn in 1715 of the boundary between the *Herrschaften* (estates) of Schattwienn and Kranichberg in Lower Austria on the one hand, and the *Stift* (church foundation) Neuberg in Styria on the other. It is at a scale of 1:7,600 and shows the various border claims and the boundary as fixed by the surveyors.[8]

5.2 The Struggle for Tax Reform

It is important to stress that the Austrian Habsburg territories were united in the person of the ruler and that forms of government varied considerably from territory to territory. Although after 1740 great efforts were made to centralize decision making, in most territories some power, notably power to levy taxes, remained in the hands of the local Estates. The *Landeskontribution* (land tax) was the main source of tax revenues for the army. Most of it was raised as a tax on immovable property. It was technically a free gift of the Estates to the ruling prince, consented to by the Diets of the various territories and levied under their supervision by individual lords. The Estates were also formally responsible until 1748, and in practice longer, for recruits for the army and for their quarters. Reform of the contribution and of army organization was thus vital to the central authorities. Reform was intended externally to demonstrate Austrian military capacity in a period when attack from the Turks and the Prussians was a constant threat, and internally in the name of the *Gottgefällige Gleichheit* or *Gottgefällige Gerechtigheit* (God-given principle of equality) to enforce the liability of all classes to a proportional share of the tax burden.[9] The provision of an effective cadaster for organizing collection of land tax was thus fundamental in the defense of the state from both external enemies and internal power groups.

5.3 Early State Taxation Cadasters

Although there is evidence that a surveyed taxation cadaster was instituted in Moravia as early as the eleventh century, the foundation for land taxation in the modern period was laid by Ferdinand I in 1527. Ferdinand needed money for a campaign against the Turks and wanted to bring in a uniform system of land taxation in all the *Erblände* (the hereditary lands of Austria and the Bohemian Crown) which he had inherited in the 1520s. The system involved self-assessment by princes and the Estates and lasted for more than two hundred years.[10] The Thirty Years' War brought such increased demand for tax and such economic dislocation that the tax system was in almost total disarray by its close. An investigation in Lower Austria, instituted in 1658 by Leopold I immediately after his succession, showed that some taxes were outstanding from before 1600 and confirmed that such tax as was collected often did not reach the central authorities but was dissipated in local bureaucracy.[11] As well as beginning a drive for efficiency which was continued by his successors, Leopold I also attempted to restrict aristocratic exemption from taxation. In 1697 he tried to subject the Hungarian nobility to taxation, but failed. His attempts to make the contribution subject to long-term agreements rather than annual negotiation were more successful. In 1691, for example, a regular, fixed contribution was agreed in Transylvania.[12]

Under Charles VI there was a "rectification" of the cadaster, which was still based on Ferdinand I's 1527 cadaster. The rectification was carried out through the Estates in Bohemia and Silesia and was aimed not at changing the existing system but at making it work more effectively. The Silesian cadaster was finished by 1740; in the new cadaster, as in its predecessors, the lords and the church paid tax. Sporadic attempts by them to claim general exemption were defeated. The new cadaster was administered by the Estates, and the lords and towns were in charge of local repartition (apportionment) of the tax in the traditional manner. In Bohemia a land tax revision was completed by order of the Bohemian Estates in 1654 and revised between 1654 and 1656. The tax roll for the cadaster was based on information provided by the taxpayers and included only *Rustikal* (peasant and town) land. *Dominikal* land (belonging to the lords and church) was omitted from the cadaster, as before the 1690s they paid tax only irregularly. From 1706 the lords and church had to pay tax, but their lands were still excluded from the cadaster. From 1710 to 1747 there was further lengthy and expensive rectification of the cadaster, although it was unclear to what end the information was being collected.[13] Charles VI's reforms in Bohemia and Silesia were an important part of a gradual process of reform which gathered momentum under his successors. The most important tax reform instituted by Charles VI was undoubtedly that in the Duchy of Milan, one of the smallest but richest of the Habsburg lands and one in which central authority was not mediated by an obstructive Diet.

5.4 The Milanese *Censimento*: The First Surveyed and Mapped Cadaster

The *censimento* in the Stato di Milano, or Austrian Lombardy, was begun by Charles VI, who was personally involved in the decision to undertake it, and was continued by his successor, Maria Theresa, both of whom saw the principle of equal taxation as crucial to their wider political and military reforms.[14] It is important for three reasons. First, it was the earliest and, for more than a century, the only fully surveyed and mapped cadaster in Austrian Habsburg territory. Second, it was the climax of the struggle of the central Milanese authorities against the autonomy and privileges of particular groups, notably the nobility and the church, and of particular territories. Third, its impact was felt far beyond Lombardy.[15]

The tax system in operation before the eighteenth-century reform was based on a land survey and evaluation instituted by Charles V. In practice the survey, which was completed in 1568, was not carried out in remote or mountainous regions or in those areas where powerful landowners were able to prevent it and no maps were drawn. The evaluation was based on only a partial sample of sale deeds, and the assessment of commerce was never finished. In 1599 arbitrary sums representing business activities were simply added to the land values, and this information was used to fix provincial and metropolitan quotas. Local repartition was left in the hands of far from impartial local administrators, rather than disinterested royal officials.[16]

The overall tax burden in Lombardy in the first half of the eighteenth century was modest by contemporary standards, but commentators held it to be insupportable, as it fell particularly on artisans and the peasantry. They began to abandon their lands and workshops as the tax burden became unbearable, and their emigration to neighboring territories significantly damaged the Milanese economy. The aristocracy and the clergy in Lombardy held two-thirds of all immovable property, but they bore only a tiny share of the tax burden and wielded great political influence. The object of the reform was to increase their share of the tax burden. Their resistance made reform difficult and complex, but in Milan, in contrast to some other areas, they did not manage to obstruct reform completely.[17]

The need for reform of taxation in Milan had been apparent from the early eighteenth century. When Prince Eugène of Savoy (1663–1736) was governor of Milan in 1707 and attempted to raise money to support his troops, he found the taxation system in disarray. As early as 1713 Charles VI sent an order to Prince Eugène in which he stressed the need to reduce church and noble exemption from taxation.[18]

On 7 September 1718 Charles VI issued a *Patent* (order) setting up the Giunta di Nuovo Censimento Milanese (Board for a New Census of Milan) to construct a cadaster to set taxation on a more equitable footing in the Duchies of Milan and Mantua. The *giunta*'s first president was Count Vicenza di Miro of Naples, who was succeeded in 1749 by the more radical Pompeo Neri (1706–76). In 1735 it was extended to cover the Duchy of Parma after this fell to Austria following the Peace of Vienna.[19]

The Milanese *giunta* proclaimed that, as well as certain "ordinary" ancient sources of state revenue, there should be three main types of "subsidiary" contribution levied respectively on immovable goods, on the rural population, and on commercial activities. It was the aim of the *censimento* to divide and extract these subsidiary taxes by clear legislation and efficient administration.[20] The following discussion is confined to the *giunta*'s reform of the tax on immovable property, the main source of state revenue.

The *giunta* began its work by gathering information from all landholders about the nature and extent of holdings and buildings and of sovereign rights over their land. Court mathematician and astronomer Johann Jakob Marinoni was dissatisfied with this system and insisted that mapping was the key to an improved taxation system. He was invited by the governor of Milan to a conference to set out his proposals for a mapped survey. Marinoni recommended that a survey should be conducted using his own improved plane table, that the common unit of measurement should be the Milanese *trabucco* (1 Milanese *trabucco* = 2.61093 m), that standard chains and poles should be issued, and that maps should be at a scale of 1:2,000 and should show property boundaries and boundaries of cultivation, communications, drainage, and settlements. The area of land parcels would be calculated directly from the maps. The cadastral maps should be used to construct smaller-scale parish maps, which in turn should be reduced in scale and used as the basis of a topographical map of the whole state.[21]

The members of the *giunta* were unsure of the merits of Marinoni's plane-table method and decided on a trial survey to compare it with the established local method of survey with a *squadro* (cross-staff). The trial took place in the parish of Melagno in the Po valley in 1720. Marinoni began first and, although he was severely hampered by the crowds of spectators, managed to survey four hundred hectares in eight days. It took a local Milanese surveyor with a *squadro* fourteen days to survey the same land, and his maps were less detailed than Marinoni's and less useful, since land area could not be calculated directly from them. The *giunta* suspected that the result might be due more to Marinoni's skill than to the intrinsic advantages of his method and ordered a second trial. This took place in Rovenna and Piazza in 1720. A pupil of Marinoni's competed against a local surveyor and was a clear winner. Marinoni's improved plane table had been shown to be quicker, cheaper, and suitable for all types of terrain. In 1720 work began according to the general method he had put forward at the 1719 meeting.[22]

The surveying for the cadaster took just over three years. The plane-table method was used without any general triangulation control, although in some places a magnetically orientated local triangulation net was drawn.[23] By 1723 the survey had covered 2,387 parishes in Lombardy with an area of 19,220 square kilometers; they were mapped at a scale of 1:2,000, and general parish maps were compiled at 1:8,000 (figures 5.5 and 5.6). By 1726 these parish maps had been joined to produce a topographical map of Milan in sixteen sheets at a scale of 1:72,000.[24]

The second part of the cadaster involved assessment for taxation. Assessors appointed by the *giunta* visited each village and divided the land according to its quality (good, average, or poor), land use, and cost of production. By 1731 the work of assessing the land and buildings was almost

Fig. 5.5 Map of the commune of Agliate at a scale of 1:2,000 from the Milanese *censimento*, 1720–23. This map is an original, drawn in the field. Source: Archivio di Stato di Milano, mappe Carlo VI, cart. 3048 (Agliate).

complete. The Polish War of Succession (1733–35) then broke out, and French and Spanish troops marched into Lombardy. Work on the cadaster stopped, and the maps and documents were taken for safekeeping to the fortress in Mantua. The death of Charles VI in 1740 and the Silesian Wars (1740–48) further delayed the work. When Maria Theresa took up the project in 1749, the assessments had to be completely redone and appeals dealt with, but by 1759 all was complete. Every parish in the duchy had been surveyed and had information about it recorded in maps, in *Grundparzellenprotokoll* (land registers), and in *Häuserparzellenprotokoll* (building plot registers). The Milan cadaster came into force on 1 January 1760.[25]

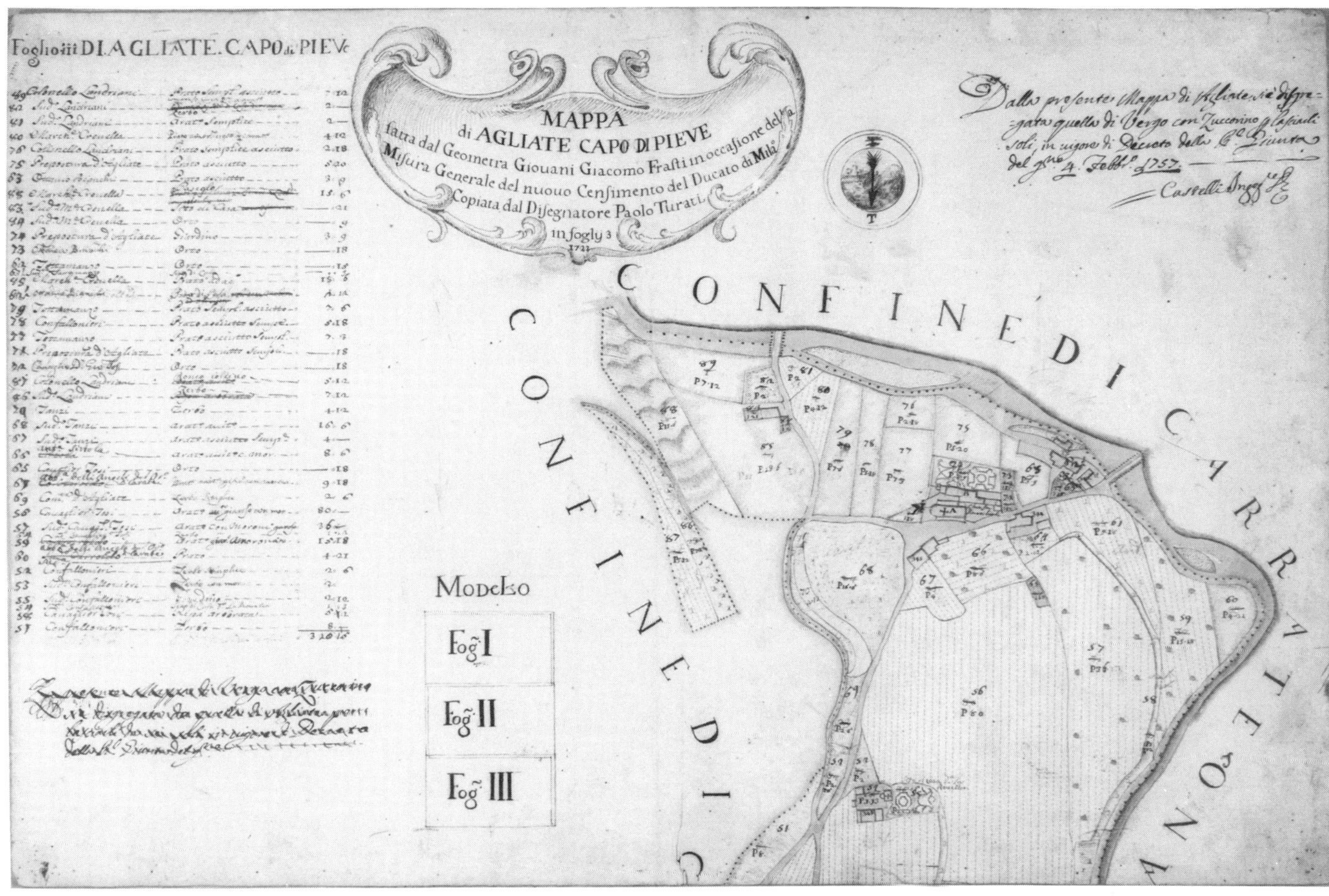

Fig. 5.6 Map of the commune of Agliate at a scale of 1:2,000 from the Milanese *censimento*, 1720–23. This map is a fair copy of the one in figure 5.5, a contemporary copy made in the levying of tax. Source: Archivio di Stato di Milano, mappe Carlo VI, cart. 3048 (Agliate).

The Milan cadaster was widely recognized by contemporaries as a pioneering cadaster. Adam Smith said: "It is esteemed one of the most accurate that has yet been made."[26] The cadaster was particularly important in that *all* land, not just productive fields, was surveyed and mapped at a large scale.[27] It has also been hailed as the first mapped cadastral survey in Europe.[28] While in fact it is predated by the Swedish *geometriska jordeböckerna* of the seventeenth century and by sixteenth- and seventeenth-century Dutch polder maps, the Milan cadaster was of undoubted significance in promoting the practice of mapped surveys and was directly emulated in neighboring states.

As a surveying and mapping exercise, the *censimento* was an unequivocal success; as a piece of taxation reform, however, it was not wholly satisfactory. The new cadaster left the overall tax burden, and hence the income of the sovereign, unchanged, but the burden was now borne by more people; in particular, most of the tax was now paid by the rich rather than the poor.[29] Another, unintended benefit of the reform was that, as the land-tax contributions were not changed through time, agricultural reform was encouraged. Those who improved and increased production from their land or enclosed common land continued to pay the old level of tax, which was thus reduced over time as a proportion of income. In contrast, those whose production fell paid relatively more.[30] The reform also successfully established the principle that an individual's liability for tax was to be determined by the central authority. The state would no longer fix overall quotas and leave repartition to local magnates.[31]

The attempts of the *giunta* to discover and limit the extent of tax privileges, however, met with only partial success, not least because Maria Theresa was reluctant to alter radically the old aristocratic society of which she and her advisers were a part. She fluctuated between setting up new centrally controlled administrative structures under the direction of foreigners like Neri, committed to attacking patrician privilege, and making significant concessions to traditional power groups. Numerous cases of exemptions from tax claimed by the aristocracy were brought to court by the *giunta,* but still more were left unresolved. In 1731 it was decided that lands for which immunity was claimed should be exempt from taxes imposed before 1599, but subject to all those imposed after that date. Further concessions were made to the patriciate in 1757 after Neri's *giunta* had been dissolved.

Ecclesiastical exemption had been even more abused than lay exemption, and many church claims of exemption were quite outrageous. The first *giunta,* under Count Miro, made little headway against them, but the second *giunta,* headed by Pompeo Neri, attacked the problem with renewed vigor. Its efforts were undermined in 1757, however, when Maria Theresa drew up a concordat with Rome which confirmed many church privileges and represented a significant concession to the church. When Joseph II, a devout Roman Catholic but one determined to end church autonomy and abuse of privilege, visited Lombardy in the summer of 1769, he became aware of the problems occasioned by the concordat and renewed the attack on church exemptions. He declared the concordat of 1757 void as from 1784, an action which finally put an end to the church's fiscal privileges.[32] The provision of accurate and precise information in the Milanese cadaster could not of itself end abuses of tax privilege, but it unquestionably helped (chapter 9.2). The *censimento* incidentally helped to extend Habsburg power in the region by defining territorial boundaries to favor Habsburg interests and compelling aggrieved parties such as Parma to plead their cases before imperial, that is, Habsburg, tribunals, and in so doing to recognize imperial suzerainty. As so often happened, cadastral surveys and maps had ramifications well beyond the purely fiscal sphere.[33]

5.5 The Influence of the Milanese Cadaster: Tuscany, Piedmont, and Spain

Reforms similar to those brought about with the Milanese cadaster were not instituted throughout Habsburg lands, not because the value of the cadaster went unrecognized, but because of the fundamental point that in most other Habsburg territories the contribution was under the control of the local Estates, agreed by them, and levied through them. Reform of the contribution did take place in all Habsburg lands in the eighteenth and nineteenth centuries, but a Milanese-type mapped cadaster would have been constitutionally inappropriate and too controversial to contemplate. The Milanese cadaster did, however, provide a model for tax reform within and particularly beyond the Habsburg lands. Within Habsburg-Lorraine territory the main influence was felt in the Grand Duchy of Tuscany when, from 1749 onward, concerted attempts were made to limit abuse of tax privilege by the nobility and particularly to end the autonomy of imperial and papal enclaves in Tuscany. A law of 16 November 1771 went further and set up a commission to investigate the extent of fiefs and to fix their boundaries. Feudal lords were required to bring documentary evidence and "an exact map" to substantiate their claims. Maps were viewed by the authorities as valuable instruments in helping them discover precisely the limits of feudal land and ensuring that tax was paid on all other land. The example of Milan was clear, and there was also an overlap of personnel. Pompeo Neri, who had headed the second Milanese *giunta,* was the secretary of the Ufficio delle Riformagioni, the body in charge of the work in Tuscany, and many Tuscan maps were drawn by surveyors who had worked in Milan.

The reform in Tuscany was successful up to a point. Some feudal lords, notably those with land in the heart of the duchy around Florence, submitted maps and documents quickly. Many sent in preexisting estate maps, mainly eighteenth century in date, but with some isolated examples from the seventeenth century. These were copied by state surveyors. The quality of some of the newly drawn maps, however, was not very high. This was partly because the better-trained state surveyors were fully occupied in copying maps, so that new mapping was done by less proficient surveyors, but chiefly because mapping was done at the expense of the lords. They obviously wanted to spend as little money as possible in carrying out a reform which could work only to their disadvantage. A more serious problem was that the feudal lords with peripheral estates far from the heart of government had managed to retain much of their autonomy and ignored the injunction entirely. The reform was thus a useful step on the path to reform but cannot be compared with the radical Milanese cadaster.[34]

The greatest influence of the Milanese cadaster was felt outside Habsburg territories: first, in the Principality of Piedmont and the Duchy of Savoy, which were ruled by the king of Sardinia and which bordered the Duchy of Milan, and second, in Spain. Because of the importance of these cadasters, especially the Piedmont cadaster, and because they were closely influenced by work in Milan, they are described in this chapter, though Spain and Piedmont were not, of course, territories of the Austrian Habsburgs. The Savoy cadaster is discussed in chapter 6.4.

In Piedmont there were tax problems very similar to those of Austrian Lombardy, and there was a comparable determination on the part of the central ruler to tackle them rigorously. By an edict of 5 May 1731 Carl Emmanuel III, king of Sardinia, brought into effect the *perequazione* (reform) of the land tax in Piedmont. This was the fulfillment of work to reform the tax system begun in 1697 by his father, Victor Amadeus II. Victor Amadeus's first attempt at a general survey of immovable property in 1677 foundered in the face of political opposition, but a second edict, on 12 June 1697, proved successful. One of the hallmarks of Victor Amadeus's reign was centralization of power, a motive evident in the instructions for the cadaster. It was to be organized centrally and carried out by surveyors and officials answerable to central government and independent of local communities.[35]

The objects of the survey were to reform tax on immovable goods so that liability was commensurate with ability to pay and to limit the tax privileges enjoyed by the church and nobility. The intention of limiting privileges was reiterated several times, but the aim was only to end the abuse of privileges and not to abolish them altogether. It was not, therefore, comparable with the principle of equal taxation established under the Napoleonic *code civil* one hundred years later. Nonetheless, the Enlightenment principle of a strong central monarch compiling detailed information about his territory in order to put through reform was clear.[36]

Work for the reform began on 14 May 1698 in the province of Cuneo and was followed by work in Mondovì, Turin, Asti, Biella, Ivrea, Pinerolo, Susa, and Vercelli. The work had two phases: the first took place between 1698 and 1730 and involved assessment and surveying of properties to apportion tax fairly among the *comunità* (communes) of the province. The second, which followed from 1730 until well into the eighteenth century, apportioned tax within each commune.[37]

The work of assessment was entrusted to a delegate and sixteen surveyors who were sent out into the provinces by the central government. They were assisted in the field by local people who could point out local boundaries. Property owners were required to submit *consegne* (depositions as to the extent and boundaries of their lands). After surveying and marking out the administrative boundaries, surveyors went on to delimit the boundaries of each type of cultivated area as a block; land was not surveyed parcel by parcel (survey *par masses de cultures;* cf. chapter 6.8). The area of fields was calculated using the *squadro* and plane table. Each commune had to provide information about the owners of church and feudal property and other property which was exempt from tax. The quality of each land-use type was then estimated based on its average productivity over the last ten years. Town houses were exempt from tax, but those in the country were taxed according to their size. Quarries and rocky ground were not taxed. All lands exempt from tax and communal lands were distinguished and listed separately.[38]

The general surveying and assessment of the territory resulted in increased revenue for central government and a considerable reduction (315,736 *giornate,* about 120,000 ha) of the amount of land classified as feudal and hence exempt from taxation.[39] However, the reform left the burden of taxation unequally apportioned both among and within communities. The survey did not take in the whole of the province of Piedmont, and the techniques used also had their limitations. Ricci considers that the most serious flaw was the failure to move from a system of gauging the overall burden according to the subject's ability to pay to one in which the needs of central government for money determined the sum raised. However, these limitations stemmed for the most part from the desire to complete the work quickly.[40]

The surveying and assessment were largely completed by 1711, despite interruptions caused by the War of the Spanish Succession, and finally finished in 1730. The results of the survey are recorded in a series of registers in which communes are listed with their areal extent, a description of the quality of their land, and agricultural information such as the crop rotations practiced.[41]

Between 1711 and 1730 the *perequazione* survey was subjected to detailed examination, and its results were compared with information from other sources, such as sales contracts supplied by the *intendente* (provincial administrator). Grave disparities were discovered, especially about the extent of allodial land (land held in absolute ownership without acknowledgment to a superior) compared with feudal and church lands. The edict of 5 May 1731 announced the findings of the *perequazione,* but it was possible to appeal against these in a special congress in which the king was the final arbiter. The congress dealt with numerous appeals in 168 sessions between 1731 and 1749. A concordat of 1727, based on the findings of the *perequazione,* limited the exemption of church lands from taxation. Various lands acquired by the church after 1620 became subject to both royal and communal taxes. The *perequazione* was also used to reduce the tax privileges of the nobility through the compilation of a special cadaster of lands which had been falsely claimed as feudal.[42]

The *perequazione* has been called the first modern cadaster in Italy. There are two reasons why this is not the case: first, it was predated by the Austrian survey in Lombardy: second, the *perequazione* itself did not involve parcel-by-parcel surveying. Such surveying, the *rilevazione,* occurred only after 1731.[43] After the central government had established through the *perequazione* the tax to be paid by each commune, it then ordered each commune to apportion tax among the individuals who owned land within it and who were not exempt from taxation. It was up to each commune to organize its own cadaster, under the supervision of the *intendente,* according to the state of its preexisting registers and maps.[44]

The cadastral work covered a large part of Piedmont, although feudal estates belonging to the emperor in the Langhe and those belonging to the pope, which were in enclaves in the territory of Savoy, were excluded, as were those communities which continued to rely on the old cadastral methods of the *consegne* or on earlier surveys.[45] A census of cadasters in the 807 communes in the province was made under the edict of 29 April 1733; that of 5 March 1739 gave instructions to be used in those communes where a new cadaster was thought necessary. Instructions for the Piedmont surveys were based on those which governed the Milanese cadaster. The surveyor was to go into the commune and, with the help of locals, delimit and map its boundary. Land parcels were then to be surveyed and mapped and registers compiled. The maps are uncolored and at a scale of 1:2,372, the same as that used in the Duchy of Savoy (chapter 6.4). The cadastral registers are extremely detailed and distinguish several types of land use and many types of building. When the surveying was finished, the valuation began. To save time, instead of assessing each parcel separately, as was done in Savoy, surveyors calculated the value of each land-use type. Individual plots were deemed to have the average productivity of the land-use type and were taxed accordingly, a system which penalized poor farmers and benefited good ones.[46]

Documents from the *rilevazione* include *il libro delle stazione,* a work journal of the survey; *il sommarione,* in which individual plots are listed in topographical order with the owner, extent, and land use; *il catasto,* in which owners are listed alphabetically with details of all their land holdings; the map of the commune; *il libro figurato,* which has a separate plan of each parcel; and *il libro delle mutazioni,* which lists alterations in ownership. The cadaster continued in operation until it was replaced by the Napoleonic survey.[47] In contrast to the plain maps of the nineteenth-century Napoleonic cadaster, those of the Piedmont cadaster are more pictorial in character. The symbols used give some immediate idea of land use, and there is some hatching to indicate relief.[48]

The influence of the Milanese *censimento* was also felt in Spain. Zeno Somodevilla, marquis de la Ensenada, who was finance minister to Ferdinand VI (1746–59), instituted a general taxation reform in Spain. Ensenada had spent some time in Italy, and his Spanish reforms were explicitly modeled on Maria Theresa's reforms in Lombardy. The problems in Spain closely resembled those of Milan and Piedmont, namely, the existence of a powerful nobility and clergy whose exemption from taxation the central state was determined to break. A new cadaster in the province of Castille was begun in 1749 and completed five years later at great cost. However, in 1754 Ensenada was ousted from his position, and the tax reform foundered in the face of insurmountable noble and clerical opposition.[49]

5.6 The Theresian Cadaster

The most important of the tax reforms introduced by Maria Theresa and her finance minister Count Friedrich Wilhelm Haugwitz (1702–65) were implemented between 1747 and 1756 as part of a gradual process of reform carried out province by province. The land-tax system had reached a crisis, since the many wars of the period greatly increased the amounts which had to be levied, and the loss of Silesia and, briefly, Bohemia in the 1740s greatly increased the burden on other territories. The problem was especially acute in Bohemia, which was required to contribute 75 percent of the whole land tax, a share which was quite out of proportion to its ability to pay. Not surprisingly, Bohemia was 3.5 million *Gulden* in arrears by 1741.[50] Reform was thus urgently needed to demonstrate Austria's military competence to her external enemies and to consolidate the power of the central authorities against provincial and aristocratic particularism.

Maria Theresa and Haugwitz drew up a program of fiscal reform for the Austrian and Bohemian lands. Hungary was explicitly excluded from Haugwitz's reforms. Haugwitz was the governor of Austrian Silesia, and he introduced a series of tax reforms between 1743 and 1748 which provided a model for fiscal reform in the other hereditary lands. The heart of the reforms concerned the funding of the army through a land tax. Under Haugwitz's reforms the Silesian Estates' liability for the supply and transport of army recruits was commuted for cash. Lay and clerical lords, peasants, and towns were required to submit tax returns and to pay a percentage of their declared revenue in tax. On 8 August 1748 the Silesian Estates agreed that the new level of contribution would be fixed for ten years, instead of being subject to frequent renegotiation. Similar reforms were subsequently instituted at the decennial recesses with the Estates of the other hereditary lands, and together they helped to consolidate central control over the army and at the same time to undermine the principles that taxation was a matter for the Estates and that the nobility should be exempt from taxation. The centrally devised reforms were, however, introduced by the negotiation of a separate agreement with each provincial Diet so that the principle of Estates' control over taxation was ostensibly respected. The reforms of land tax and indirect taxation enabled the state to increase tax revenue significantly.[51]

The revision of the contribution is known as the Theresian cadaster, although the state undertook systematic cadastral surveying only in the Tyrol.[52] The patent of 9 October 1748 and instructions of 24 May 1749 required the compilation of separate tax returns for peasants' land (*Rustikalfassionen*) and nobles' land (*Dominikalfassionen*); land tax was levied on the basis of the *Rektifizierten Dominikalakte und Bekanntnuss-Tabellen der theresianischen Steuerfassion* (tables drawn up from the submitted tax returns). Although there was no official mapping outside the Tyrol, nobles in other areas used the abundant material of the tax returns to commission surveys of their own and the peasants' land. Between 1745 and 1752 Karl Anselm Heiß drew twenty-eight maps and Ferdinand Erdlanger fifty maps of St. Florian in Upper Austria at a scale of 1:21,600. At the same time, Wolfgang Josef Schnepf (b. 1721) was compiling the atlas of the estates of Weinberg and Wartburg, an elegantly bound collection of nineteen maps.[53]

In the Tyrol, surveying for the Tyrolean *Peräquationssystem* (reformed tax system) was started under Maria Theresa and continued by Joseph II. The reason behind the survey is the subject of much debate, but it is clear that the Tyrolean taxation system was particularly complicated and unfair and that all previous attempts to reform it had foundered on the rock of noble opposition. Maria Theresa's own early attempts to reform taxation in 1746, 1771, and 1772 all failed, but those of 1774 and 1777 succeeded, and the reformed tax system came into force in 1784. The system was similar to that which Joseph later introduced in the rest of Austria and in the Bohemian lands (section 5.7) in that it was based on surveying carried out by the landholders themselves, who were also required to fill in tax returns. It proved difficult and expensive to survey in this mountainous area, but costs to the state were small, as the land was surveyed and information gathered at the landholders' expense. Informers were given substantial rewards to expose those who failed to declare taxable property, and the property in question was confiscated, so there was considerable incentive to fill in returns accurately. In most cases maps were not drawn; one exception is a map of Vomp near Schwaz which was drawn in 1796 as part of an appeal against the level of tax assessment; it is probably the earliest cadastral map of the Tyrol. The tax system in the Tyrol remained in force until 1882, although this had not been the original intention. Plans to keep it up to date came to nothing.[54]

Maria Theresa's reforms went some way toward taking land taxation out of the control of the provincial aristocracy and establishing a firm financial base for the state, but much remained to be done. The contribution remained unfairly apportioned. Bohemia was still overburdened, and the peasantry continued to bear an unfair proportion of the tax. Injustice led to inefficiency, and in the 1770s ad hoc adjustments had to be made to the level of taxation in Bohemia, Moravia, and Silesia in response to famine, destitution, and even revolt.[55] There remained much for Joseph II to accomplish.

5.7 The Josephine Cadaster

Joseph II was committed to far more radical social and political reform than was his mother, Maria Theresa.[56] As early as 1783 he made known his intention of reforming the cadaster and the system of land taxation, but there were problems of both principle and practice to be resolved before such reform could be implemented.[57] The principle which Joseph wanted to establish was that of *one* tax (on land) to be paid by *all*. The similarity of his reform principles and the ideas of the Physiocrats is striking, although Joseph himself denied that his ideas had originated from any one source. He wrote: "We can pay no attention to the customs and prejudices which have become established over the centuries. The land and the soil, gifts of Nature to Mankind, are the source of all value. From this comes the undeniable truth that the state's need for money must be met from the land alone and that there can be no differences amongst the possessions of men, no matter what their estate."[58]

As might have been predicted, the nobility opposed such objectives. They considered the principle of taxing land alone unjust, since it left bourgeois wealth untouched. The principle of a tax on all, they considered a contravention of their traditional privileges. Joseph was also concerned that all provinces should contribute their fair share of tax revenues to end the overburdening of Bohemia, but this was less controversial.

The most important practical problem was how to survey the land, as there was a serious shortage of trained surveyors throughout the Habsburg lands. Grünburg claims that 36 percent of productive land had escaped record in Maria Theresa's cadaster of 1756, so there was an inadequate basis on which Joseph's surveyors could build.[59]

Joseph was, however, set on achieving reform, and on 20 April 1785 issued a *Patent* ordering regulation of the land taxation system. The original *Patent* applied only to Vienna, Upper and Lower Austria, Styria, and Carinthia, but it was extended to cover Hungary and Transylvania in the following year. Work got underway, but the shortage of surveyors proved increasingly problematic. Joseph addressed the problem first by assigning military surveyors to the project and second by getting peasants to carry out some basic surveying. Peasants had recently been involved in surveying as part of Joseph's scheme to abolish *Robot* (labor duties) on state land. Peasants were now to be used for cadastral surveying where land was fairly level and offered no unusual complications. They worked in teams under the supervision of a lord's bailiff or estate official. They were given some basic instruction and used a newly simplified system of measurement units. They were excused other services, such as *Robot,* as compensation for the time spent on the survey. Upland areas and those that posed special problems were surveyed by trained surveyors, who also checked the work of the peasants. General surveying to fix and mark parish boundaries was undertaken under the supervision of the 264 army officers assigned to the project.[60]

There was considerable debate as to whether gross or net income from land was to be taxed. Joseph favored taxing gross income, not least because it would have taken too long to work out net income. Eventually a compromise was reached: gross income was to be taxed, but it was to be reduced by arbitrary amounts to take into consideration factors such as the distance of a piece of land from the market. Land was divided into nine land-use types, and the holder of each piece of land was to declare the income received from it over the past ten years. These declarations were made public in the hope that this would encourage honesty, and rewards were offered for information about false declarations. Previously the land tax had been collected by *Herrschaftsbezirken* (areas based on estates), but this was cumbersome and reminded the lords of their traditional right to repartition and collect the land tax. Joseph's reforms replaced the *Herrschaftsbezirk* by the *Steuergemeinde* (tax parish) as the unit of taxation. Within the *Steuergemeinde* taxes were to be collected by agents of the state, a change which removed the most significant power of the Estates.[61]

It was not intended to undertake comprehensive mapping, although some maps of *Dominien* (domain lands) were produced. The general aim was to produce results cheaply and above all *quickly*. Joseph said: "The benefits which a cheap tax reform must surely bring are too important for us to allow any delay or dilatoriness to interfere with it."[62] He had hoped to finish the survey in six months; although this proved impossible, 207,370 square kilometers were surveyed in the space of four years. This very considerable achievement would have been impossible if the surveyors had had to produce maps as well.[63] The surveyors did, however, produce *Brouillons* (field sketches), but most of these were discarded once the land parcel in question had been surveyed and assessed.[64] Some private mapping was done in the period 1785–88 in connection with the Josephine reforms, as had happened with the Theresian tax reforms.[65]

Joseph then embarked on his most radical reform, which was to combine the new tax law with a law to limit lords' rights to exact payments and services from the peasantry and to allow peasants to pay remaining dues in cash rather than kind. Joseph feared that if such obligations were not codified and regulated, the nobility would simply offset their new state land-tax liabilities by extracting more dues and services from the peasantry. What Joseph proposed was a radical intrusion of the state into serf-seigneur relations.[66]

Agrarian and tax reforms were combined in the *Steuer- und Urbarialregulierung* (Regulation of Taxation and Feudal Obligations), which was to come into force on 1 November 1789. The regulation limited the liability of the peasant to the state to 12⅔ of his gross income and to the lord and church to 17⅞ of his gross income. The new land tax, or *Grundsteuer,* replaced the old *Kontribution,* but taxes on income from urban property and indirect taxes remained, so that the idea of one tax on land alone was not realized. It is not clear how attached Joseph was to the principle of one tax, but in any case he could not do without the income from the other taxes, since he had pledged not to increase land tax.[67]

The preparatory work for the reconciliation of agricultural and tax reforms proved difficult and contentious. Joseph was forced to allow a period of grace before it came into operation to hear petitions from lords who would have suffered under the new system. The delay provoked rural riots by the peasants. The lords were concerned at the material hardship which the new law would bring and in particular feared that the peasants' payments would become a rental fee rather than a token of the subjects' submission to the lord.[68]

Joseph's concern with speed proved well founded. He succeeded in finishing the surveying and gathering of information, but these were just the preconditions for the cadaster, and he was too ill to see the system well established. His cadaster was introduced in Bohemia and Austria from 1789 but was never completed in the Hungarian lands.[69] Joseph died in 1790 and was succeeded by his brother Emperor Leopold II, who disbanded the Josephine cadastral commissions, declared Joseph's reforms null and void, and ordered a substantial return to the old Maria Theresian tax lists.[70] Joseph's reforms were certainly short lived, but he was a pioneer among European tax reformers, and his ideas were in large measure realized in the nineteenth century.

5.8 The *Josephinische Landesaufnahme*

The Seven Years' War exposed Austria's lack of reliable military maps. Maria Theresa tried to rectify this situation, and in 1763 a survey of Silesia marked the first phase of a project known as the *Josephinische Landesaufnahme,* begun by Maria Theresa and completed by Joseph II. By the end of Joseph's reign in 1790, Silesia, Bohemia, Moravia, Marmaros, the Banat, Transylvania, and all of Austria had been surveyed. It was primarily a military survey, carried out by surveyors from regiments stationed locally who worked under military orders and produced topographical maps at the normal military scale of 1:28,800. The *Josephinische Landesaufnahme* was, however, not just a narrow military and strategic mapping exercise, as surveyors were instructed to identify all houses, to record the number of livestock, to describe woods, rivers, and roads, and to indicate the nature of terrain on the maps. Joseph II said of it: "If one is to rule countries well, one must first know them exactly."[71]

In Marmaros, the Banat, and Transylvania the aim went further, and military mapping was accompanied by cadastral mapping. The survey of Marmaros had military, economic (taxation), and general geographic aims. The work was carried out from 1766 to 1768, but it proved very difficult to reconcile the many, often conflicting demands made of this survey. Military maps were produced at a scale of 1:28,000, and economic maps were made at varying scales including 1:7,200 and 1:4,800.[72]

The Banat survey was undertaken from 1769 to 1772. A special request was made that an economic survey be carried out "because the land plots have never been surveyed and property disputes are a constant burden." The request was granted, and cadastral and military survey were carried out side by side. Military maps were later drawn from the economic maps, although this proved difficult, as there was no general triangulation. Cadastral maps were produced at a scale of 1:7,200 and were used as the basis for taxation until 1819. Stavenhagen considers that the Banat survey did not justify its expense but that the mapping of Transylvania, completed in 1774, was more successful in that it produced a significant increase in tax revenues to the state.[73]

5.9 The *Stabile*, or Franciscan, Cadaster

Like Joseph II, Emperor Francis I of Austria was convinced of the need for taxation reform. Discussions as to how this was to be achieved began early in his reign, and the Treasury began preparatory work in 1806. The Napoleonic Wars delayed progress, but the central point at issue was whether the cadaster should be based on the Milanese model (i.e., comprehensively surveyed and mapped by professional surveyors) or the Josephine model (i.e., by requiring landholders to survey but not to map their own land). Francis himself favored the Milanese model, but he was not convinced of the need to base the survey on triangulation. His technical advisers, by contrast, saw the construction of a triangulation net not only as indispensable but also as eminently feasible, since a military triangulation had been underway since 1806. Francis was persuaded, and in 1817 trial surveying began to produce maps at a scale of 1:2,880, exactly ten times larger than the normal military scale. On 23 December 1817 Francis issued a *Patent* which set in train the reform known as the *Stabile,* or Franciscan, cadaster; as its name suggests, it was intended to provide a stable or long-term solution to the problem of land taxation.[74] Some of the maps drawn for the trial survey used shading to indicate relative relief and slope. One such was that of Berchtoldsdorf, drawn in 1817 by Wilhelm Reiche (figure 5.7). The technique was judged inappropriate for a cadastral survey and was discontinued after the trial.[75]

Francis, unlike Joseph, could afford to base his taxation reforms on mapping, not least because the principle of reform had been established so that speed was no longer so important. In recognition that a mapped cadaster would take a long time to effect, provisional orders were issued in 1819 to specify the systems which were to operate while work for the new cadaster was carried out. In the Tyrol the Maria Theresian *Peräquationssystem* was to remain in force; in Vorarlberg, recently under Bavarian rule, the Bavarian system was to remain in place; and in part of Illyria the existing French system was to remain.[76]

Detailed instructions were issued to govern the surveying. The original intention had been to use the military triangulation and to add a secondary triangulation to control the cadastral survey. This proved impossible, as the military work was not far enough advanced, nor of high enough quality in its early stages. Cadastral surveyors constructed their own triangulation net, and detailed cadastral maps were constructed by plane-tabling.[77] Each tax parish was surveyed and mapped individually where practicable. Josephine tax parishes, later known as *Katastergemeinde* (cadastral parishes), were adopted with comparatively few changes, and great efforts were made to ensure that these units consisted where possible of one settlement and its lands. Only where these were too small were settlements grouped together.[78] Parish boundaries were fixed before detailed surveying started, and landowners were obliged to mark the boundaries of their properties with stones or staves.

Fig. 5.7 Manuscript map of Berchtoldsdorf (Perchtoldsdorf), Austria, drawn in 1817 for the Franciscan cadaster at a scale of 1:2,880. This map was drawn as part of the trial survey and, unlike later maps, indicates relief by hill shading. Source: Bundesamt für Eich- und Vermessungswesen, Vienna, 16 121 Sheet V.

It was hoped to make the *Stabile* cadaster maps available for a wide variety of state and private uses, and as early as 1818 an institute was set up in Vienna to reproduce the maps using Alois Senefelder's new process of lithography. This was done with the active cooperation of Senefelder, who was simultaneously supervising reproduction of cadastral maps in Bavaria. Lithography was not completely satisfactory, but it was far cheaper and quicker than the alternative of copper engraving.[79] Lithographic reproduction also helped to standardize map presentation. Legends were lithographed and then hand-colored. They were issued in 1820 and periodically revised (figure 5.8). Their complexity reflected the great diversity of land use and language to be found in the areas under Austrian rule.[80]

When surveying in a region was finished, a *Grund-Ertragsschätzung* (land assessment) was carried out. Under the supervision of a *Katastral-Schäztungsinspektor* (cadastral and valuation inspector), valuation officers and their assistants worked through the tax parishes, helped by people from the parish and local officials. Forest areas were valued by special commissioners.[81]

Before the start of the valuation proper the most experienced and responsible landowner of each parish was sent a very detailed questionnaire which he was required to complete with details of the physical qualities of the land, its use, farming practices, the population, and so on. This information helped in the compilation of a *Steuerschätzungsoperat* or *Katastralschätzungselaborat* (taxation assessment) for each locality. This was an extremely detailed description of the area, together with its tax assessments. When the assessment process was completed, the documents were translated into the official languages of each province.[82]

The Franciscan cadaster began in 1817 in Lower Austria, and the Austrian part of the survey was completed in the Tyrol in 1861. Its quality is variable. By the time the surveyors reached the Tyrol and Vorarlberg in 1855, a second and improved military triangulation had been completed on which cadastral surveyors could base their work. In the Austrian part of the empire a total of 30,556 cadastral parishes with a ground area of 300,082 square kilometers divided into nearly 50 million land parcels were surveyed. This was an astonishing rate of progress, achieved because the surveyors worked from sunrise to sunset, six days a week, and because much routine work was done by assistants.[83]

By the time surveying was completed in the Austrian part of the empire in 1861, the pricing system on which the assessors based their valuation was completely out of date and valuation was impossible. By 1861 there were five different land-tax systems in operation in the Austrian lands. In the Tyrol the 1784 *Peräquationssystem* was used; Galicia had a system based on the Josephine reform; in Bukovina an unsatisfactory system introduced in 1832 was used; Vorarlberg had a system similar to that of Bavaria; and in the rest of Austria the *Stabile* system operated. In addition to these difficulties agricultural reforms and the construction of railways had produced radical change in landholdings in some areas.[84] Pressures such as these led to the law for the regula-

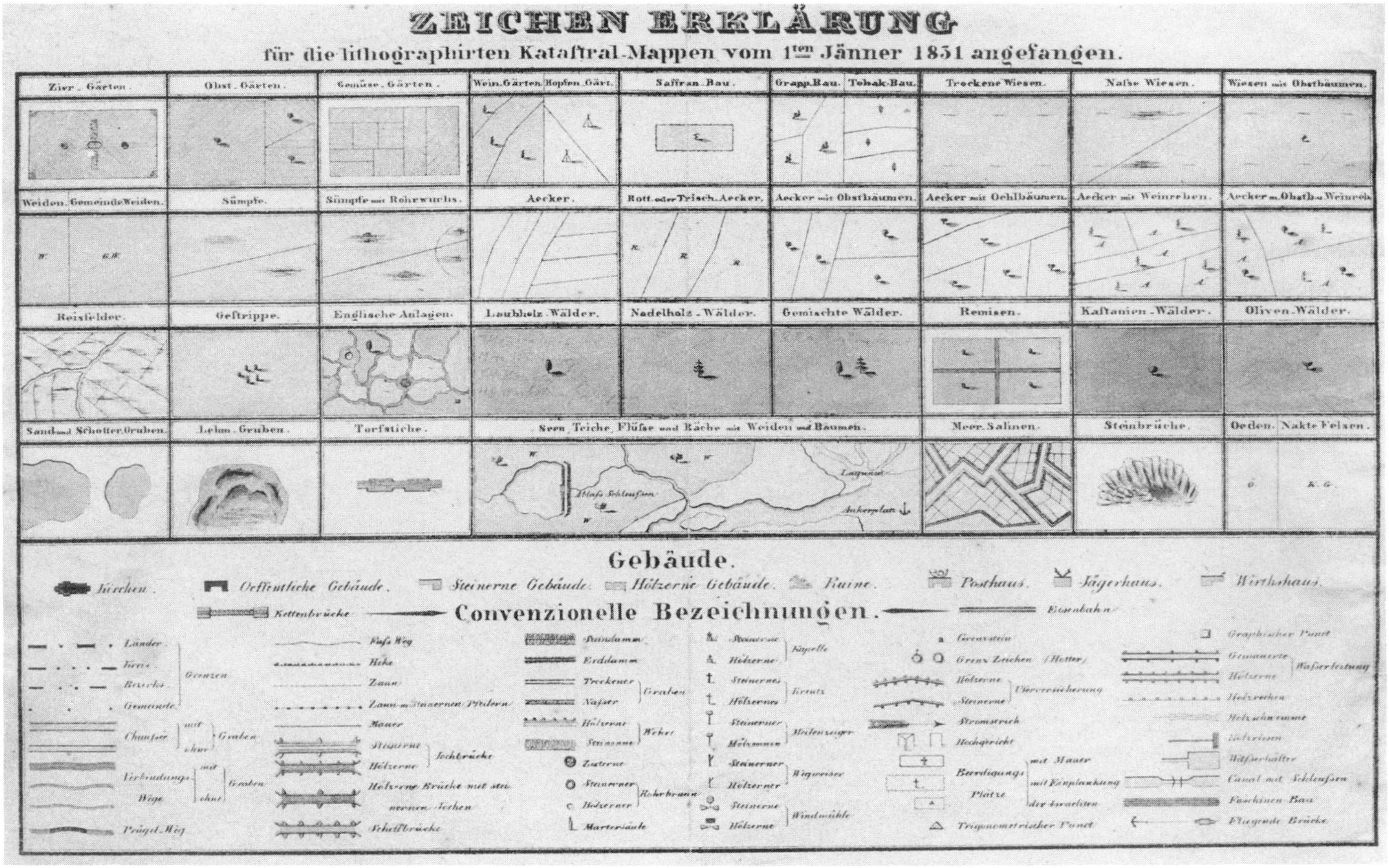

Fig. 5.8 Lithographed model legend for the Franciscan cadaster, 1831. Source: Bundesamt für Eich- und Vermessungswesen, Vienna.

tion of land tax of 24 May 1869. This reform, or *Reambulierung,* led to the drawing of new maps where numerous changes had taken place since the *Stabile* survey, or to alterations to original maps where changes were fewer. A significant improvement was the marking of triangulation posts in the field.[85]

In 1849 the *Patent* for the *Stabile* cadaster was introduced in the Hungarian part of the empire, and work started in the western part of Hungary, now Burgenland. When the so-called *Ausgleich* took place in 1867, under which the Austrian Empire became the Austro-Hungarian monarchy, the western part had been mapped by the Austrian administration. In the eastern part of Hungary the work was carried out after 1867 by the Königlich Ungarischen Kataster (Royal Hungarian Cadaster).[86]

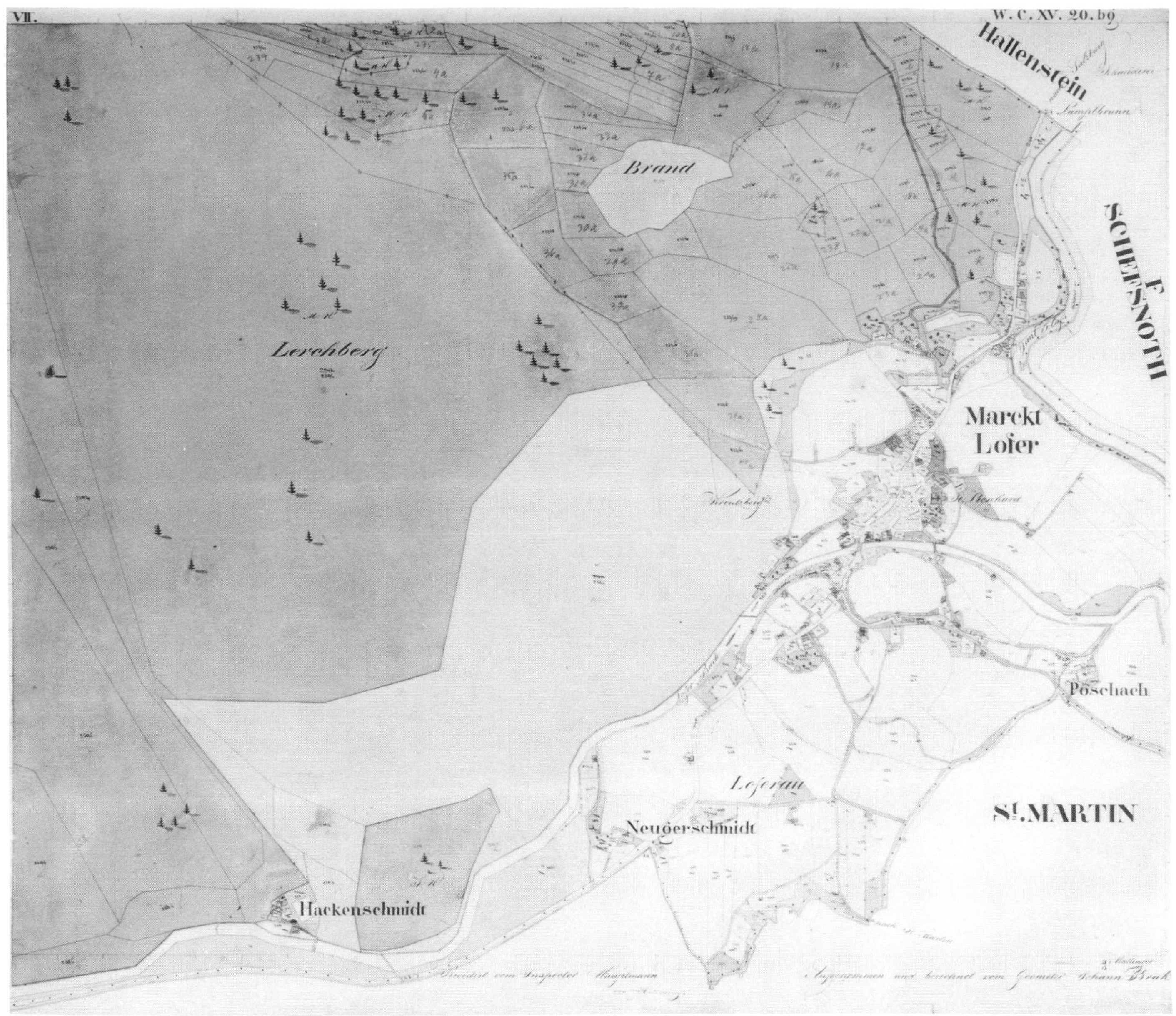

Fig. 5.9 Manuscript map of Lofer, Austria, drawn in 1830 at a scale of 1:2,880 for the Franciscan cadaster. Source: Bundesamt für Eich- und Vermessungswesen, Vienna, 57 117 Sheet VII.

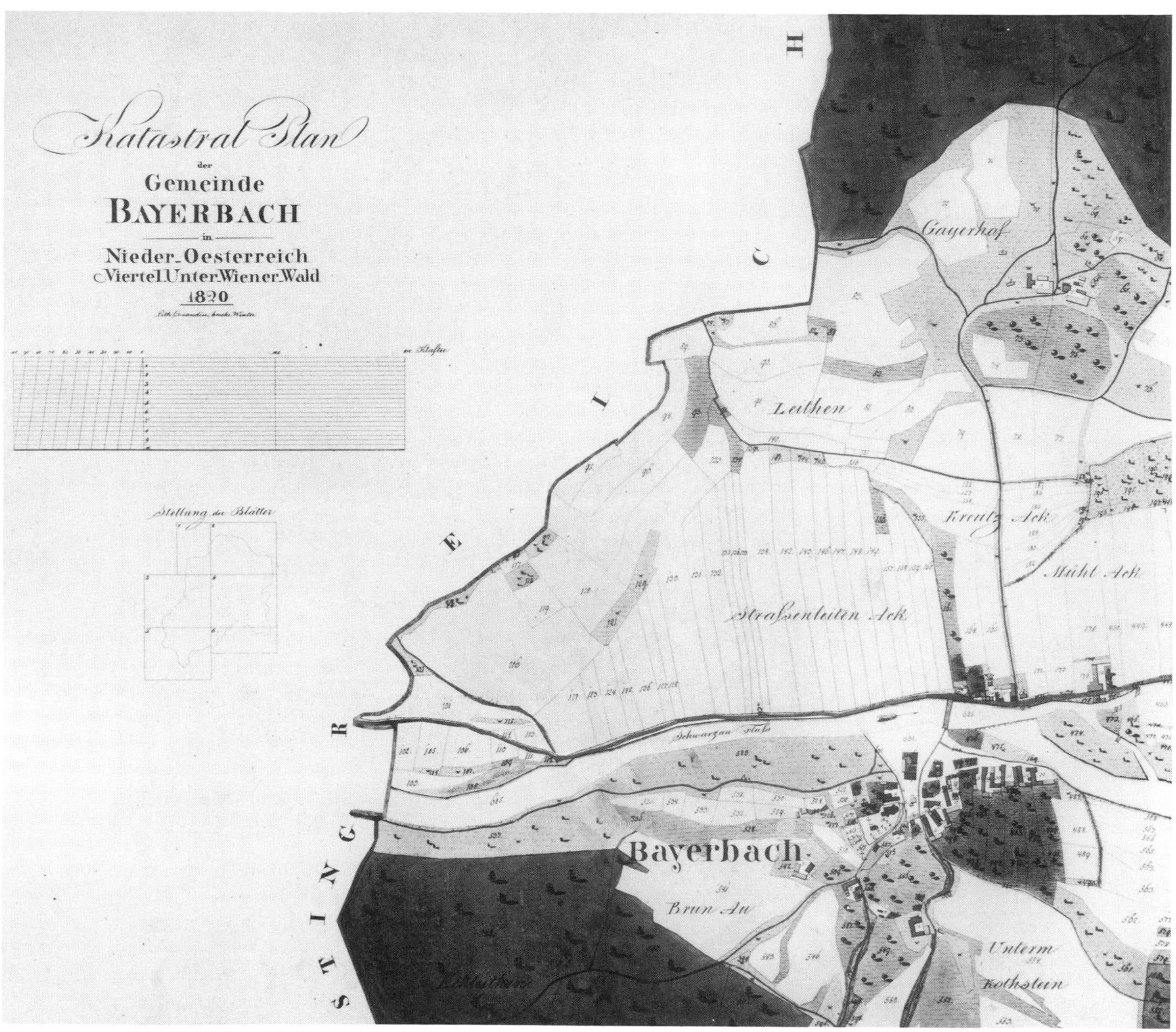

Fig. 5.10 Manuscript map of Bayerbach (Payerbach), Austria, drawn in 1820 for the Franciscan cadaster at a scale of 1:2,880. The map contains subsequent alterations to show a new water supply system, a railway line, and new house building. Source: Bundesamt für Eich- und Vermessungswesen, Vienna, 23 129 Sheet III.

The Franciscan cadaster consists of several elements, the most important of which is the *Katastralmappe* or *Parzellenplan* (cadastral map).[87] These were normally at a scale of 1:2,880, although a scale of 1:1,440 was used where land was fertile and plots were small and 1:5,760 where land was poor and properties large (figures 5.9 and 5.10). In all towns and some other settlements *Bauparzellpläne* (maps of building plots) were drawn at scales of 1:1,440 or 1:720. Provincial archives were established to house the maps, and from 1833 a central archive was set up in Vienna to house copies of all the maps.[88]

Each map is accompanied by a *Grundparzellenprotokoll* (register of land plots) and a *Bauparzellenprotokoll* (register of building plots). The *Grundparzellenprotokoll* lists for each land parcel its reference number, the name of the landholder, the land use, and its area in Lower Austrian *Joch* and *Quadratklafter* (1 Viennese *Klafter* = 1.896 m). There is also a section in which changes of ownership, subdivision of plots, and changes of land use are noted. At the end of the *Grundparzellenprotokoll* all the landowners of the locality are listed in alphabetic order and their occupations stated. There are also aggregate statistics of the area of each land use.[89]

The *Bauparzellenprotokoll* lists the function of each building, its ground area, whether it had one or more storys, the building material, and sometimes details of its state of repair. In some areas the number of stalls for animals and even the number of animals which could be housed are given. Finally the cadaster has an alphabetic list of all landholders resident in the locality, a description of its boundaries, and additional documents which show alterations made in ownership or land use until the 1860s, when the *Reambulierung* was begun. Such comprehensive detail permits modern reconstruction of changing patterns of land use and ownership.[90]

It was not immediately recognized that it would be necessary constantly to revise the *Stabile* cadaster. Revisions and alterations were not made systematically, and in some areas the maps rapidly went out of date. The first comprehensive revision of the cadaster was begun in 1869, when it was ordered that all boundary and ownership changes be entered on the maps. However, it was the order of 23 May 1883 which really began the process of constant revision using modern surveying techniques.[91] The revision overlapped with agrarian *Operationen* (reforms) which were carried out following legislation of 7 June 1883 and which affected field distribution, drainage, and communications. This often entailed revision of property boundaries, and the officials in charge of agrarian reform were obliged to give information to cadastral commissions who were carrying out the *Evidenzhaltung* (revision of the cadaster).[92] Where necessary, new mapping was carried out at a scale of 1:2,880 in Austro-Hungary and at 1:12,500, 1:6,250, and 1:3,125 in the occupied territories.[93]

5.10 Later State Surveys

A new survey of 1878 followed the Congress of Berlin of that year in which Austro-Hungary gained administrative rights over Bosnia and Herzogovinia. Work was begun in Bosnia to create a triangulation net joined to that of the Monarchy. Work on the topographical map was started but abandoned in favor of a cadastral survey which the administration considered vital to establish and record property boundaries. Work on the cadastral survey began in 1880 under military leadership. Nearly 3½ million land parcels were surveyed and mapped at a scale of 1:6,250, while scales of 1:3,125 and 1:1,562 were used for towns and Sarajevo respectively.[94]

The *Stabile* cadaster had reformed the system of land taxation, but imperial finances remained in a fairly constant state of crisis throughout the mid-nineteenth century. These crises, usually the result of war, were in general met without reform of the tax system but by means of long-term borrowing, money creation, and the sale or lease of state property. In the nineteenth century direct tax accounted for a decreasing share of all tax revenue. In Austria direct taxes contributed about 30 percent of state revenue in the mid-nineteenth century, and this fell to about 25 percent by the early twentieth century. In Hungary the proportion contributed by direct taxes was higher, but it fell from about 40 percent in 1867 to about 28 percent by the early twentieth century. The land tax was only one, though an important one, of a number of direct taxes, and rates and exemptions very much favored rural property at the expense of urban property.[95] The reduction in the burden of tax on rural land from the 1820s and especially after 1896 has been described as "a signal triumph for the agrarian interests."[96] It is ironic that at the very time when the collection of the land tax was put on a sound scientific footing, the tax had ceased to matter as much. The coincidence is not fortuitous: only when the Crown had conceded that the rural interest should not bear the brunt of taxation and when the provincial nobility had conceded that such tax as remained should be collected by the crown did the issue become uncontroversial enough for a scientific mapped cadaster to be introduced.

5.11 The Cadastral Map in Austria

Austria's role in the development of topographical mapping befits its position as a major European power. There is evidence from the end of the seventeenth century onward that the rulers of Austria were aware of the military and administrative advantages which the possession of precise maps and a skilled body of surveyors could bring.[97] By contrast, Austria's role in the development of cadastral mapping is slight. Cadastral mapping was instituted in Austro-Hungary as a whole only in the nineteenth century with the *Stabile* cadaster, and although this covered a remarkably large area and was carried out efficiently and with great determination in widely varying terrains, it was not innovative in technique or organization. Perhaps its most innovative aspect was the very early recognition of the potential of lithography to reproduce maps cheaply and abundantly. The cost-benefit ratio of cadastral mapping alters greatly if cheap reproduction allows more people to benefit from it and hence spreads the initial high cost of surveying and mapping. The failure to institute a mapped cadaster earlier is, however, striking and highlights the controversial nature of such cadasters. The Austrian Habsburgs did not introduce a mapped cadaster earlier because they would not have been able to. The power which the possession of such detailed and comprehensive information would have given the central authorities was widely recognized by the church and nobility. They resisted, albeit unsuccessfully, the acquisition of such knowledge in Milan. They succeeded at least in part in obstructing its collection in Tuscany, and they ensured that it was not proposed in the remaining territories. After two centuries of struggle by the central authorities to limit and then to reduce the power and autonomy of the nobles, they succeeded in introducing the *Stabile* cadaster in the nineteenth century. By then it was a nonissue, but only because the burden of taxation had swung decisively away from the rural interest. Perhaps more than in any other country in Europe we see in the Austrian experience how far the map is from being a neutral and apolitical instrument.

Acknowledgments

We should like to acknowledge the financial support of the Travel and Research Fund of the Oxford University Sub-Faculty of Geography and the generous help of Ms. Maggie Baigent, Mr. Peter Barber, Prof. Peter G. M. Dickson, Dr. Gerhard König, Prof. Dr. Ingrid Kretschmer, Dr. Colin Thomas, Prof. Vladimiro Valerio, and the staff of the Bundesamt für Eich- und Vermessungswesen, Universitätsbibliothek Wien, Niederösterreichische Landesarchiv, Niederösterreichische Landesbibliothek and Österreichische Nationalbibliothek, and the Taylorian Library, Oxford.

6

FRANCE

Among all the nation-states of Enlightenment Europe, France occupied a preeminent position in both science and culture (figure 6.1). French administrators, notably A.-R. J. Turgot, Louis XVI's controller-general of finances, used science and systematic knowledge to develop policies to strengthen the power of the French monarchy.[1] Among the questions Turgot had to address in the latter part of the eighteenth century was whether the monarchy could ever again achieve financial solvency and whether government could deploy the power to "make the public interest prevail over the constitutional prerogatives of the privileged orders and corporate bodies, ending or abating their exemption from direct taxation."[2] It was for an instrument to effect this latter that the French state turned to cadastral mapping.

In 1929 Marc Bloch wrote: "Tax reform was one of the *raisons d'être* of the [French] revolution; to base taxation on land in a manner as equitable as possible, topographic surveys were absolutely essential."[3] A new land tax was introduced based on size of properties and nature of land use (*par masses de cultures*); from 1807 a new cadaster (now known as the *ancien cadastre* to distinguish it from later surveys) was compiled and recorded in large-scale *plans parcellaires*.

This chapter reviews the eighteenth-century and earlier antecedents of this state involvement in land survey and mapping, the debates about cadastral surveys conducted by the revolutionary government, and the implementation of the post revolutionary surveys in France and territories now included in Belgium, the Netherlands, and Luxembourg. The *cadastre* both contributed to and was affected by political and economic reforms, while in the field of cartography, it was to shape the character of the national map survey beyond the end of the Napoleonic First Empire.[4]

6.1 Mapping Individual Properties and Communes

In France the mapping of individual properties or *seigneuries* for practical purposes in association with agricultural improvement was rare before the eighteenth century. Though some parts of

France, notably the north and the Paris basin, were the scene of agricultural reform, the sixteenth and seventeenth centuries in France did not witness the sort of fundamental technical changes that were affecting agriculture in, for example, England and the Netherlands. Charles Estienne's and Jean Liébault's *L'Agriculture et maison rustique* (1564), the first agricultural treatise in the French language, was repeatedly reissued during the sixteenth and seventeenth centuries, as was Olivier de Serres's *Théâtre de l'agriculture,* first published in 1600.[5] The next French text of importance in this field was Duhamel de Monceau's *Traité de la culture des terres,* a title not published until 1750. As Neveux, Jacquart, and Le Roy Ladurie comment, this vacuum in the agricultural literature is a sure sign that there was no demand to change the time honored ways of doing things.[6] Such a picture is also in accord with the comparatively late arrival of *plans terriers* in the French archival record.[7]

Furthermore, it can be argued that in feudal France, when land was important to landowners not so much in its own right in terms of acres or precise physical extent but rather for the seigneurial rights and dues that accompanied it, there was no need of maps when the written word could precisely identify rights, obligations, and taxes associated with landownership. This was particularly the case in that greater part of France (all except Provence and Languedoc) where taxation was based on the person, not land. In the far south of France, in "land-tax country," written registers (*matrices cadastrales* or *compoix,* as they were termed locally) had been in use throughout the Middle Ages (section 6.3).[8] As a general rule even the most recent *compoix* are not accompanied by plans but list the names of proprietors, the land use of each parcel, its area in local measure, and its location by describing abutting parcels.[9] The exceptions are few but include communes such as Saint-Geniès-des-Mourges in the diocese of Montpellier. Here the *compoix* was renewed in 1787, and the changes were recorded on a map.[10] Similarly, recording and regularizing the multiplicity of changing *reconnaissances féodales* (feudal obligations) and *compoix* were the reasons for mapping Cambon-les-Lavour in 1632 and its recopying in 1708.[11] Frêche lists some twenty plans of communities in the Toulouse region and in Languedoc compiled during the eighteenth century.[12] Legislation concerning *compoix* never insisted on plans because of their high cost. Figure 6.2 is an example of one that was compiled. Dated 1772, it is entitled "Plan de divers terroirs avec application des titres et compoix" and covers the area of present-day Chaussadenches in the Vivarais. It is at a scale of 1:400 and was compiled by a local notary who wrote into each parcel changes in ownership and cultivation since the fourteenth century, providing a record of cadastral characteristics probably unique in France.[13]

In early modern France, as elsewhere in Europe at this time, litigation about matters such as tithe payment and property boundaries resulted in the production of maps for use as evidence in courts of law.[14] In France, these are known variously as *plans d'arpentage, plans de mesurage, plans de délimitation,* or *plans de bornage.* An example of a map associated with tithe payment is the

Fig. 6.1 France: provinces, *pays*, and places noted in the text. Sources: Droysens 1886; Muir and Philip 1927; Shepherd 1930; Kinder and Hilgermann 1974–78; Darby and Fullard 1978.

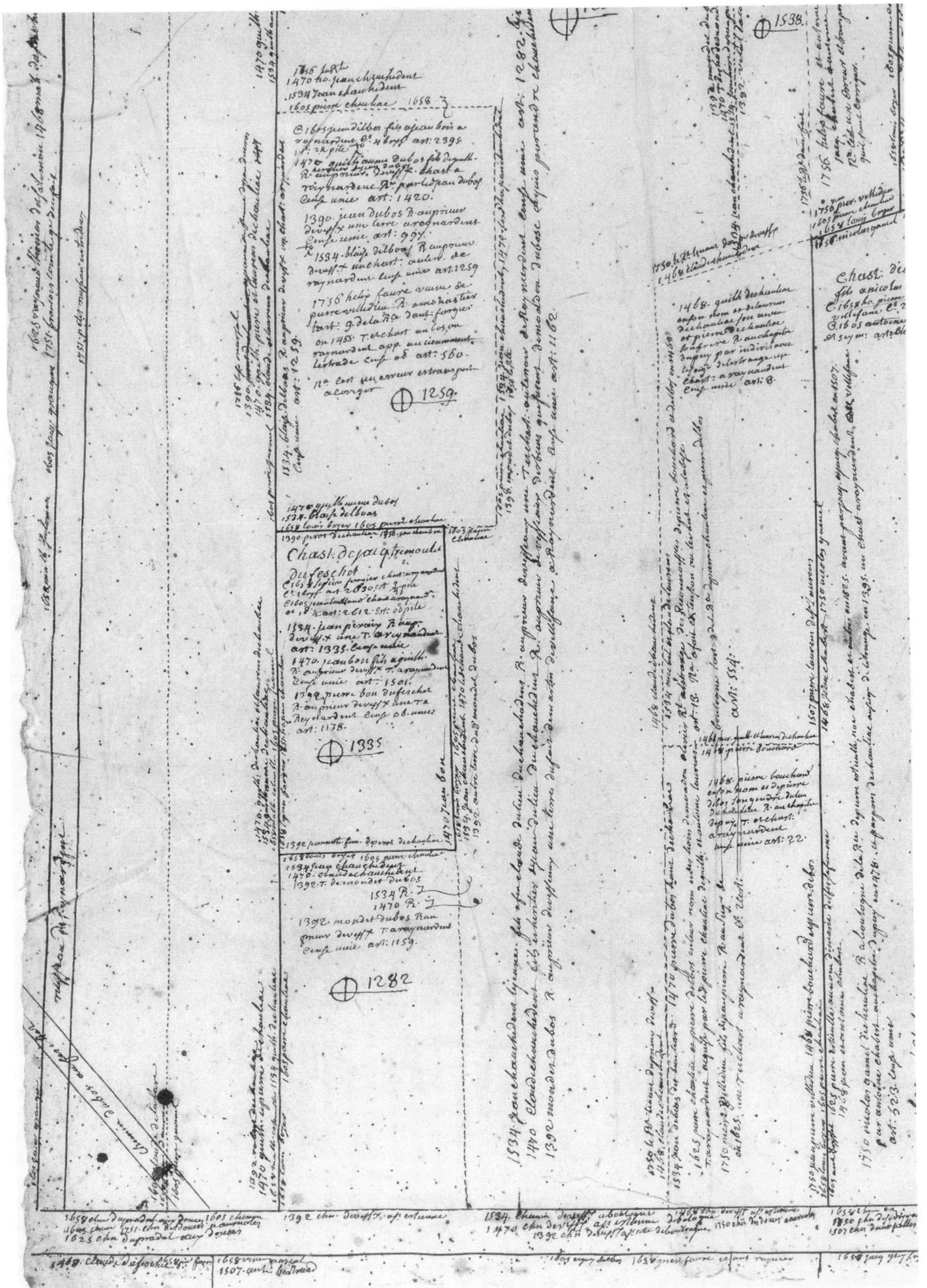

Fig. 6.2 "Plan de divers terroirs avec application de titres et compoix" of present-day Chaussadenches in the Vivarais. Source: Archives Départementales de l'Ardèche, 3E 187.

Fig. 6.3 "Plan des dîmes de Champeaux" (Seine et Marne), a fifteenth-century map indicating the division of tithe dues among a number of parishes. Source: Archives Nationales, Paris, L898 n. 52.

fifteenth-century "Plan des dîmes de Champeaux" (Seine et Marne). This was a multichurch village, and the plan indicates the division of tithe among the various parishes (figure 6.3). Such tithe plans are relatively rare today, not least because of destruction when tithe payment was abolished during the Revolution.

From the sixteenth century are found the first schematic maps of the boundaries of *seigneuries* and of *censives* (tracts of land held by individuals to whom the *cens,* the fundamental feudal rent which acknowledged subservience to the *seigneur,* was due).[15] From about 1650, true *plans parcellaires* are found accompanying written terriers and *censiers* (lists of those from whom the *cens* and other payments, such as tithe, were due). Guerout considers that their arrival can be explained largely by the fact that the compounding of several centuries of fragmentation of holdings as a result of inheritance practices eventually rendered the written terrier an inadequate instrument for checking on the rights and dues of land ownership. Examples of such maps are those c. 1670 of communities in the *censive* of St. Germain des Près in Paris and the Île-de-France.[16]

By the eighteenth century, when state agencies in France were actively considering mapped cadastral surveys as instruments for assessing and recording land taxes, *plans parcellaires seigneuriaux* became much more common as a result of two main processes. First, in the economic arena, what Guerot, following the argument of Marc Bloch, calls the *réaction féodale,* which encouraged the more rigorous management of estates.[17] A second factor is related to the wider scientific enlightenment, in which the precision of a map fitted more comfortably than an approximate, verbal description. The role of maps in this developing environment of exactitude was made explicit by E. de la Poix de Fréminville in his *La Pratique universelle pour la rénovation des terriers* (1752) when he posed the question: "Can one draw up a revised terrier without compiling a plan of the lands?" His answer was straightforward: "It is impossible!"[18] Similarly, in his *Instructions pour les seigneurs et leurs gens d'affaires* (1770), the Toulouse lawyer Joseph Rousselle wrote that unequivocal property descriptions required a survey and map.[19] Furthermore, though there was no equivalent of the eighteenth-century English enclosure movement in France, in a number of communes enclosures and reallocation of properties were effected which, in Jean Guerout's opinion, "quite naturally resulted in plans representing the new state of affairs."[20] Among these may be counted the maps produced in association with eighteenth-century *remembrements* (enclosures and land redistributions) in Lorraine.[21]

Lastly, the question may be posed as to how many individual landed estates had been privately mapped by 1789. Babeuf in his treatise of 1789 advocating a national property cadaster estimated that about two-thirds of *seigneuries* in the whole of France had estate plans by this date.[22] As Bloch comments, Babeuf was exaggerating by a very considerable amount.[23] In those regions such as the Île-de-France, where estates were large and wealthy and attitudes sufficiently open to allow "rational" economic management, such a figure might at least approach the truth. In other areas, and the Midi especially, attitudes were very different and maps very few.

6.2 Forest Maps

One category of land that was of particular economic, military, and recreational importance to the aristocratic rulers of *ancien régime* France was the forest. Forest mapping shows that from the seventeenth century, government, like private owners, was becoming aware of the value of cadas-

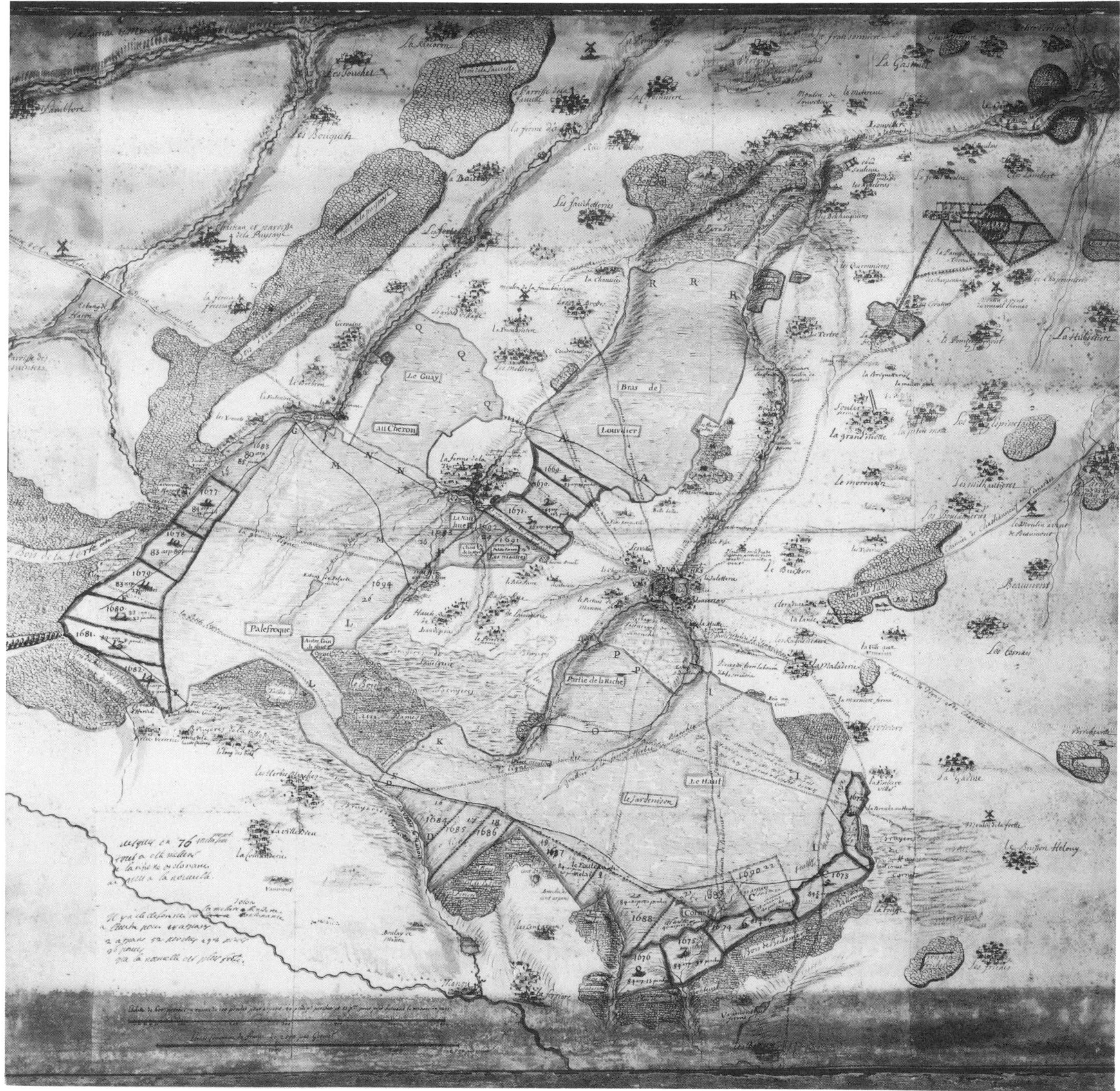

Fig. 6.4 "Plan de la Forêt de Senonches," late seventeenth century. Source: Archives Nationales, Paris, N II Eure-et-Loir 8 [No. 916].

tral mapping. A number of royal forests had already been mapped by the second half of the sixteenth century; maps of the Forêt de Chantilly and that of Nontron in Périgord dating from the end of the fifteenth century are among the earliest.[24] Many of these forest maps use color tints and symbols to distinguish timber and coppice trees, old trees, and young plantations to provide an inventory of the forest resource.[25] During the seventeenth century, woodland reserves were further diminished in France and their precise extent rendered uncertain by fires, military destruction, and illicit felling. When Colbert was given charge of Louis XIV's forest, he instituted in 1662–63 a wide-ranging program of forest reform to ensure there was enough timber for naval ship construction. One element of this was the compilation of a complete cartographic inventory of the royal forests of France. These maps are "an exact description of the complete extent of forests, specifying their area in *arpents,* and detailing their lines of subdivision, the nature of the trees with which each is planted—whether with timber or coppice—and noting their age and whether of strong or weak, stunted growth."[26]

The maps were used to regulate the felling and sale of timber; their continued use into the eighteenth century is attested by the many hundreds of tracings, reductions, copies, and recopies which were made and which are still extant.[27] The original plans might be counted in the thousands (figure 6.4).[28] Many, though, have been lost: for example, large-scale, seventeenth-century surveys at 1:10,000 of the royal forests of the Nord survive for only the forests of Rihoult and Tournehem.[29] Louis de Froidour and a team of thirty surveyors alone made 1,008 plans of 257,000 hectares of woods around Toulouse.[30] At the other end of the scale, the woodlands of the Île-de-France, Brie, Perche, and Picardie were accommodated on just 44 plans.[31]

6.3 Taxation "Cadasters" in France before the Eighteenth Century

Although state-sponsored taxation surveys accompanied by cadastral maps are a product of the eighteenth and nineteenth centuries in France, in a number of quasi-autonomous regions local written lists of dues were produced. Reference has already been made in section 6.1 to the *compoix* of Languedoc. In 1269 Louis IX (Saint Louis) ordered a survey of buildings to be recorded in a series of *livres d'estimes,* which served as the basis of taxation until 1491. In 1359 Charles V ordered a revision of these records in the Dauphiné province by surveyors entitled *sapiteurs.* Then in 1491 Charles VIII envisaged a project to conduct a general cadastral survey over the whole of the four regions of France—Languedoc, Languedoil, Outre-Seine, and Normandie—though in fact it failed to extend the use of written terriers beyond Languedoc.[32]

After some similar operations in the *généralités* (Crown administrative districts) of Toulouse, Montauban, Aix, and Bordeaux in execution of an edict of Francis I in 1535, in the Agenois in

1604, and in the Condomois in 1664, Colbert, then Louis XIV's minister of finances, again attempted to establish a cadaster over the whole kingdom in 1679, only for it to be abandoned in the face of fierce opposition from the nobility and the clergy, who had the most to lose from a taxation system based on a proper survey of property.[33] Such interests with their jealously guarded privileges of tax reductions and exemptions continued to oppose general taxation reform until they themselves were "reformed" by the Revolution.

What all of these early attempts at nationwide taxation reform underline is that some necessary political conditions had to be fulfilled before government could ever be successful in this area of policy. First, the political and administrative unification of the country had to be attained; second, the equality of each citizen in law had to be recognized; and third, equality of each individual in the realm of taxation had to be achieved. These political conditions were not attained until after the Revolution of 1789, and even then perhaps more in rhetoric than in reality. However, some advances were made before this in Savoy and the three *généralités* of Limoges, Riom, and Paris. These developments toward a full survey of France are examined in the following two sections.

6.4 The Savoy Cadaster, 1728–38

The first state-sponsored mapped cadaster on a commune-by-commune basis covering part of the territory of modern France is that of the Duchy of Savoy conducted between 1728 and 1738.[34] Written cadasters were being employed as taxation instruments in some communes in Savoy from the sixteenth century, particularly those mountain areas which at that time were richest and most advanced.[35] These documents resemble the *compoix* of prerevolutionary France and describe the boundaries and extent of parcels, their quality as established by the valuers, and the level of tax payable. The end of this system of assessing and recording tax liability in this part of western Europe was presaged by the adoption of mapped cadasters in the Kingdom of Piedmont, in Austrian Lombardy from 1718 (chapter 5.5), and then in the Duchy of Savoy from 9 April 1728.

Savoy was divided up into *départements* (administrative districts), corresponding in the main to natural regions, each headed by a *délégué* who controlled a group of surveyors and their assistants. The survey work was carried out by a relatively small number of surveyors and valuers, perhaps a hundred, most of whom had gained experience of such work in mountain areas in the Piedmont tax reforms (chapter 5.5).

Men who knew the limits of a parish were nominated *indicateurs*, and the community provided two *estimateurs* to value each parcel of land and assign it to one of three categories (good, average, or poor). To avoid systematic undervaluation, an *estimateur d'office* was recruited from a neighboring parish. Parcels were delimited in the presence of their proprietors or their repre-

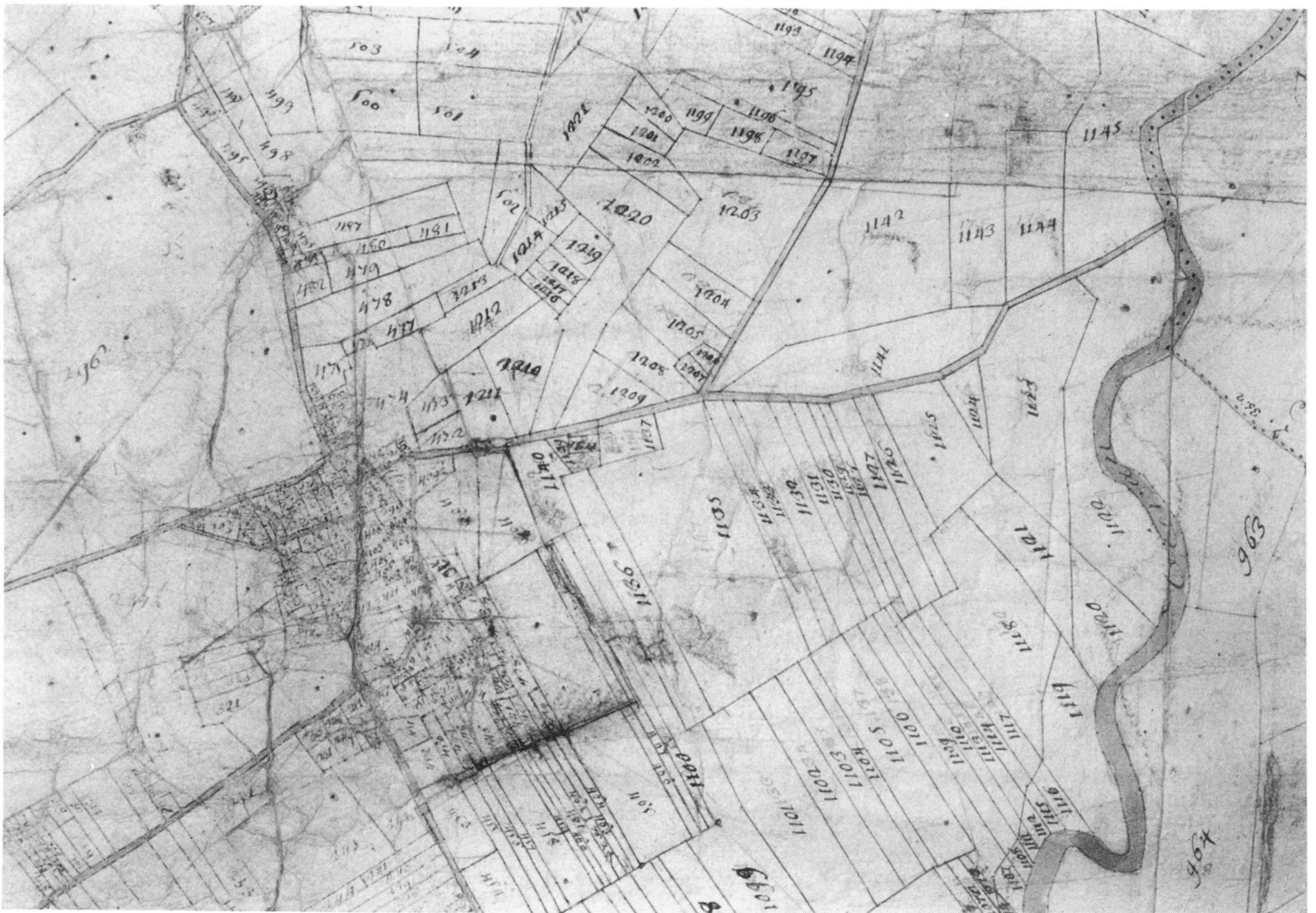

Fig. 6.5 Savoy cadaster. Part of the original map of the commune of Metz portraying the village and fields of the village of Fessy, 1730. Source: Archives Départementales de la Haute Savoie, *cadastre sarde* no. 253.

sentatives, who had to bring evidence of ownership and any rights to tax exemption. Thus the cadaster served three purposes for government administration: it provided a fair and rational basis to use for assessing tax liability, it provided a precise record of that liability, and it also served as a record of landownership. It was "the product of enlightened despotism . . . instituted primarily to identify the number of minor aristocrats who were avoiding dues, but also to extract fair taxes from all land owning peasants."[36]

Surveying and mapmaking moved ahead quickly and were effectively complete by 1733. As surveyors were liable to pay the costs of rectifying any discovered errors, there was a built-in

Numeros relatifs à la Mape.	Qualités des Piéces avec les Noms & Sur-Noms des Particuliers, par ordre Alphabetique.	Mas.	Degré de Bonté.	Mesure de Piémont. Journaux	Tables	Pieds	Mesure de Savoye. Journaux	Toises	Pieds	Estimati en Arg par Jour de Savo liv.	s.
	Bachet .... Jean Claude										
143	Champ . à Champagny		3	„	82	5	„ 1	25	„	19	6
702	Paturage aux vignes de metz		3	„	3	4	„	17	1	0	15
737	Vigne .... aud.		3	„	8	8	„	44	5	~~23~~	~~12~~
784	Broussailles .. aud.		3	1	2	5	1	128	1	0	3
974	Pré ... au metz ...		1	„	10	2	„	52	3	13	10
1011	Maison Cour .. aud.		2	„	4	4		22	3	10	10
1040	Champ ... au Burnon		1	1	19	11	1	218	3	27	0
	Bachet Jean ... Pierre										
144	Champs .. à Champagny		3	„	69	„	„	355	6	19	6
157	Champ .... aud.		3	.	51	8	.	266	3	19	6

Fig. 6.6 Savoy cadaster. First page of the *tabelle* for the commune of Metz, which lists proprietors and their landholdings in alphabetic order. Source: Archives Départementales de la Haute Savoie, Icd 1499.

Fig. 6.7 Savoy cadaster. Part of the copy map of approximately the same area of the commune of Metz, 1732, as in figure 6.5. Copy maps indicate land use by a combination of color tints and conventional symbols. Source: Archives Départementales de la Haute Savoie, *cadastre sarde* no. 253, copie.

mechanism to encourage accuracy. In each survey team a *géomètre-inspecteur* checked work before it was despatched to the office of the *délégué général* at Chambéry. Each draft map and accompanying table of land parcels and proprietors was deposited in its parish for fifteen days so that residual errors could be identified by landowners. Checking and rectification of draft cadasters occupied another five years until 15 September 1738.[37]

There is no indication of scale on the maps or in the instructions published for their specification. Measurements undertaken in the early nineteenth century, when the Napoleonic cadaster was being produced, reveal a uniform scale of 1:2,372 (cf. 1:3,000 for the Piedmont cadaster). The original maps, though carefully drawn, are quite plain but carry a date, orientation, and names of settlements (figure 6.5). Watercourses and roads are often tinted in blue and green, but there is no other detail in the parcels other than their boundaries and reference numbers corresponding to the *tabelle alphabétique définitive* (figure 6.6). This is a list of land parcels in alphabetic order of proprietors. Each parcel is identified by its map reference number and is described and named, its quality is assessed according to the three-category scale, and its area in both Savoy and Piedmont measure and its taxation assessment are recorded. In addition to the map and terrier (and drafts of these), the cadastral documents include the *journalier* and the *livre de transport,* which record post-1738 revisions. These contain changes in ownership through to 1860, when Savoy was incorporated into France and the French cadastral system, but with an interruption in the Revolution and First Empire period. While the 1738 map is no longer the basis for land taxation, the *tabelles* and the *livres* are still recognized as providing legal proof of landownership.[38]

The *cadastre sarde* of 1728–38 survives intact with the exception of one map for the commune of Gaillard near Annemasse. Most maps were made as three copies. The original map and a contemporary copy signed by the surveyors and the Crown's representative were deposited in the archives of Annecy and Chambéry. These copies are exquisitely drawn and use tints to distinguish the land use of each parcel (figure 6.7). Another copy was made at the same time for parish use and might still be found in the local archive. These copies have suffered from usage and, while the crossings-out and corrections may make them difficult to use as historical sources, they are a certain indication of the role that these maps played in the years after 1738 in matters of tax liability and rights and boundaries of properties.[39] Indeed, later whole or partial copies of the original maps were made for such purposes at the expense of local inhabitants.

The *cadastre sarde* was of importance in a context much wider than that of the tiny Duchy of Savoy. The imagination and effort expended in the cadastral survey of Savoy was much admired by contemporaries, especially by Physiocrats and political writers in France.[40]

6.5 *Plans d'Intendance* in the *Généralités* of Limoges, Riom (Auvergne), and Paris

In the second half of the eighteenth century, in three separately governed *généralités* of the part of France characterized by liability to personal taxes, attempts were made to convert these to taxes related to landed property (cf. the *compoix* of land-tax France, namely, Provence and Languedoc). The role that maps played in each of these regions is different, but between them they provided models for almost all the components of the later revolutionary and Napoleonic cadasters. They are known collectively as *plans d'intendance* for reasons of their advocacy by the *intendants* (Crown representatives) who governed them. Each system was based on a detailed survey of land extent and quality to determine tax liability; large-scale maps were instruments which served as a record of that liability for all time.

Why did these eighteenth-century initiatives at mapped cadasters take place in *généralités* where tax was personal rather than, as might be expected, in those provinces where taxation was already related to land? First, in the two remote *généralités* of Limousin and Auvergne the *intendants,* Turgot and Trudaine, were at the forefront of the scientific enlightenment in France. The former was a powerful advocate of employing scientific methods in government administration and a sympathizer with Physiocratic concepts. Turgot accepted the primacy of the land tax and the need to spread its burden more widely and fairly through French society, having written his *Réflexions sur la formation et la distribution des richesses* in 1766. Second, the *généralité* of Paris was the one most thoroughly infused with modernizing seigneurial management and with Physiocratic mores; it was also where population and landownership change rendered the unreformed system perhaps the most inequitable of the whole country. Third, the personal tax yielded so little by the eighteenth century. Administrators and political activists alike campaigned for its replacement by a new tax founded on a "general cadaster" of the whole of France. One such proposal, albeit for a simple estimate and written survey of possessions, was published in 1763. Other schemes, while not addressing the crucial difficulty of how to abolish the manifold exemptions from personal taxation enjoyed by some classes, attempted at least to rationalize the basis for collection from those who were liable to pay. Some communities went ahead and surveyed and valued the resources. They included the landholdings of each taxpayer and made the level of tax properly proportional to real wealth, so that in some parishes by the eighteenth century there was what became known as *taille proportionelle* or *taille tarifée.*[41]

Attempts to legislate even for these limited commutations of personal taxation for the whole of France, however, failed in the *ancien régime.* In a country still far from fully unified politically, reforms by "general law" encountered too many obstacles. Enclosure reformers had demonstrated that it was politically more expedient to restrict legislation to particular areas or regions; this was the only practicable way forward for the tax reformers as well with the semiautonomous *généralités* as the appropriate units. The establishment of a proportional personal-wealth tax did not of itself require the production of cadastral maps, but for Fougères "their use was a natural consequence."[42] This process in the three *généralités* of Limoges, Riom, and Paris is charted briefly below.

Limoges

The first attempts to establish a *taille proportionelle* in the *généralité* of Limoges were made between 1730 and 1743 when Tourny was *intendant.* His first scheme was to allow taxpayers to declare the value of their possessions, including each parcel of land, using a generalized (by comparison with the sophisticated *compoix* of Languedoc) two- or threefold classification of value depending on yields. It has the typical failings of a voluntary system, as taxpayers submitted underestimated returns. Tourny was prepared for this problem and had always planned to use it to push through a more radical reform. To remove the inequalities occasioned by purposeful underestimation, communities were required to pay for an independent expert, a surveyor, to measure and value their land. By about 1779, some three-quarters of the parishes in this province had been surveyed but not, at this stage, mapped. Maps were not seen as an integral part of this exercise. They were introduced under the *intendance* of Anne-Robert Jacques Turgot by the Limoges administrator, Pierre Cornuau, who was a trained surveyor and mapmaker who had earlier worked on the Cassini project to compile a topographical map of the whole of France. One parish, Sainte-Claire de Soubevas, was used by Cornuau to illustrate the potential of a map-recorded cadaster. Even if a generous allowance is made for subsequent losses, it is clear that only a relatively small number of maps were made. Cornuau's plan was always running up against opposition, both from within the administration as a result of personal rivalries, and from wealthier individuals who had the most to lose from a truly proportional wealth tax.[43] When Turgot moved on from the Limoges, the project was deprived of one of its strongest supporters.

Riom

A *taille tarifée* was introduced into this part of the Auvergne in 1733 by *intendant* Daniel Trudaine. Initially, as in the Limousin, this was based on declarations of proprietors, but in the 1740s in the Haute-Auvergne, the properties of some proprietors were submitted to detailed survey. According to Fougères, the basis of their selection is not clear, but it led quite quickly to surveys being widely adopted.[44] A difference with Limousin was that the cadaster was recorded in a terrier where parcels were gathered together by name of owner (*matrices cadastrales*) rather than by the topographical ordering (*états de sections*) of Limousin. As in Limousin, detailed surveying of properties was opposed by the wealthy, who had for a long time been accustomed to paying much less than an equitable proportion. It was not unusual to see them threatening or even setting upon surveyors in the field.[45] In contrast, the less well off, comparatively numerous in this poor upland area, pressed hard for a general cadastral survey of the whole Auvergne and in 1786 twice petitioned the national government in Paris to this end. As in Limoges, no doubt some plans have been lost, but from the small numbers which survive it seems safe to say that in many communes no plan was envisaged by the surveyor on which to record the *taille tarifée*. Conversely, some of the accompanying maps are much more than just a record of field boundaries, since they use color tints to distinguish land use.

The popularity of the Auvergne surveys with peasant proprietors was not without its influ-

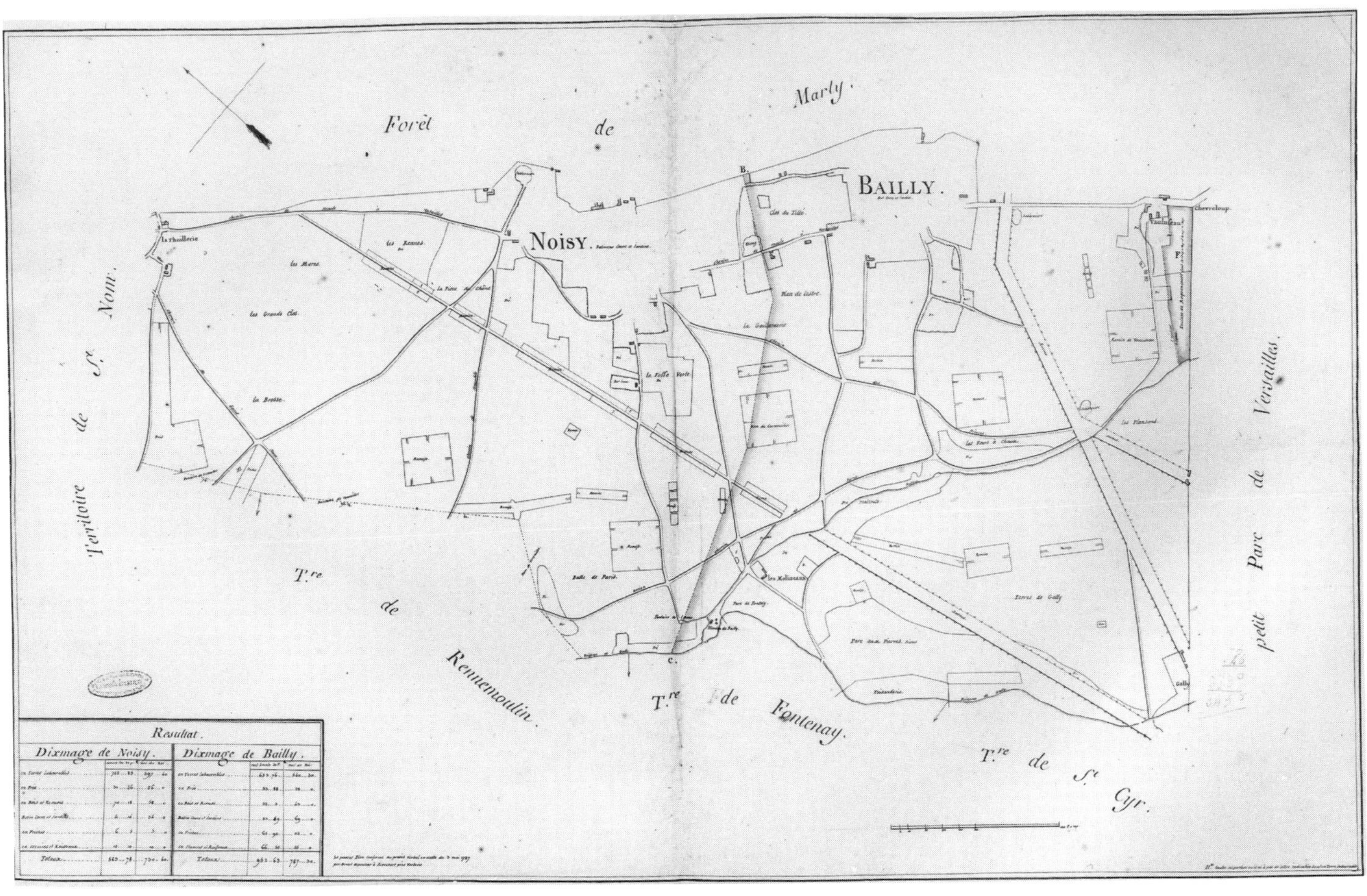

Fig. 6.8 "Plan d'intendance de Noisy-le-Roi," by Devert, 1787. Source: Archives des Yvelines et de l'Ancien Département de Seine-et-Oise, C97/44[750016].

ence on revolutionary policies. The law of 28 August 1791 gave communes the right, at their own expense, to commission a plan of their lands, more particularly a *plan parcellaire* after 23 September of that year. In France as a whole this option was taken up by only a small number of communities, though in the Auvergne itself the numbers were not inconsiderable. Fournier's list of twenty-two communes in Auvergne is itself incomplete.[46]

Paris

In the *généralité* of Paris a *taille tarifée* was not introduced until 1776, following its promotion by *intendant* Berthier de Sauvigny. An important difference by comparison with Limoges and

Auvergne was that plans were not optional but were a required part of the cadaster. They are not, however, *plans parcellaires* but rather land-use plans with states of cultivation indicated by color to provide not a record of the liability of individual parcels but a means of calculating tax based on land use (figure 6.8). There is a direct link between this survey and the *plans par masses de cultures* of the first Napoleonic cadaster.[47]

6.6 Cadastral Mapping and Resource Evaluation in Corsica, 1770–96

At the same time as the government was deliberating tax reform by cadastral survey in revolutionary and postrevolutionary France, the newly acquired island of Corsica was mapped. This provided both a firsthand model to which reference could be made and a training ground for cadastral surveyors, not least for Jean-François Henry de Richeprey, the so-called father of the French cadastral survey.

After a series of military coups d'état, invasions, and counterinvasions from the end of Genoese rule in Corsica in 1729, the island was finally purchased by France under the terms of the Treaty of Versailles on 15 December 1768 and formally annexed on 15 August 1769 (coincidentally the date of birth of Napoleon Bonaparte in Ajaccio).[48] The following year a royal edict set in train a cadastral survey of the island as a means to evaluate its land resources and to serve as a basis for the economic development of the island and especially its agriculture toward the end of the Physiocratic era.[49]

The question may be posed as to why this particular instance of territorial aggrandizement was accompanied by a cadastral survey of the newly acquired lands. Part of the answer has to do with the Corsican practice of partible inheritance, which had produced an extremely fragmented landownership structure. In the words of de Bedigis, one of the directors of the survey, taken from a memoir dated 1 July 1772, "Continued equal division of properties amongst children means that there will come a time when Corsica's land holdings are so infinitely small that the boundaries between each, if marked out on the ground, would cover the entire surface!"[50] This extreme point had not been attained by the eighteenth century, but property ownership had become so obscure that in practice dues derived from many estates that had long been subdivided out of existence. Little had been achieved by way of rectification of this problem under Genoese rule, but from 1770 the French government was much occupied with the question. An anonymous prospectus conserved in the Archives Nationales indicates that the "Plan Terrier" of Corsica was to be compiled to satisfy two main objects. First, it was to establish the actual situation of landownership and in particular to define that land owned by the Crown, that held in common by particular communities, and that in private ownership. A second aim was to compile a detailed

Fig. 6.9 *Plan terrier* of Corsica. Part of the map of Communauté de Castellare. Source: Archives Départementales de la Corse du Sud.

inventory of the economy and resources of the island. Maps were required to ensure that "the government would know the whole island as well as an individual does his own domain, and bringing everything together in one view on a map would facilitate effective management."[51] Indeed the *Instruction sur la levée des plans* published in 1771 begins by defining a *plan terrier* as "an exact image of the land for all conceivable uses."[52]

The cadastral maps were a marked technical advance over those of Savoy, for example, as they were constructed within a triangulation of the whole island. By 1783, the two engineer-directors of the survey, Testevuide and Bédigis, had covered Corsica with a network of 91 major

✣ I. ✣

DÉPARTEMENT DE CORSE

DISTRICT de Bastia | CANTON de Casinca

COMMUNAUTÉ de Porri

TERRIER GÉNÉRAL

TOPOGRAPHIE | EXPLICATION

DESCRIPTION

DÉTAILLÉE DE LA COMMUNAUTÉ

de Porri

DANS L'ORDRE TOPOGRAPHIQUE

L'ORDRE DES MATIÈRES

ET L'ÉXÉCUTION DE L'OUVRAGE

M. DCC. XCIIIII.

L' AN quatrieme DE LA RÉPUBLIQUE.

NUMÉRO Du District	NUMÉRO de la Communauté	NUMÉRO Des Rouleaux de Plans
1.er	67.	11.

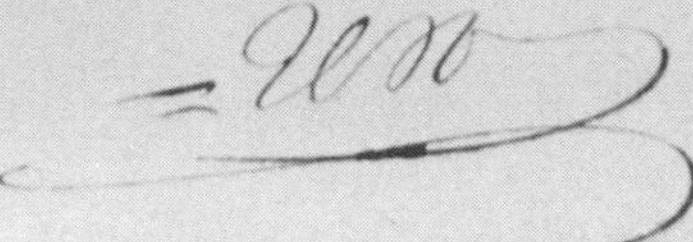

Fig. 6.10 *Plan terrier* of Corsica. The *terrier général* for the Communauté de Porri. Source: Archives Départementales de la Corse du Sud.

triangles and 386 secondary, or cross-checking, triangles. An exceptionally thorough system of checking the accuracy of completed maps was also employed. Both the triangulation and the topographical content were checked on the ground. In the office, measures recorded in the terrier were examined against those calculated from the maps. As Antoine Albitreccia comments: "These instructions for checking, 122 manuscript pages in length, are a model of their type. Through them we can perceive the intentions of the survey directors, know something of their method, appreciate their high ideals and understand the motives which guided their work."[53]

Testevuide's and Bédegis's contract with the French government required them to provide two sets of cadastral plans not later than June 1794 in bands of two miles width at 1:10,800 and descriptions of each commune on the island (figure 6.9). For this part of the job they had a team of draftsmen and clerks.[54] The maps consist of thirty-seven rolls, and each commune has a file containing topographical and statistical descriptions (figure 6.10). From these were compiled summary evaluations for the two *départements* which made up the island. Work was effectively completed by 1796 but was little used until 1810, when Napoleon ordered the reduction of the cadastral surveys to 1:86,400, the scale of the Cassini topographical map of France. The scale was eventually decreased to 1:100,000, and the map published by the Dépot de la Guerre in 1824.[55]

6.7 Jean-François Henry de Richeprey: "Father of the French Cadaster"

The *plans d'intendance* illustrate the ways in which some local administrators were thinking about and using maps as part of taxation reform programs in the decades immediately prior to the Revolution. Each of the three main schemes contains elements recognizable in the nineteenth-century general cadaster of France. One man who put together all the elements of this system during the eighteenth century and so has been dubbed the father of the French cadaster was Jean-François Henry de Richeprey, a political economist, advocate of social reform, and prolific topographical writer. In his youth he had worked on the *plan parcellaire* of Corsica and the Milanese cadaster (section 5.4). In the 1780s he was employed as one of a team in the Haute-Guyenne brought together to direct fiscal reform in the province.[56] By that time there was nothing novel in France about using mapped surveys as a means of evaluating and/or recording land-tax liabilities. What de Richeprey contributed was the development of a rigorous and definitive formula. The general cadaster of France undertaken from 1807 is identical with that proposed by de Richeprey. The method of production and the scale of the *plans parcellaires* are identical, the *Recueil méthodique des lois, décrets, règlements, instructions, et décisions sur le cadastre de la France* of 1811 reproduces word for word the *Projet de règlement pour les ingénieurs-géomètres* of de Richeprey published in 1782.[57] However, before this elaborate system was adopted for the whole of France an attempt was made to effect the reform using maps more generalized, simpler, and cheaper than *plans parcellaires*. These were the *plans par masses de cultures* reviewed in the next section.

6.8 First Napoleonic Cadastral Survey with *Plans par Masses de Cultures*

The closing years of the *ancien régime* in France saw both practical examples of tax reform by cadastral survey and a plethora of didactic works advocating wholesale reform, among which can be counted books by Munier, Dutillet de Villars, and Lamy, and that by de Richeprey discussed in the previous section.[58] Lamy's ideas were successfully tested in the commune of Hornoy near Amiens. P. F. Aubry-Dubrochet in his tract *L'Exécution du cadastre général de la France* . . . expressed his conviction of the need of a map-based cadaster and included a model of the type of map that he envisaged based on his local parish of Villers (figure 6.11).[59] Babeuf in his *Projet de cadastre perpétuel* of 1789 was even more radical in proposing a redistribution of properties into equal-sized lots, so radical in fact that in the opinion of Herbin and Pebereau it set real reform back by several years.[60] Though proposals differed in detail, everyone agreed that a general cadaster was the only practicable solution to the question of inequality of taxation burdens which had festered for so long. The law of 1 December 1790 abolished all the old taxes and replaced them by a single property tax which had to be divided equally among all properties on the basis of net productivity, with but few exceptions required in the interests of agricultural development.[61]

Municipal councils were first required to draw up a list of the various parts of their administrative territories and the properties which they comprised. Landowners or their tenant farmers then either declared the type and extent of their properties or had them valued. Council officers then assessed the net revenues of each property and drafted the *matrice de rôle* (cadastral registers). Plans were not required by law; an instruction of 23 November 1790 said only that commissioners could make use of existing cadastral maps together with any other maps, terriers, or documents which they could obtain.[62]

This law thus set down the first foundations of a national cadaster. Gaspard-François de Prony, director of the Ecole des Ponts et Chaussées, was appointed director of the cadastral department in 1791. As he conceived it, the cadaster was "a highly centralized, rigorously scientific operation designed to permit the piecing together of maps of local areas and comparisons between them."[63] There were thus to be two distinct components to the compilation of the cadaster as Prony saw it: first, the geodetic survey; second, the local field surveys to assess land valuations. In terms of precedent, this is clearly closer in spirit and purpose to the survey of Corsica than to the *cadastre sarde* or the *plans d'intendance*. Mounting costs, reducing budgets, political uncertain-

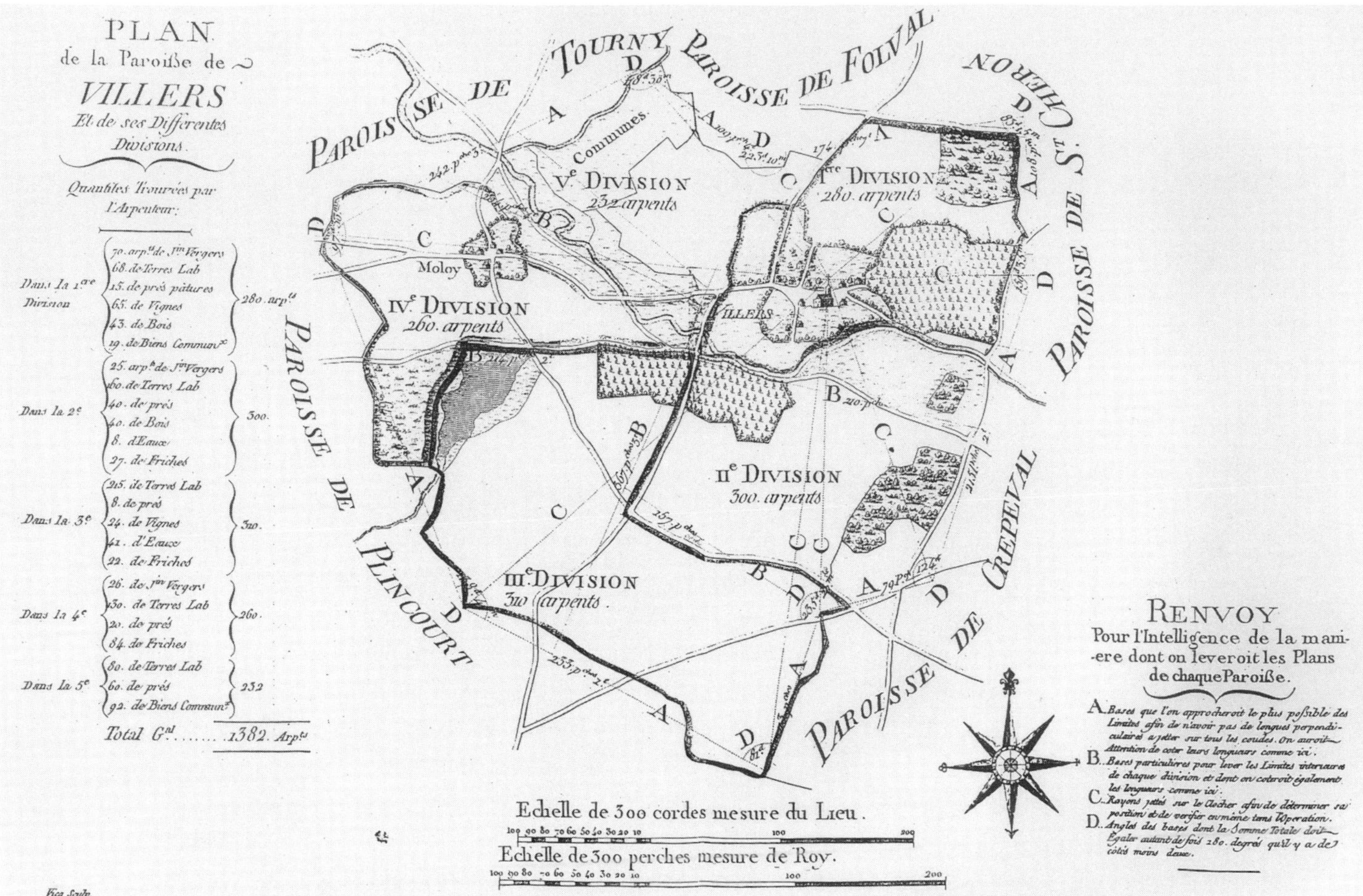

Fig. 6.11 "Plan de la paroisse de Villers et de ses différentes divisions," which accompanies P. F. Aubry-Dubrochet's *L'Exécution du cadastre général de la France* (1790).

ties, and delays in reforming systems of measurement and training personnel meant that little progress toward taxation reform, for which the cadaster was to be an instrument, was effected in the 1790s. Moreover, although *ancien régime* inequalities in rates of tax were so extreme, there was much dispute about what was to succeed them. Difficulties in determining individual property revenues produced a whole host of complaints, claims, and counterclaims.

A centralized panel of official commissioners was set up in 1797 to try to resolve the continuing problem of establishing a fair apportionment of taxes. This took direct responsibility for reorganizing the compilation of the *matrices* under a law of 22 January 1801 which ordered land-

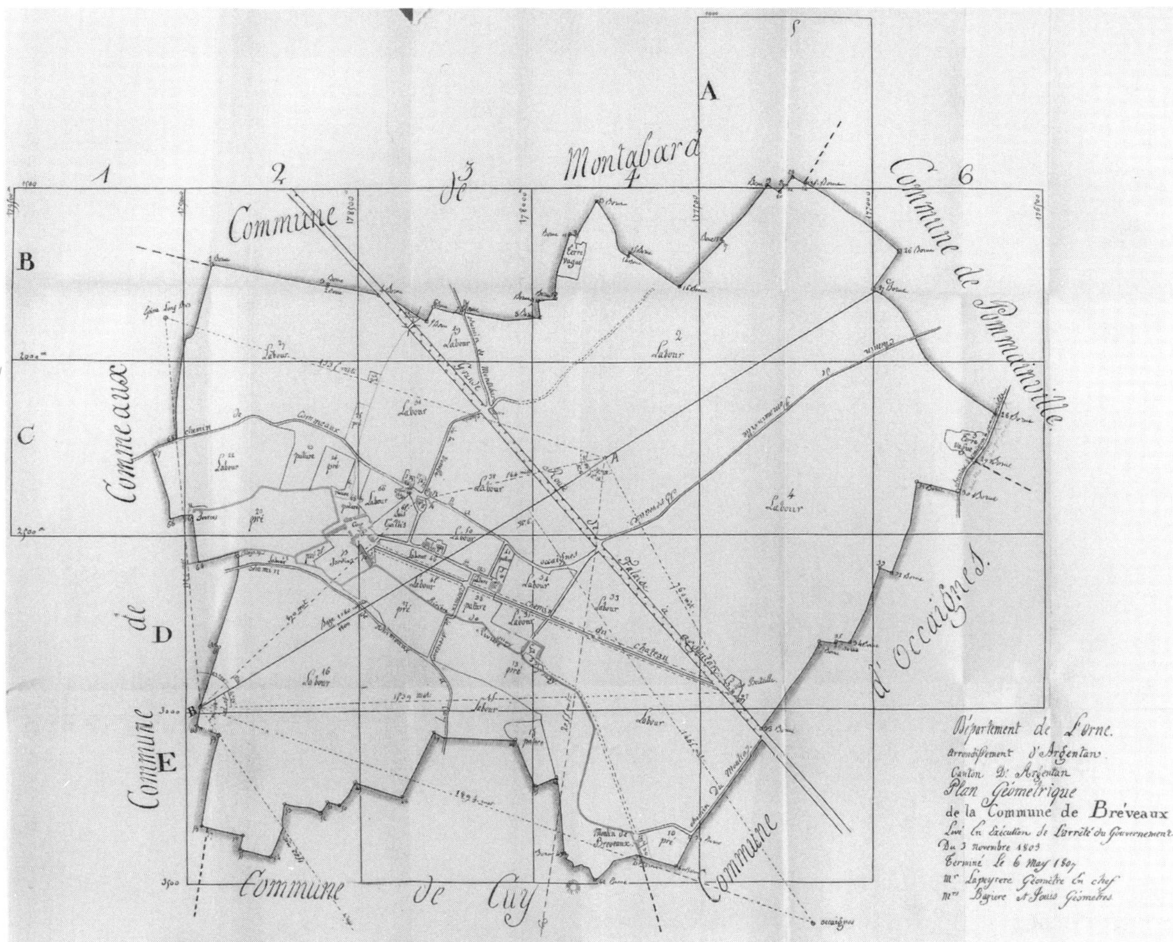

Fig. 6.12 "Plan géometrique de la commune de Brévaux, Département de l'Orne . . . levé en exécution de l'arrêté du gouvernement, du 3 Novembre 1803." Source: Archives Départementales de l'Orne.

owners to declare their incomes. As always, however, landowners underestimated their properties. The organizing authorities were convinced of the need to survey the whole country to determine the contents and income of each property, but they were daunted by the cost both in money and time of an operation such as Prony had envisaged. Cheaper and quicker means were explored—in particular, the idea of a *cadastre général par masses de cultures.* This was tried in 1,915 communes by a government decree of 3 November 1802.[64] For each of these places, a map at 1:5,000 was produced on which were plotted from two to eight categories of land use, such as arable, grass, vines, and chestnuts, distinguished by color tints (figure 6.12). The area of each land-use zone was calculated from the map, as was the total extent of the whole commune. Proprietors then declared the content of the fields and land parcels they owned in each of the land-use zones. Comparison between the total declared area and that calculated from the maps produced a weighting factor which could be applied to increase declared areas in other, unmapped communes in each *arrondissement* (administrative district), thereby establishing the total tax liability for apportionment. The problem was that, as the maps did not show field boundaries, it was not possible to take any account of variations in yields, and therefore revenue, between fields in the same land-use zone. Complaints came from all sides—the government, municipal councils, mayors of communes, and landowners—all agreeing that a parcel-by-parcel survey was the only practicable solution. From October 1805 assessments were in fact made on this basis, although the maps continued to be produced delimiting the land-use zones only. The land parcel was clearly the most appropriate unit for the cadaster, as it was occupied by one individual and usually was devoted to one type of production over its whole extent.[65]

6.9 The *Cadastre Parcellaire* of 1807

In 1807 Napoleon said to Mollien, his finance minister: "Half measures are always a waste of time and money. The only way forward is to survey all the land in all the communes of the empire, property-by-property. This *cadastre parcellaire* will complement my legal *code* in matters of landownership. It is imperative that the plans be sufficiently accurate and complete to confirm property boundaries and to suppress litigation."[66]

Enshrined in this manifesto are the two components common to other tax-based cadastral surveys in nineteenth-century Europe: that a parcel-by-parcel survey will provide a fair method of apportioning global amounts, and that large-scale maps with field boundary and ownership data will provide a permanent record of that apportionment and so prevent future litigation.

Such a system was set in progress in France by the budgetary law of 15 September 1807. Its preamble states that its objects are:

> to survey an area of more than 7,901 square *myriamètres* [= 790,100 km^2] containing more than a hundred million land parcels . . . ; to compile plans for each commune on which these 100 million parcels will be delimited, to classify each according to soil fertility, to assess the production which each might bear, to aggregate under the name of each proprietor all his scattered parcels, to determine by the aggregate of their productivity his total revenue, and to make this total revenue the basis of his land tax liability.[67]

A commission met on 7 November of that year in the finance ministry to work out the principles and practice of the mapping. It was chaired by Delambre, permanent secretary of the

Académie des Sciences. They worked out a set of regulations which were approved on 27 January 1808.

France was divided into twelve regions, each headed by an inspector-general of the cadaster charged with supervising operations in the constituent *départements.* By 1814, some 9,000 communes containing 12 million hectares and about 37 million individual land parcels had been surveyed. The political events of 1814–15 which brought the First Empire to an end held up the process, but mapping continued after the fall of Napoleon at a rate of between 300 to 400 communes a year until 1821. A law passed in July of that year removed the cadaster from state responsibility and made it a *département*/commune operation, with the national government retaining a general supervisory role. From the point of view of the maps, the most important change was that the supervising engineers who previously had checked the *cadastre* were replaced by *géomètres en chef* nominated by the prefect, so that checking in effect passed from the hands of expert government agents to those of valuers chosen from among the landed proprietors themselves.[68] Nevertheless, after 1822 progress did accelerate and was particularly quick from 1826 to 1840, during which period more than 21,600 communes, almost two-thirds of the country, were surveyed and mapped. By 1838, work was complete in eighteen *départements,* and by about 1850, almost the whole of continental France had been surveyed and mapped.[69] The exceptions were Savoy and the Comté de Nice, which were not formally annexed until 1860.

The *cadastre* of a commune consists essentially of three related documents. First there is the *atlas cadastral* or *plan cadastral* at a scale varying from 1:5,000 to 1:500 depending on the fragmentation of fields and holdings (figure 6.13). Second, there is a register known as the *état de sections* which lists all the parcels in order of their reference numbers on the plan. Third, a complementary register, the *matrice cadastrale,* groups the same information by landowners arranged in alphabetic order. All the laws, decrees, rules, and instructions relating to the compilation of the general cadaster were published in 1811.[70] This document, which runs to no fewer than 1,444 clauses, has been translated into several languages, and many of its components are enshrined in the cadastral legislation of other countries.[71]

A fundamental problem with this *ancien cadastre* as a practical instrument for assessing and apportioning land tax was the fact that only the accompanying written *matrices cadastrales,* not the maps or the *états de sections,* were annually updated to take account of alterations to properties. Through the nineteenth century, general developments in the rural economy, the coming of railways and other new forms of communication, together with urban and industrial encroachments into the countryside radically transformed the pattern of landholding in some places, so that the value of the maps decreased with each passing year. The *cadastre* was not finished before it was necessary to think of bringing the maps up to date.

A number of projects for updating were considered in 1828, 1830, and 1836, and special

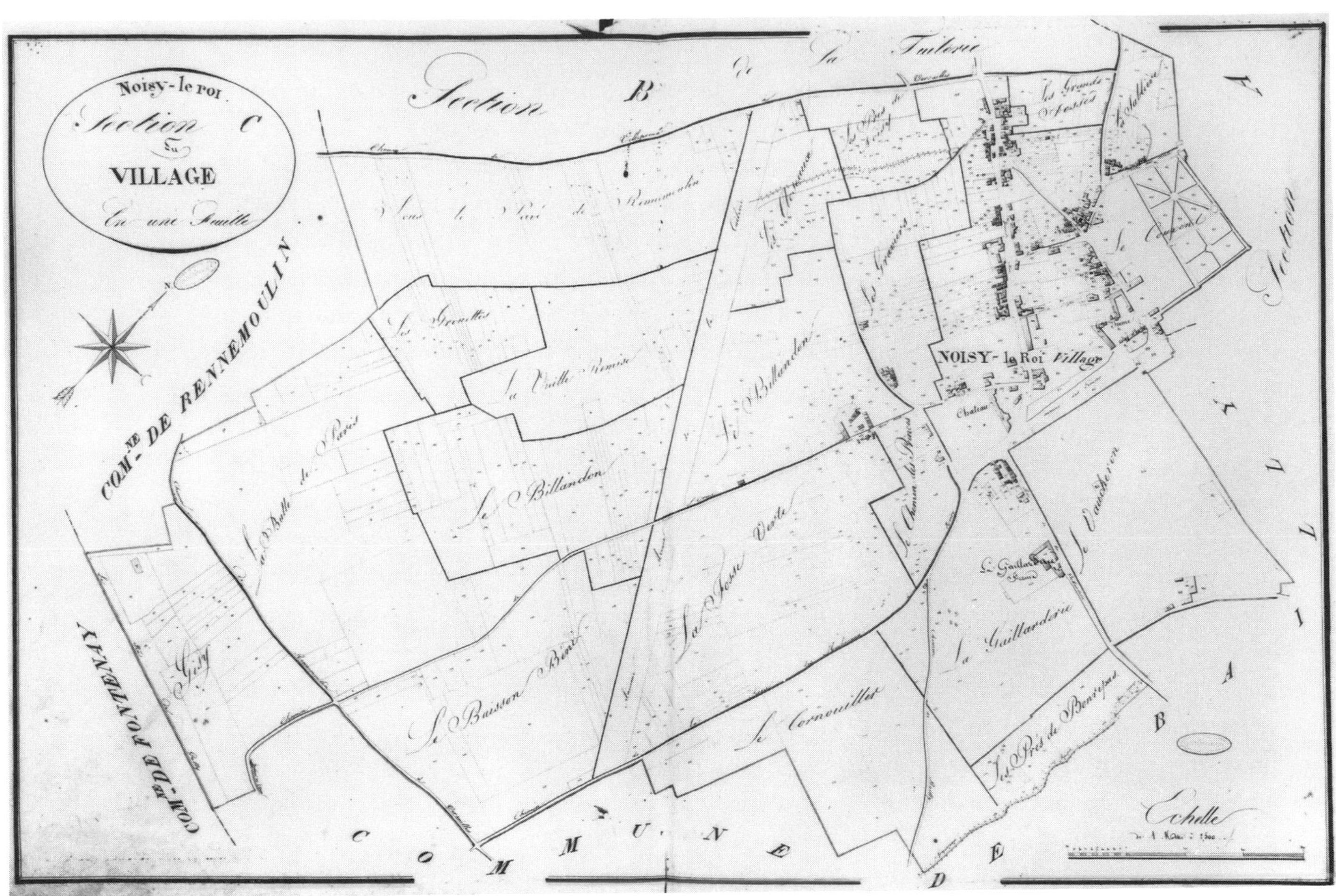

Fig. 6.13 Cadastral map of Noisy-le-Roi, by Donnet, 1819. Section C depicts the village area. Source: Archives des Yvelines et de l'Ancien Département de Seine-et-Oise, 3P3 Noisy le Roi, Plan cadastral.

commissions were set up in 1837 and 1846 charged with investigating means and methods. The consensus of proposals was that annual changes be recorded not only in the *matrices cadastrales* but also on the *plans parcellaires* and in the *états de sections*. By annual revisions of the plans, the *cadastre* would have become an effective register of title to land as well as a fiscal instrument. Tests were carried out in a number of communes in thirteen *départements* to confirm the practicality of proposed methods.[72] Various pieces of enabling legislation were enacted, for example, in 1850 and 1898, to permit communes to undertake a resurvey, but the costs of so doing meant that little was done in the nineteenth century other than in *départements* such as the Nord, where urban and industrial developments had produced profound field and ownership change. The *cadastre révisé. cadastre refait,* and *cadastre rénové* are essentially twentieth-century matters.[73]

6.10 Compiling the *Ancien Cadastre*

For the three *départements* of Calvados, Orne, and Maine-et-Loire, there are excellent reconstructions of the process of compiling the nineteenth-century cadasters, the local organization, surveyors, and progress through the revolutionary, postrevolutionary, empire, and restoration periods.[74] The first task was to identify the lands belonging to each commune and to record these in a document known as a *procès-verbal de délimitation,* attached to which is a sketch map of the boundaries. This work was undertaken by a surveyor in conjunction with the mayors of contiguous communes assisted by a number of local men drawn from the body of substantial landowners. The whole commune was then divided into a number of *sections* (subareas) to help locate particular properties in the documents and also to avoid an overlong series of reference numbers to relate parcels on the map with associated written documents. Boundaries of the *sections* followed natural features as far as possible and usually extended for between two hundred and four hundred hectares with the settlement itself being accorded a separate *section. Sections* are identified by a letter and a name beginning with "A" in the north and proceeding in an eastward spiral ending with the central *section* of the commune.

The *Recueil méthodique* of 1811 specified the triangulation of each commune, not only to ensure accuracy, but also to provide a means of checking the completed work. During the period 1808 to 1827, not all maps were constructed according to these strict triangulation instructions because of lack of experience or instruments. Some networks were supposedly fitted in after the plan was done. From 1827 a specialist surveyor was entrusted with the task of establishing the cadastral networks of each *département.*[75] Nevertheless, the triangulation of the *ancien cadastre* possesses the critical fault common to all such surveys based on the use of a separate triangulation for each map, namely, that individual maps cannot be joined accurately to form a general map. Some *géomètres en chef* took it upon themselves to carry out local triangulations or even, as in the Puy-de-Dôme, to proceed with a *département*-wide triangulation.

Once the principal features of a commune had been fixed by triangulation, surveyors then proceeded to fill in the detail of parcel boundaries, establish the owners of each, and record the state of cultivation—for example, as arable, meadow, or wood. A land "parcel" was defined for these purposes as a piece of land with no internal subdivisions and forming a single ownership unit.

Draft plans were tested by the *géomètres en chef.* Generally, if differences between ground and map exceeded 1 in 200, the plan was considered defective and rejected. Up to 1837, maps were drawn at scales of 1:5,000, 1:2,500, and 1:1,250; from 1837, the scales were 1:4,000, 1:2,000, 1:1,000, and, in exceptional instances, 1:500. An index plan at 1:5,000, 1:10,000, or 1:20,000 indicates the various topographical sections and cadastral map sheets which cover a commune. In addition, the major lines of communication, watercourses, and principal settlements are indicated on these index plans. Various colors are used to indicate these features but the methods were not standardized.[76]

Areas of parcels were calculated from the plans after the testing, and each proprietor was given a schedule of parcels entered under his name. Disputed parcels were remeasured, and proprietors had the right to request a complete resurvey at their own expense if the first plan was found accurate, but at the expense of the original surveyor if his work was proved defective on resurvey. As Anthony Lewison comments in his meticulous reconstruction of the cadastration of the commune of Cipières in the Alpes-Maritime, 1831–42, justice had not only to be done but also to be seen to be done. Participation of peasant proprietors in the process was explicitly required by the 1811 *Recueil méthodique*.[77]

Once the plan, list of proprietors, and extent of parcels were agreed to the satisfaction of all parties, the land-tax registers were compiled, and the proportion of global land tax apportioned to each parcel was calculated by reference to its area and the tax per hectare which applied to that particular type of land. As noted in section 6.9, this taxation register is presented in two forms: as an *état de sections,* with parcels arranged according to the topographical subdivisions or *sections* of the commune, and as a *matrice cadastrale,* in which land parcels are listed by alphabetic order of landowners. Two copies of the map, *état de sections,* and *matrices cadastrales* were produced, one for the commune and one for the land-tax office.

The *Recueil méthodique* envisaged regular updating of the cadastral registers to take account of changes in landownership. These annual revisions applied only to the registers; the maps remained unaltered because of the magnitude and cost of the work this would have entailed.[78]

6.11 Value of the *Ancien Cadastre*

To some contemporary administrators and to at least one modern historian, the 1807 *cadastre* is a seriously flawed survey: individual maps were constructed without reference to regional or national triangulation. For example, Gaspard-François de Prony, director of the cadastral mapping department in the 1790s, was later charged with evaluating a proposal from *ingénieur-géographe* Laprade that the individual commune maps of the *cadastre parcellaire* could in fact be copied and reduced to obtain at minimal cost a general map of France that would, in terms of both scale and topographical content, be better suited to the needs of civil and military engineers than the existing Cassini maps. This way of obtaining a general map was considered by Prony's inquiry to be no more possible in early nineteenth-century France than it was in mid-nineteenth-century England, when the Ordnance Survey was evaluating the possibility of constructing a general map of England by joining together parish tithe maps (chapter 7.3). In France, as in England, the new general map was produced not from existing cadastral surveys but by the army, which established a new geodetic grid between 1818 and 1827 and then mapped France at 1:80,000 (the *carte de l'état major*). The *cadastre* proceeded independently of this but with the army having access to the cadastral material, a procedure which increased the total cost of the topographical maps and the cadastral maps by comparison with that of a truly joint exercise.[79]

Though the combination of general map and cadastral survey had been discussed by director Prony in the 1790s, by the time of the 1807 *cadastre,* cadastral needs took precedence over general map requirements in the survey specification. After so many failed attempts, the problem of inequitable taxation levies, on the one hand, and the empire with its burgeoning budgetary needs, on the other, had still not got an efficient and rigorous revenue base.

To its authors, the *cadastre parcellaire* of 1807 was to serve three distinct purposes. First, it was to be an instrument of government fiscal policy to ensure the fair calculation and apportionment of land tax commune by commune and property by property. Second, it would provide a documentary record of much wider value, providing, for example, a large-scale map of landed property in all the communes of France. Third, the *cadastre* of 1807 was to be accorded the status of a legal document to prove title to land.

With the first of these aims, the *cadastre* was an undoubted success; it did introduce what was generally recognized as an equitable distribution of tax liability. On the second point, maps of the *ancien cadastre* have indeed been much used by public bodies such as the war, highways and bridges, mines, and forest and water departments; by later land surveyors; and by countless thousands of individuals for a whole host of purposes, not least as a source of historical data.[80] As a legal document, however, the *cadastre* has not lived up to the role of *grand livre terrier de la France* accorded to it in the 1811 *Recueil méthodique* because of a lack of independent verification of property boundaries. Thus, courts decline to consider the *cadastre* the final arbiter in property disputes.[81] Despite this, plans and registers of the *ancien cadastre* have been widely used as a basis for property transactions. For most such purposes, the areas of parcels are stated with sufficient accuracy. Nor was the influence of the *cadastre* limited to the narrow confines of the present territory of France. In both Switzerland and the Republic of Geneva a *cadastre parcellaire* was begun under Napoleonic rule.[82] This same process in those parts of present-day Belgium, the Netherlands, and Luxembourg annexed to France after the Revolution is discussed in the following section.

6.12 Napoleonic Cadasters in Belgium, the Netherlands, and Luxembourg

In 1795 the present territory of Belgium was incorporated into the Republic of France and became subject to French cadastral legislation (chapter 2.7). The first cadastral plans *par masses de cultures* were completed in the *département* of Jemappes, Antwerp was surveyed in detail from August 1807, and in 1808 a parcel-by-parcel survey was initiated in the *département* of Dyle (with Brussels as its capital).[83] The fall of Napoleon delayed but did not prevent the completion of the cadastral operation which was continued under Dutch rule by a decree of 30 September 1814. By 1834, the Brabant surveys had been accomplished, and some five years after Belgian independence the *cadastre parcellaire* was complete except for Limburg and Luxembourg, for which the *matrices cadastrales* were compiled in 1843,[84] Although the general methods used were identical to those employed in France proper, two characteristics of the completed surveys combine to make these Low Country surveys unique. First, more than half of them were reduced in scale and published during the nineteenth century. Philippe Vandermaelen, founder of the Etablissement Géographique de Bruxelles, made a start on this by publishing the maps and *matrices* of the 137 communes of Brabant between 1837 and 1847.[85] The task was taken up by P.-C. Popp, former director of the *cadastre* at Bruges, in his *Atlas cadastral parcellaire de la Belgique*. . . . By the time of his death in 1879, this atlas covered the provinces of Antwerp, Brabant, Flanders, Hainaut, and Liège.[86] A second distinction is that, although the original cadastral maps were not updated as parcels changed (thus following French practice), these alterations are recorded in a series of supplementary plans known as *croquis complémentaires*.

As in France proper, in those parts of the empire where cadastral mapping followed closely on the French model as set out in the 1811 *Recueil méthodique,* the need to establish a just system for levying land tax took precedence over subsidiary objects of the cadaster, such as its intended role of providing a means of registering title to land. As in France, the Low Countries' cadasters are not recognized in courts of law as definitive evidence of landownership. Nevertheless, cadaster-derived data have been used as critical arbiters in a number of legal and administrative situations. For example, a law of 1854 used the proportion of fiscal due apportioned to a particular piece of land as a basis for valuing properties for compulsory purchase.[87] Cadastral maps were used as one source for compiling the large-scale *Carte topographique de la Belgique* (1846–54).[88]

6.13 The Cadastral Map in France

Large-scale maps of properties were employed in France, as elsewhere, in increasing numbers from the seventeenth century as evidence in property disputes and to assess and develop strategic resources, such as forest timber. The mapping of individual estates for general managerial purposes was not widespread in France until the eighteenth century. However, by the close of the First Empire, a massive, nationwide *cadastre parcellaire* was well advanced.

The reform of land taxation was an important objective of the French Revolution; a mapped cadaster was the prime instrument by which a more equitable taxation regime was brought about. This was also extended to territories such as Belgium, the Netherlands, Luxembourg, and the Republic of Geneva, which were annexed to France. It was also held up as a model and exemplar for taxation reform elsewhere, for example, in Switzerland and Prussia.

Taxation reform was much discussed and written about in eighteenth-century France, not least by the Physiocrats. By the time of the Revolution, there was a clear consensus that this objective could be advanced only by means of a surveyed cadaster to reveal and record each individual's properties. Experiments in the Duchy of Savoy and by *intendants* in Limoges, Riom, and Paris had earlier pointed the way, but opposition from the nobility and clergy, who had the most to lose from a levy based on actual quantities and actual values of land, precluded national reform until the Revolution removed these vested interests from power and influence. The *plan terrier* of Corsica, compiled in the 1770s and 1780s to distinguish Crown, common, and individual landownership and to assist with the agricultural development of this newly gained island, also confirmed to government the potential value of a cadaster beyond the immediate realm of fiscal reform. In a practical sense, the Corsican project also provided a training ground for men like de Richeprey who were to play leading parts in establishing the actual method of the Napoleonic *cadastre*, notably the triangulation of communes and its eventual parcel by-parcel nature. With this last, as with the architectural, engineering, and urban planning projects of the First Empire, the role of Napoleon himself in instituting and molding the *cadastre* was by no means insignificant. He saw a *cadastre parcellaire* as a natural adjunct in matters of landownership to the new legal code of postrevolutionary France.

As with the tithe survey of England and Wales, there was discussion right from the start about whether individual cadastral maps of communes could form the basis of a general map of France. Neither happened, but rather the military authorities in both countries started again using national triangulation to control a new survey. This apart, the *cadastre* of France can be seen as a successful instance of the two-way relationship between map and state: in this case the map was conceived initially as an instrument for fiscal policy but was then accorded wider roles in land transfer and as a general land data-base.

Acknowledgments

We are pleased to acknowledge help from the following colleagues: Prof. Paul Claval, Dr. Mark Cleary, Prof. Hugh Clout, Prof. Josef Konvitz, Dr. Piet Lombaerde, Mme Mireille Pastoureau, Prof. Jean-Robert Pitte, and Dr. David Siddle.

7

ENGLAND AND WALES

7.1 Surveys of Church and Crown Lands Sequestered during the Civil Wars

Experimentation with map-based cadasters in Sweden, French government maps of forest reserves, and Dutch maps of polder holdings indicate that by the seventeenth century, some European governments and provincial rulers were conscious of the utility of maps not only for plotting national strategy and the organization of fortification and warfare but also for recording inventories of property, including its ownership, resources, and liabilities. That a cartographic approach was not yet an inevitable concomitant of government-sponsored surveying of real property can be illustrated by reference to the Crown and parliamentary surveys of seventeenth-century England.

James VI of Scotland, who inherited the throne of England in 1603, acquired a burden of debt that could be annulled only by the sale of Crown lands. In the same year his manorial stewards were instructed to make a return of leases, values, and rents of estates. From 1604, surveyors were engaged to survey the estates which from 1607 included forest land. They compiled what is now known as the Great Survey, an undertaking that was never finished and includes few mapped surveys.[1]

In the seventeenth-century civil wars, first estates of church bishops were sequestered in 1646, then those of the deans and chapters of cathedrals, and finally Crown lands in 1649. Parliament required that these be surveyed prior to sale for the benefit of the government. By 1649 an elaborate administrative procedure had been established with a surveyor-general at its head and teams of surveyors assigned to each county.[2] However, very few maps were constructed; rather, surveyors were required to identify individual land parcels by describing their abutments. Speed was an important consideration, and for this reason detailed survey measurements were forbidden.[3] It was also rather more of a valuation than a surveying exercise, as a main purpose was to establish the level of fine payments for which Royalists could regain possession of their estates. After the wars much confiscated land was returned to prewar proprietors, but some remained in the hands of its new owners. If government in England was still wedded to the idea of survey by written description and valuation, the Civil War upheavals may have contributed to the output of local maps by their effect on the land market, much as the dissolution of the monasteries had done a century or so earlier.

7.2 Cadastral Maps and the Enclosure of Open Fields, Commons, and Waste

Enclosure is the process whereby land that was exploited collectively, or over which there existed common rights, was divided into parcels owned in severalty, with each proprietor "exchanging his share of common rights over the wider area for exclusive rights in part of it."[4] The purpose of enclosure as perceived by some contemporaries was to introduce more efficient farming to open-field parishes by making farms compact, to encourage improvement of livestock management and crop rotation, to remove the burden of tithes, to convert common and other rough grazing land to more profitable uses by expanding the area of land under the plow, or, as was the case with some of the earliest enclosures in the Tudor period, to do the reverse—to lay down arable land to pasture for sheep farming.[5]

The earliest government involvement in the process of redistribution and enclosure of communally held land in England was in 1604, when an act of Parliament was obtained for the enclosure of Radipole in Dorset (figure 7.1). It is still largely unclear why by the eighteenth-century enclosure by act, known as parliamentary enclosure, became the norm, when in preceding centuries communities had found local agreements sufficient. At the most pragmatic of levels, an act enabled a majority of proprietors in a village community who advocated enclosure to constrain a minority who were opposed to it. The number of private or public acts of enclosure and enclosures under the terms of General Enclosure Acts, passed by Parliament in 1801, 1836, 1840, and 1845 to obviate the need for a separate act for each community, totals some 5,250, covering more than 3 million hectares, or about a quarter of England.[6] The period 1755–80 saw 38 percent of the total of all parliamentary enclosures, and the Napoleonic War years, some 43 percent. Though private enclosure acts were petitioned for individually, the wording of their clauses is stereotyped, differing only, for example, in the amount of common to be allotted to the lord of the manor in lieu of his rights, or of land to the tithe owner in return for a commutation of tithe dues.[7] The first General Enclosure Act was passed in 1801 during the Napoleonic Wars to facilitate enclosure and thus the general increase in agricultural production which was an expected concomitant. This provided model clauses, including the requirement that a cadastral plan be

Fig. 7.1 England and Wales: counties and places noted in the text.

made and verified. Communities could insert such model clauses into their own individual enclosure bills. A truly general act was passed in 1836 for enclosing open arable fields, which was extended to other categories of communal land in 1840. If two-thirds of local proprietors who together held two-thirds of the land agreed, enclosure commissioners could be appointed, and enclosure could proceed without any reference to Parliament. In common with earlier private acts, enclosure under this public act would be recorded in a formal enclosure award deposited in the parochial documents and another copy enrolled with the respective clerk of the peace. To protect the interests of small proprietors, a further General Enclosure Act in 1845 established a central Enclosure Commission and a body of assistant enclosure commissioners and surveyors who were entrusted with the local administration.[8]

The provisions of each enclosure act were normally put into effect by enclosure commissioners, usually three in number, to represent the interests of the lord of the manor, the ecclesiastical or lay tithe owner, and the other proprietors. After their appointment, these men in turn recruited a clerk and a surveyor. A select committee of the House of Commons set up to investigate the nature of parliamentary enclosure prior to the first General Enclosure Act, reported in 1800 that:

> Acts of Inclosure commonly require a Survey to be made, either by the Commissioners, or by some person employed by them, and a Map to be prepared from it; both which is generally done by a Surveyor, especially appointed for the purpose . . . a fresh Survey and Map are often ordered though there may have been One of each in Existence, fully or nearly adequate to the Purpose; and that in some Counties a Practice has prevailed of employing Two Surveyors; one to take a General, the other a Particular Survey. In some Instances another Description of Persons is appointed by the Act, called Quality-Men, whose Business it is to value the land.[9]

As the commissioners made their decisions about each of the new allotments which they judged a fair equivalent "in full and perfect satisfaction" of preexisting open lands and common rights, these were set out on the map to replace the preenclosure cadaster (figures 7.2 and 7.3). Draft enclosure plans representing the preenclosure cadaster survive more rarely from before about 1830 but may be found with the other enclosure documents, as may copies of earlier estate plans, as the commissioners had a duty to investigate the evidence of old surveys when framing their award.[10]

Enclosure maps are attested by their surveyors on oath as true and accurate. Parcels depicted on the plans are linked by reference numbers to the enclosure award. Not all enrolled (officially deposited) copies of awards have maps, as commissioners were in law required to enroll only the award and, probably for the sake of economy, sometimes adhered strictly to this limited require-

Fig. 7.2 Pewsey, Wiltshire, enclosure map. This map of 1775 accompanies the award to enclose the common fields of this parish. Source: Wiltshire Record Office, 1634/34.

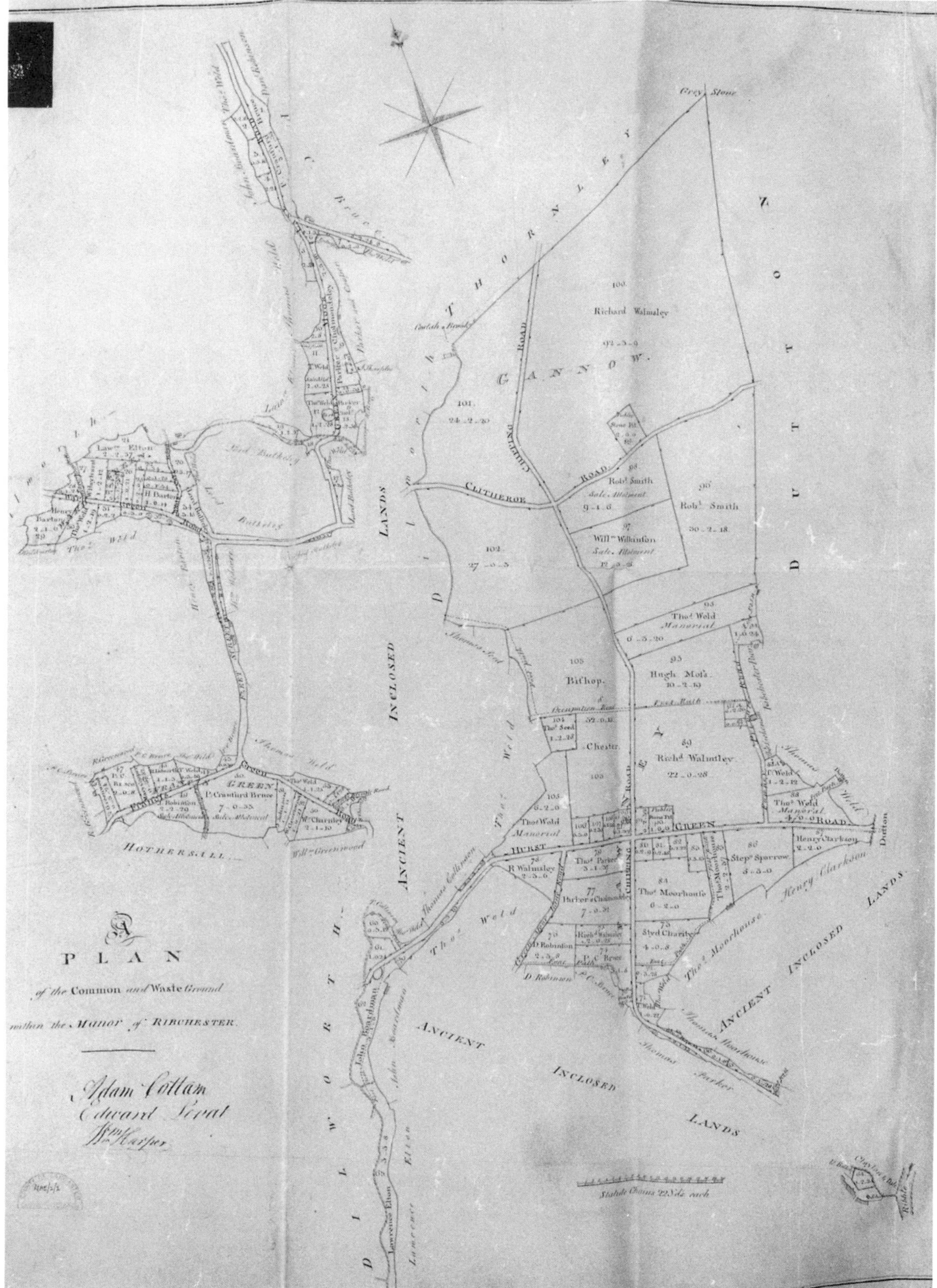

Fig. 7.3 part of the enclosure map of Chipping, Milton and Ribchester, Lancashire, of 1812, showing some of the common and waste which was enclosed by act of Parliament in 1808. Source: Lancashire Record Office, AE 2/2.

ment.[11] Nor have all maps survived. For parliamentary enclosures before 1770 the survival rate is less than 33 percent; for those enacted after 1810 it is more than 90 percent; for the period as a whole it is just over 70 percent.[12]

To a twentieth-century observer the utility of a map to record the new cadaster resulting from the exchanges and consolidations decided by the enclosure commissioners seems very obvious. So it did also by the 1790s to promoters of bills, to witnesses before select committees, and to agricultural commentators such as William Marshall. For example, Sir John Sinclair's 1797 bill *For Enabling any Person or Persons Entitled to any Waste, Uninclosed and Unproductive Lands, Common Arable Fields, Common Meadows, or Common of Pasture, . . . in England* required that "the Surveyor or Surveyors shall, with all convenient Speed, make an exact Survey of all such Common Arable Fields, Common Meadows, Common Pastures, Wastes, or Commons."

William Marshall, an outspoken commentator on many agrarian topics, was a convinced advocate of the value of cadastral maps in estate management. In his *On the Landed Property of England,* published in 1804, he discusses the utility of maps for reorganizing inefficiently disposed farm holdings on an estate, and the value of general maps for "promptly exhibiting the several farms and fields, as they lie . . . it is to him [the improver] what the map of a country is to a traveller, or a sea chart to a navigator."[13] Maps head his list of "office furniture" for the modern estate office.[14] On the narrower question of enclosure, Marshall was questioned by House of Commons committees on two enclosure bills and published his views in 1801 as a pamphlet entitled *On the Appropriation and Inclosure of Commonable and Intermixed Lands* to inform the House of Lords committee appointed to consider the general bill in that year. Marshall's proposals for a general act of enclosure included the preparation of "a map or maps, of such lands, with a number or other distinguishing marks,—the quantity in acres, and the estimated rental value, by the acre, marked upon each of the several parcels."[15] This did not necessarily demand a new survey, as "the surveyors chosen . . . shall measure and map (if sufficient surveys cannot otherwise be procured)."[16] Maps were needed so that the commissioners "may be able to conduct the business of appropriation with facility and due effect."[17] Two maps were to be produced. The first was of the existing state of the land, and the second of the land divided into parcels.[18] A similar advocacy of the cadastral map is evident in the successful 1801 *Act for Consolidating in One Act Certain Provisions Usually Inserted in Acts of Inclosure:* "A Survey, Admeasurement, Plan and Valuation of the Lands etc. to be inclosed shall be made, and kept by the Commissioners, which shall be verified by the Persons making them. . . . Maps made at the Time of passing Acts may be used, without making new ones, if the Commissioners shall think fit."[19]

A generation earlier in 1766, H. S. Homer had published his *Essay on the Nature and Method of Ascertaining the Specifick Shares of Proprietors upon the Inclosure of Common Fields* at a time when the activity of parliamentary enclosure was accelerating markedly. He discusses in detail the surveying procedures to be followed upon enclosure; mapping at this date seems also to have been a matter of routine, as he makes no special pleading for the utility or importance of recording enclosure on maps. Though he details the method, the need to prepare a map is taken for granted. First of all the yearly value, number of furlongs, and plots composing the common land of a district should be ascertained.

> When this Point is settled, the next Thing to be procured is a Plan and Survey of the Number of Acres in each Division; which, when compleated, is called the General Survey: Afterwards the known Property of every owner in each Division is separately to be measured, and this when finished, is called the Particular Survey. By means of the latter every Proprietor's Estate is reduced first into its Measure or Number of Acres, and then by a Comparison of the Measure of the several parts with the Value of each by the Acre, as before fix'd, it is reduced into Money, or its whole annual value, unconnected with the Common Ground. . . . when the value of every Proprietor's Estate is settled according to this Method, the General Survey and Plan, which points out also the respective Dwelling-Houses of the Proprietors, easily shews the natural Situation of the several future Estates; from which there will be no Occasion to vary.[20]

As to the method of survey:

> The most approved Method of surveying and planning any Tracts of Land, is to measure the Outside thereof from station to station with a chain, taking Offsets from the straight line at convenient Distances, where there is any Irregularity in the Figure; and at each Station also to mark the Angles or Bearings between the Lines; which is done either by means of a graduated Instrument, as a Theodolite, Circumferentor, or Semicircle; or otherwise it is done by means of a Telescope and Plain Table.[21]

He recommended that all the working lines be left on the plan so that the maps could be tested.

Parliamentary enclosure in the eighteenth and nineteenth centuries, as noted above, affected perhaps a quarter of England, which means that an overwhelming majority of land was enclosed by private agreements in the sixteenth to eighteenth centuries though agreements do extend beyond these centuries in both directions.[22] All this was done without maps. Moreover, few maps are found accompanying parliamentary enclosure awards before the last quarter of the eighteenth century.[23] Why, then, did enclosure by award *and* map become a desideratum thereafter?

There seems no clear answer to this question. What is certain is that the making of a map for enclosure was an already established practice under private acts by the end of the eighteenth century, so the inclusion of clauses requiring maps and specifying their nature in the 1801 General Enclosure Act simply enshrined the current practice into an act of Parliament. A possible explanation for the growth in the use of maps among enclosers in the eighteenth century is that the use of maps in private enclosures arose out of the by then established tradition of private estate mapping but was not a product of the institution of enclosure itself. This would certainly help explain the written nature of enclosure agreements vis-à-vis the map-recorded parliamentary enclosure activity, which was itself accelerating at precisely the time of burgeoning output of private estate maps.[24]

7.3 The Tithe Commutation Surveys

When the British government passed its Tithe Commutation Act in 1836, it set in motion a survey which was unprecedented in England and Wales in terms of area covered, detail of inquiry, and exactness of the record.[25] In scope and content, the English and Welsh tithe surveys marked a new departure. Their employment of cadastral maps to record the survey was, however, by no means new or innovative. In England some thousands of private enclosure acts were passed in the eighteenth and early nineteenth centuries, and these acts generally required that the new cadastral pattern of a parish be described on a map.

Customarily, tithes represented a tenth of the annual increase of the produce of the soil and for about a thousand years were the heaviest direct tax on farming and, from their mode of collection, the most repugnant.[26] After the Napoleonic Wars, crippling taxes, rates, and tithes as well as uncertain grain prices, a succession of poor harvests, and a decline in household-based industry brought poverty and distress to many farmers and hunger to many laboring families. William Cobbett, a social commentator and advocate of political reform, repeatedly reminded farming audiences that the privileges enjoyed by the clergy could not last much longer: "There must be a *settlement of some sort;* and that settlement never can leave that mass, that immense mass, of public property, called 'church property', to be used as it now is."[27]

When protest and violence erupted in the Kent countryside and in other southern counties in the autumn of 1830, it was directed not only against farmers' property and the workhouses but, in many parishes, against their parsons.[28] From being a source of local irritation, the "tithe question" had become a matter of concern for national government. Between 1833 and 1836 no fewer than four attempts were made to secure the passage of bills to commute tithes; on 13 August 1836 Lord John Russell's Tithe Commutation Act received its royal assent and substituted a monetary tithe rent-charge that was index-linked to the price of grain for all customary payments and for tithe in kind.

The implementation of the Tithe Commutation Act progressed swiftly and without much resistance. Between 1836 and 1845 the tithe commissioners completed about 90 percent of their work, and almost all of the rest was finished by 1855. Three documents were prepared for each

No. 5.—London: Published (By Authority,) by G. Routledge, 11, Ryder's Court, Leicester Square.

LANDOWNERS.	OCCUPIERS.	Numbers referring to the Plan.	NAME AND DESCRIPTION OF LANDS AND PREMISES.	STATE OF CULTIVATION.	QUANTITIES IN STATUTE MEASURE.			Amount of Rent-Charge apportioned upon the several Lands, and to whom payable. PAYABLE TO VICAR.			PAYABLE TO Impropriator			REMARKS.
					A.	R.	P.	£.	s.	d.	£.	s.	d.	
Adams Jane	William Good		Small Bridges											
		514	Nayland End Wood	Wood	28	1	39							
		575	Six Acre Ley	Arable	7	0	31							
		576	Nine Acre Ley	do	9	2	19							
		594	Part of 20 Acres	do	4	2	23							
		596	Eighteen Acres	do	20	1	23							
		597	Twelve Acres	do	13	2	1							A Prescription of 32 Shillings per Annum in lieu of Hay and Clover and Small Tithe is paid by the Tenant of this Farm the whole of which is covered thereby
		598	Thirteen Acres	do	14	1	12							
		599	Long Meadow	Pasture	5	2	15							
		600	Charity Field	Arable	17	2	12							
		601	Cross Path Field	do	18	0	11							
		602	Tare Field	do	10	1	37							
		608	Five Acre Meadow	Pasture	4	1	23							
		609	Meadow Field	Arable	6	3	6							
		610	Three Acre Meadow	Pasture	2	3	26							
		611	The Slipe	do	2	0	22							
		612	Six Acre Meadow	do	5	3	26							
		613	Cottage and Garden		.	.	29							
		614	Great Walnut tree Close	Pasture	9	2	5							
		615	Bures Green	do	6	0	7							
		616	Weston's Green	do	0	3	4							
		617	do	do	1	1	35							
		618	Barns Yards &c		0	3	0							
		619	Paddock	Pasture	1	1	34							
		620	Boat House Mead	do	7	2	38							
		621	House &c		3	1	13							
					203	1	11				42	9	2	
Adams Jane	Simon Seaman		Hole Farm											
		520	Hop Ground	Arable	2	1	21							
		522	Little High Field	do	7	1	39							
		523	Third Back Meadow	Pasture	2	1	37							
		524	Second do	do	1	0	38							
		525	Great High Field	Arable	12	3	37							
		526	First Back Meadow	Pasture	1	1	36							
		527	Chapel Field	Arable	11	0	23							
		562	Barn Field	do	11	0	1							
		563	Church Field	do	7	0	3							
		564	Barn Field Bottom	do	2	3	7							
		565	Barn bottom Mead	Pasture	3	1	7							
		566	Farm House, Barns Yards &c		1	3	19							
		567	Church field bottom	Arable	2	1	18							
		568	Pass Brooks	do	5	0	10							
		569	Pass Brooks bottom	Pasture	2	0	27							
				Forward	76	3	2							

Fig. 7.4 Extract from the schedule of tithe apportionment for Bures St. Mary, Suffolk, in 1838. Source: Suffolk Record Office, Bury St. Edmunds, T60/1.

tithe district: a tithe apportionment, a tithe map, and a tithe file. The most important part of a tithe apportionment is the "schedule of apportionment," in which each tithe area or piece of tithable land (usually an open-field strip or an enclosed field) is listed under the name of its owner and occupier, it is named, and (where not a field) it is described (figure 7.4). Its "state of cultivation," or land use, is entered for purposes of valuation; the statute acreage of each tithe area as established by a field-by-field survey is recorded in the apportionment, as is the monetary amount of the global, or district, rent-charge apportioned to that particular tithe area. Each tithe area is also allocated a unique reference number in the apportionment; these numbers link the tithe apportionment to the tithe map.

The tithe apportionment is the legal instrument in which the commutation occasioned by the 1836 act is recorded. Why, then, did government require there to be a map of each tithe district? A field-by-field *survey* was certainly needed to establish a number of facts. First, a parish had to be valued to establish its agricultural productivity for comparison with levels of tithe actually collected in the recent past. Second, this rent-charge had to be shared out or apportioned among the individual tithe areas of a district according to the land use and size of each. Neither activity needed a map; indeed, tithe in about a quarter of English and Welsh parishes had been satisfactorily commuted before 1836 without maps.

The Tithe Commutation Act itself contained contradictory clauses on the nature of maps it required. Clause 35 states that

> the Valuer or Valuers or Umpire may, if they think fit, use for the Purposes of this Act any Admeasurement, Plan or Valuation previously made of the Lands or Tithes in question of the Accuracy of which they shall be satisfied; and that it be lawful for the Meeting at which such Valuer or Valuers shall be chosen to agree upon the Adoption for the Purpose aforesaid of any such Admeasurement, Plan or Valuation, and such Agreement shall be binding upon the Valuer or Valuers; provided always, that Three Fourths of the Land Owners in Number and Value shall concur therein.

But clauses 63 and 64 of the act required the Tithe Commissioners to sanction the accuracy of every map by affixing their official seal as a final duty when confirming the commutation of tithes in a district. "If the Commissioners shall approve the Apportionment they confirm the Instrument of Apportionment under their Hands and Seal, . . . and every Recital or Statement in or Map or Plan annexed to such Confirmed Apportionment or Agreement for giving Land, or any sealed Copy thereof, shall be deemed satisfactory Evidence of the Matters therein recited or stated, or of the Accuracy of such Plan."

Clearly there was some ambiguity here. On the one hand, the act entitled landowners to adopt any plan of their choice; on the other hand, the tithe commissioners were required to certify the accuracy of every map before confirming an apportionment. Maps were certainly required by the act, but it was left to the Tithe Commission and their map adviser, Lieutenant Robert Kearsley Dawson, to define the precise purpose and nature of tithe maps.

On 29 November 1836 Dawson advised the commissioners that in his opinion tithe maps should portray both tithe district and tithe area boundaries with absolute accuracy and at a scale large enough to enable quantities to be computed from the map.[29] It was necessary to know the exact area of a district to enable a fair and just apportionment of rent-charge; levels of rent-charge also varied quite considerably between contiguous parishes, so it was important to know precisely where parish boundaries ran. However, the accurate representation of field boundaries was not at all necessary for the immediate purposes of tithe commutation. It was, though, a vital element for a larger, more ambitious project for which Dawson sought the Tithe Commission's support. This was to extend the tithe mapping to produce, in Dawson's terms, a "General Survey or Cadastre" of the whole country.[30] This proposal found immediate favor with the Tithe Commission, who saw it not only as a way of ensuring a set of conformable, accurate tithe maps but also as a possible means of devolving some of the costs of tithe commutation on the government. The argument used to support the case for accurate, large-scale cadastral maps for tithe survey purposes was one of the *permanence of record* that such maps would provide. The Tithe Commutation Act was designed to produce a settlement once and for all of the tithe question. In order to avoid future litigation, it was important that fences be accurately portrayed so that particular tithe areas and their rent-charge liability could always be identified.

On 4 January 1837 the commission's secretary wrote to Dawson outlining its opinion on the role of tithe maps.

> Considering the permanent nature of the record which it is meant that the instrument of apportionment, with the Plan which is to be annexed to it, should constitute, it appears to the Commissioners to be their duty to require such Plans as will ensure a possibility, and, as far as may be, a facility for ascertaining precisely the extent and position of lands to be declared subject to or free from the rent-charges to be created under the Tithe Act, when the present boundaries of those lands have been altered or displaced. The Commissioners are of the opinion that the Scale recommended in your Report [3 chains to an inch], and the lines of construction proposed to be left marked on the Plans, are essential for this purpose, and they authorize you to communicate this opinion to all parties with whom you are, or may hereafter be in correspondence or communication on the subject.[31]

The commissioners were quite clear about the type of map necessary for tithe commutation. Tithe maps were to be first-class, accurate plans at a scale of 26.7 inches to a mile (1:2,373), constructed according to a system of internal triangulation (figure 7.5). Such maps would have been ideally suited to form a basis for a full cadastral survey of the nation as well as forming a permanent map record of tithe liability. Cadastral maps had been used in, for example, the Netherlands, France, Denmark, Austria, and some German states to record the tax liability of particular pieces of land. As Dawson told the tithe commissioners, "In many of the States of Continental Europe, Cadastral Surveys have long been in progress, at an annual expense commensurate with the importance attached to the possession of such documents."[32]

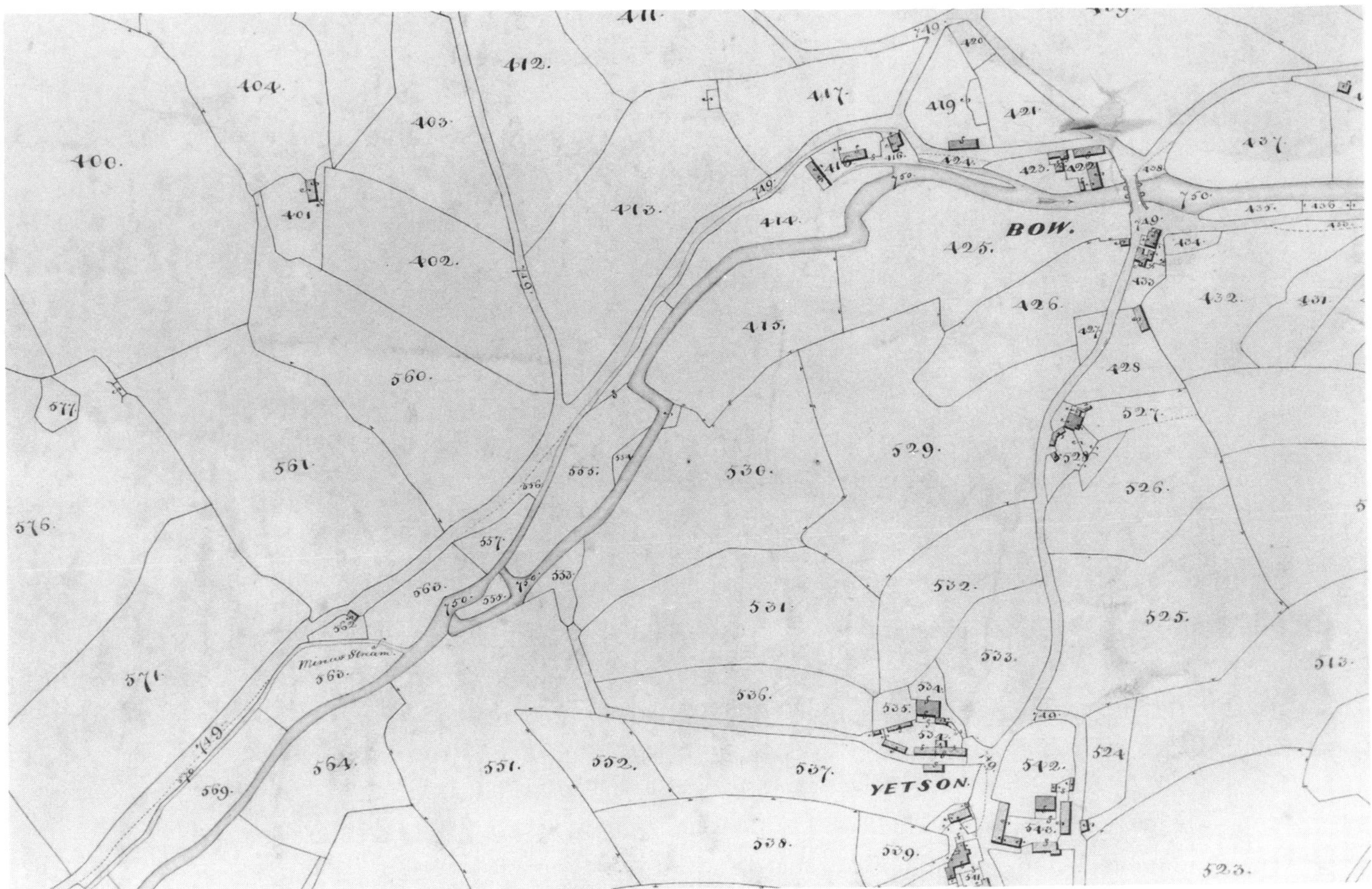

Fig. 7.5 Ashprington, Devon. First-class tithe map drawn at three chains to an inch (1:2,373) in 1843. Source: Devon Record Office, Ashprington tithe map.

In the same letter he set out an impressive list of advantages to be derived from a similar survey of England and Wales. These included the resolution of boundary disputes, easier transfer of real property, and identification of the best lines for new roads, railways, and canals. The government would also obtain an accurate statement of the "real capabilities of the country" and would be able to decide where investment in improvements might be most beneficial. He concluded this review by saying that "the necessity which now exists, for Surveys of nearly the whole country, for a specific purpose presents means for forming a General Survey or Cadastre, at such a cheap rate, that the opportunity cannot be lost without exposing those who ought to present its importance, to the certainty of future censure, if they fail to perform that imperative duty."[33]

The tithe commissioners remained convinced that accurate maps were necessary for an effective and lasting commutation for tithes. By pressing for accuracy, however, they were increasing the already considerable cost of commutation for some landowners. In a letter to the chancellor of the Exchequer, the commissioners reiterated Dawson's arguments for combining tithe surveys with a general survey of the whole country in the hope that some of the cost of cadastral accuracy might be defrayed from the public purse.[34] They also outlined other tax collection advantages that would accrue from a full cadastral survey. In particular they referred to maps required for assessing Poor Rates under the New Poor Law. The Poor Law commissioners strongly supported both the idea of a general survey and the method of producing parish maps described by Dawson.[35]

The tithe commissioners concluded their appeal to government by asking whether "such maps shall be attained at enormous expense at some future period, or whether the large sums of money which must now be expended on the maps, good or bad, supplied for the purposes of the Tithe Act, instead of being wasted for all other public purposes, shall be so expended as to be the means, as far as it goes, of supplying all the wants of the Nation as connected with surveys."[36] These arguments persuaded the chancellor of the Exchequer to appoint a House of Commons select committee on 16 March 1837, "to consider the best mode of effecting the Surveys of Parishes for the purpose of carrying into effect the Act for the Commutation of Tithes in England and Wales."

The committee had to decide whether accurate maps were needed for tithe commutation, maps which might be combined into a general survey and then be reduced, engraved, and published. The committee soon realized that for the immediate purposes of commutation and apportionment it was not necessary to have a map at all. Even the chairman of the tithe commissioners, William Blamire, stated that an accurate schedule would suffice; it was the future confusion that might result from the adoption of such a system that concerned the tithe commissioners.[37]

After weighing the evidence, the select committee reported in May 1837 their recommendation that precise, accurate maps were not required for successful tithe commutation. Its conclusion was influenced by matters of cost to landowners and by the spirit of the 1836 act, which was to encourage voluntary commutations wherever possible. The committee recommended some relaxation of clauses 63 and 64 of the Tithe Commutation Act to relieve the commissioners of the duty of sanctioning map accuracy.

In June 1837 a bill to amend the Tithe Commutation Act was brought before the House of Commons and became law on 15 July. An opportunity for a cadastral survey of the full kingdom was lost. The maps accepted by the tithe commissioners are on a variety of scales and made at a variety of dates (figures 7.6 and 7.7). As a body, the tithe maps were quite unsuitable to form the basis for a national cadastral survey. The country "lost its chance of a cadastral system on the continental pattern, with all that means in terms of cheap and simple property transfers."[38]

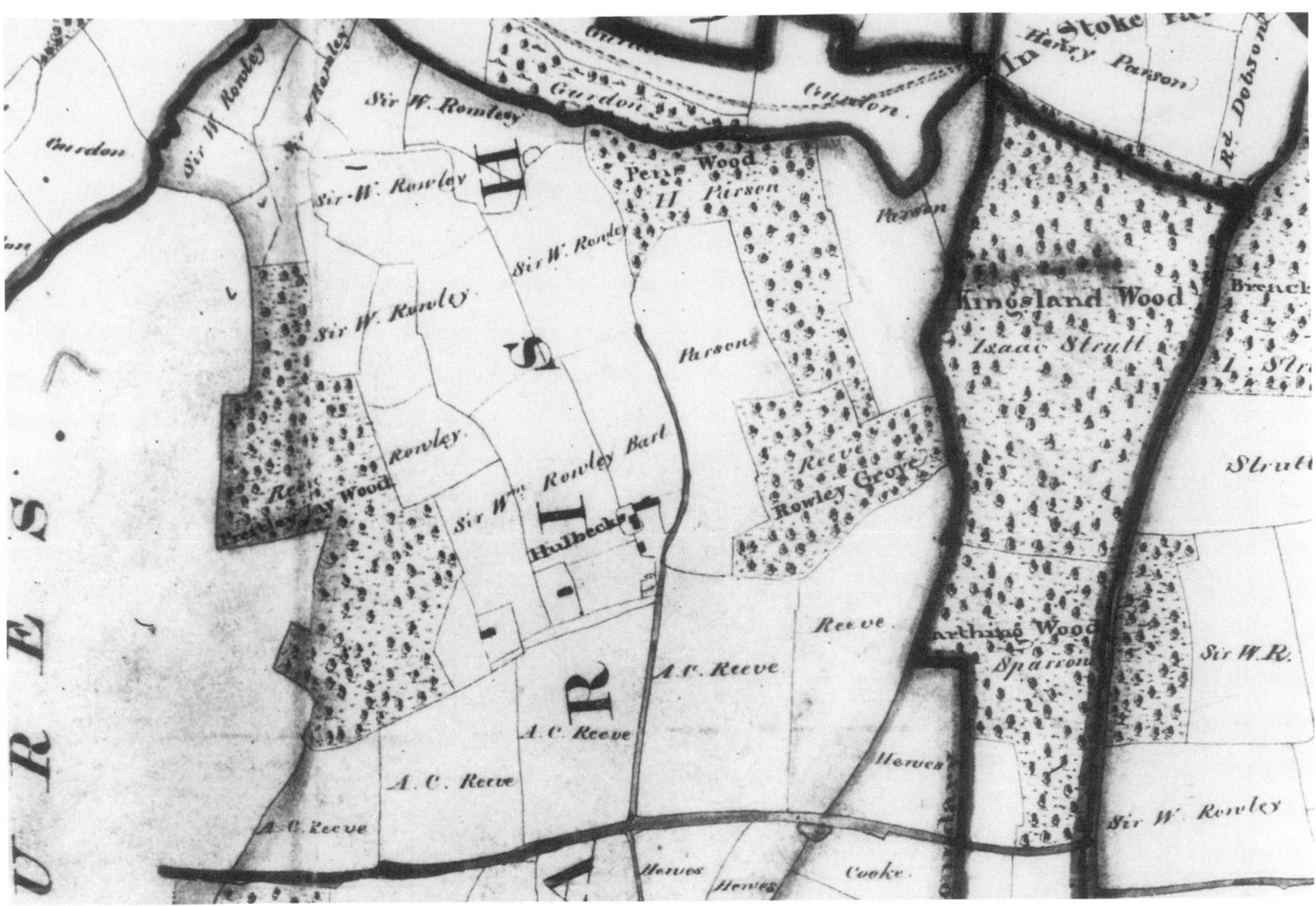

Fig. 7.6 Enclosure map of Stoke by Nayland and adjoining parishes in Suffolk, constructed in 1817 and upon which the later tithe map of Wiston is based. Source: Suffolk Record Office, Bury St. Edmunds, Q/RI/34.

The first section of the amended Tithe Act relieved the commissioners of the need to certify the accuracy of every map and accompanying apportionments and permitted them to establish two classes of tithe map. "First-class maps" are those which the commissioners considered sufficiently accurate to serve as legal evidence of boundaries and areas and can be identified by the certificate of accuracy which they bear and also by the presence of the commission's official seal (figure 7.8). All first-class maps and accompanying field books were carefully checked by Lieutenant Dawson's staff in London. The technical specification that Dawson prepared before the 1836 act was amended still applied to first-class maps, with the exception that the commissioners no longer considered it essential that a uniform system of conventional symbols be used. "The maps

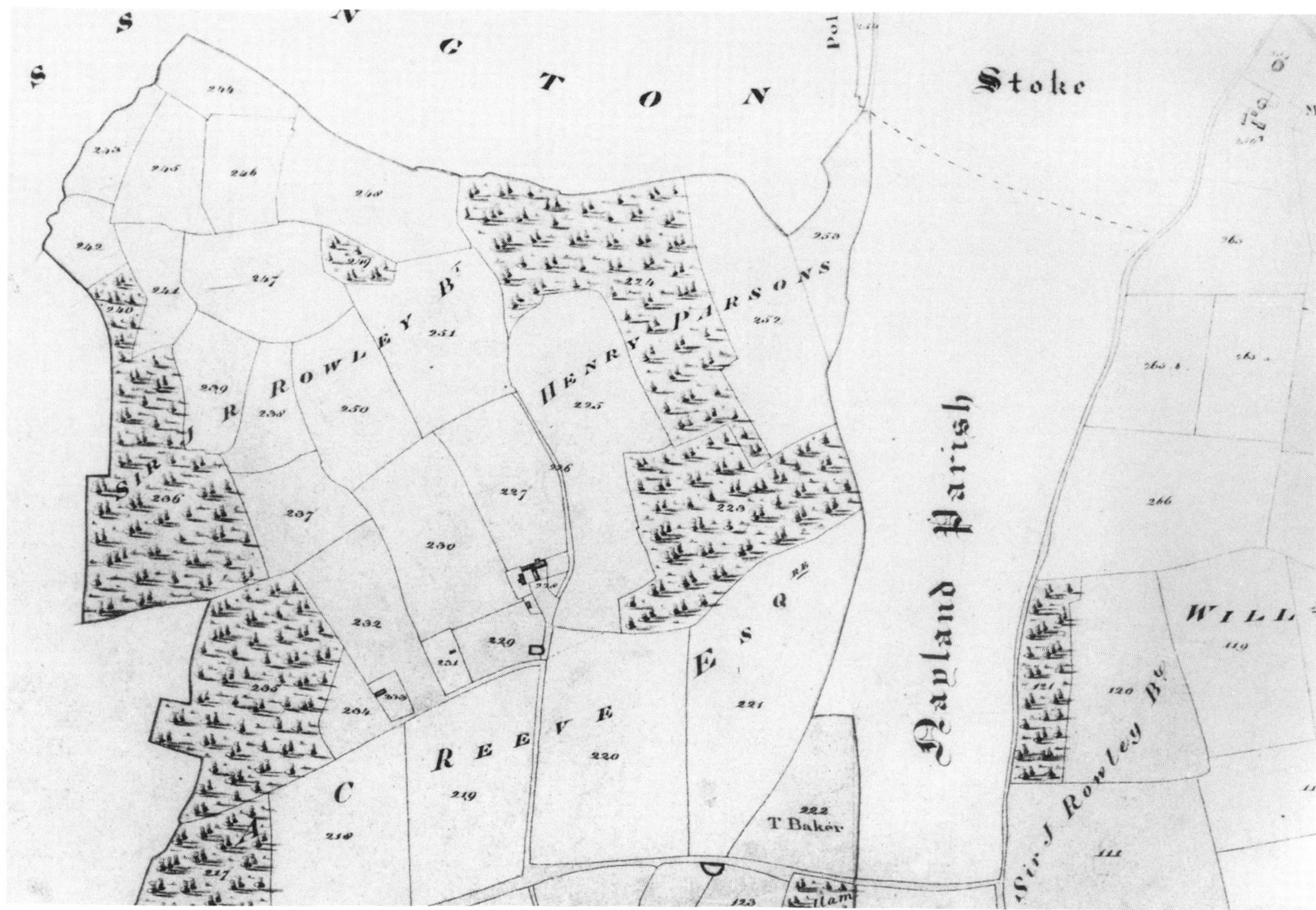

Fig. 7.7 Wiston, Suffolk, tithe map of 1837, derived from the enclosure map pictured in figure 7.6. Source: Public Record Office, London, IR30/33/470.

which will be most acceptable to the tithe commissioners are the plain working plans, with the lines of construction, names and reference figures shown upon them, and with no other ornament or colour whatever; and the most ready way of obtaining the seal of the commission will be to send up the actual working plan."[39]

The category of second-class maps includes those which were intended to be first-class but which failed Dawson's tests and were not subsequently corrected to the commission's satisfaction. By far the larger number are those maps "which three-fourths of the landowners are desirous to use, but which the parties do not mean to submit to the test of the Commission."[40]

All tithe maps, whatever their origin, show the boundaries of the tithe district. The tithe

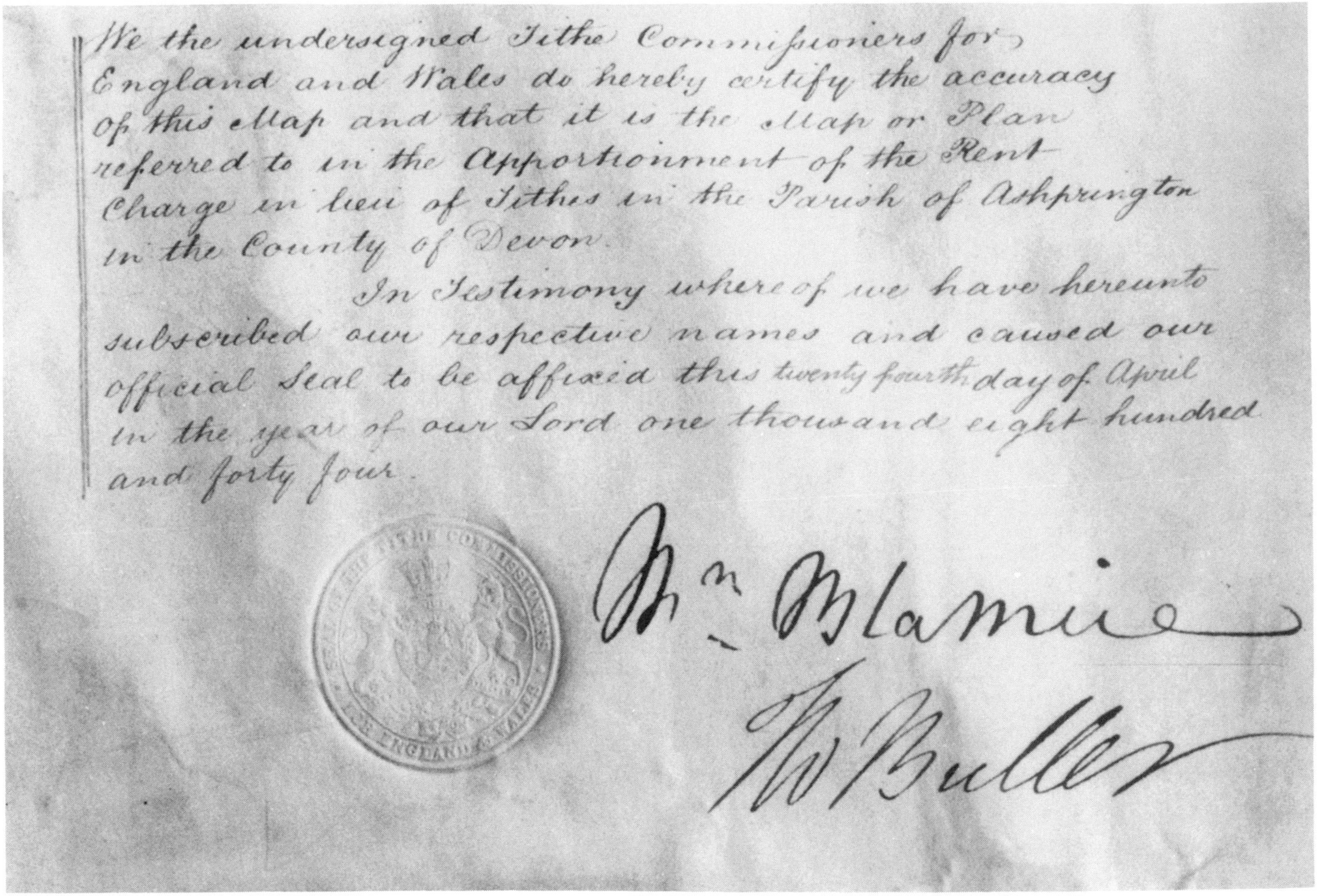
We the undersigned Tithe Commissioners for England and Wales do hereby certify the accuracy of this Map and that it is the Map or Plan referred to in the Apportionment of the Rent Charge in lieu of Tithes in the Parish of Ashprington in the County of Devon.

In Testimony whereof we have hereunto subscribed our respective names and caused our official Seal to be affixed this twenty fourth day of April in the year of our Lord one thousand eight hundred and forty four.

SEAL OF THE TITHE COMMISSIONERS FOR ENGLAND & WALES

Wm Blamire

TW Buller

Fig. 7.8 The tithe commissioners' official seal, which identifies a "first-class map." Source: Devon Record Office, Ashprington tithe map.

areas usually correspond to fields, but in some instances they constituted whole farms or, more rarely, a whole township. On most maps the boundaries of enclosed fields are represented by continuous lines and those of unenclosed fields by dotted lines. Occasionally, hedges, fences, and gates are also portrayed. The amount of other detail shown on tithe maps varies considerably. Most maps mark the courses of streams, canals, ditches, and drains as well as the outline of lakes and ponds and the lines of roads and paths. Some, such as the second-class map of Gittisham, Devon (figure 7.9), use conventional symbols recommended by Dawson in the first edition of his "Instructions" to identify different types of land.

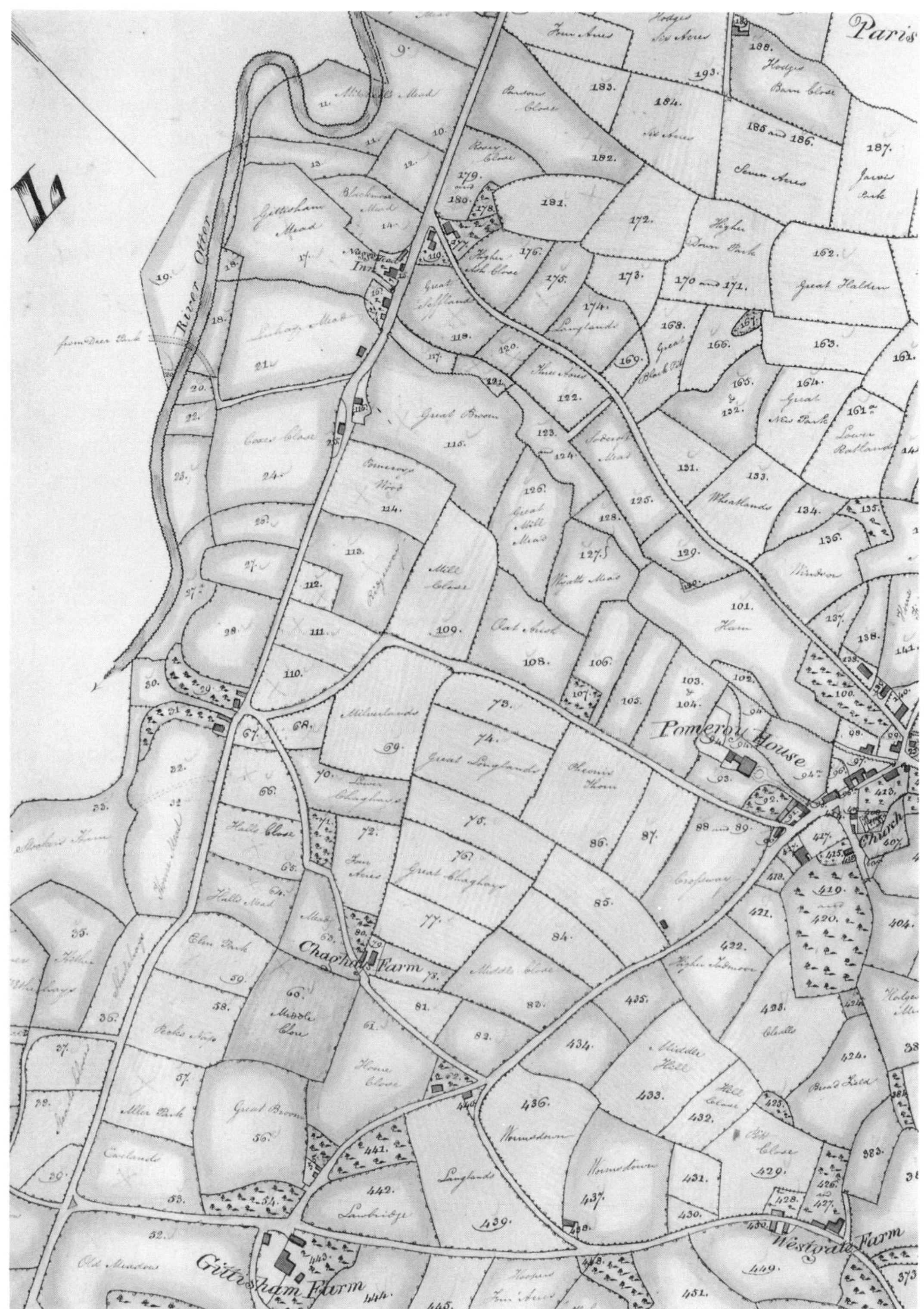

Fig. 7.9 Gittisham, Devon. Second-class tithe map drawn at a scale of six chains to an inch (1:4,746). Arable and pasture fields are distinguished by color washes, and woodland and other features by conventional symbols. Source: Devon Record Office, Gittisham tithe map.

The linchpin of the scheme for obtaining accurate maps from unsupervised private surveyors was the application of a system of rigorous checks to the completed work. The work of testing maps was a prodigious effort. By 21 March 1844 the London office had tested 2,017 maps, the work of 405 surveyors. They had approved 1,399 of these maps, had rejected 521, and were still considering 97. To expedite this operation, and presumably also to ensure consistency, examiners were provided with printed examination papers which provided a checklist of potential faults and a convenient schedule for organizing their comments.

7.4 Registration of Title to Land

When the British government received a report from its Commission on the Law of Real Property in 1830, it marked almost exactly two centuries of attempts to suppress the inconvenience, expense, and potential for fraud associated with transferring ownership of land by private conveyance (by reference to private deeds with no public verification or deposit). Dowson and Shepherd, in their international review of land registration, define the key defect of private conveyancing as the total absolution of the vendor from responsibility over title. The vendor can quite properly say: "You must take the land as I have described it, whether such description is right or wrong. With regard to title deeds and evidence of facts, you must accept what I happen to have in my possession—(which may be nothing at all). The rest you must search and pay for. The title you can accept or reject."[41]

Title deeds are drawn up as a single copy; if one is lost or destroyed, all proof is gone. A legal examination of the title deeds is made, but private conveyancing does not require an examination of the actual property for comparison with descriptions in the deeds, so purchasers are seldom aware of the limits of their property in any sense that would satisfy a court.

From about 1830 there was a great deal of parliamentary discussion of a national system of land registration and whether this should be based on public registration of previously private deeds or, more radically, by registration of title on cadastral maps in a public registry as Sir Robert Torrens introduced to the Australasian colonies (see chapter 8.13). That this largely failed to occur in any form is evidenced by remarks made in 1913 by His Majesty's land registrar, Sir Charles Fortescue-Brickdale:

> During the last thirteen years over 150,000 conveyances have passed through the Land Registry and it is our experience that about one in every four descriptions necessitates a visit to the ground to clear up some question of doubt . . . properties are described by names long disused, by occupiers long since departed; wrong dimensions are inserted, wrong positions, wrong boundaries, no boundaries; half a house is conveyed as if it were

> the whole, six houses as if they were five, north appears for south, long for short, straight for crooked, a house on one side of a street is conveyed by mistake for one on the other.[42]

In some restricted parts of the country, local acts had required the registration of deeds for some time: for the Bedford Level in the Fens since 1663, in Yorkshire from the early eighteenth century, and in Middlesex from 1708. In practice, these registers did not greatly reduce the cost of conveyancing. The registers signaled the presence of deeds but not what they contained, so large numbers of transactions might still need to be searched, while in places where the same surnames occurred frequently, the registers were well nigh useless.

In 1830 a parliamentary commission recommended this inferior system of registering deeds rather than titles for want of suitable cadastral maps to form a general index of titles. A series of bills to introduce such registration of deeds was presented to Parliament, but by 1850 none had become law. By the 1850s the inadequacies of even the most comprehensive registration of deeds were clear to the conveyancing reform movement, and reformers turned their attention to registration of title, which in turn brought the need of a proper cadastral survey to the attention of the British government for the first time since 1836–37, when Robert Kearsley Dawson had argued that the tithe commutation surveys might be extended to form such a cadaster (section 7.3).

In 1850 the registration and conveyancing commissioners were still recommending a system based on the registration of deeds, but with in addition "an index referring to the land itself founded upon a public map."[43] A bill was drafted based on the recommendations of the commissioners' report. This was debated in Parliament and referred in 1853 to a select committee which recommended that yet another royal commission be appointed, this time to examine specifically the question of registration of title. They reported in 1857 and firmly rejected registration of deeds in favor of registration of titles on general index maps.[44] Cadastral maps were to play a pivotal role in this scheme, and their nature was being discussed by the British government at exactly the same time that Robert Torrens was steering the Torrens Act, which provided a map-based system of land title registration, through the South Australia legislature (chapter 8.13). A critical difference, however, between the Torrens Acts and the Land Registry Act which finally passed through the British Parliament in 1862 was that in Australia and New Zealand *all* land granted by the Crown had to be registered; in England, registration was completely voluntary, as at this time Parliament still objected to compulsory registration as a matter of principle.[45]

The 1862 act was very much a dead letter. After five years only 507 applications to register title to land had been received, and just 327 effected. By 1875, only 650 titles had been registered. Setting aside the opposition of many sellers and purchasers to registration on ideological grounds, the practical problems with the system related to the requirement that the Land Registry should confirm and guarantee the boundaries as recorded on the maps deposited as part of the description of each title. The 1862 act contained no provisions to produce maps, so 1:2,500 scale Ordnance Survey maps (i.e., those produced by the national mapping agency) could be used in those limited areas of the country where they were extant, and tithe maps, enclosure maps, and even private estate maps were adopted elsewhere. The cost of new survey was sufficient to deter most owners from registering their properties.[46] When land was registered, property boundaries shown on the deposited maps were checked on the ground by staff from the Tithe Commission's map department. This was an expensive and often inconclusive business, involving the appointment of a surveyor who tried to arbitrate between owners of adjoining properties and then recorded the agreed boundary. Agreement was very often difficult to reach.

A fourth royal commission on land registration was set up and inquired into the abject failure of the 1862 act. Its recommendations led to the 1875 Land Transfer Act, which was again purely voluntary and again an absolute failure, with only forty-seven titles registered after three years. Its object was to make land registration easier and thus more attractive to landowners.[47] It tried to do this by simplifying and reducing the cost of the index maps on which boundaries of all registered properties were to be recorded. Boundaries could be "general" boundaries, a euphemism of the day for indefinite boundaries. All the act required was that "registered land shall be described in such a manner as the Registrar thinks best calculated to secure accuracy, but such description shall not be conclusive as to boundaries or extent of the registered land." The Land Registry was still not empowered to make its own maps, so tithe maps continued to be used where Ordnance Survey 1:2,500 maps had not been published. As the head of the Land Registry, Sir Charles Fortescue Brickdale, said in 1900: "Had it not been for the triumphant proof furnished by [Australian] experience that the system really meets the needs of persons who want to sell and mortgage their land, there can be little doubt that the discredit inflicted upon it by the unfortunate experiments made in England . . . would have sufficed to destroy all faith in it as a practical possibility."[48]

Compulsory registration within defined districts was first effected in England by the 1897 Land Transfer Act in the counties of London and Middlesex and a small number of then county boroughs. However, the overwhelming majority of the country was left with optional procedures. At all stages and at all times the Treasury was reluctant to vote money for the general cadastral survey of the country which was necessary for effective registration.[49] The parsimony of the British government in this respect has already been demonstrated in section 7.3 in connection with the extension of the tithe surveys to form a general map of the nation. Why did the British government continue to be such an obdurate opponent of a mapped cadaster which other national governments considered very much a sine qua non by this time?

Thomas Colby of the Ordnance Survey map office spoke of land registration in terms of a "lurking spectre" and was aware that "a considerable prejudice is supposed to exist against any measure which might facilitate a Register of Property."[50] On what was such prejudice founded?

To Dowson and Shepherd, the answer is to be found in the customary familiarity of landowners in private conveyancing, despite all its acknowledged faults, and in the fact that, despite all the potential for fraud, lawyers operating the system worked with professional care.[51] The habitual practice of primogeniture avoided the fragmentation of properties on death, and the general stability of the countryside once enclosed set "barriers to petty encroachments and acts as a powerful corrective to untrustworthy or unintelligible plans."[52] To these should doubtless also be added the lobbying weight of the legal profession, which, in the nineteenth century as now, saw conveyancing as the bread and butter of its practices. Furthermore, in those parts of the world such as France and India, where general registers of property had been long established at the time when the British Parliament was still debating their merits and demerits, registration was desired as much to enable fair and adequate land taxation as to facilitate land transfer.

7.5 Land Taxation and Cadastral Mapping

In continental European countries one of the main motives for a centrally organized mapped cadaster was the desire to improve the efficiency with which land taxes were levied. In England there was considerable debate from the late seventeenth century to the late eighteenth century about the best way to raise the huge amounts of money needed to finance the country's role in European wars. This debate, however, did not include a discussion of a cadaster. In the nineteenth century the British government was to use cadastral maps to abolish one particular land tax, the tithe, but the general land tax was collected every year between 1692 and 1798 without a surveyed or mapped base or even a comprehensive written register.

The English land tax was levied *on* the landed classes *by* the landed classes. In many parts of continental Europe, such as the Austrian Habsburg lands and the Duchy of Cleve, the aristocracy had traditionally collected but not paid the land tax, and a mapped cadaster proved a vital weapon for the monarch in his battle to subject the nobility to taxation.[53] In England the propertied classes and central government also clashed over land taxation. When they reached a consensus, it was one in which it was in no one's interest to compile a mapped cadaster.

Government revenues in England came from three main sources: customs duties (levied on foreign goods and not discussed here), excise duties levied on home-produced goods, and assessed taxes on the home population, normally a property tax, sometimes combined with a poll tax. The levy which we know as "the land tax" was introduced in 1692 under 4 William III, c. 1, to finance "a vigorous war against France." In theory it was a general tax on income from real and personal property and the profits of office, but in practice only income from immovable property was taxed, and the levy thus became known as the land tax. The owner of real property was taxed on the income which would have accrued from that property had it been let "at rack rent," a notional assessed rent, not an actual rent. The tax officials compiled returns which list the owner and/or occupier of each property, its rack rent, and the amount of tax paid.[54] The returns did not always give the name of the property and never indicated its acreage or boundaries. Still less was there any surveying or mapping.[55]

Having granted the land tax in 1692, the English landed classes, as members of Parliament, blocked attempts by the government in the seventeenth and early eighteenth centuries to extend the scope of the excise duties and decrease or even abolish the land tax.[56] They did so because they wanted to keep control of taxation. They administered and hence controlled the land tax, while the excise was levied by excise officers, professional government servants whose numbers expanded greatly in the eighteenth century.[57] Excise officers had extensive powers of search and control over units of production in industries subject to the excise. Their powers were resented, and they themselves were despised by the landowning "country" members of Parliament, who prided themselves on their independence of government. To limit these officers' power, landowners favored the land tax over the excise. They regarded the excise as a permanent charge which, once granted, was outside their control. The land tax, by contrast, was intended to last only for the duration of the war.[58] As Samuel Pepys commented when the same question was discussed in the 1660s: "The true reason why the country-gentleman is for a land Tax and against a general Excize, is because they are fearful that if the latter be granted, they shall never get it down again; whereas the land tax will be but for so much, and when the war ceases there will be no ground got by the Court to keep it up."[59]

Landowners were also highly suspicious of borrowing as the alternative to immediate taxation, which seemed to them to increase the power, wealth, and influence of financiers at the expense of the landowning classes.[60] Thus landowners prefered the land tax to the excise and to borrowing because they had more control over it, although they may well have paid more in tax as a result.[61]

While most landowners favored the land tax, they were divided as to how it should be imposed. The tax was not levied equally. In particular, landowners in the north of the country paid little compared with those in the south, and merchants and the bourgeoisie paid little compared with the landed gentry, since the former held most of their wealth in stocks and bonds, which in practice were not taxed. Changes which would have made the tax more equitable, its yield greater, or its collection more efficient were debated but in general not implemented, since they were too contentious.[62] The landed classes rejected any changes which would have diminished their control over the tax and increased that of the executive. A mapped or surveyed cadaster would necessarily have been in the hands of professional surveyors under the control of central government and would have removed control of the tax from the local landowners and as such was unthinkable.

Thus the land tax remained, and remained largely unchanged. The king and his government were content that the tax produced a large amount of money reliably, uncontroversially, and with little effort on their part.[63] In fact the land tax declined in real terms in the eighteenth century, since land was not revalued. At the same time, government expenditure rose rapidly, but the government avoided controversial interference with the land tax by borrowing money and by levying customs and excise duties and other assessed taxes. Although the landowners had successfully resisted the *general* excise, the customs and excise duties increased steadily as a proportion of government revenue.[64]

The government and the king were content with the political stability the land tax brought. It was introduced against a background of continual political turmoil. The Civil War and the Interregnum had brought no lasting political settlement, nor did the Restoration of the Stuart monarchs in 1660, since James II was forced to abdicate less than thirty years later, and William of Orange was invited to the throne in 1688.[65] Against this background the land tax was essential to the achievement of domestic order and political stability. It was operated locally, not centrally or bureaucratically; in administering it, the political nation was attached to the apparatus of the state and not, as had happened so many times in the recent past, alienated from it. The assessors and collectors of the tax were people of some substance, often yeomen or graziers in the countryside and prosperous tradesmen in the town. They were not professional government servants but did the tax work to confirm their status in the locality. The land-tax commission in each area was the preserve of the local gentry. Central government was represented by the land-tax agents, but in practice central government intervened very little, and the administration was left largely in the hands of assessors, collectors, and commissioners. This confirmed their local authority and their place in the political nation as substantial men, free of the corrupting taint of government but active in establishing and maintaining the political settlement.[66] This political consensus would have been prejudiced by the kind of heavy-handed government intervention in the tax administration which a cadaster would have demanded.

The central government was thus content with a tax which brought in substantial sums of money and bound the nation together. The landowners were content with the land tax, to which they and the clergy contributed but which they controlled and which was never so onerous as to be unsupportable. The common people were content with the tax, since in general they did not pay it. In some areas they were recorded as ratepayers and actually handed over the money to the collectors, but it was the landowners who finally footed the bill.[67] There was thus no agitation from below for reform of the tax, as was the case in Cleve and Milan.

Not the central government, nor landowners, nor the common people had any incentive to introduce a surveyed or mapped cadaster for the land tax. Such a cadaster would have aroused the opposition of landowners without benefiting the government or the common people. It would have been unjustifiably expensive, when there was no wholesale evasion of the tax and when the king was desperately short of money. England thus stands apart from the general European trend of levying land taxes on the basis of cadastral surveys in the eighteenth and nineteenth centuries.[68]

7.6 The Ordnance Survey's "Cadastral" Maps

In June 1855, Sir Henry James, the newly appointed director of the Ordnance Survey, said in evidence before the House of Lords Select Committee on Agricultural Statistics that the Survey "should be empowered to map every property in the country . . . nothing would be more simple."[69] In writing the Ordnance Survey annual report for 1860, Sir Henry urged the resurvey of southern Britain at a large scale of 1:2,500, for which he coined the name "Cadastral Survey." This is a term extensively used in the period 1861–63 by those involved in considering the 1:2,500 survey. In 1862, for example, a House of Commons select committee reported on the "Cadastral Survey."[70] After about 1875, the year in which James retired from the Ordance Survey, the "Cadastral Survey" is not much referred to.

Despite this use of the term, a cadastral survey in the sense of a survey of *properties* was never envisaged by the Ordnance Survey, and it was never done. The Ordnance Survey 1:2,500 scale maps identify each parcel of land by parcel numbers, but these refer not to registers of owners and occupiers but to books of reference in which the acreage and, until about 1880, the land use of each parcel were recorded.[71] The ownership of parcels, the hallmark of a true cadastral survey, is not recorded in these books. Though comparable in scale with the English tithe maps and private estate maps and also with continental European full cadastral surveys, it is with the similarity of scale that the "cadastral" likeness ends.

Robert Kearsley Dawson had urged the completion of a full cadastral survey in 1836–37 in association with tithe commutation (section 7.3), and both Parliament and the Ordnance Survey were well aware of what was meant by "cadastral" surveying and mapping on the Continent. W. S. Farr of the General Register Office attended the 1853 International Statistical Congress in Brussels and reported to the Treasury on the congress's cadastral mapping debates; his views are recorded in the parliamentary inquiries published in the 1850s into the most useful scales for Ordnance Survey maps.[72] The 1862 select committee report "on the Cadastral Survey" includes extracts from the address by the Portuguese delegate, M. Davila, who acted as rapporteur of the cadastral survey section. He set out the congress's manifesto of advantages which a country could obtain from a cadastral survey, which was received with acclamation by those present at the time.[73]

England, though, was to benefit from none of these putative advantages of a map-based national cadaster.

It can be argued, as does Richard Oliver, that the Ordnance Survey's use of the word "cadastral" was just a convenient, non-scale-specific label to describe its proposed resurvey of southern England in debates where opinion within the Ordnance Survey was divided between advocates favoring revision of the one-inch series of maps and those, including director Sir Henry James, who fought for the principle of resurvey at the large, twenty-five-inch (1:2,500) scale.[74] James's use of the word "cadastral" probably also helped his advocacy of the resurvey to the Treasury by stressing its general usefulness to landowners and others who had previously commissioned large-scale maps from private surveyors or had them redrawn from the aging and imperfect body of tithe maps.[75]

7.7 The 1910 Finance Act and the Maps of "Lloyd George's Domesday"

Through the nineteenth century the area of land owned by the largest landed proprietors in England with estates of four thousand hectares or more steadily increased until the great landed families attained their greatest territorial power in the 1880s. Surveys such as the parliamentary *Return of Owners of Land, 1872–73* and analyses of this such as John Bateman's *Great Landowners of Great Britain and Ireland* (1883) confirmed this concentration of landed wealth in fewer and fewer hands. Political groups such as the Land Reform Movement sought to institute a free trade in land to break up these large landed estates and the political system which they fostered and which exacerbated the polarization of wealth in English society.[76]

A solution tried by Prime Minister Lloyd George's new Liberal government to redistribute this wealth, if not the ownership of land itself, was to tax increases in property values over time. The Finance Act of 29 April 1910 provided for the valuation of all land in the United Kingdom, so that increases in value since that date could be taxed when properties were transferred or sold.

The valuation was conducted by identifying and separately numbering hereditaments (each piece of separately occupied landed property) in "income tax parishes" (civil parishes or amalgamations of civil parishes). Once identified on lists, properties were valued in the field, and the boundaries of each hereditament were recorded on 1:2,500 or larger-scale Ordnance Survey maps. Hereditament numbers recorded on the maps linked the maps to valuation books in which owners' names and addresses and the property values were recorded. The legislation of 1910 finally gave England its first and only comprehensive mapped cadastral survey covering every individual hereditament (figure 7.10). These maps and the valuation books are still extant, but the increment value duty itself was but shortlived; it was repealed just ten years later.[77]

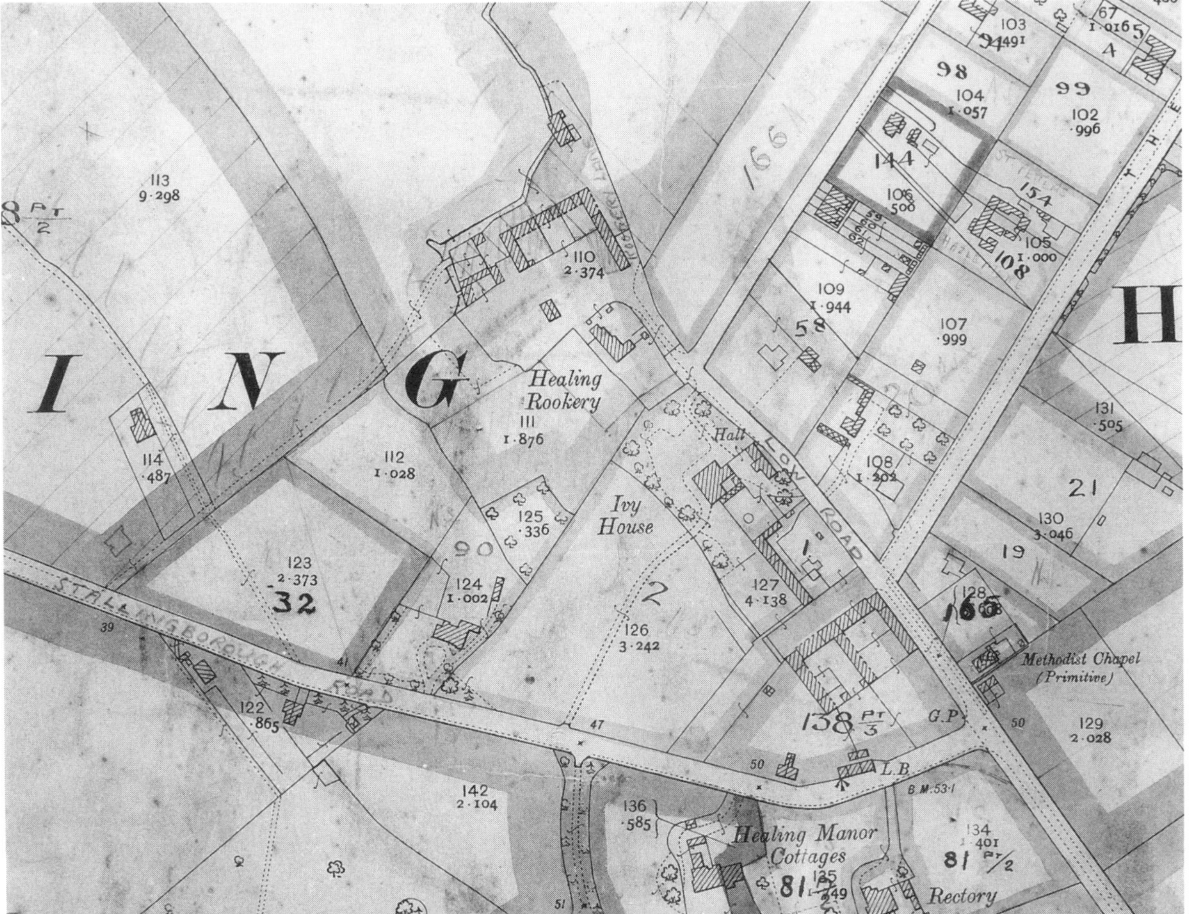

Fig. 7.10 An extract from a 1910 valuation map sheet. Source: Public Record Office, London, IR 130/3/178.

7.8 The Cadastral Map in England and Wales

Chapter 1.2 noted the precocious development of private estate mapping in England in comparison with the position in, say, France or the Austrian Habsburg territories. This can be related to differential timings of the transition from feudalism to capitalist modes of agrarian production. In contrast, the recency of comprehensive national cadastral mapping in England stands in stark contrast with both France and most other continental nations and states. Whereas all the land of metropolitan France has been recorded in periodically revised cadastral surveys since the Revolution, that position (setting aside the short-lived 1910 Finance Act survey) has even today not been fully achieved in England and Wales.

The Crown and parliamentary surveys of seventeenth-century England are notable for their explicit lack of mapping provisions. In the sixteenth and seventeenth centuries, private enclosure agreements in a majority of English parishes were achieved satisfactorily without recourse to maps. The private enclosure acts of the accelerated phase of enclosure of the later eighteenth century invariably required maps, though it is difficult to find an explanation of this within the process of enclosure itself. To us the usefulness of maps in enclosure and redistribution of strip holdings is obvious. It is argued in section 7.2 that the appearance of maps as an intrinsic part of the enclosure process in the eighteenth century can be related to the fact that the use of maps was by then well established in the management of private estates. Enclosure swept away some of the last vestiges of feudal society; enclosure maps also symbolize the new capitalist landed society which supplanted it.

The tithe maps of mid-nineteenth-century England and Wales have been hailed as a high point of Western cadastral achievement. Those who drafted the successful tithe commutation bill clearly believed that cadastral maps ought to play some part in the reform, a fact with which Parliament concurred by passing those clauses into the Tithe Commutation Act of 1836. However, no sooner was the legislation in effect than the practice of tithe commutation revealed ambivalence about the need for maps. Without a clear understanding of what role maps were to play, there was uncertainty about the precise nature of the maps required.

The superintendent of tithe commutation surveys, Robert Kearsley Dawson, argued long and hard, but in vain, for the widening of tithe mapping to effect a general cadaster of England and Wales. Evidence was presented to the government of both the internal advantages to be gained and the profit which a number of continental European countries derived from their established national cadasters. Questions of cost were adduced to counter these, together with a fear of subverting the essentially voluntary nature of tithe commutation by making the mapping more onerous and costly. There was also opposition from the Ordnance Survey, which feared the advance of private surveyors into its national mapping domain.

Vested interests—in this case, the legal profession, the bulk of whose income derived from property conveyancing—were also instrumental in deferring efficient, map-based land registration on either continental or Torrens models for the whole of the nineteenth and much of the twentieth centuries.

A further explanation of the reluctance of the British government to adopt cadastral maps in the eighteenth and nineteenth centuries is the way in which the land tax was levied. Notional income from land was taxed as part of a general income tax imposed by the landed classes on themselves. Thus there was not the pressure to reform the basis of its collection in England as there was in postrevolutionary France, in the Austrian Habsburg territories, or among some German states. The new increment value land tax introduced by Prime Minister Lloyd George's government in 1910 *was* founded on the principal of assessing individual land parcels, or more specifically, their annual increases in value. This legislation did indeed produce England's first, and to date only, fully comprehensive mapped cadastral survey. As an active instrument of government policy, this survey, though, lasted no longer than the First World War. Within ten years the taxation was repealed, and the survey on which its levy was based and recorded became a historical document. Today, there is compulsory registration of real property in England and Wales, but since registration occurs only when ownership is transferred, it will still be some time before every hereditament is included in the comprehensive, map-based cadaster which is being built up.

Acknowledgments

We should like to acknowledge the help of Miss Geraldine Beech, Dr. John Chapman, Dr. Richard Oliver, Dr. Michael Turner, and Dr. Sarah Wilmot.

8

COLONIAL SETTLEMENT FROM EUROPE

8.1 Maps and Early Colonial Settlement

In the early years of European settlement in the New World in the seventeenth century, whether in the Liesbeeck River valley east of Cape Town in South Africa (figure 8.1), or on the Atlantic seaboard of North America, land surveying and the production of cadastral maps became established as a concomitant of colonial settlement.[1] Land availability, if not the only lure of migrations from Europe, was a most important influence in the individual decision to migrate. As Sarah Hughes comments in the context of Virginia: "Immigrant colonists gazing at a wilderness envisaged its taming and imagined new markers bounding the edges of their own fields and meadows. The men who could measure the metes and bounds of those fields held the key to transforming a worthless, uncultivated territory into individual farms."[2]

The actual process of settlement and the relationships between cadastral surveying and settlement form varied from one colony to another in North America, but two broad types reflect radically different attitudes to political and social control. These can be called the New England and Virginia methods. The Puritan colonies exerted social control over land by the orderly, planned granting of blocks of land to town communities.[3] These were surveyed first and divided into farm tracts and peopled second. In colonies south of Pennsylvania, landholdings were larger and granted to individuals who were free to pick and choose the best land as they perceived the realities of the physical environment. Land tracts were claimed first and surveyed second by metes and bounds methods (figure 8.2). In contrast to the concept of community living which governed New England political geography, the "Virginia system" was strongly individualistic and competitive and reflected colonial authorities' lack of concern for ensuring orderly and equal access to resources or with shaping the pattern of settlement.

Though the position of land plats (New World terminology for cadastral plans) in the settlement process differed, the method of survey in North America by compass or circumferentor (primitive theodolite) and chain was much the same, whether of individual tracts in Georgia,

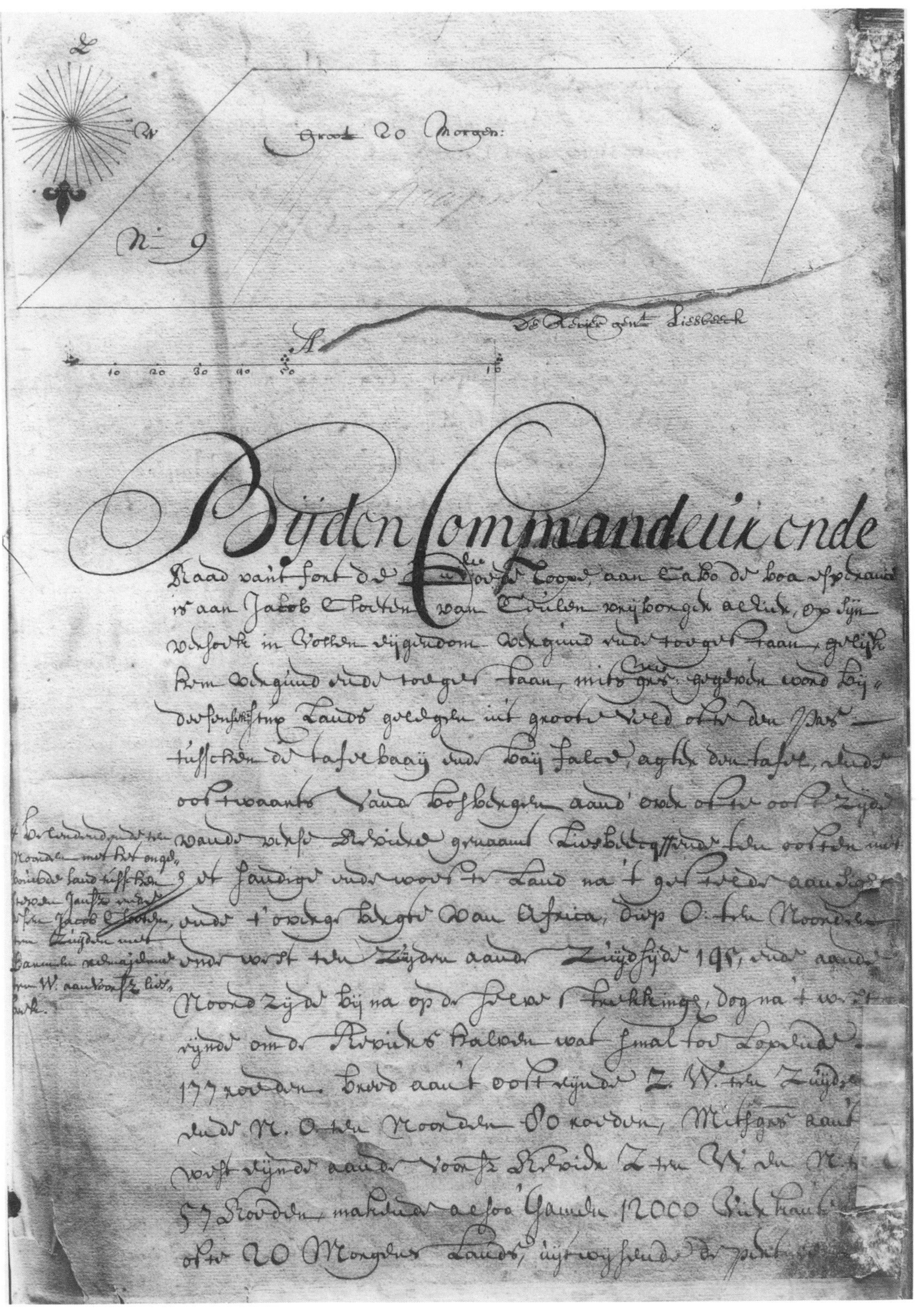

Groot 20 Morgen:

No 9

de Rivier genl Liesbeeck

Bijden Commandeur ende

Fig. 8.1 Grant of land to Jacob Cloete on the Liesbeeck River, South Africa, 1657. Source: Cape Archives Depot, M 117 (after Christopher 1976a: frontispiece).

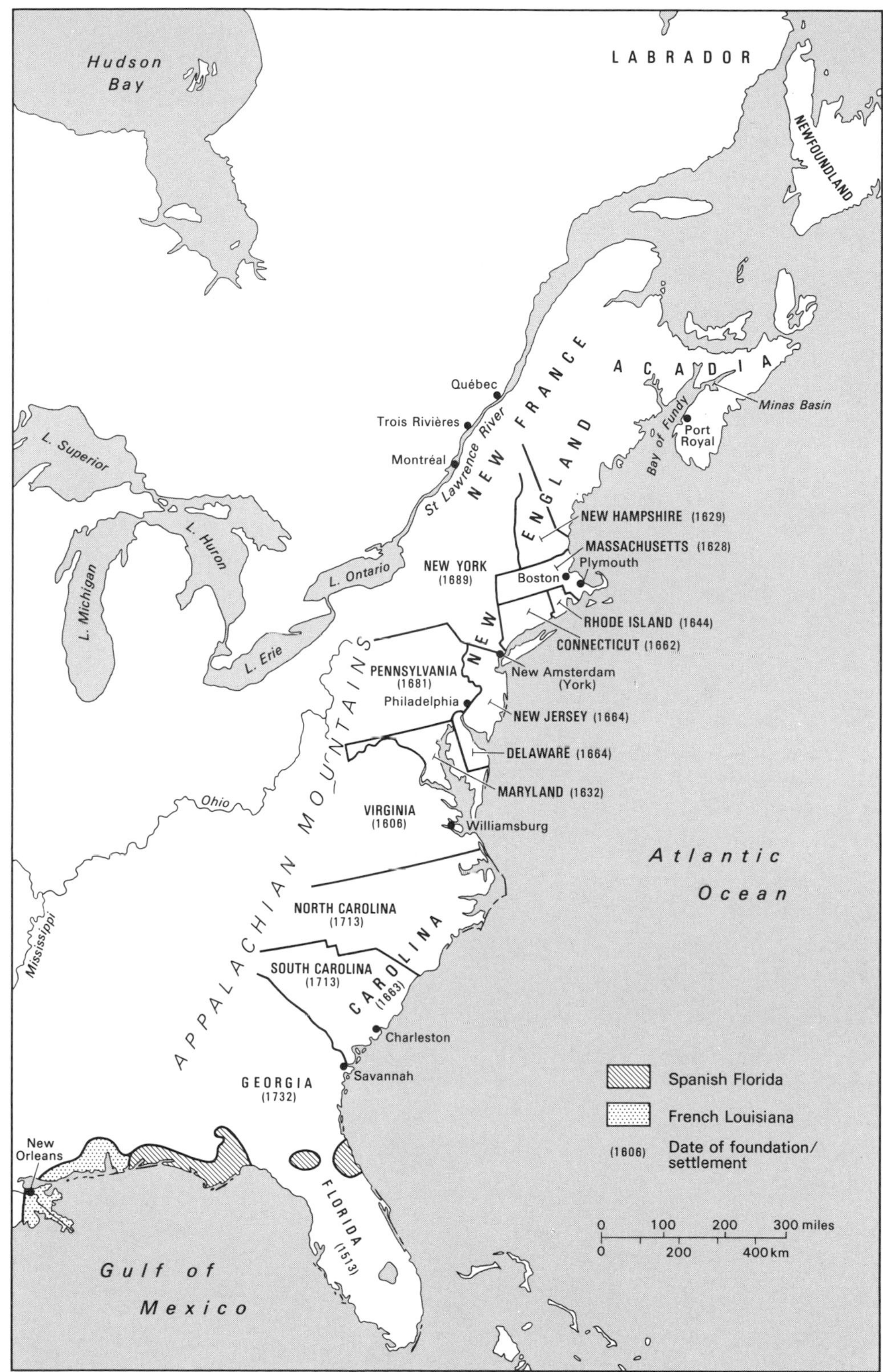

Fig. 8.2 The eastern colonies of North America. Sources: Droysens 1886; Muir and Philip 1927; Shepherd 1930; Kinder and Hilgermann 1974–78; Darby and Fullard 1978.

townships in New Hampshire, or *seigneuries* in Quebec.[4] In 1610 the English surveyor William Folkingham had dedicated his *Feudigraphia* to "all undertakers in the plantations of Ireland or Virginia."[5] Didactic treatises published in Europe and in the colonies adapted Old World surveying and mapping techniques to colonial needs, advocating in particular the compass traverse for surveying new lands.[6] One of the reasons for John Love's writing his *Geodaesia; or, The Art of Surveying and Measuring of Land Made Easie* (1688) was that he had earlier seen "young men, in *America*, often nonplus'd so, that their Books would not help them forward, particularly in *Carolina*, about Laying out Lands, when a certain quantity of Acres has been given to be laid out five or six times as broad as long."[7]

Whether texts such as Folkingham's or Love's were actually read by surveyors in the colonies is another matter. There was certainly a need of advice and instruction on instruments and techniques, for early seventeenth-century surveyors commonly estimated angles by eye and distances by pacing, by counting the footsteps of their horse, or even from a slow-moving boat.[8]

From the second half of the eighteenth century, the increasing number of surveying textbooks published in America testifies to both the continuing need and, by then, a market for works of instruction. Thomas Abel's *Substensial Plain Trigonometry . . . And this Method Apply'd to Navigation and Surveying* of 1761 and John Carter's *Young Surveyor's Instructor; or, An Introduction to the Art of Surveying* of 1774 are examples of pre-Independence texts published in Philadelphia. In 1785, the year in which the presettlement mapping of the federal public domain began (section 8.8), the fourth edition of Robert Gibson's *Treatise of Practical Surveying* appeared, "Adapted to the Use of American Surveyors; Some Parts of the Work have been abridged, and other Parts totally omitted, as being of little or no Use to the American Surveyor." The prodigious survey and mapping activity which the 1785 and later land ordinances generated was matched and serviced by more texts in the following quarter century, for example, those by Wall in 1788, Clendinin in 1793, and Moore in 1796 (who has a section on the recovery of lost boundaries when original survey markers have disappeared); those of Dewey, Jess (whose book was "adapted for the easy and Regular instruction of Youth in our American Schools"), and Little in 1799; Flint in 1804; texts by Conway and by Eliot in 1807; Gummere in 1814; and treatises by each of Antony and Day in 1817; and Fairlamb and Hanna in 1818.[9]

8.2 Bermuda

After the right to colonize the Somers Islands (Islands of Bermuda) was granted to the Virginia Company of London, the first party of settlers set out for the island in 1612.[10] In 1614 the company employed Richard Norwood, an English mathematician, navigator, and surveyor, to conduct a reconnaissance survey to determine the extent of land available for settlement. This revealed that there was enough potential agricultural land for each of the company's adventurers to receive a share of twenty-five acres. During 1616 to 1617, Norwood made a second survey and map (subsequently published) on which the island is divided into eight tribes or parishes, each of which is further subdivided into fifty shares, or farms of twenty-five acres. The survey fixed farm boundaries and thus laid down a basis for the orderly settlement and development of the plantation. Within five years of Norwood's cadastral survey, each share had been settled, and the principle of settlement effected through survey and mapping had been established in an English colony. As Wesley Craven comments in his introduction to Richard Norwood's *Journal*, "In the first survey Norwood supplied the information necessary for the drafting of an exact plan of settlement; in the second he performed [what was to become] the normal role of the American surveyor in the conversion of a claim into a title."[11]

8.3 The "Virginia Method" of Settlement before Cadastral Survey

In Bermuda, survey preceded European settlement, and the position there was thus much closer to that obtaining in the Puritan New England colonies than in the Virginia Company of London's mainland colony. The tendency in all colonies from Pennsylvania to Georgia was for the individual to locate and settle land prior to survey and also for tracts to be patented in relatively large blocks. The granting of land on "headrights" (allocations to heads of households with additional land according to the number of their dependents) became the principal basis for obtaining title to land in seventeenth-century Virginia. Its origin was in the Virginia Company's 1618 charter, which included the provision: "that for all persons . . . which during the next seven years after Midsummer Day 1618 shall go into Virginia with intent there to inhabite if they continue there three years or dye after they are shiped there shall be made a grant of fifty acres for every person upon a first division."[12]

From 1642 it was a legal requirement in Virginia that all surveyors "shall deliver an exact plott to each parcell surveyed and measured" (figure 8.3).[13] The 1642 enactment coincided with a period of contentious disputes over boundaries; from the 1640s a series of government measures was designed to tighten up survey and land patent procedures. In Maryland early seventeenth-century surveying of tidewater land grants was similarly slipshod. Carville Earle recounts how, in Allhallows Parish, surveyors simply measured off a desired distance along an estuary shore, projected the other three lines to produce a rectangular lot, and then calculated its area.[14] Field measurements of distance, direction, and area were also very approximate. In Earle's opinion, surveyors who measured distance with a Gunter's chain before about 1670 were few and far between. Indeed the Maryland assembly of 1674 recognized the imprecision of surveys and

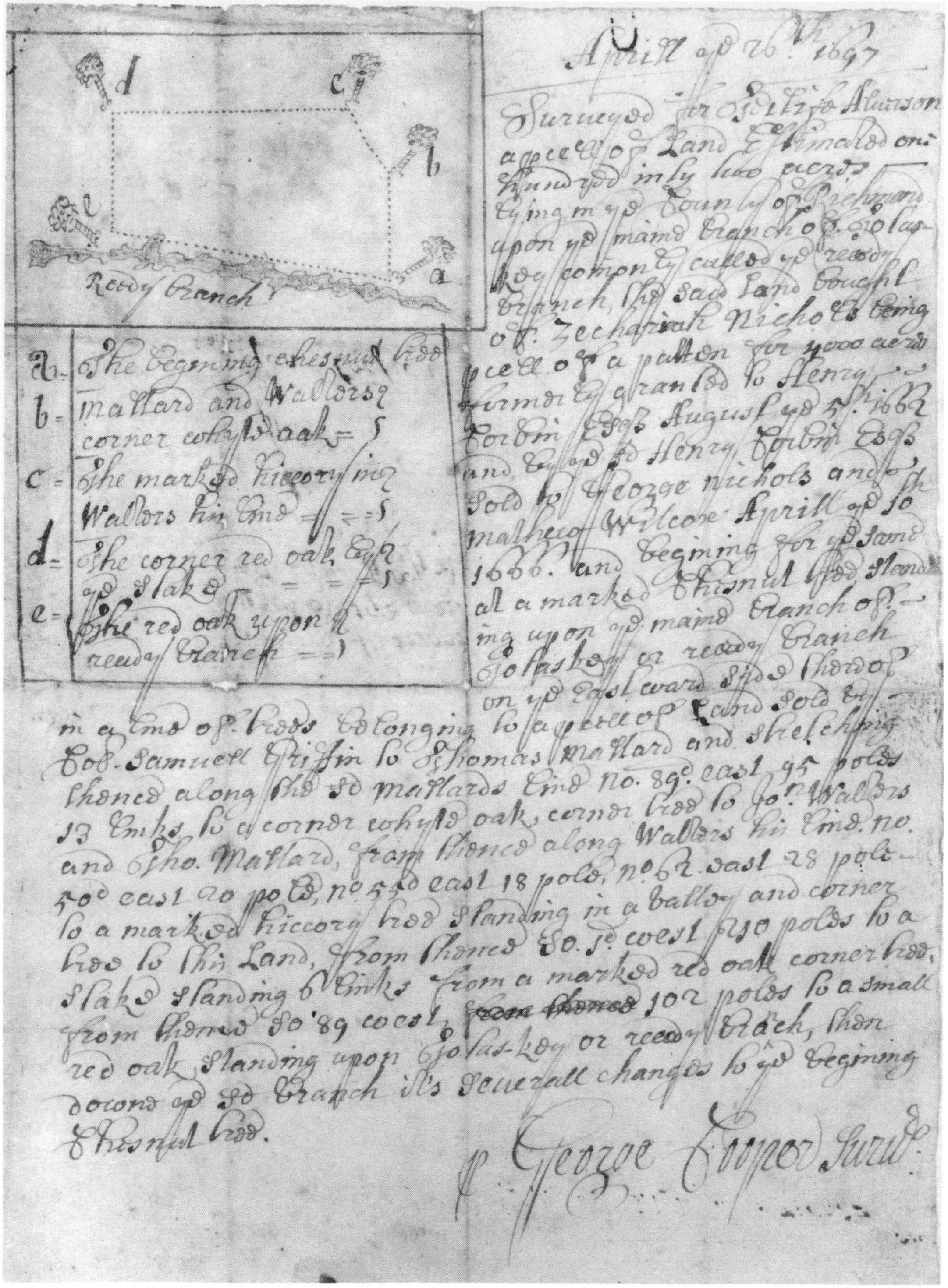

Fig. 8.3 Survey of Teleife Alverson, a grant of 192 acres in Richmond County, Virginia, 1697. Source: Virginia State Library, Richmond, Va., Northern Neck Surveys Richmond County, Box 40.

passed a law which provided a doubling of fees for surveyors who used chain and circumferentor. The eighteenth-century resurveys in Allhallows Parish, which Earle has exhaustively analyzed, reveal errors ranging from a 150 percent surplus to a 42 percent deficiency. In 1723 the Maryland assembly provided a procedure for the "Ease of the Inhabitants in Examining evidences relating to the Bounds of Land" whereby disinterested freeholders could be appointed commissioners to examine and arbitrate in boundary disputes. They were not required to commission new surveys, and the procedure proved inexpensive and effective. As Earle comments, "By this act, the Assembly tacitly acknowledged that cadastral boundaries were based on customary bounds as proven by reliable local residents and not on the inaccurate descriptions given by incompetent seventeenth-century surveyors."[15]

Once land had been improved by clearance and settlement and was recognized by Europeans as a valuable resource, careless surveying created multifarious problems. When land was thought of as limitless, there was little incentive to accuracy. The recording of survey work on a plat for public witness was but one symptom of a developing capitalist attitude to land in the tidewater colonies as social and economic changes enhanced the value of survey and provided a role for property maps in the historical process of land settlement. Some Virginia land plats, with their ornamentation, color coding, and use of conventional signs, seem also to have acquired some of the symbolic meaning of English estate maps of this time as icons of the capitalist landowning system transplanted from an England where wealth and social standing could be measured by acres of land. Sarah Hughes cites James Minge's 1701 plan of William Byrd's Westover Plantation as a fine example of this genre (figure 8.4).[16]

In the eighteenth century, the presentation of these cadastral plans became more standardized, with the plat and written description on the same sheet of paper. Henceforth, all written descriptions included the date of survey, name of client, total acreage, a descriptive location in a county and a location by reference to marked topographical features, a history of the patent and changes in ownership, metes, and bounds, the names of adjoining property owners, and the surveyor's signature.[17] In addition, between 1713 and 1749, surveyors were required to note how much of each land grant was, in their opinion, suitable for cultivation.

By the middle of the eighteenth century, attempts were being made in other states to regulate the manner of surveying and mapping. In 1747 the surveyor-general of New Jersey issued instructions for his surveyors which included an example of how the cadastral map should be drafted. The map was to be drawn to scale and watercourses and distances were to be properly noted as well as the boundaries of adjoining properties and all natural features and roads. Surveyors were also required to submit composite cadastral maps of all the land allocations in the county in which they worked:

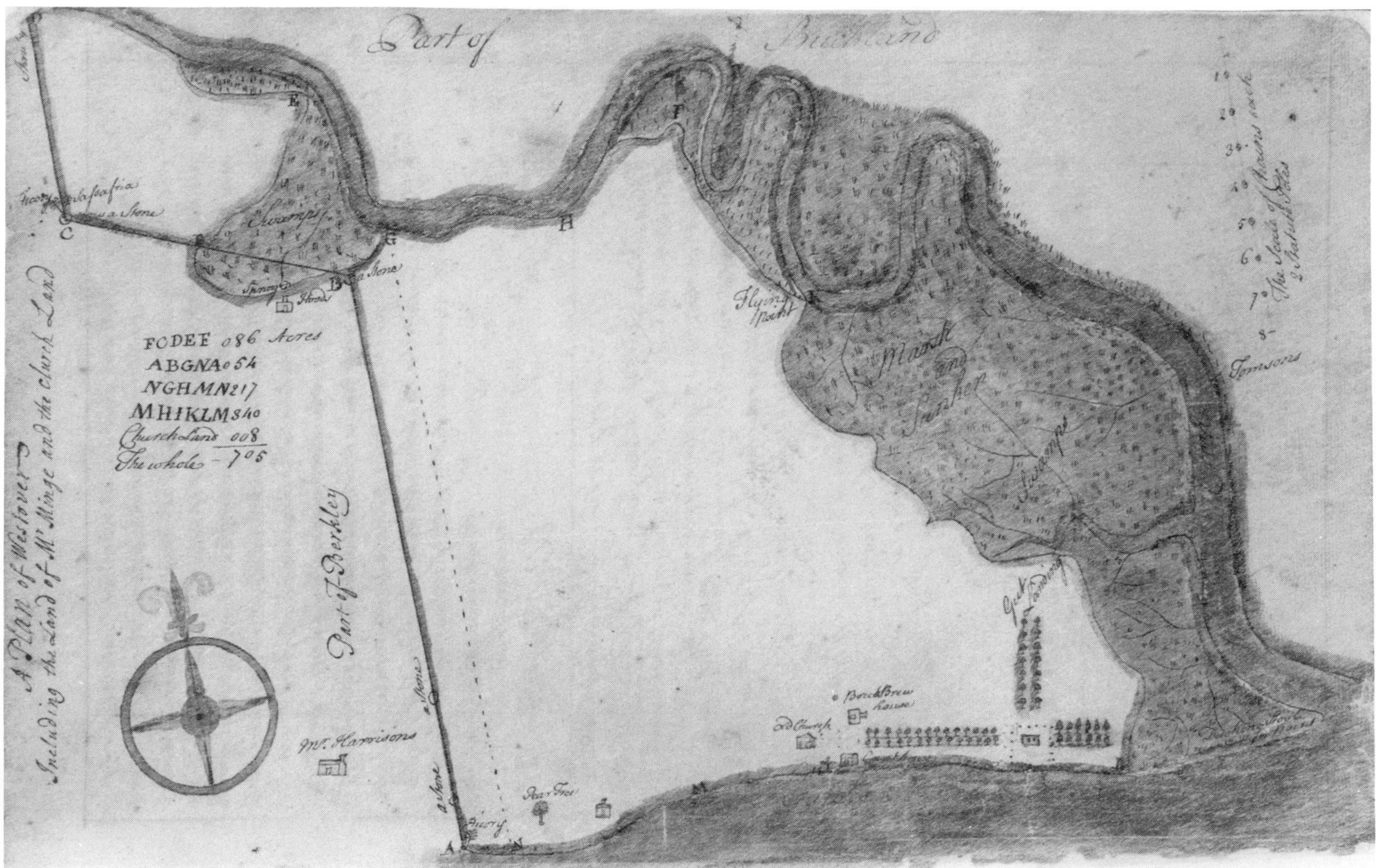

Fig. 8.4 James Minge's Westover Plantation, Virginia, 1701. Source: Virginia Historical Society, Richmond, Va.

As far as in you lies, for your own Satisfaction, and the Good of the Country, endeavour to make one Map of the Land appropriated in the County wherein you Survey; adding thereto, the several Tracts of Land you shall survey, with their proper Situation in respect of one another, to the best of your Judgements; let the Scale by which you do this, be one Mile, or eighty Chains to one Inch: But this will make some of the Tracts appear too small, that they will not admit of writing in them the Quantity, nor the Person to whom they were surveyed; therefore let the Tracts be numbered on the Map 1, 2, 3, 4, 5, and so on, and put a Table on the Margin of your Map. . . . And send me yearly a Copy of your Map and Table, in order for preventing the Confusion that otherwise the Country may fall into, by Surveyors encroaching one upon the other.[18]

As Peter Wacker comments, had such a set of maps been forthcoming from New Jersey surveyors, it would have done much to alleviate subsequent problems of overlapping claims. Despite entreaties throughout the eighteenth century, this particular instruction was not heeded. Not until 1788 were deputy surveyors in New Jersey required to keep a book of all unlocated lands and to return all surveys to the surveyor-general's office within one month—an attempt, albeit too late, to obviate overlapping land titles. Yet even these rules were not adhered to; not all deputies kept books, and many continued to survey where others had surveyed before.[19]

The output of cadastral maps in colonial America rose as colonization proceeded and more and more land titles had to be proved. This mapping activity was added to by the growing number of disputed boundaries which were commonly resolved by resurvey and mapping. For example, between 1607 and 1700, just five thousand square miles of tidewater Virginia had been surveyed and settled; by the eve of the Revolution, this had been extended by forty-five thousand square miles as settlement pushed westward with the surveyor following in its wake.[20] Survey and mapping accelerated during the eighteenth century, and in 1745 the surveyor Thomas Lewis surveyed thirty-one tracts around the present town of Lexington totaling 7,142 acres in thirty-nine days.[21]

This increased activity was fueled also by enactments such as the Virginia Land Law of 1705. This enabled a citizen of the colony to obtain up to 500 acres of virgin land for himself plus an additional 200 acres on headrights for every slave or servant, up to a maximum of 4,000 acres. In the colonies of the Deep South, such systems of headright grants made to the heads of families and bounty grants made to reward services rendered in the Revolutionary War were instituted as the colonies competed for immigrants. Public officials tried to make it as easy as possible to transfer land from public to private ownership.[22] By the end of the colonial period, the headright system was common practice in all the southern states as each tried to stimulate immigration and development. As George Burrington, from 1731 first governor of the Crown colony of North Carolina, asserted: "Land is not wanting for men in Carolina, but men for land."[23] In 1777 the Georgia legislature passed its land act, under which "every free white person, or head of a family, shall be entitled to, allotted, and granted him, two hundred acres of land, and for every other white person of the said family, fifty acres of land, and fifty acres for every negro, the property of such white person or family."[24] Headright grants were easily acquired by first obtaining a warrant for a specified number of acres. The county surveyor then laid out and marked the corners by running lines by compass before calculating the area of the tract and producing a plat. Finally, the plat and accompanying description were lodged in the county courthouse and with the state's surveyor-general (figure 8.5).

Though the headright system was attractive to the authorities, who thought it encouraged settlement, it was wide open to abuse by land speculators. The defects inherent in a metes and

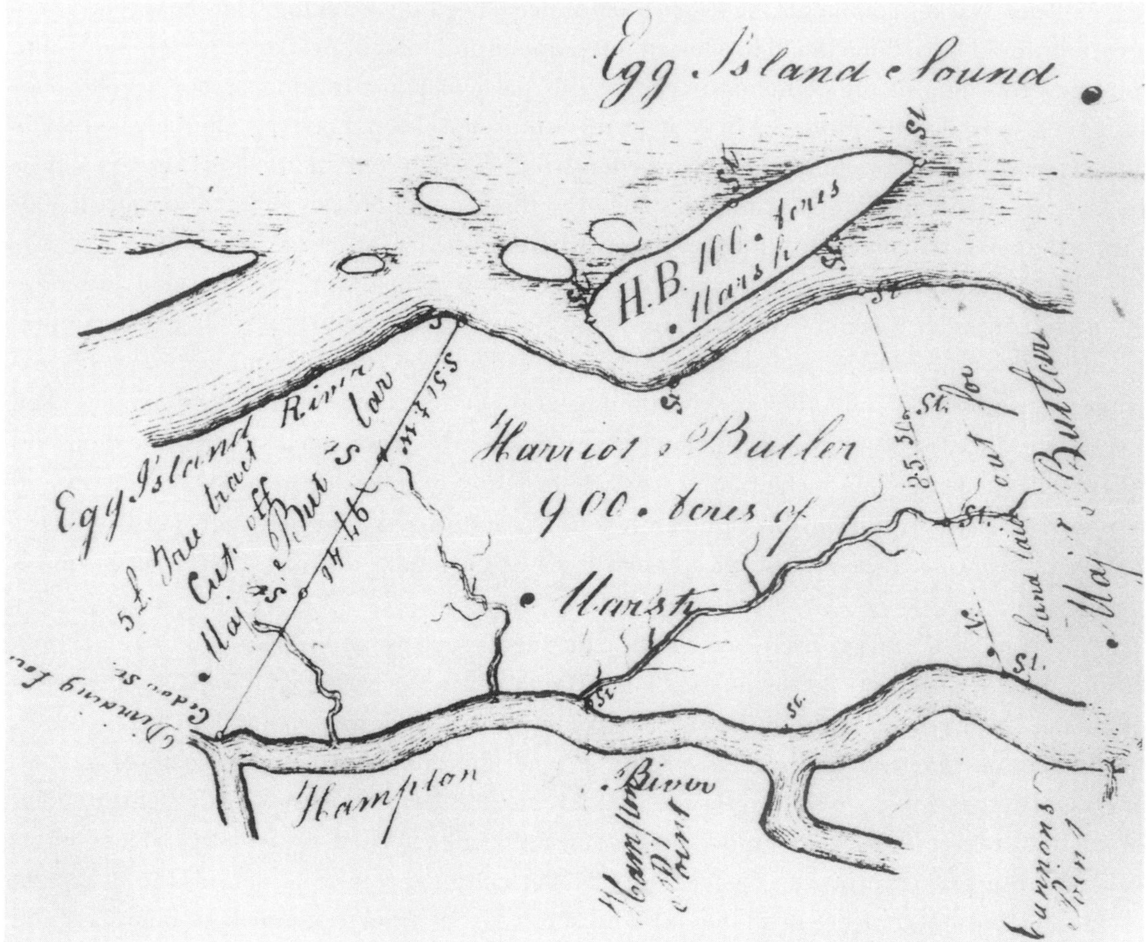

Fig. 8.5 Headright grant to Harriot Butler in Glynn County, Georgia. Source: Surveyor General Department, Georgia Department of Archives and History, Atlanta, Ga., Plat Book, DD p. 528.

bounds system of surveying were frequently compounded by ambiguous references to poorly defined physical features. Some surveyors, as Governor Dobbs of North Carolina asserted in a letter to the Board of Trade in 1755, merely established what kind of timber grew on the land they were commissioned to survey: "And at the fire side laid down their plan, if not joined to any neighbouring Plantation then named an imaginery Tree, a pine red white or black oak or hiccory etc and so enter beginning at a hiccory and so name imaginery Trees at any angle and conclude as usual so on to the first station. . . . You may judge what confusion that has and does create."[25] In South Carolina, headrights could be claimed for slaves. It was not uncommon for such claims to be made, the slaves sold, and then for the new master to claim further headrights on the same slaves.[26]

An excellent account of the actual practice of metes and bounds survey in the field in the eighteenth century is given by William Roome in his history of land surveying in the colony of East Jersey:

> The Deputy-Surveyor, when on a land expedition, would sally out on horseback, his compass duly boxed, and his chain in his saddle-bags behind him. Arriving at the point where the land lay which was to be surveyed, and where by previous appointment his assistants were to be found, the first step was to cut a straight stick . . . for a "Jacob's staff." . . . All things being now ready, the compass was placed at the beginning point and the bearing decided on, taken, and the flag-man sent ahead and placed in position. The compass was then removed to the flag-man's point, but no back sight left at the point it was removed from. The bearing was here again taken at the same degree as before (by the needle's point) and the flag-man again sent forward, and this was continued until the survey was completed—the chain-bearers following directly after the surveyor as nearly on the same line as practicable. . . . Often no flag was used ahead at all. The surveyor would set up his compass, and with the remark that a certain tree, rock, or some other object ahead was on the line, or within a rod, or a rod and a half of it, perhaps on either side, as the case might be, he would direct his chain-bearers to "go for it" and they would all start for the place designated. . . . When it was considered that the chain had been carried "too crooked" an allowance would be made by guess "to make the distance about right." And this mode would be continued until the lot was completed.[27]

Imperfect surveys resulted in overlapping claims, and patents were often issued on land that was not vacant. Surveys rarely contained less land than called for in the description but often contained a surplus because surveyors would be held to be liable if at any time a shortfall was proved. Hilliard recounts how in eastern Georgia, where the maximum land grant was 1,000 acres, more than a hundred individuals put in multiple headright claims for in total more than 100,000 acres.[28] Thousands of such fraudulent claims were approved; in Franklin County, 5 million acres were granted, more than ten times the county acreage! The surveyor-general of Georgia reported in 1839 that "the twenty-four counties existing in 1796 contained actually 8,717,960 acres of land, whereas the maps and records in the Surveyor-General's office show that in these counties there had been granted 29,097,866 acres."[29] Few of these grants were ever proved by survey but were simply sold on paper to out-of-state speculators. Hammon discusses the private and court settlement of such disputes in Kentucky, where the only positive outcome was the decision of Congress to revise the system completely by creating one-mile-square sections.[30] No longer would the corner of a person's property be defined by such as the infamous point of origin quoted by C. E. Sherman in his report on Ohio land subdivision: "beginning at the old crow's nest on the north fork of the Kentucky river."[31]

8.4 New France: Acadia and Quebec

In 1605 the town of Port Royal was founded by French settlers and became the focus of the colony of Acadia in the area of present northern Maine and the three Maritime Provinces of Canada (figure 8.1 above). Not until the 1650s, however, was a residential French population well established here.[32] French settlement in Canada was successful earlier in what was to become the heart of New France, the Quebec colony along the St. Lawrence. In 1641 Jean Bourdon was appointed engineer of the Company of New France, a body established in 1627 by Cardinal Richelieu in an attempt to extend the power of the French state to North America by ensuring the transplantation of the feudal system of *seigneuries* and *censitaires* (chapter 6.1), by which, in theory if not in practice in New France, the lowliest subject of the state farming his *roture* (holding) was linked by dues and obligations to his sovereign king.[33] Bourdon mapped the infant French colony in 1641 (figure 8.6) to record seigneurial boundaries and also to indicate the potential for settlement expansion in the Quebec area.[34]

From the first land grants made in Quebec in the 1620s, through to the 1670s, almost all grants were made along the two hundred miles of the St. Lawrence River from Quebec to Montreal, despite Colbert's attempts to concentrate settlement into contiguous blocks.[35] Trapezoidal *seigneuries* were divided into ranges of strip *rotures*, which enabled the greatest number of settlers to have access to navigable river courses or good road access and a share in different ecological zones back from the river. A ratio of width to length of about 1:10 was common.[36] The boundaries of these landholdings were surveyed and marked on the ground but not at first on maps. From the 1630s, a simple survey line perpendicular to the river, a *rhumb de vent*, was used as the long axis to define most seigneurial concessions.[37] Nor were plats really needed to record boundaries when all properties emanated from a base line along the river. Two posts to mark the end of the base line, a *rhumb de vent* perpendicular to this, and a note of the names of adjoining *seigneuries* were perceived to be sufficient. However, maps of the overall landownership pattern were compiled and sent back to France in response to requests from the government on the state of *seigneuries* in New France. For instance, in 1712 Gédéon de Catalogne sent back a *Mémoire sur les plans des seigneuries et habitations des gouvernements de Québec, les Trois-Rivières, et de Montréal*, the maps for which were made in 1709 by Jean Baptiste Doucagne based on de Catalogne's surveys (figure 8.7). A cadastral map of the Île Jésus was prepared in 1749 by the Séminaire de Québec.[38] Simi-

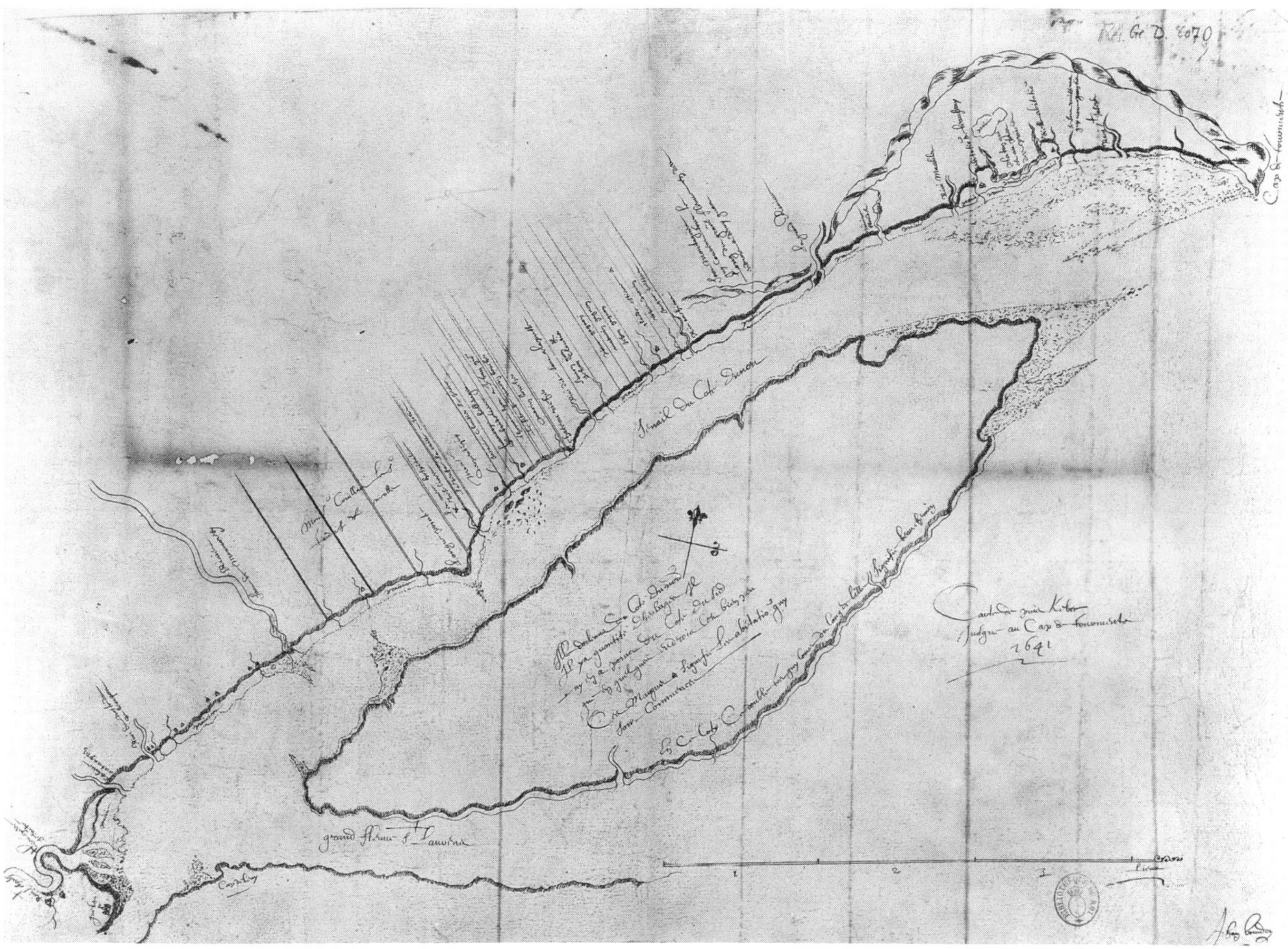

Fig. 8.6 "Carte depuis Kébec jusques au cap du Tourmentes," by Jean Bourdon, 1641, which depicts *seigneuries* along the St. Lawrence River. Source: Bibliothèque Nationale, Paris, Département des Cartes et Plans, D338.

larly, there is evidence that a map showing the boundaries of Acadian *seigneuries* in 1700 was prepared in Acadia and sent to Versailles, though it is now lost.[39] However, it would seem that, as in Quebec in this period, cadastral maps were not essential to prove title to land in Acadia.

Permanent British control of the Acadian colony began with the capture of Port Royal in 1701 and was confirmed by the Treaty of Utrecht in 1713. Land titles were confused when the British took over, both rents and seigneurial dues were nominal, and away from the longer-established settlements, even these token amounts were not rendered.[40] Major Lawrence Arm-

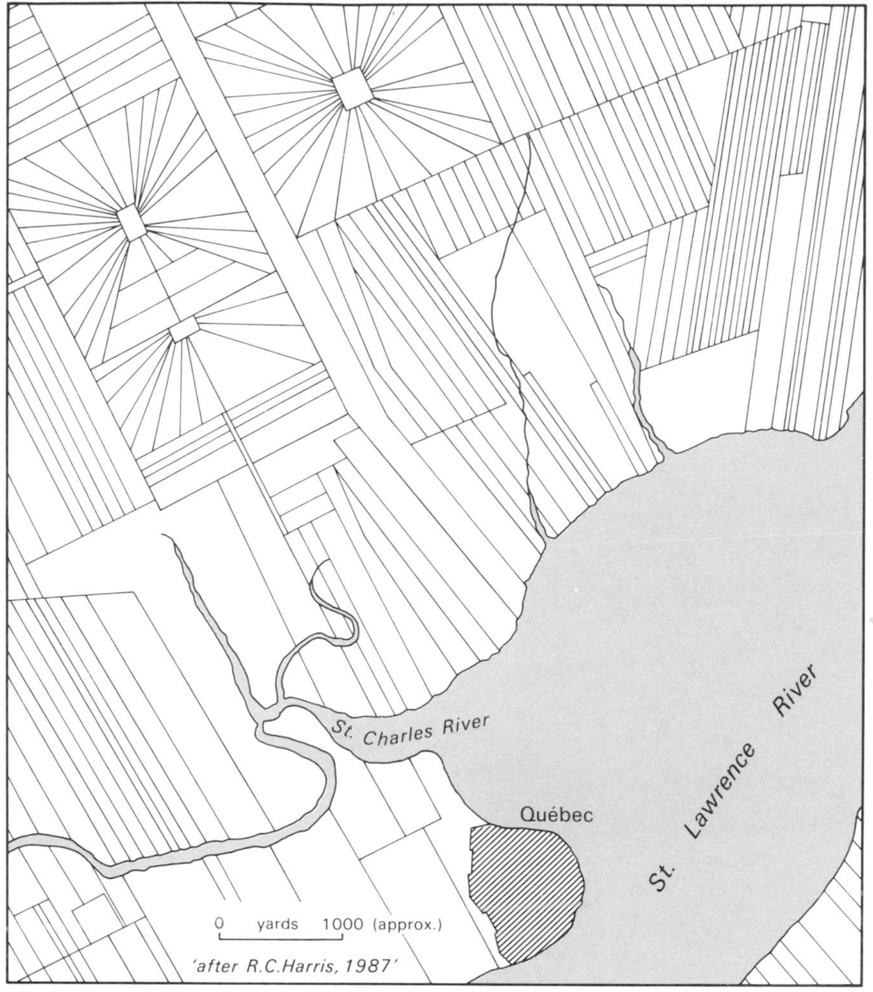

Fig. 8.7 *Seigneuries* along the St. Lawrence, after the survey of de Catalogne, 1709–12. Source: Harris 1987.

strong of the British administration proposed "the necessity of having the ffrench Inhabitants Estates survey'd and Measur'd . . . to lay before Your Lordships any just plan of this Country, for it is said that some, if not all, of them possess and claim Greater Tracts than they are anyways Intitled to."[41] In November 1733 Armstrong sent "an exact plan" by George Mitchell of properties along the Annapolis River to the Board of Trade, and the following year Mitchell was instructed to extend his mapping to the other French settlements.

These, though, were all post hoc surveying and mapping exercises undertaken on the instructions of governing authorities concerned to verify preexisting cadastral patterns. In the 1750s, with the forced removal of the one-time French population of Acadia and the influx of settlers from New England, the first formal surveys prior to settlement in Maritime Canada were carried out when in 1753–54 the Lunenberg settlement farm lots were laid out. From the 1760s, a typical New England township system of land grants was instituted here.[42] Nova Scotia townships contained on average about 100,000 acres, the external lines of which it was the government's responsibility to establish, while the internal subdivision into lots was left to the settlers themselves; lots thus vary in size and shape considerably from township to township. Surveyors who laid out internal lot lines were responsible to local township committees, not the government, so their plans and field notes have often been lost or destroyed.[43] One such loss is of the original survey map of Cornwallis Township, which tradition has it was burned in a house fire. The Register of Deeds for this township contains lots numbered 1–10 located in divisions numbered 1–15; it is assumed that these numbers refer to a missing map.[44]

8.5 Louisiana, Texas, and the Florida Parishes: France, Spain, and England

In the French Louisiana colony, the characteristic French "long-lot," or *arpent*, system of land grants was used and subsequently adopted by the Spanish administration.[45] Instead of completely unsystematic tract selection as in east coast southern colonies, plots were granted with a frontage along streams in order to allocate equitable amounts of the very productive Louisiana alluvial land. The counterpart in Quebec (section 8.4) was the seigneurial system but with the essential difference that in New France the landownership unit was the *seigneurie*, or large estate with tenanted farms, while in Louisiana, farm lots were granted to individuals and set out with long axes perpendicular to the river (figure 8.8). As in Quebec, survey of these Louisiana holdings often meant just the measurement of the river line and river width of lots, with backlines left unsurveyed for many years. Indeed, many lots remained unsurveyed until the Louisiana Purchase in 1803 resulted in the extension of the United States rectangular survey system into this area.[46] In legal terms, owners who did not perfect their title by survey and cadastral map remained as squatters, and new settlers could and did encroach on their claims, which all fueled land disputes in the eighteenth century. Even after 1803, settlers were reluctant to allow the federal surveyors on to their land for fear of offending the Spanish authorities, whom they expected to return to power, and, more directly, from fear of being discovered with more land than their "order of survey" stipulated. Under Spanish law, those who informed on persons encroaching beyond their allotted acreage could be awarded half the selling price of such illegally held land. As J. W. Hall comments, "It is not surprising that the United States deputy surveyors were sometimes watched over gun barrels as they attempted to survey private claims."[47]

In Texas (figure 8.9), the first riparian long-lots were set down in 1731 as irrigated farms, or

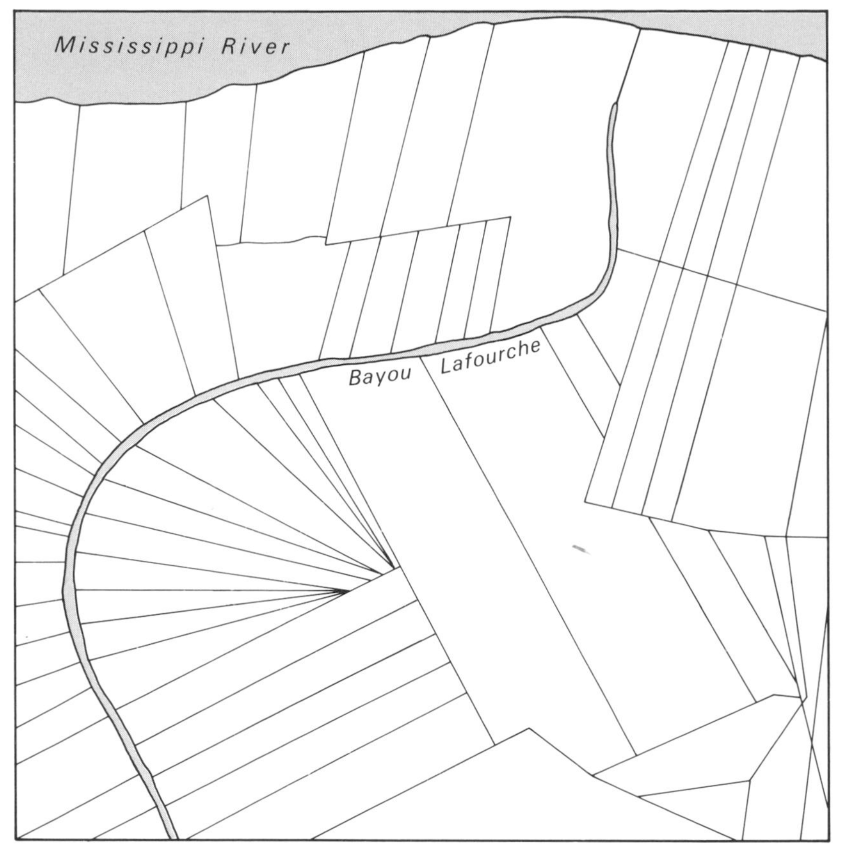

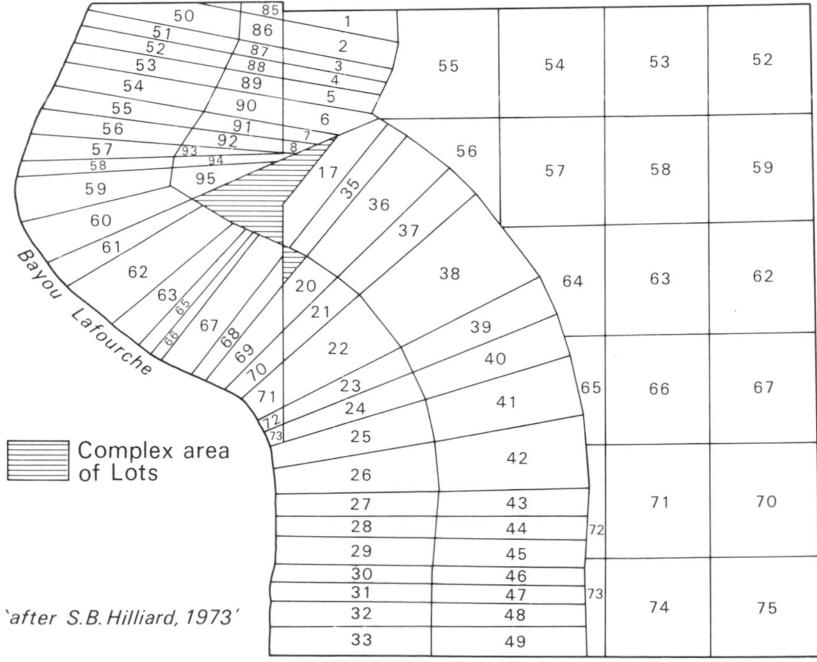

Fig. 8.8 Long-lots in Louisiana.
Source: Hilliard 1973.

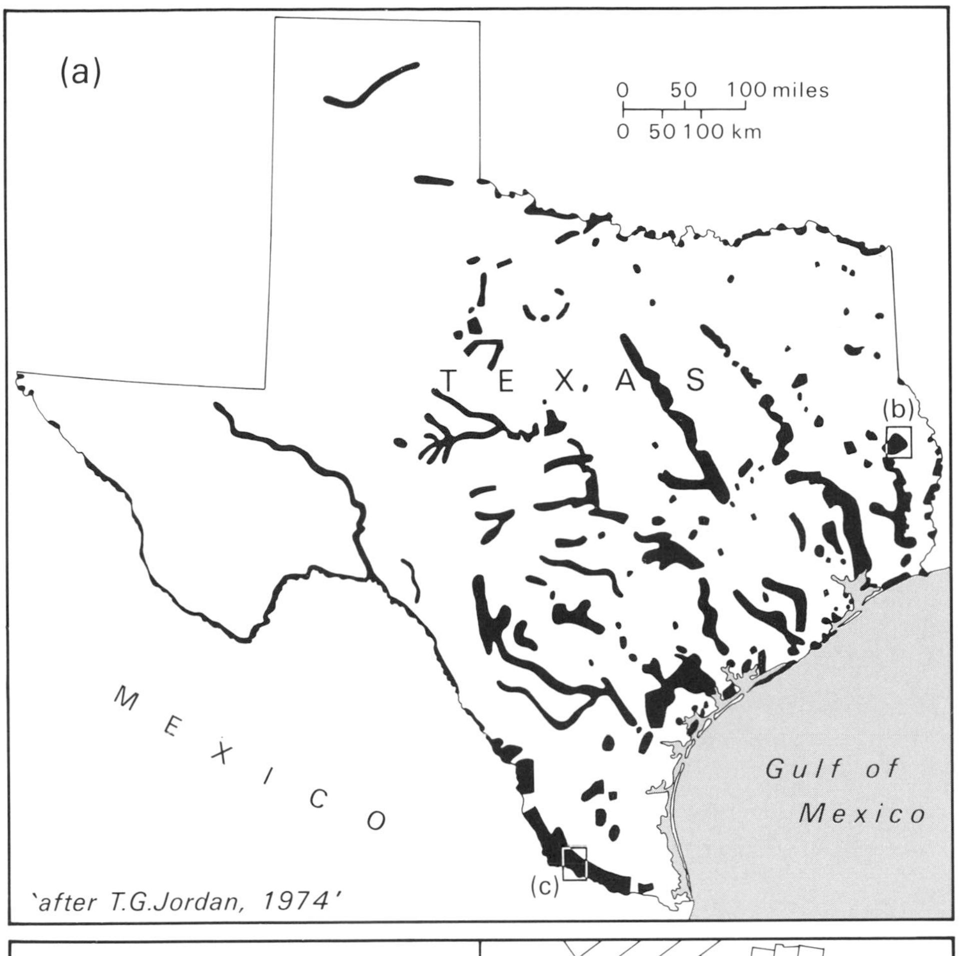

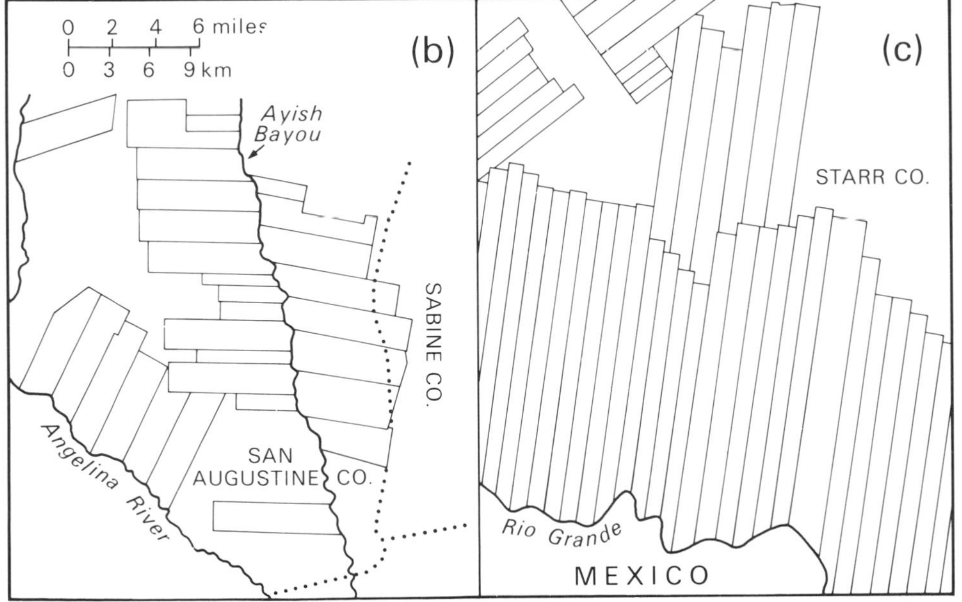

Fig. 8.9 Long-lots in Texas: (*a*) distribution; (*b*) San Augustine County, eastern Texas, surveyed in the Mexican period, 1834–35; (*c*) Starr County, southern Texas, surveyed in the 1760s during the Spanish period. Source: Jordan 1974.

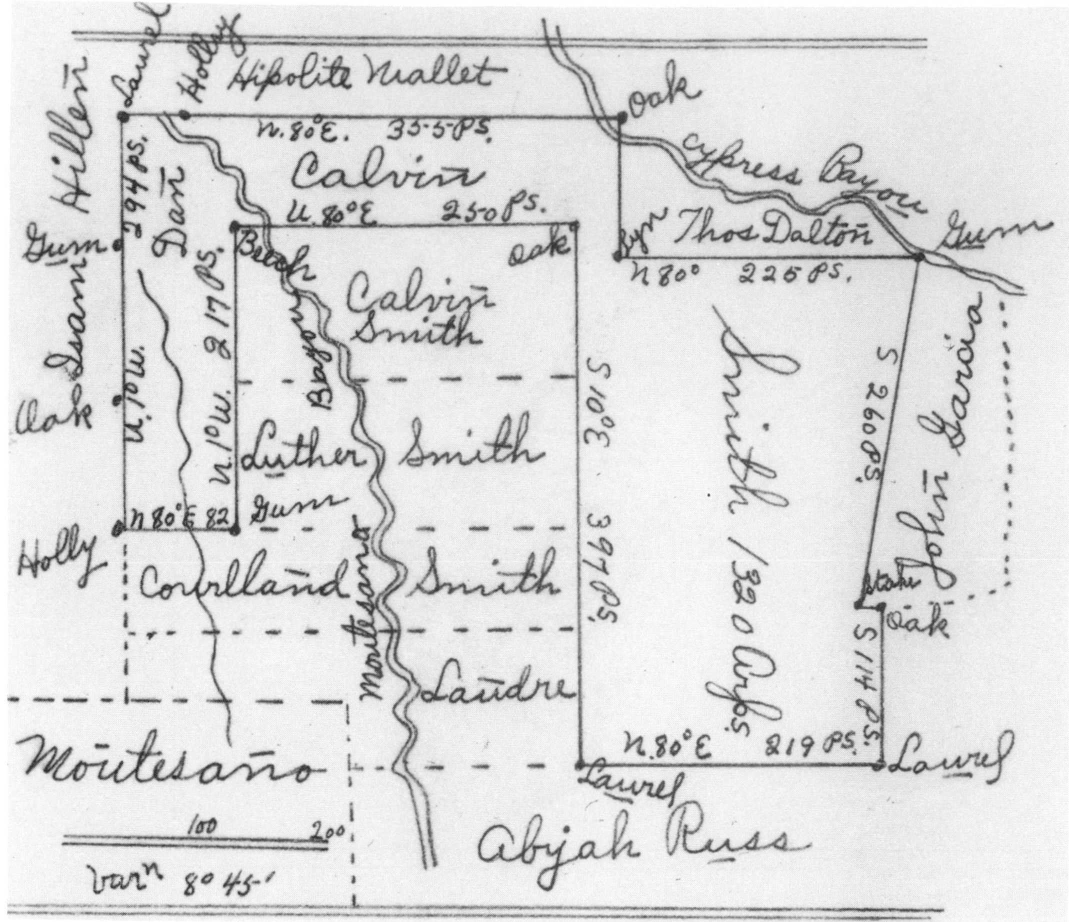

Fig. 8.10 Plat of the Spanish patent to Calvin Smith, 1799, Greensburg, Louisiana. Source: Hill Memorial Library, Louisiana State University Baton Rouge, Special Collections (copy) and Greensburg District Land Office.

suertes, around San Antonio. From the 1760s and into the Mexican period, they were surveyed in large numbers along the Rio Grande. These last measure about one kilometer wide and sixteen kilometers in depth, with access to the river here being for cattle drinking water, rather than for irrigation. Though introduced in the Spanish period, it is Jordan's opinion that the prime influence, certainly after 1767, was the earlier use of long-lots by French settlers to the east in Missouri and Louisiana with which there was Spanish-Mexican cultural contact.[48]

In those inland areas of Louisiana in Spanish control, settlement and survey systems were quite different. The lands suited to cattle ranching west of the Mississippi and into Spanish Texas were granted in huge, rectangular tracts (figure 8.10). These *sitio* land grants had been estab-

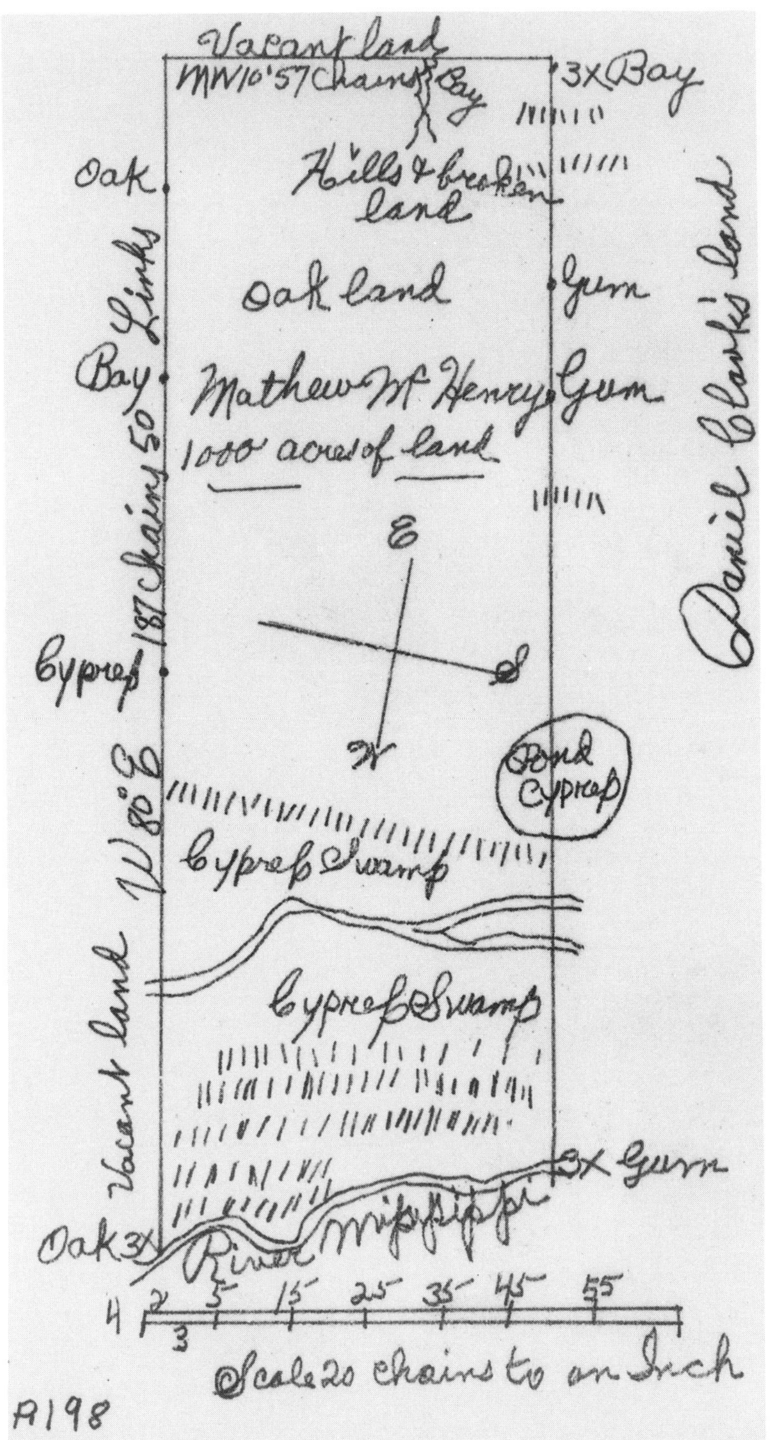

Fig. 8.11 Plat of British grant to Mathew McHenry, 1768. Source: Hill Memorial Library, Louisiana State University Baton Rouge, Special Collections (copy) and Greensburg District Land Office.

Fig. 8.12 Vincente Pintado's map of landholdings c. 1805 on the east bank of the Mississippi near Thompson's Creek, Louisiana. Source: Library of Congress, Washington, D.C., Pintado Papers, MMC 3361.

lished first in the Spanish Kingdom of the Indies in 1680. Though differing in scale, a similar procedure was followed to perfect title as with *arpent* strips. A survey was needed, but this might be conducted *a pasos de caballo*, with distances calculated later from the particular length and number of steps taken by the surveyor's horse.[49]

In the Florida parishes of Louisiana north and east of the Mississippi, cadastral survey and settlement differed yet again.[50] This territory had similarly passed from France to Spain but was in British control when, following the Treaty of Paris in 1763, effective settlement began. As in other British colonies in the South, headright grants and unsystematic grant location were the rule (figure 8.11).[51]

The territory of Spanish West Florida, (i.e., the east bank of the Mississippi and the Pensacola area), which was in Spanish hands in 1803, was not included in the Louisiana Purchase from France. In the first decade of the nineteenth century, a Spanish surveyor, Vincente Pintado, compiled composite cadastral and topographical maps for the Districts of Manchac, Baton Rouge, and Feliciana for Don Juan Ventura Morales, who was *intendente* (the Crown's representative) in Spanish West Florida and controlled land distribution. Figure 8.12 is one of these maps of the landholdings on the east bank of the Mississippi c. 1805, which identifies the landowners and those lands which Pintado considered needed verification or resurvey.[52]

8.6 The "New England Method" of Cadastral Survey before Sale of Land

The colonies from Virginia southward exhibit a great variety of settlement systems, but in the colonial period, each was based on the concept of grant *prior* to the cadastral survey which proved title to claimed tracts. In the New England colonies the process was reversed: grants to individuals generally *followed* cadastral survey. By 1630 three nuclei of English settlement had been established on the New England coast, with the Massachusetts Bay Colony destined to be dominant in the later stages of the region's development. This area was the destination of the great Puritan migrations of 1630–40 and the point of dispersal for groups of dissenters who left Massachusetts Bay during the religious disputes of that decade. From then on, the original conception of a single settlement was supplanted by a system of six nodes of English settlement (figure 8.2 above) on Massachusetts Bay, Providence and Rhode Island, coastal New Hampshire–Maine, the Connecticut Valley, New Haven, and Plymouth.[53] In the first instance it was the intention of the Massachusetts Bay Company to grant land to each individual shareholder and settler as had been done in Virginia, but this idea was rapidly, if not completely, replaced with grants to groups of settlers acting together as "towns."[54]

A town consisted of a tract of land with defined legal status granted to a *group* of settlers, so

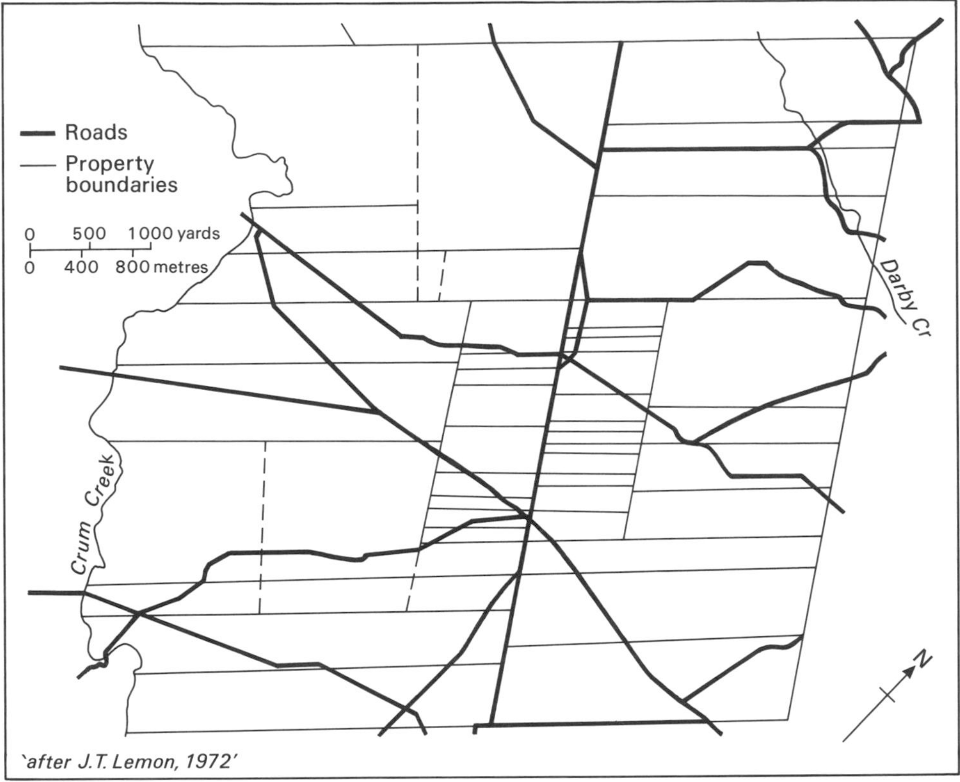

Fig. 8.13 The layout of a Pennsylvania township: Newtown, Chester County. Source: Lemon 1972.

that New England settlement, at least initially, was a cooperative, communal venture. Towns were established by would-be settlers who first petitioned the colonial authority for a new settlement: if this was agreed, a committee of proprietors was appointed to organize the settlement and, in particular, the distribution of land. English land tenure in its many varieties was the model for actual land subdivision, although, because of the large size of most original towns, land outside the immediate village lots was only gradually transferred to private ownership. Towns of eight or ten square miles are not uncommon, but with experience it was found that six square miles was about as much land as a community could manage, and this, or its approximate equivalent acreage, became the prevalent size of the New England township.[55] New England townships are not all square and consistently oriented but are normally bounded by straight lines. Though they were usually set out prior to settlement, the internal units were not themselves characterized by uniformity in size or shape.[56] The variety of townships set out in early eighteenth-century New Hampshire is illustrated by James Garvin.[57]

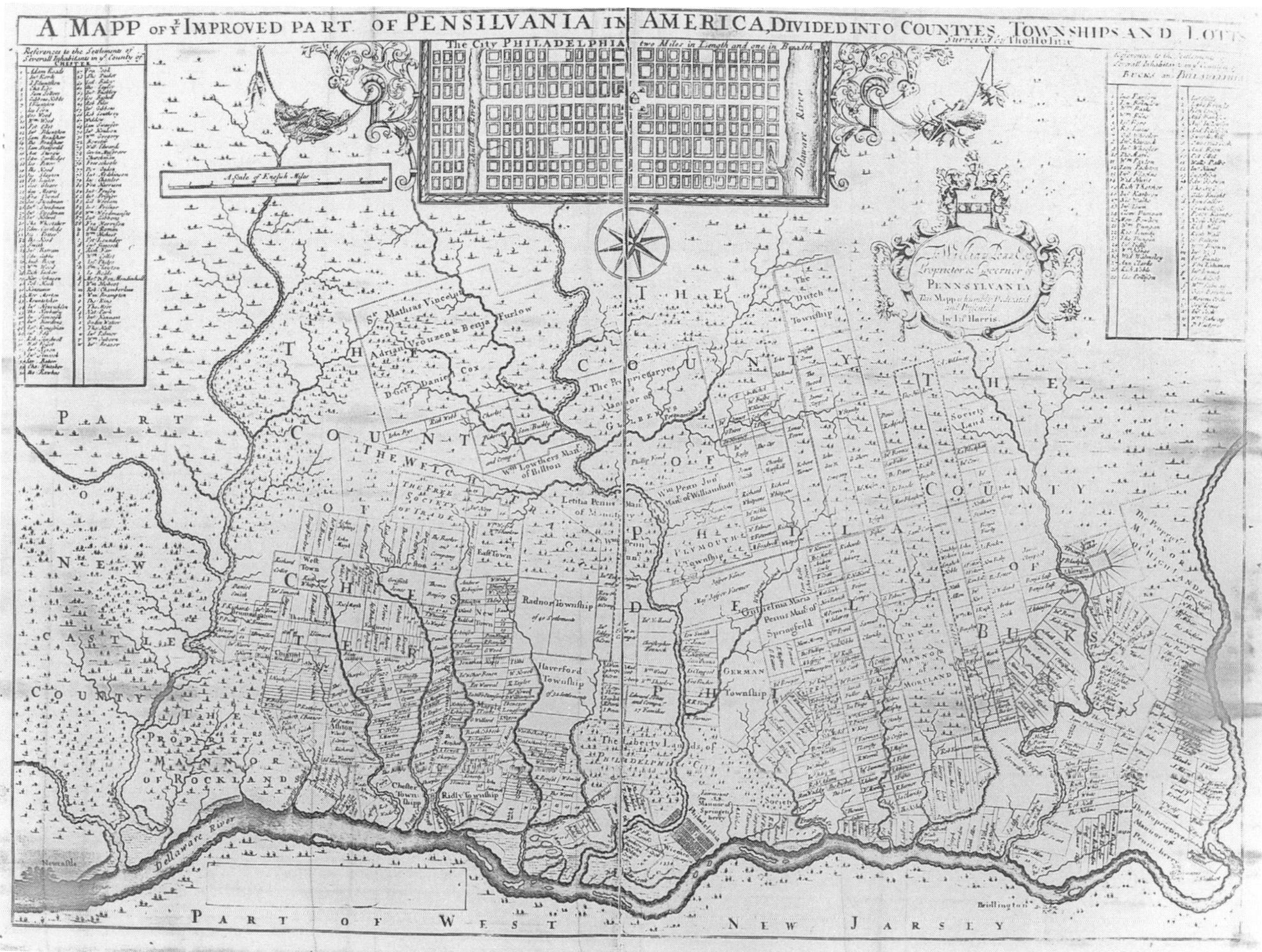

Fig. 8.14 Thomas Holme's "A Map of ye Improved Parts of Pennsylvania," 1681. Source: Pennsylvania State Archives, Harrisburg, Pa.

William Penn, who in 1681 was granted the territory now known as Pennsylvania, had clear ideas of how his colony should be developed, from the layout of his chief town, Philadelphia, to the regulation of rural settlement. Cadastral mapping was a fundamental instrument of his settlement policy. In 1685 he said, "We do settle in the way of Township or Villages each of which contains 5,000 acres, in square, and at least Ten Families; the regulation of the Country being a Family to each five hundred Acres " (figure 8.13).[58] Penn envisaged sets of contiguous tiers of townships surveyed prior to settlement, with the Delaware River as the baseline.[59] A land office was established, and about 860,000 acres of land sold to "First Purchasers," mostly English Quakers. The prime role of survey was confirmed by the fact that it was the surveyor-general who was responsible for implementing Penn's settlement program. Thomas Holme's map (figure 8.14) indicates that contiguity and a degree of regularity had been maintained in the first two decades of settlement. Thereafter, the land commissioners found it difficult to assemble like-minded groups of people and easier to deal with individuals and sell them the acreage they required, a course of action precipitated by the need to raise revenue. This procedure was firmly established as government policy after some 60,000 acres were granted in 1699 to the Pennsylvania Company of London and located in a number of widely scattered parcels. Thus regularization gave way to the sale of individual warrants with patents validated after survey.[60] In a mid-eighteenth-century report to the Pennsylvania proprietors, a surveyor in Lancaster County observed that "the habitable land lyes so in patches among the mountains and barrens that one tract canot always (nor often) lye contiguous with another, for that reason alone it will be very much to your loss if every man be alowed to take what quantity he thinks fit. . . . And this I fear will often be the case if the warrants are issued before the tracts are survey'd."[61]

His fears were borne out; warrants issued before survey produced irregular boundaries and noncontiguous holdings as in other mid-Atlantic and southern colonies. In 1766 Robert Levers, a Pennsylvania surveyor, commented in a letter to a colleague that "the survey of Broadhead and Scull was an iniquitous survey, . . . it is so contrived as to leave out every place that has the least appearance of being bad."[62]

8.7 West and East Jersey

The proprietary colonies of West and East Jersey (after 1702, the Crown colony of New Jersey) display in microcosm the whole variety of relationships between cadastral mapping and land settlement in the colonial period by virtue of their frontier position between New England/Pennsylvania and southern practices and because the Dutch introduced both large *patroon* estates with irregular boundaries set out by metes and bounds and long-lots arranged along navigable waterways.

The Dutch were defeated by the English in 1664, after which the intention was to settle by towns, with village sites platted before settlement. In addition to townships settled by New Englanders and the settlements of English Quaker groups, whose land division reflected their ideology of cooperation and community, individuals and small groups were also granted land in the proprietary period with freedom to seek out the best land.[63] Seventeenth-century surveying was technically deficient, and however good the systems of registering land were intended to be, the period here, as in the other Atlantic colonies, was highly litigious such that in 1695 the Council of Proprietors of West Jersey ordered a general resurvey as "sundry persons do suffer ye Lines of their Surveys and marks of their line trees to Decay and become indiscernible whereby many Contests and Suits of Law are likely to arrise and Accrew."[64]

8.8 The "Land of the Western Waters" and the Origins of the United States' Public Domain

In the period before the American Revolution, land grants and titles in trans-Appalachia were issued by the proprietary governors or appointed officials of the king of England. Title to these western lands became a pressing matter at the time of the Declaration of Independence in 1776. The six former colonies of Massachusetts, Connecticut, Virginia, North and South Carolina, and Georgia had claims to all the unsettled land in the old Northwest, while New York had title to a large area west of the Delaware.[65] During the war Congress urged these states to cease granting western lands, and in February 1780 New York defined its boundaries and ceded its western claims to the United States. Other states followed after the war and ceded most of their claims, retaining reservations of land principally so that they could discharge their obligations to soldiers who had provided military service for land in lieu of cash. One such was the so-called Military District of Virginia between the Scioto and Little Miami rivers in the present-day state of Ohio.[66] This was granted using the Virginia method, with land claimed in advance of the surveying which was necessary to perfect title. Not all trans-Appalachian land was disposed of in this unsystematic way.[67] Both land companies and state agencies, for example, tried to ensure that land was divided into roughly rectangular lots in the new states of Kentucky and Tennessee.

Once the territory north of the Ohio River and west to the Mississippi had been ceded to Congress, a system had to be devised to alienate these now-public lands to avoid the problems inherent in the indiscriminate system of claiming land first and then surveying for title second. The shortcomings of previous methods were well known to contemporary politicians and administrators. The House of Representatives debated the whole issue at the First Congress in Philadelphia on 27 December 1790. Elias Boudinot from New Jersey was doubtless speaking from more than his own state's experience when he said that "more money had been spent at law in disputes arising from that mode of settlement in New Jersey than would have been necessary to purchase all the land of the State."[68] As Congress discussed this matter in 1790, the debate polarized between (1) the generally New England, democratic advocates of land sale in small parcels to individual settlers to create a landowning democracy, and (2) conservative, southern, plantation-owning interests who argued in favor of large grants at low prices to companies or wealthy individuals who would then organize settlement by the resale of land.[69] A congressional committee of five with Thomas Jefferson as chairman was charged with the job of preparing a survey ordinance which would govern the survey and division of the public lands prior to sale, settlement, and development and so bring "some order to the frenzied scramble among a welter of avaricious interests, large and small, local, national and international, to reap some profit out of this vast national domain."[70] The instruments which government was to use to control land alienation were seen from the very first draft ordinance announced on 7 May 1784 to be land surveys in rectangular units recorded on large-scale, cadastral maps or land plats.

The principal authors of the Ordinance Establishing a Land Office for the United States of 1784 are generally held to be Thomas Jefferson and a North Carolina member of his committee, Hugh Williamson. They proposed that the public domain:

> shall be divided into Hundreds of ten geographical miles square, each mile containing 6086 feet and four tenths of a foot, by lines to be run and marked due North and South, and others crossing these at right angles, the first of which lines, each way, shall be at ten miles distance from one of the corners of the state within which they shall be. . . . These Hundreds shall be divided into lots one mile square each, or 850 acres and four tenths of an acre, by lines running in like manner due North and South, and others crossing these at right angles.[71]

A regular map of political units for administration was thus intended, and a coherent system of decimal measures was to be adopted based on the nautical (i.e., Jefferson's "geographical") mile. Paradoxically, however, indiscriminate location of individual holdings was envisaged within the regular political framework. Land would be chosen by warrant, and warrant holders were to be completely free to locate their claim, even far ahead of the frontier of settlement. But title would be granted only for properties of a size and shape controlled by the grid; no longer could a pioneer select only the prime parcels and exclude poor land from his grant boundaries. This law would thus have controlled, rather than abolished, southern practices of land allocation.[72] Compass and chain surveys were envisaged, trees were to be marked by blazes, while land was to be described "on a plat marking water courses, mountains and other remarkable and permanent things."[73] Land plats of both New England towns and southern plantations had been essential to illustrate and support written descriptions of land and so ensure security of title. The 1784 Land Ordinance envisaged extending this role for cadastral maps to the public domain.[74]

Though Congress's financial state was parlous and the sale of these public lands was intended to reduce the debt burden, the draft ordinance was not accepted. George Washington, for one, wanted to promote contiguous settlement, discourage wide dispersal, and halt the accelerating process of illegal squatting on United States government lands. As Washington wrote in 1784 to President of Congress Richard Henry Lee, "The spirit for emigration is great, people have got impatient, and tho' you cannot stop the road, it is yet in your power to mark the way; a little while and you will not be able to do either."[75]

The national debt was even more of a burden in 1785 when the 1784 draft was discussed again. Early amendments substituted the statute mile for the Jefferson "geographical" mile and townships of seven miles square for those of ten miles square. Land was to be surveyed prior to sale, descriptions of land made available, and land sold outright rather than by warrants, so that it could be purchased by those unable to select a tract on the spot. Washington proposed the idea of sale by whole townships, which he thought advantageous in New England.[76] Further discussion reduced the seven-mile townships to six miles, and a compromise between southern and New England systems of land alienation was achieved by selling alternate townships entire and by lots. The agreed Land Ordinance of 1785 provided that the survey should begin where the western boundary of Pennsylvania met the Ohio River some ten miles west of the Revolutionary War outpost of Fort McIntosh and near the confluence of the Ohio with the Little Beaver Creek. The federal surveyors were to start off westward from the north or right bank of the Ohio River (figure 8.15) and set out townships numbered from south to north in seven "ranges" numbered from east to west.[77]

Eight surveyors began work on these first seven ranges in September 1785, but skirmishes with Indians and insufficient manpower and military cover meant that after two years only four ranges of townships were prepared for sale.[78] Subsequent land auctions were not a great success either and did not produce the hoped-for income. Nevertheless, the basic principle of rectangular survey prior to sale had been established and was resuscitated in the Act Providing for the Sale of the Lands of the United States in the Territory Northwest of the River Ohio (1796), discussed and adopted by Congress after the Battle of Fallen Timbers removed the danger of Indian attack in the near West. The act reestablished the rectangular system, and the idea was to proceed systematically in an east-to-west sequence. As Lola Cazier clearly demonstrates in her history of surveys and surveyors in the public domain, in practice the Federal Land Survey was executed piecemeal as cessions were obtained and after investigation of the legality of private land claims made prior to the survey.[79] These last were particularly prevalent in the Southwest, Louisiana, and the Mississippi Territory.[80] These was also the complication of huge areas such as Texas, where land was never ceded to the United States, with the result that its cadastral surveys and land alienation system display a variety of long-lots, *sitios*, metes and bounds, irregularly aligned rectangular surveys, and rectangular surveys by ranges (figure 8.9 above). In California, there were many Spanish and Mexican land grants to be checked and incorporated.[81] Early Mexican petitions for land grants in California were accompanied by a *diseño* (rough map) of the property, but this practice fell into disuse; later grants were specified by written descriptions only. When United States deputy surveyors came to survey these holdings with chain and compass so that

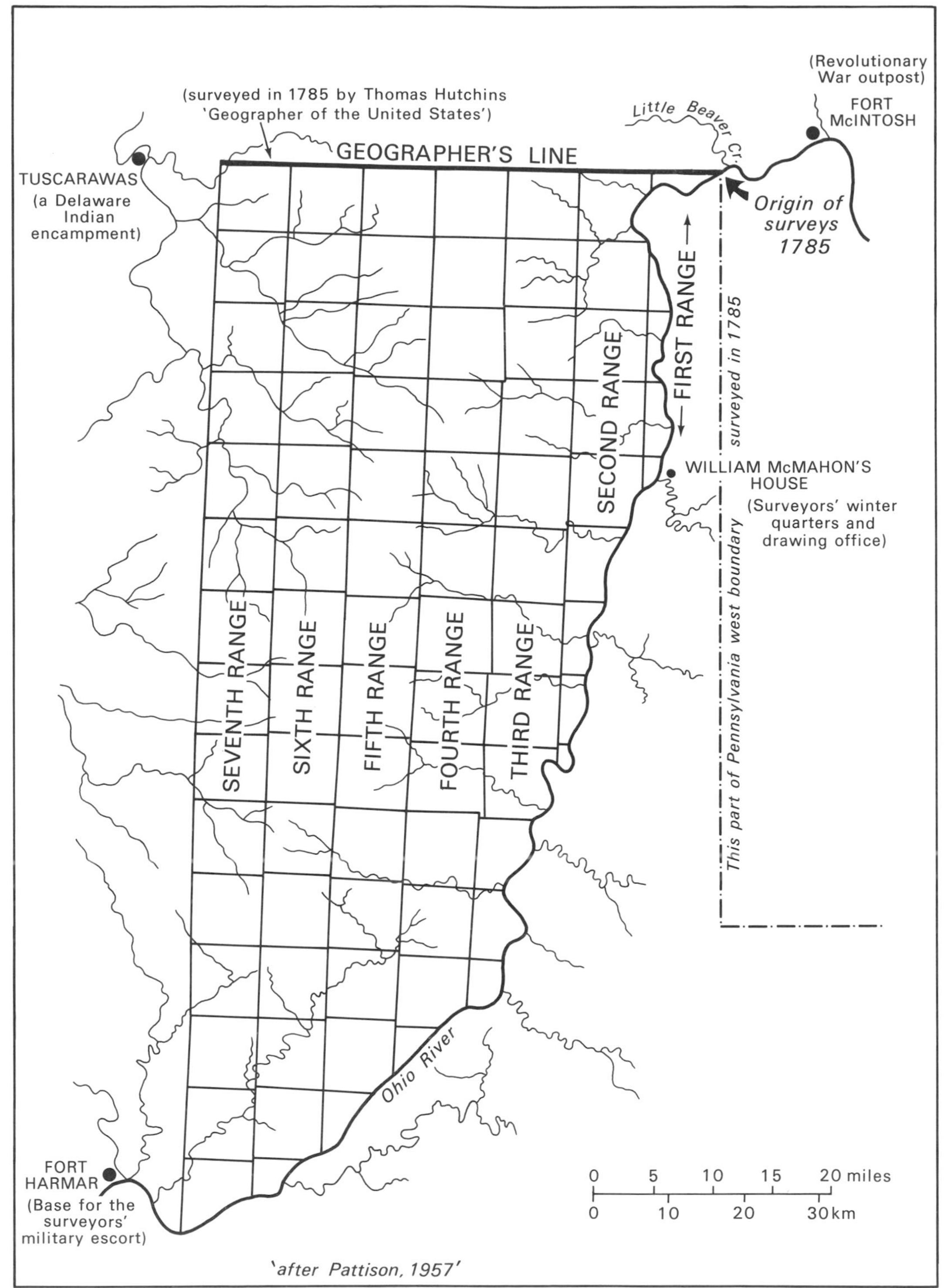

Fig. 8.15 The "Seven Ranges." Source: Pattison 1957:129.

United States title could be patented, their difficulties were compounded not only by the uncertainty of directions and distances but also by the way measurement was recorded, for example, by reference to lengths of a lariat.[82]

In the same year as the 1796 Land Ordinance, Samuel Moore, "a practical surveyor of more than thirty years," published one of the first wholly American textbooks on surveying: *An Accurate System of Surveying* . . . . Of the early surveys he thought: "The cheapness of lands superseding the necessity of precision, but afterwards opens a door for endless litigation . . . should those unappropriated lands be accurately surveyed in the first place, it would undoubtedly prevent after generations from that multiplicity of disputes, respecting boundaries and division lines, which too often harrass our courts of judicature in these States."[83]

It took, though, more than fifty years to establish the consistency and accuracy that was to carry the Federal Land Survey across the continent (figure 8.16). Ohio, the first state to be created from the public domain, was the scene of most of the experiments.[84] By the time the survey reached states such as Kansas and Nebraska (1854–95), a well-defined system of surveying and mapping with its own bureaucracy of contract deputy surveyors working under surveyors-general and supported by register clerks and draftsmen based in a network of local land offices had been established.[85] In 1855 a manual entitled *Instructions to the Surveyors General of Public Lands of the United States . . . Containing also a Manual of Instructions to Regulate the Field Operations of Deputy Surveyors, Illustrated by Diagrams* was produced by the General Land Office as a codification of practices enshrined in earlier manuals relating to particular states and territories.[86] Figure 8.17 is a part of "Diagram B" from the 1855 manual, which indicates the mode of laying out sections and quarter sections on a township plat.

The United States Land Office in Washington, its district offices, its surveyors in the field, and its staff of plat drafters succeeded in processing sufficient land for the 4.5 million people who moved into the lands west of the Appalachians in the half century from 1787 to 1837. The public domain was enlarged by the Georgia cession in 1802, the purchase of Louisiana from France in 1803, the acquisition of West and East Florida in 1811 and 1819 respectively, the Oregon settlement of 1846, the Mexican cession of 1848, and the Texas, Gadsden, and Alaska purchases of 1850, 1853, and 1867 respectively (figure 8.16).[87] All these areas in turn fell under the rectangular survey system with six-mile townships, as the enlarging public domain acted as a magnet for one of the greatest mass migrations in human history.[88] A uniform system of surveying and platting regular-bounded townships and sections was an important element in this process, even if some of the ground surveys proved faulty and progress was not as fast as government might have wished.

Though the survey system itself became increasingly uniform through the nineteenth century, the landownership cadaster represented on the plats lacked overall uniformity. From initial

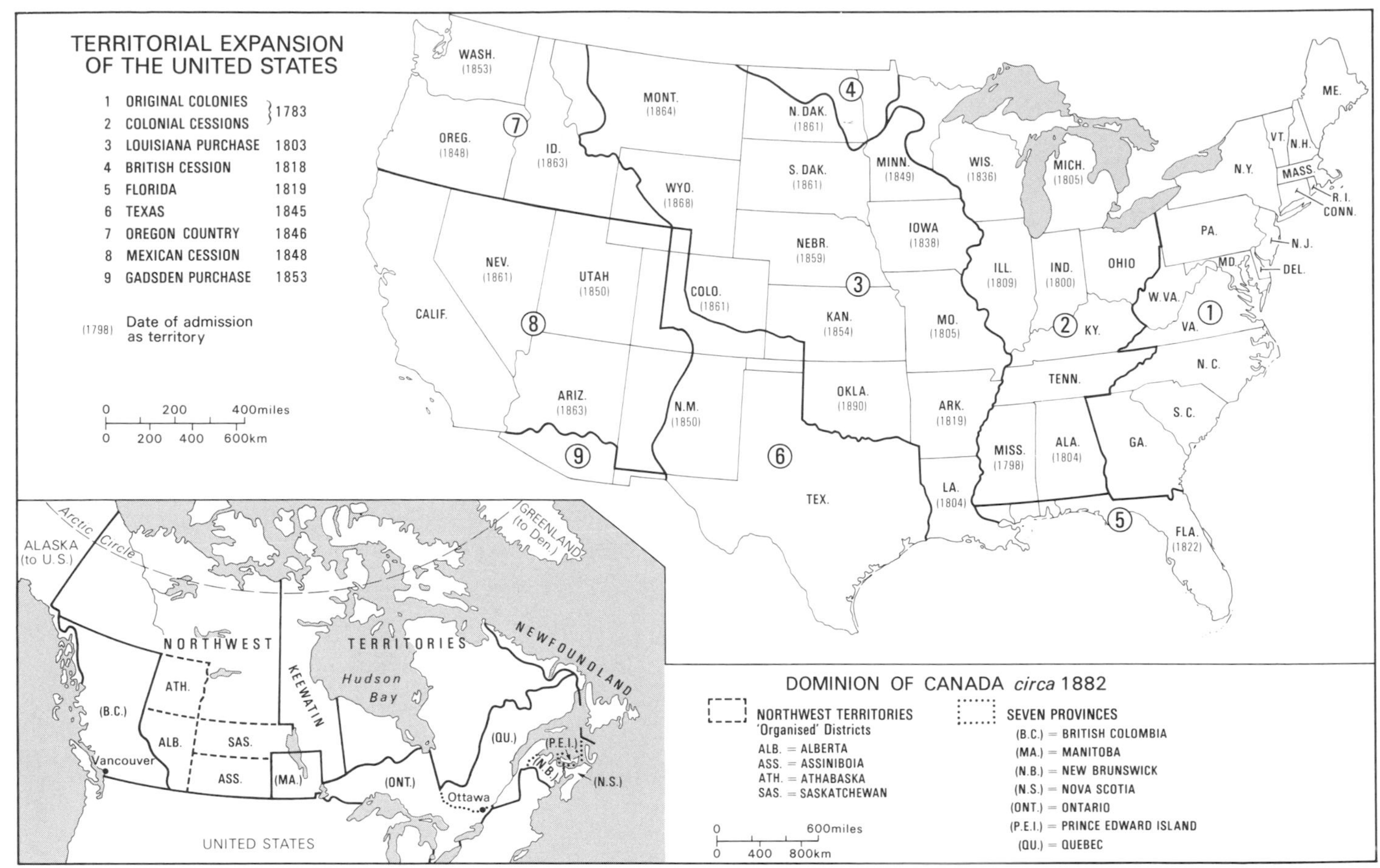

Fig. 8.16 The territorial expansion of the United States in the nineteenth century and the Dominion of Canada about 1882. Source: Droysens 1886; Muir and Philip 1927; Shepherd 1930; Kinder and Hilgermann 1974–78; Darby and Fullard 1978.

sales of whole sections of one square mile, land was progressively made available to purchasers in half sections from 1800, in quarter sections from 1804, in half quarter sections from 1820, and in quarter quarter sections of just forty acres from 1832. With a land purchase unit as small as forty acres, it was perfectly possible to buy up a farm holding of a few hundred acres with an extremely irregular shape so as to maximize the content of good land and minimize that of inferior.[89]

However, the role of cadastral maps in the developing process of land alienation went on unchanged. Plats and accompanying notes provided, first, a flow of information on land quality and resources to potential purchasers to support the sale of land. Second, they were part of the

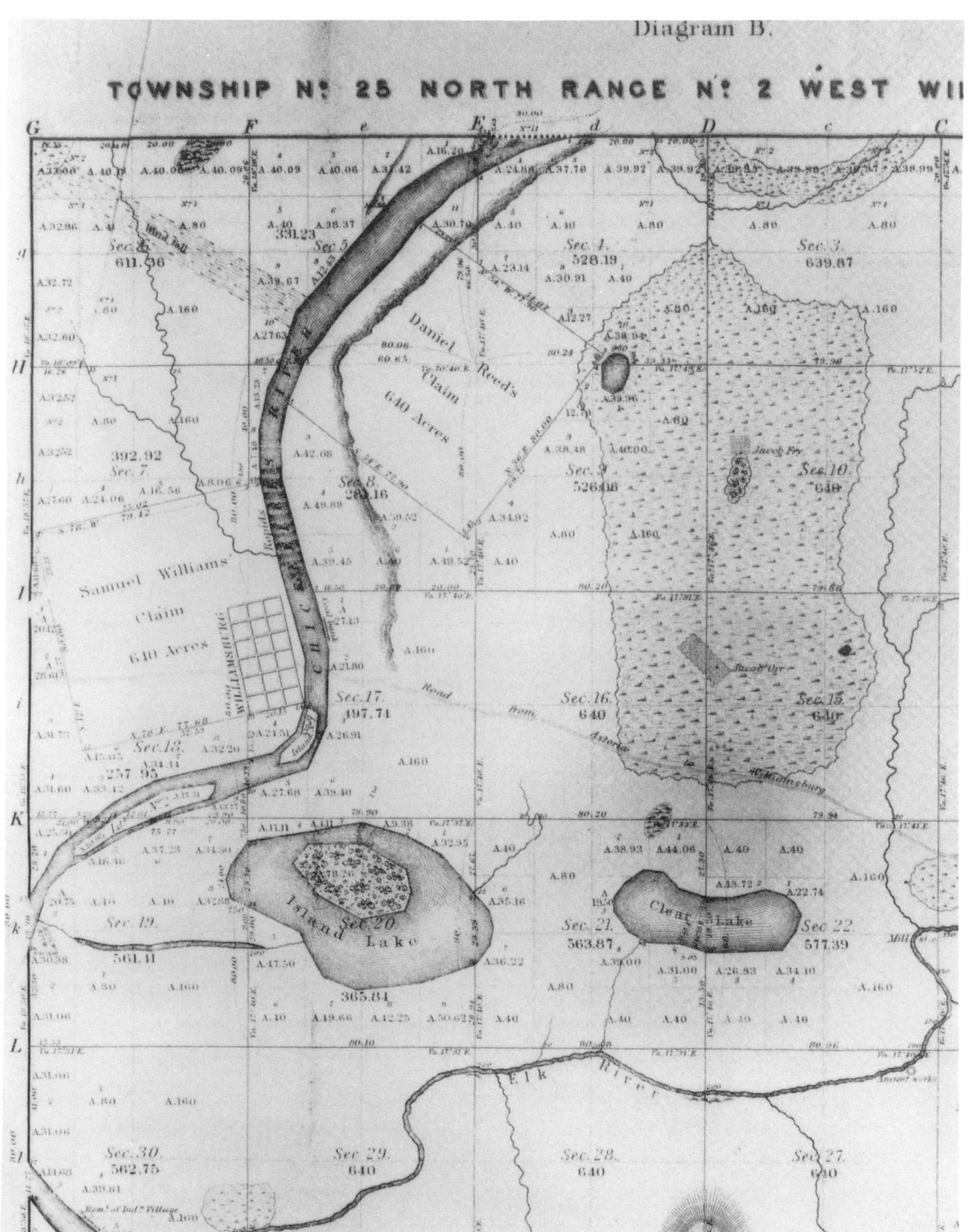

Fig. 8.17 "Diagram B" from the 1855 *Manual of Instructions to the Surveyors General . . .*, which indicates the mode of subdividing townships and sections. Source: White 1983.

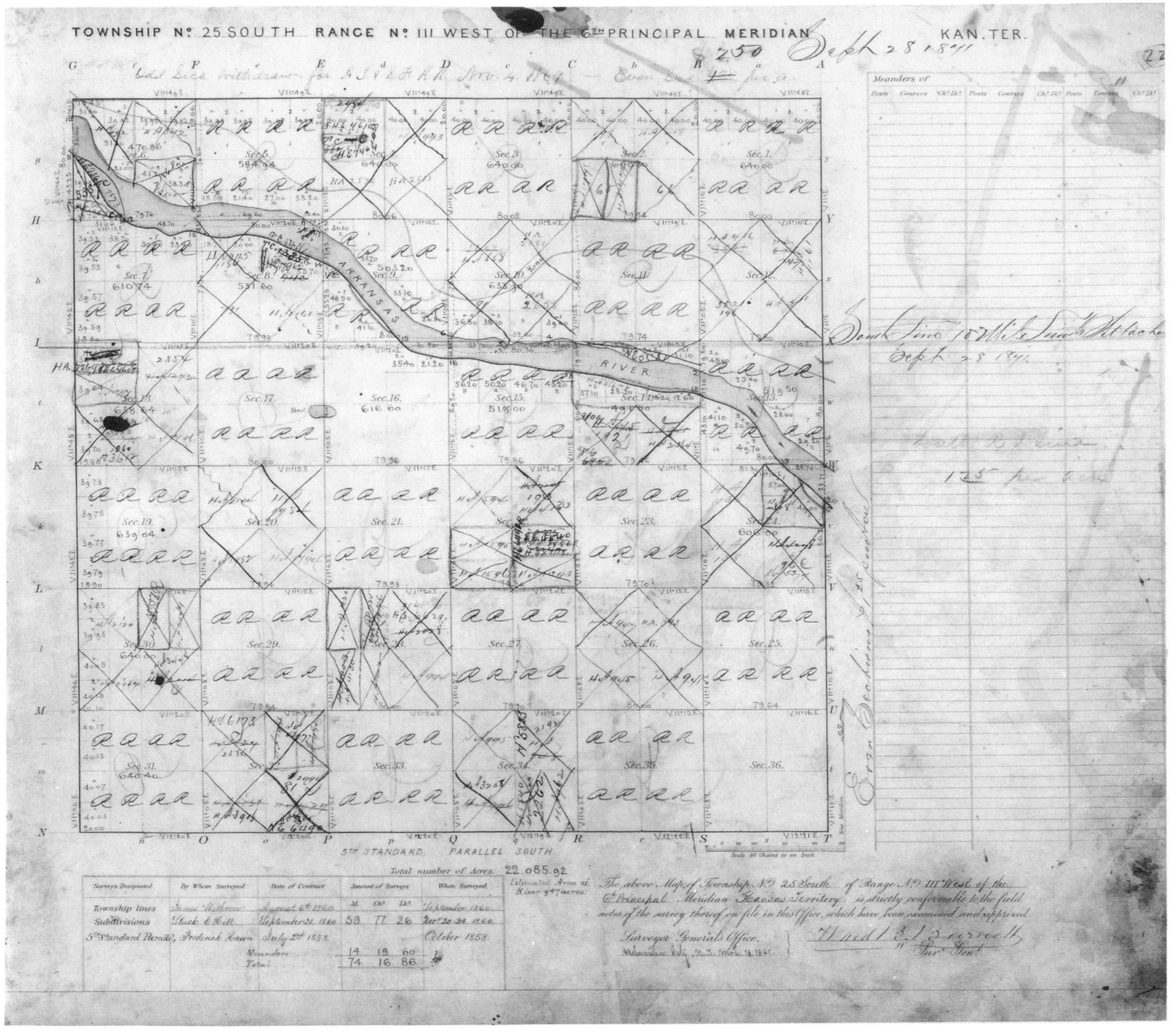

Fig. 8.18 The local office plat of township 25, south range 3 west, Kansas, annotated with the names of landowners. Source: United States National Archives, Washington, D.C., RG49 Township Plats Kansas, 6th PM T25S/R3W.

permanent record of the boundaries and subdivisions of public lands. A modern edition of the government's *Manual of Instructions for the Survey of the Public Lands of the United States* explains that the physical evidence of these property boundaries "consists of monuments established in the ground and the record evidence of which consists of field notes and plats."[90] Three copies of the plats were drawn up by draftsmen in the office of the surveyor-general from the deputy surveyors' original field notes: one plat was for the General Land Office in Washington, D.C., one for the regional office, and one for the local land office where the land was to be put on sale.[91] These third copies provide a graphic record of the General Land Office's initial disposal of the land, as when land was sold these plats were annotated with owners' names (figure 8.18). Land could be sold only after it had been delineated in the plat books.

Third, and perhaps of most crucial importance to the financially hard-pressed government authorities of the day, the uniformity of the land survey and alienation system "provided a quick way to get land on the market in a mode perfect for speculation. Such absolute standardization of units, each efficiently and exactly defined and registered, made the buying and selling of land simple, safe, and fast."[92] In this respect, the cadastral survey and map were the means for converting land into a market commodity. That the system worked is evidenced by the fact that the number of farms in the United States increased from some 1.5 million in 1850 to about 5.75 million by 1900. Had title been granted by metes and bounds with concomitant infinite potential for odd-shaped, overlapping parcels, then conflicting claims and litigation would doubtless have been the harvest.[93]

Fourth, the rectilinear base of the Federal Land Survey also ensured its democratic appeal by making it possible for all settlers, employing only the simplest of equipment, to verify the contents of their land purchases. "When the directive for rectilinear partitioning was added to provisions for prior survey and for sale at private auction . . . the *quondam facilitator* of a dream of democratic rationality became a device in the service of government as vendor."[94]

Fifth, the plats and notes constitute a basic reference source for county surveyors and the Bureau of Land Management (the successor to the General Land Office), which has the duty to retrace lost or obliterated lines. The United States Geological Survey field engineers used them as a source of section and township lines for topographical maps. Federal and county courts call for the evidence of the original survey to help settle property disputes.[95] In addition, the plats and descriptive notes of the United States Federal Land Survey have been used by modern historians and historical geographers to reconstruct the pre-European landscape of parts of the nation.[96]

8.9 The British in Canada

The Treaty of Paris in 1763 provided for the cession of Canada to Britain and created the province of Quebec (figure 8.16 above). In December of the same year the British government sent Governor Murray instructions for granting land. These specified that settlement should be by 20,000-acre townships surveyed prior to settlement, that surveyors should make returns noting the resource potential of each township, and that this information should be returned "with a plot or description of the lands so surveyed thereunto annexed."[97] These instructions emphasize the role of surveying and mapping in the development of Canada's agricultural land resources. Though early intentions in Quebec may have been to change the pattern of settlement from one of long-lots inherited from the French to townships, the long-lot form survived. Figure 8.19 reveals this continuity as late as 1851, when the area around Lake St. John, some one hundred miles north of Quebec City, was settled by farmers from the St. Lawrence valley.[98]

When hostilities between Britain and the United States ceased, emigration by United Empire Loyalists began. In 1783 fleets of transports sailed from New York carrying some thirty thousand refugees to Nova Scotia, and in total some 1.5 million acres were escheated to the Crown for Loyalist settlement between 1783 and 1788. Land was granted by warrant and had to be surveyed prior to legal title being obtained, but regulations in the 1780s did not explicitly require a mapped survey.[99]

Empire Loyalists also looked to settle in the territory which became Upper Canada under the Constitutional Act of 1791. The pressure to get on to the land was thought to be too great to permit the complete survey of farm lots prior to settlement, so Surveyor-General Samuel Holland devised a scheme whereby only the frontages of townships bordering lakes or rivers were laid out by survey and only the front corners of lots were physically marked. Townships were constructed afterward on paper with lot lines initially left unsurveyed.[100]

The specific pattern for settlement in what is now known as Ontario is enshrined in a set of surveying instructions issued by Governor-General Haldimand in September 1783. He directed that the "method of laying out Townships of six Miles square I consider the best to be followed, as the people to be settled there are most used to it, and will best answer the proportion of lands I propose to grant to each family, viz., 120 acres. . . . You will begin your survey by a township on each side of the bay."[101] Townships were the prime unit of settlement along the Upper St. Lawrence, as the *seigneurie* had been in French Canada. John Collins, Samuel Holland's deputy surveyor-general, developed a model for township subdivision from these first Ontario township surveys (figure 8.20). In their interior subdivision, with long narrow farm lots, these Haldimand townships (named after the governor-general) departed radically from their New England coun-

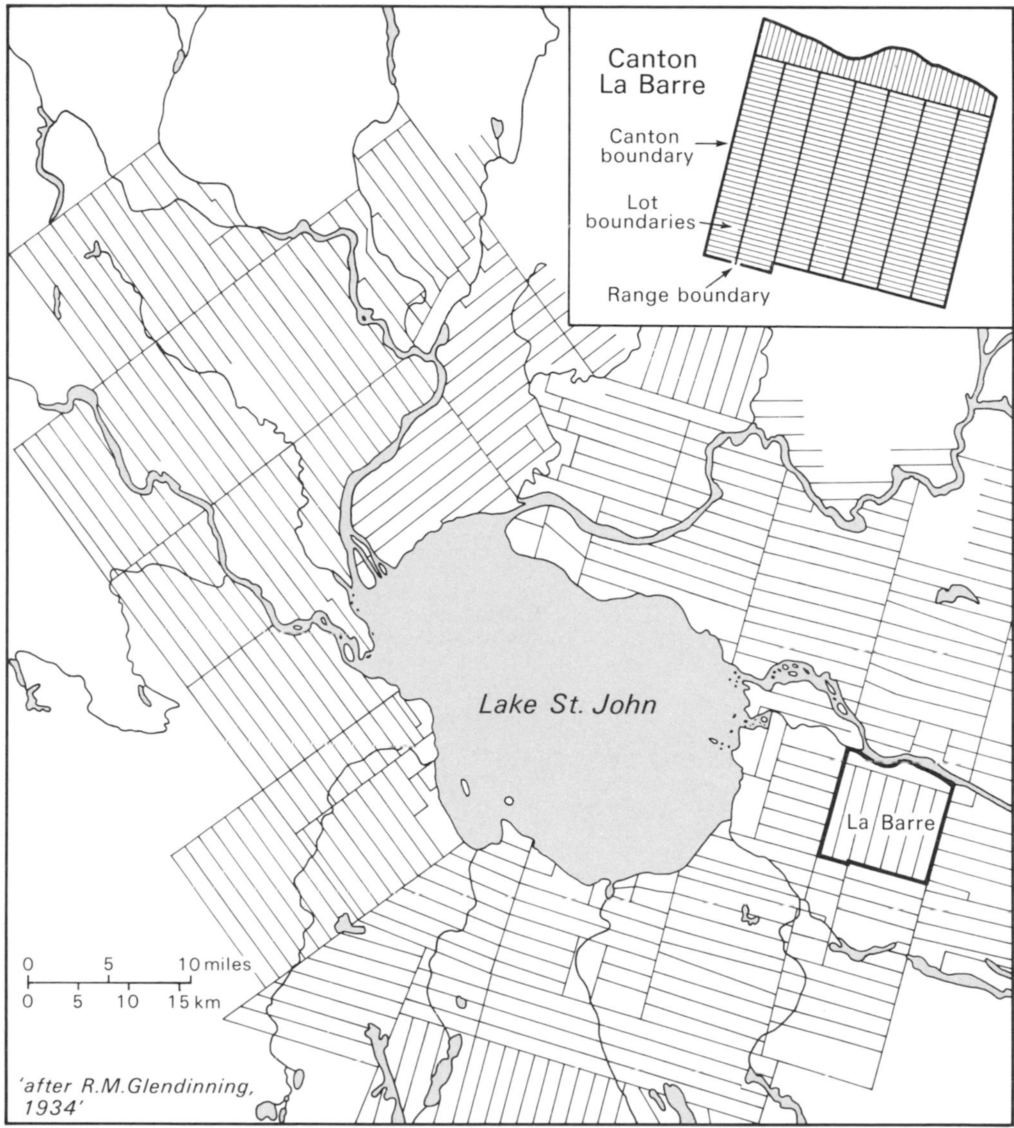

Fig. 8.19 The cadastral pattern around Lake St. John. Source: Glendinning 1934.

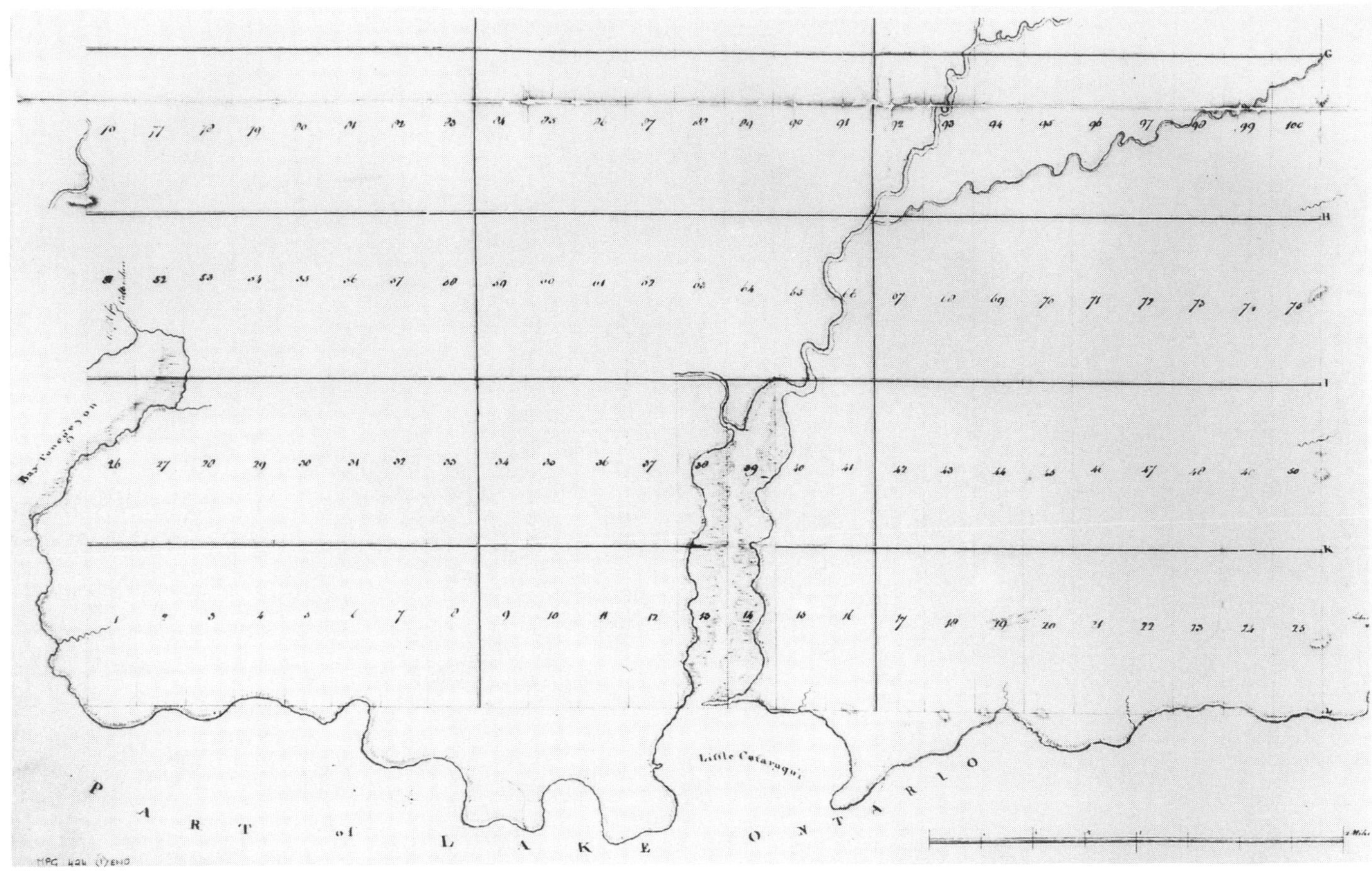

Fig. 8.20 John Collins's Ontario township model. Source: Public Record Office, London, MPG 424.

terparts. Indeed, for several years they were often referred to as *seigneuries*.[102] After 1784, the size of townships was increased to about nine miles front by twelve miles deep to take account of the larger land grants owed for the war service of soldiers.

In the following year the Ordinance concerning Land Surveyors and the Admeasurement of Lands (1785) required surveyors to keep field books and to make plats of their surveys on which the ownership of land was recorded. These were then filed in the surveyor-general's office. In this spirit, D. W. Smith, deputy surveyor-general, wrote in 1796 to tell one of his deputy surveyors that "it will be necessary that you prepare . . . separate four copies of all the Townships within your trust; the surveyed part of each to be ruled in black and the remaining unsurveyed part to be projected in red ink. . . . The Lots are to be left vacant for the grantees names to be inserted here, as the Plans are intended for a Record."[103]

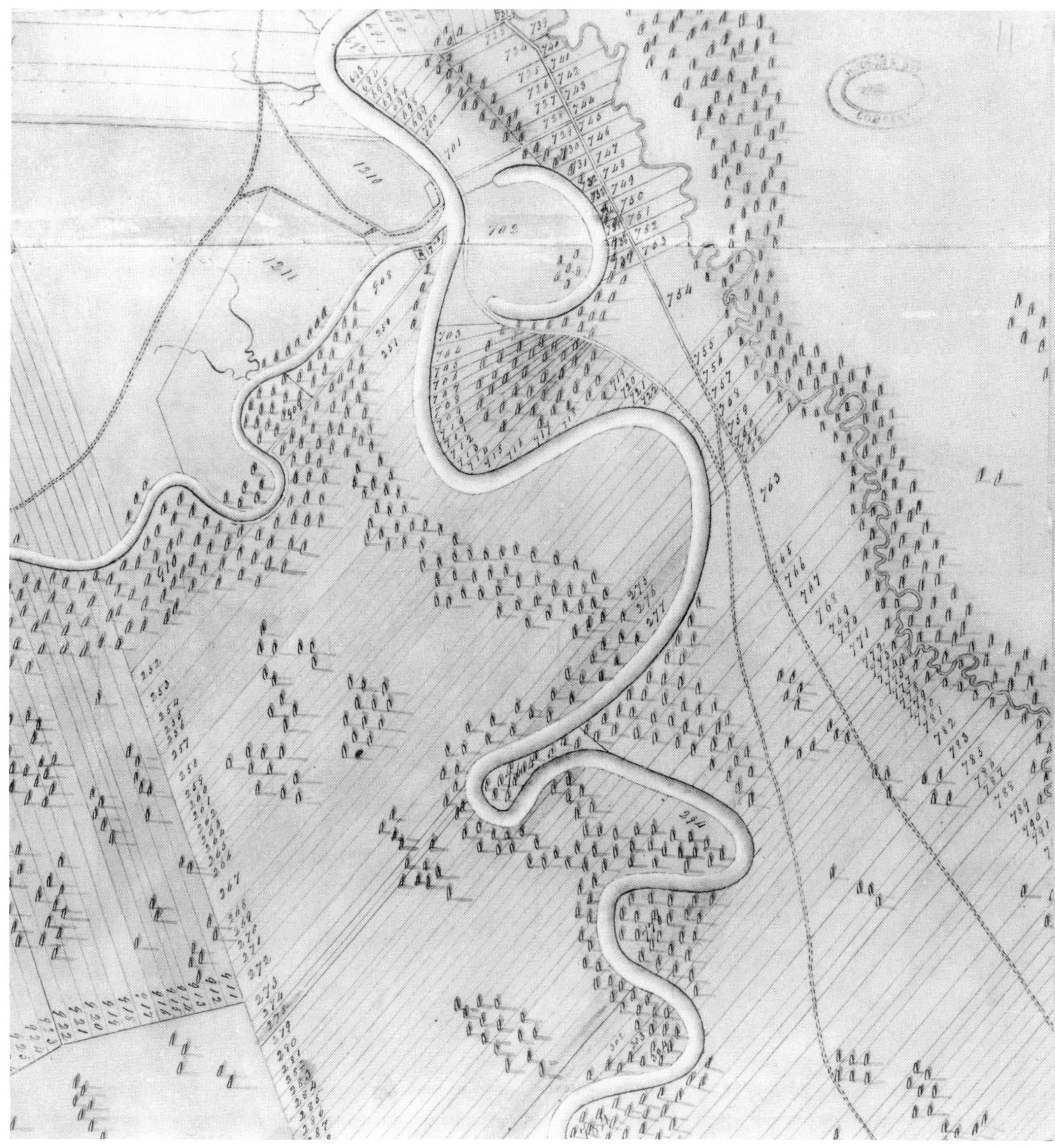

Fig. 8.21 Plan of the Red River Colony, surveyed in 1836 by George Taylor. Source: Hudson's Bay Company Archives, Provincial Archives of Manitoba, Winnipeg, E6/14.

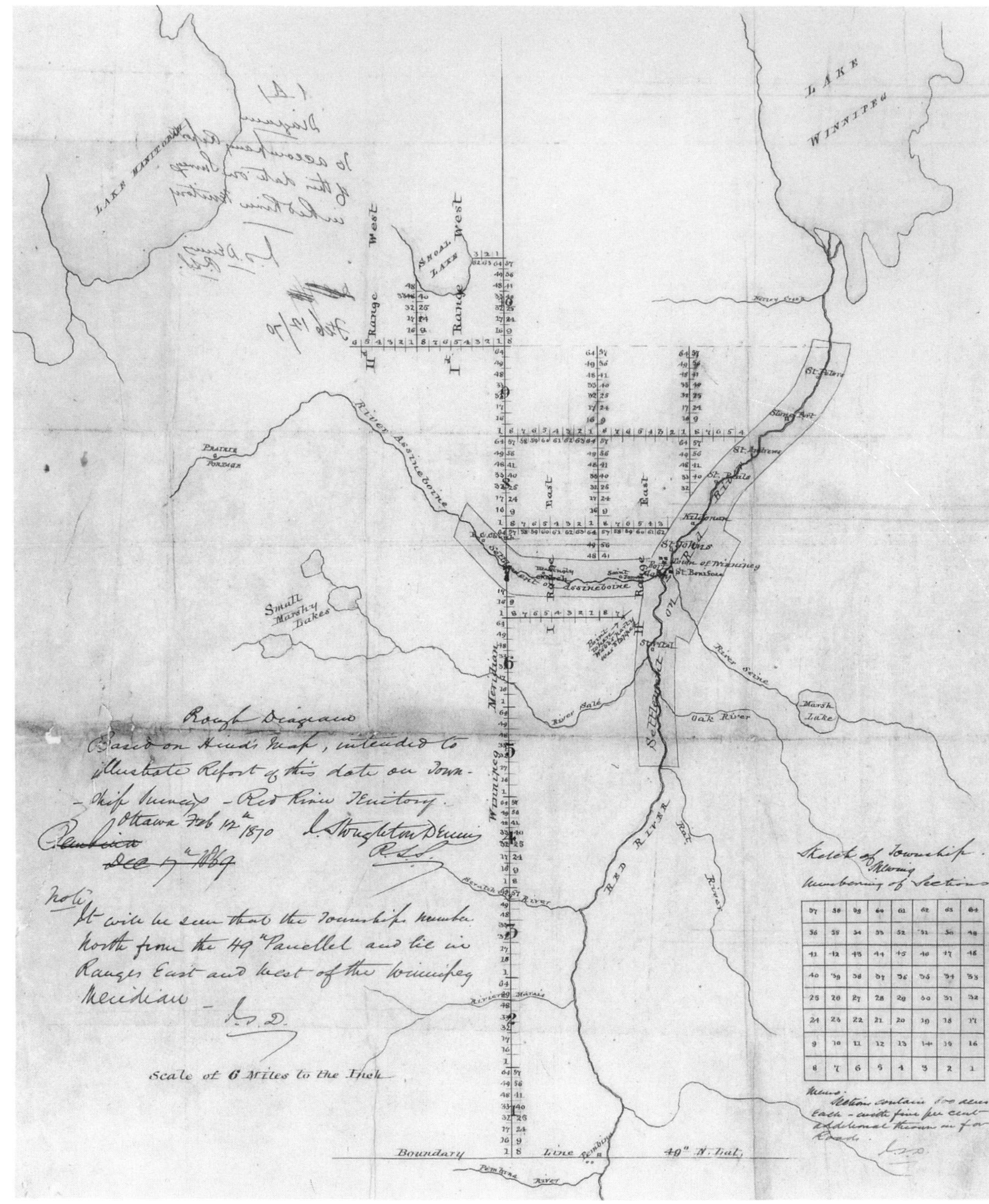

Fig. 8.22 Lt. Col. J. S. Dennis's sketch map of the Winnipeg meridian and the arrangements for section land survey, 1870. Source: National Archives of Canada, Ottawa, NMC 7064, H12/740/Red River Settlement/1870.

Besides providing a record of land grants, this survey information probably also influenced potential settlers in their choice of lots. Certainly in some offices, plans of lots were made available for consultation prior to land auction sales.[104] "On your plan it will be requisite you lay down the swamps, rivers, creeks, mill seats, lakes and minerals and note the same also in your field notes, together with the quality of the soil, timber and everything else worthy of remark."[105]

Several settlement schemes in interior Canada were proposed in these early years by private companies. One such was the earl of Selkirk's vision of a Red River Colony for the Hudson's Bay Company.[106] Settlement began in 1812 from Scotland and Ireland, and Swiss, French, and English immigrants were also recruited. Peter Fidler surveyed the first lots in May 1813, but no map of this date has survived. The river controlled the general pattern of settlement, and Governor Miles MacDonell's detailed layout follows the basic form of earlier French-Canadian riparian long-lot settlement but set out with somewhat wider frontages (figure 8.21).[107]

In the United States to the south, survey and settlement pushed ahead in the nineteenth century based on the six-mile-square township, but in Canada various sizes and internal divisions of townships were tried. Governors Lord Dorchester and Simcoe had both tried to standardize the size and design of townships and had ideal models drawn up, but little standardization was achieved until after the 1867 British North American Act established the Dominion of Canada stretching from the Atlantic to the Pacific and from the United States border to the Arctic (figure 8.16 above).[108] In 1869 the territory of Rupert's Land, which included all the western prairies, was purchased from the Hudson's Bay Company, and its settlement placed under dominion control. In August of that year a system of land survey prior to prairie settlement was instituted consisting of approximately nine-mile-square townships and 600-acre sections, with a baseline or principal meridian running north from the forty-ninth parallel ten miles west of the Red River.[109] Townships were set out (figure 8.22) in ranges east and west of this first, or Winnipeg, meridian. From 1871 the size of townships was reduced to six miles square so as to conform with the United States system, particularly as the bordering state of Minnesota and the territory of Dakota were about to be set out at this time. One-square-mile (640-acre) townships had also been adopted in northern Ontario, where this smaller size was thought better suited for administering forestry and mining country.[110] In May 1871 a manual of instructions to surveyors which followed closely that for the United States was produced, and the land subdivision of all the prairie provinces and the resurvey of the Red River long-lots was begun. In 1883 alone, as a response to demand generated by the Canadian Pacific Railway, 1,221 townships with more than 170,000 farms totaling some 27 million acres were surveyed in advance of incoming homesteaders, an achievement probably unequaled in the survey history of any country.[111]

8.10 County and State Maps and Atlases Based on Land Plats

As the settlement of the Atlantic colonies advanced, individual land plats, the prime purpose of which was the registration of land titles, were used as sources to compile general cadastral maps. Some were commissioned by colonial governments as instruments to assess and regulate their settlement programs. An early example of such a compiled map is illustrated in figure 8.23. The English proprietors of the East Jersey colony requested that "an exact Map of the Country" be sent to them to indicate proprietors' land tracts. In 1686 "Mapp of Rariton River, Milstone River, South River, Raway River, Bound Brook, Greenbrook and Cedar brook with the Plantations thereupon" was produced for them by John Reid.[112]

Printed compilations of township land plats were first published for public sale on the Atlantic seaboard in the early nineteenth century. During these early years of the republic, the county was important as both an administrative and a cultural unit. One of the first printed county maps was Charles Varlé's "Map of Frederick and Washington Counties, State of Maryland," published in 1808. However, Varlé depicted only a small sample of landholdings. The first comprehensive, published cadastral map was Jason Torrey's "Map of Wayne and Pike Counties, Pennsylvania," published in 1814.[113]

By the end of the American Civil War (1861–65), most potential agricultural land in the eastern half of the United States had been subdivided and recorded on cadastral maps. After the war, the manuscript land plats of federal surveyors were increasingly used as sources for an expanding series of published, smaller-scale county and state cadastral wall maps and atlases.[114] Both the atlases and maps contain the names of property owners. This presentation of cadastral information was the commercial rationale: property owners were potential subscribers. Most of these maps and atlases carry a source acknowledgment such as "Compiled from the United States Surveys exhibiting the Sections and Fractional Sections" (figure 8.24).

There was a limit to the number of commissioned illustrations of landowners' properties which could be fitted around a sheet map; an atlas format with one page per township greatly extended the space available for such material and also permitted inclusion of personal portraits and biographies and other material paid for by sponsors at so much per word or per square inch. Atlas production was also a response to the unwieldiness of large-scale county wall maps. The first atlas, Lawrence Fagan's *Map of Berks County, Pennsylvania*, appeared in 1861. Within a few years, atlas production outstripped that of wall maps as the main way in which cadastral maps were prepared for sale. The application of lithography to map production after the Civil War also made atlas publication financially attractive.[115] Populous rural counties of more than ten thousand inhabitants were identified as good prospects; perhaps one in twenty landowners would subscribe to a forthcoming atlas. Cadastral atlases were published chiefly in the Midwest, with

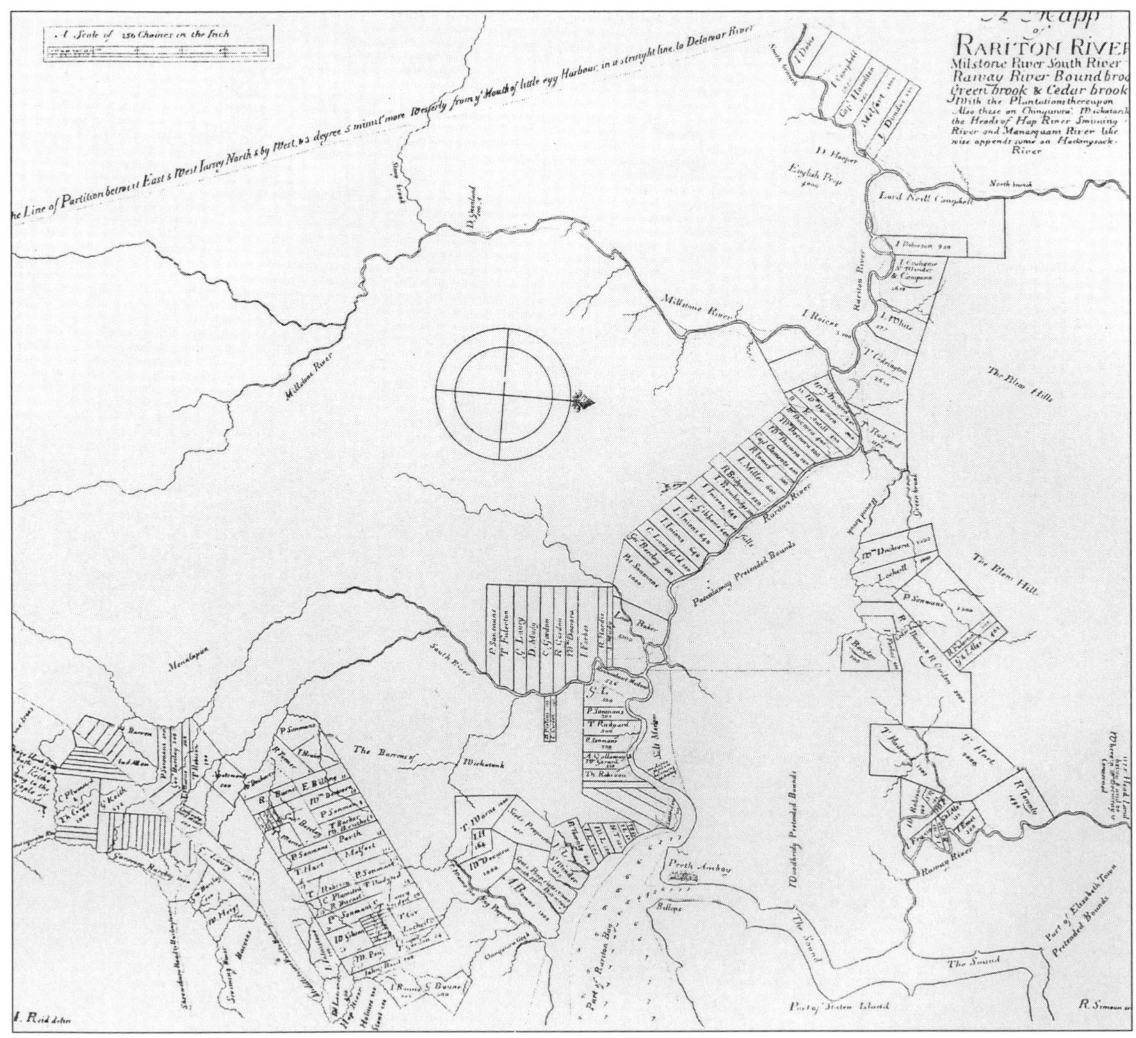

Fig. 8.23 "A Mapp of Rariton River . . . ," by John Reid, 1686. Source: New Jersey Historical Society, Newark, N.J.

almost none from the Deep South. For the southern farmer, defeated and impoverished in the Civil War, a cadastral atlas was hardly a necessity, while because individual plantations were so large, there were fewer potential subscribers per county. It was in the Midwest and to a lesser extent on the West Coast that there were enough relatively prosperous landowners in the second half of the nineteenth century to ensure a market for an atlas.[116]

Cadastral data from land plats also figure in state atlases. Alfred T. Andreas of Chicago, one

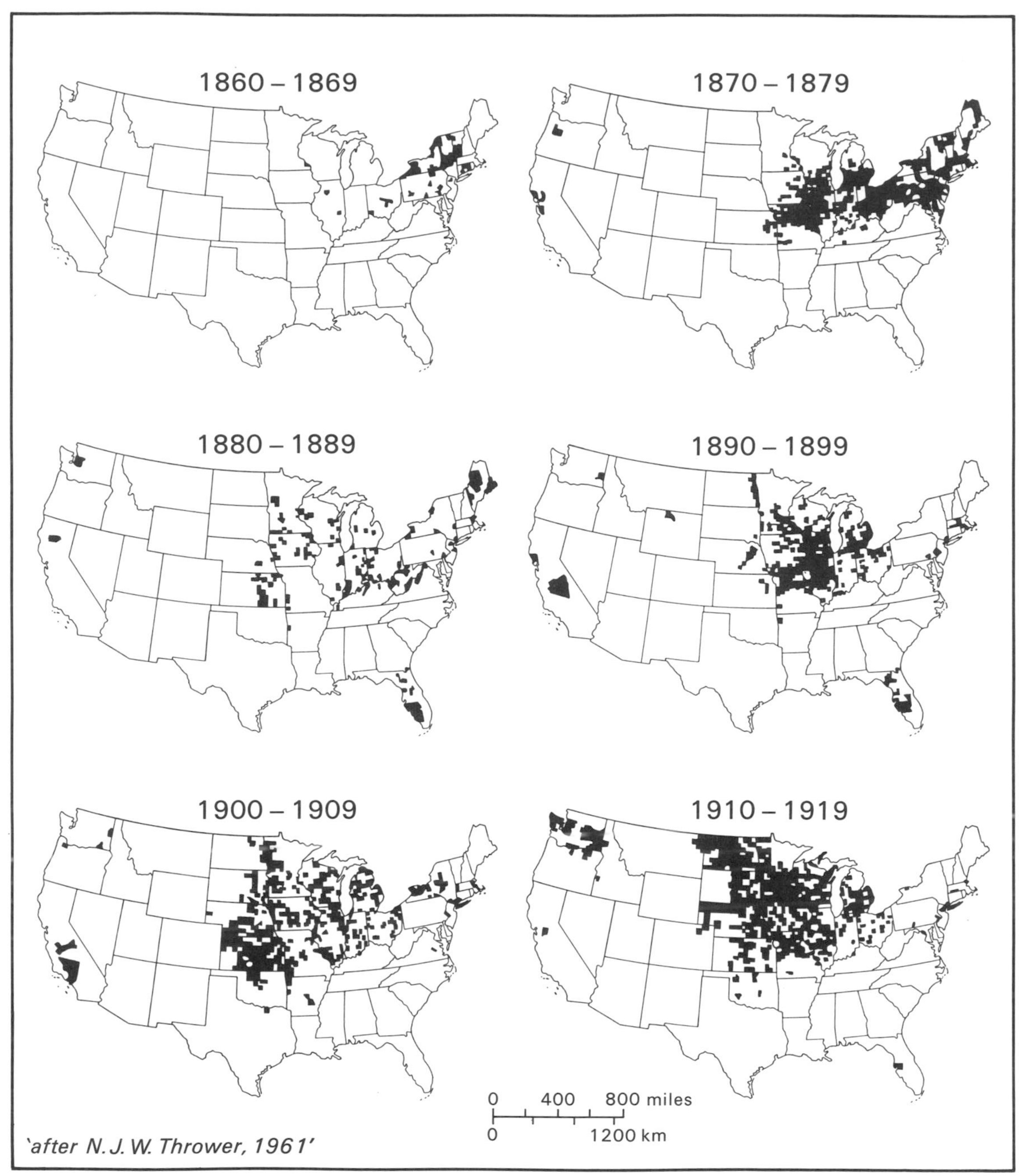

Fig. 8.24 Publication of county atlases in the coterminous United States, 1860–1919. Source: Thrower 1961.

of the most prolific and successful atlas publishers of the post–Civil War period, published his state atlas of Minnesota in 1874 with seventy pages of county maps and maps of cities and towns. His Iowa atlas followed in 1875 and that for Wisconsin in 1878.

The production of county maps and atlases was also actively pursued north of the border in Ontario in the later nineteenth century. As in the United States, most of the data for these maps derive from surveyor-generals' original cadastral maps of townships, and only a minimum of new fieldwork was undertaken to check and revise the original township maps.[117]

Cadastral maps in atlases or in the form of wall maps contributed to public awareness of the cadastral grid and associated pattern of landownership, as not only were they subscribed to by quite large numbers of individuals (one thousand copies might be a typical print run), but they were also hung in public places such as general stores and banks. To landowners in the new states, the presence of their property clearly identified on a map in a published county atlas confirmed their stake in the new nation. It was the prospect of owning land that had probably fired people to drive their wagons and later to buy their railroad tickets out west. The reverence with which these symbols might be treated is nicely instanced by Ross Lockridge's character Mr. Shawnessy in his novel *Raintree County*, who "had bought a copy of the *Atlas* for ten dollars and put it on the parlor table with the Family Bible and the Photograph Album."[118] The cadastral survey plats were thus exploited for private profit while also etching the cadaster into the public mind.[119]

8.11 New South Wales

The system of land alienation and the role that cadastral surveying and mapping played in that process originated and evolved separately in each of the Australian colonies (figure 8.25). New South Wales and Tasmania were settled with little effective overall control; in Victoria a degree of order was imposed in an attempt to stem the land plundering of extensive pastoral farmers; in Western Australia huge tracts were granted in unregulated fashion, while the South Australian (and New Zealand) "systems" were in a sense a measured reaction to all of this. Even after Australian Federation in 1901, there was no single federal system of land alienation comparable to that which governed land settlement in the United States public domain in the nineteenth century.[120] Victoria and Queensland were not partitioned from New South Wales until 1851 and 1859 respectively, and many early surveyors in Van Diemen's Land (Tasmania) came from New South Wales.[121] Within this diversity, it was New South Wales practice which affected by far the largest part of the settled continent for much of the nineteenth century.

In the year following the establishment of the convict settlement at Botany Bay in 1788, a time-expired convict became the first settler in Australia when he took up a 30-acre farm at Parramatta in the emerging colony of New South Wales. By 1804 a policy for the control of land

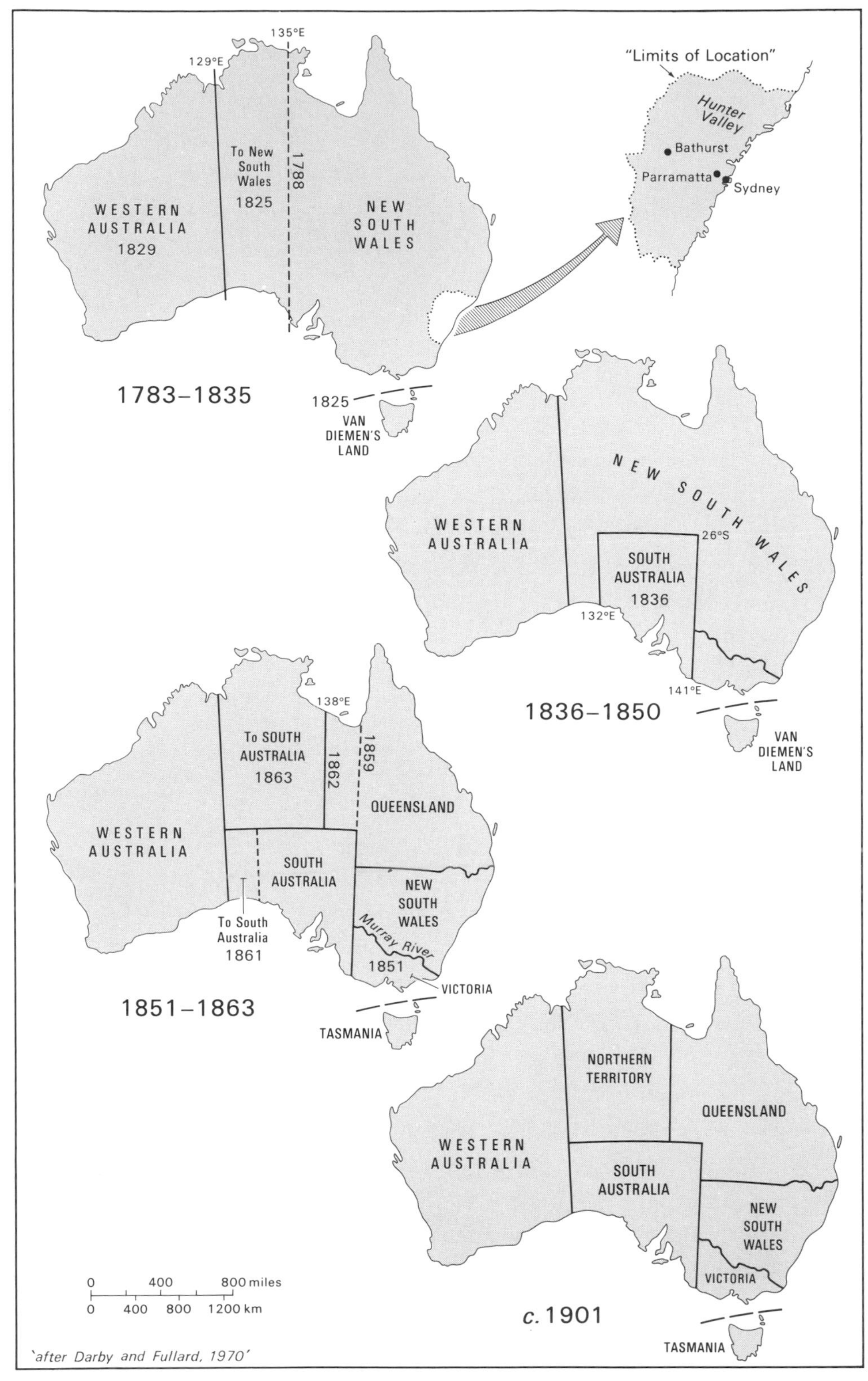

Fig. 8.25 Colonial Australia, 1788–1901.

settlement was being devised with a principle of grouping settlers in townships of up to 30,000 acres. In 1820 about 325,000 acres had been granted, but control was distinctly lacking, with settlers preceding surveyors, whose preoccupation with Admiralty work, exploration, and road surveys had enabled them to survey only about 145 farms.[122] Nor was there much control over the acreages allotted. Under the terms of settlement instructions issued in 1788, convicts were supposed to receive 30 acres, with 20 acres more if they were married and 10 acres for each child. Private soldiers were to be allotted 80 acres, and free settlers up to 100 acres. Governors were empowered to make increased grants, and some large allotments of 5,000 acres upward were made. In return for land grants, government obtained revenue through quitrents (small payments in lieu of services).

In 1821 Governor Sir Thomas Brisbane set his surveyors to experiment demarcating six-mile-square townships following American models, and he issued instructions identical to those used in the United States. The six-mile-square townships were to be divided into thirty-six sections of one square mile. Some townships were surveyed, subdivided, and mapped in this fashion along the Macquarie and Hunter rivers.

Governor Darling, who succeeded Brisbane in 1825, brought with him from London a new set of instructions which abandoned townships and introduced an "English" hierarchy of 40-mile-square counties, 10-mile-square hundreds, and 25-square-mile parishes to try to ensure contiguous and close settlement. New parishes were not be created beyond the then limits of settlement (figure 8.25).[123] Parishes were to be divided internally by rectangular survey into square-mile lots, at first offered for sale, but later given by grants subject to quitrents. Though grid subdivision was employed, there was no common baseline for this or a prime meridian for the colony, so, with surveying beginning at a number of widely separated points with reference to magnetic north and no allowance made for compass deviations, a series of regional grids, rather than one colonial pattern, was established. Section lines were also modified along watercourses, producing narrow but deep lots to maximize holdings with vital water access.[124]

Governor Darling's instructions had ordered that land survey should not be allowed into "such Districts of Our Territory as lie beyond the range of any actual settlements, but shall from time to time be extended into the parts thereof which are at present unsettled, as the cultivation of Our said Territory may progressively advance."[125] Though survey was supposed to precede selection of land, in practice squatting without survey was a usual means of obtaining the large acreages of land needed by pastoral sheep farmers as settlement extended away from the coast.

The extreme case of profligate, effectively unregulated land alienation in the first half of the nineteenth century occurred in the Swan River Colony of Western Australia, where land was granted in huge tracts of 100,000 acres and more, such that, within five months of the settlement's beginning, all the fertile land along the Swan River had been claimed. In 1830, of the 1.2

Fig. 8.26 New South Wales cadasters: (*a*) a parish with early, regular grid alignment of parcel boundaries; (*b*) a later parish in which boundaries were set out at the time of land sale. Source: Jeans 1975:6.

million acres alienated, only 160 acres were cultivated.[126] In New South Wales, 4 million acres had been chosen piecemeal over an area six times that extent, and there was unrecorded settlement even beyond this zone. The requirement to survey before settlement in New South Wales was a virtual dead letter by 1826, not least because of the shortage of surveyors and a buildup of arrears of survey. Settlement expanded rapidly in the 1820s around Bathurst and into the Hunter valley. "Not a cow calves in the colony," complained Governor Brisbane, "but her owner applies for an additional grant in consequence of the encrease of his stock." [127]

In 1829 an expedient of drawing a line around the then settled area of New South Wales, known as the "Limits of Location," was adopted to overcome the time lag between survey and settlement. Land could be occupied within this area without prior survey, but with a guarantee of confirmation of title in the future.[128] Beyond the line, no roads were made, and towns were established only in special circumstances; within the line, a settled community was created with security of tenure.[129] The supposed stricter enforcement of this policy after 1831 did little in practice to restrict settlement, nor did it limit the size of holdings. Therefore, reserves were set aside specifically for division into small farms of less than 320 acres, a policy institutionalized by the Crown Lands Act of 1842, by which land within five miles of an established or projected town was to be sold as suburban lots to small-scale agriculturalists.[130]

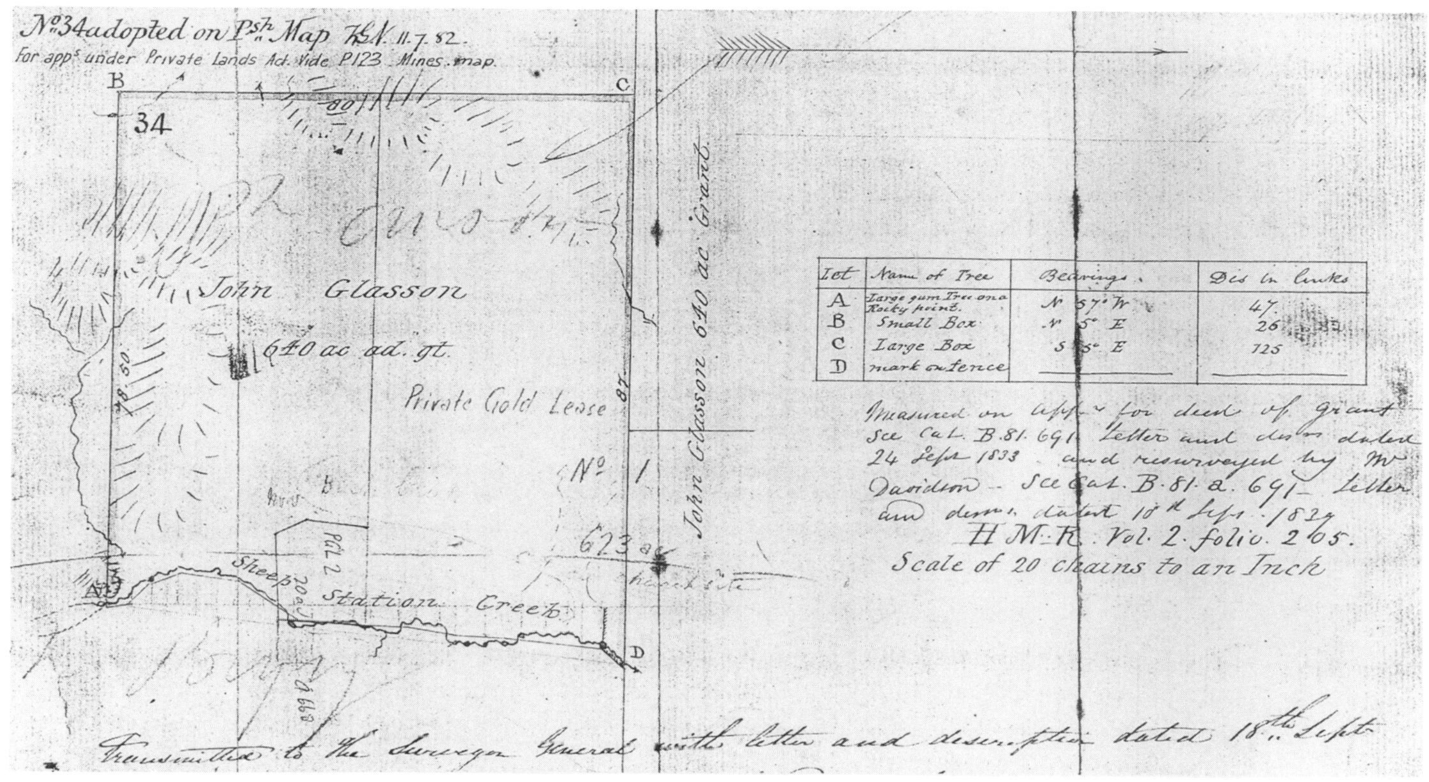

Fig. 8.27 Plan of a 640-acre "additional grant" to John Glasson, county of Bathurst, parishes of Byng and Anson, 18 September 1839. Source: Department of Lands New South Wales, Sydney, 81a.691.

The area known as the Settled Districts within the Limits of Location was surveyed using a rectangular grid, at first on the ground, but after 1830, with the paucity of resources and increasing demand, the cadastral grid was drawn on maps from which it was transferred to the ground as and when land was alienated. The contrasting regularity is shown in figure 8.26. As Roberts comments: "So dilatory was the survey . . . that more than half the settlers in the Australian colonies had no claim to their lands, even though they were within the official boundaries of settlement."[131]

The general practice of this rectangular survey was very similar to that of the United States Federal Land Survey. The surveyors' plans, known as portion plans, record the species or type of tree used to mark the corners (though rock marks or stakes were also used) (figure 8.27).[132] The 1848 *Regulations for Licensed Surveyors* issued by Deputy Surveyor-General S. A. Perry specified that:

> The scale on which plans are to be drawn is for country lots 4 inches, and for general survey of features 2 inches to the mile. . . . on the corner of each plan there must be a reference to the marks or marked trees upon the ground, which shew the corners of the surveyed lands. The surveyed lines and stations, with their lengths and compass bearings, must be shewn on the plans in red color, and a proof line, run diagonally, must also be shewn in the same color. Boundary lines must be black, as well as the outlines of existing tracks or features, if surveyed. If merely sketched they must be shewn in black dots, as indicative of uncertainty. . . . Every plan and description, or set of plans and descriptions, transmitted to head-quarters, is to be accompanied by a letter of transmission and concise report. . . . Payments will be made upon the Certificate of the Surveyor-General . . . that the work has been examined and found to be correct.[133]

In New South Wales in 1864, new guidelines entitled *Regulations for Guidance of Licensed Surveyors Connected with the Survey Department of New South Wales* were issued, requiring that surveyors conduct an inventory of the colony as they laid out tracts of land for alienation. They were to continue to mark corner trees on their portion plans but also "the boundaries of swamps, forests, plains, land liable to inundation . . . the geological and mineralogical character of lands measured, their suitability for towns, to building or cultivation purposes, and in country portions to agricultural and pastoral occupation—the supply of water and indigenous produce as timber, grass etc."[134] These details are usually found written across the portion plans rather than in an accompanying note as in the United States. In New South Wales there seems to have been no systematic use made of land resource information in the process of land selection.

Proposals to connect all Crown land grants into a general survey (cf. R. K. Dawson and the English tithe surveys in chapter 7.3) were made as early as the 1820s but remained largely unrealized. In Victoria, a geodetic survey was begun in 1858 with primary meridians and parallels at one-degree intervals, but on the whole authorities considered that the system of isolated surveys was effective for its prime purpose of alienating land in times of rapidly increasing demand. Neither surveyors nor legislators were anxious to modify a system that, in the short term, might delay alienation of land.[135] This cadastral mapping in Australia was no triumph of geometry over geography.[136]

Notwithstanding its efficacy for alienating land, the survey and mapping work was at least as inaccurate as that of the United States federal land surveyors. Even in the nineteenth century, distances were sometimes measured by a wheeled perambulator rather than chained, and as in the United States, a compass or circumferentor was used for angle measurements rather than a theodolite.[137]

In 1885 a royal commission was set up to investigate the work of the Victoria survey department. It reported that:

> thousands of certificates of title have been issued which do not represent what the land owners possess; and a very large number are in existence in which the land is so vaguely defined, that no surveyor could possibly, by any known process, from the information contained therein, define the position on the ground of the allotments shown in the certificates. And the effect of all this has been the creation or development of endless complications and difficulties, from the apparent overlapping of boundaries, excesses and deficiencies in dimensions and areas, the clearing up of which adds materially to the work of the drafting branch, and is also the cause of great trouble and expense to the owners of land.[138]

8.12 South Australia

In nineteenth-century Australia, attitudes to land became sharply polarized between the interests of commercial pastoralism, which required extensive landholdings, and that of the colonial administration, which envisaged a closely settled country with land held in small parcels by "civilized" agriculturalists and which held that "More and Smaller Is Better."[139] It was against this background that Edward Gibbon Wakefield and his colonial reformers formulated their theories of settlement organization, which envisaged, first, a "sufficient price" for land as the regulator to ensure land was alienated in proportion to the labor it could attract, and, second, the employment of systematic cadastral surveys to ensure order in the whole process.[140] "Order" involved promoting contiguous settlement, which was expected to promote a sense of community. The prospect of landownership for immigrants who built up their capital with some years of laboring would ensure a "permanent interest" of the Australian population in the Australian nation. It was a recipe designed to reproduce England's capitalist society overseas and to avert the scenario Wakefield described in 1835.

> The process by which a colony goes to utter ruin, or is reduced to misery, and then gradually recovers, has been witnessed over and over again. The colonists proceeding from a civilized country, possessing capital, divided into classes, skilful, accustomed to law and order, bent on exertion and full of high hopes; such a body of people reach their destination and then what happens? The society, which at the moment of its landing consists of two ranks, bearing towards each other the relation of master to servant, becomes instantly a dead level, without ranks, without servants or masters. Every one ob-

> tains land of his own. From that moment no one can employ more capital than his own hands will use. The greater part, therefore, of the capital which has been taken out necessarily wastes away. The seeds, ploughs, and other tools, with materials for building, rot on the beach; the cattle stray and perish. In a few months, nothing in the shape of capital remains beyond such small stocks as one isolated person can manage. But those of the society who have not been used to labour cannot, with their own hands, manage even that small stock, so as to increase or even preserve it; while those who have been accustomed to nothing but labour, and to labour in combination with others, finding themselves each one alone in a vast wilderness, are unable to use with advantage such small stocks as they begin with; and thus both classes (or rather the whole body, for now there are no classes) soon fall into a state of want. Despair follows of course, with mutual reproaches, and then rage, hate, plunder and fighting; and in the end, unless aid comes from without, all the people die of hunger. This is not an overcoloured picture.[141]

The South Australia colony founded in 1836 was the first real trial ground for Wakefield settlement. For at least the first two decades of its development, a landscape of townships subdivided into small farms was produced. The prime instrument which government employed to try to bring this about was a tightly controlled land survey and mapping system.

Right from the start, settlement regulations specified that cadastral surveys should precede sales, land was to be divided into eighty-acre sections and sold by sections for not less than twenty shillings an acre (later reduced to twelve shillings), and holdings of more than eighty acres had to be composed of contiguous sections.[142] Every purchaser was entitled to rent large tracts of "common," and purchasers of more than £4,000 worth of land could request "special surveys" outside the then contiguous sections. If the purity of Wakefield philosophy was thus diluted, settlement in South Australia was still strictly controlled by comparison with the grant-quitrent system of New South Wales.

It would take an immigrant agricultural laborer perhaps five years to accumulate the necessary £80 stake to purchase the minimum holding. As Michael Williams puts it: "An ordered, hierarchical society could be ensured, consisting of capitalist farmers employing those who were labouring to accumulate capital. However one may view the philosophy implicit in this arrangement, the result was initially successful and in marked contrast with the systems prevailing elsewhere in Australia."[143]

The first surveys were organized by Colonel William Light, surveyor-general of the colony, on a strict trigonometric basis. In this manner 150,000 acres were surveyed around Adelaide (figure 8.28) before the London directors of the South Australia scheme insisted on a cheaper and quicker rectangular running survey. They were advised by the English assistant tithe com-

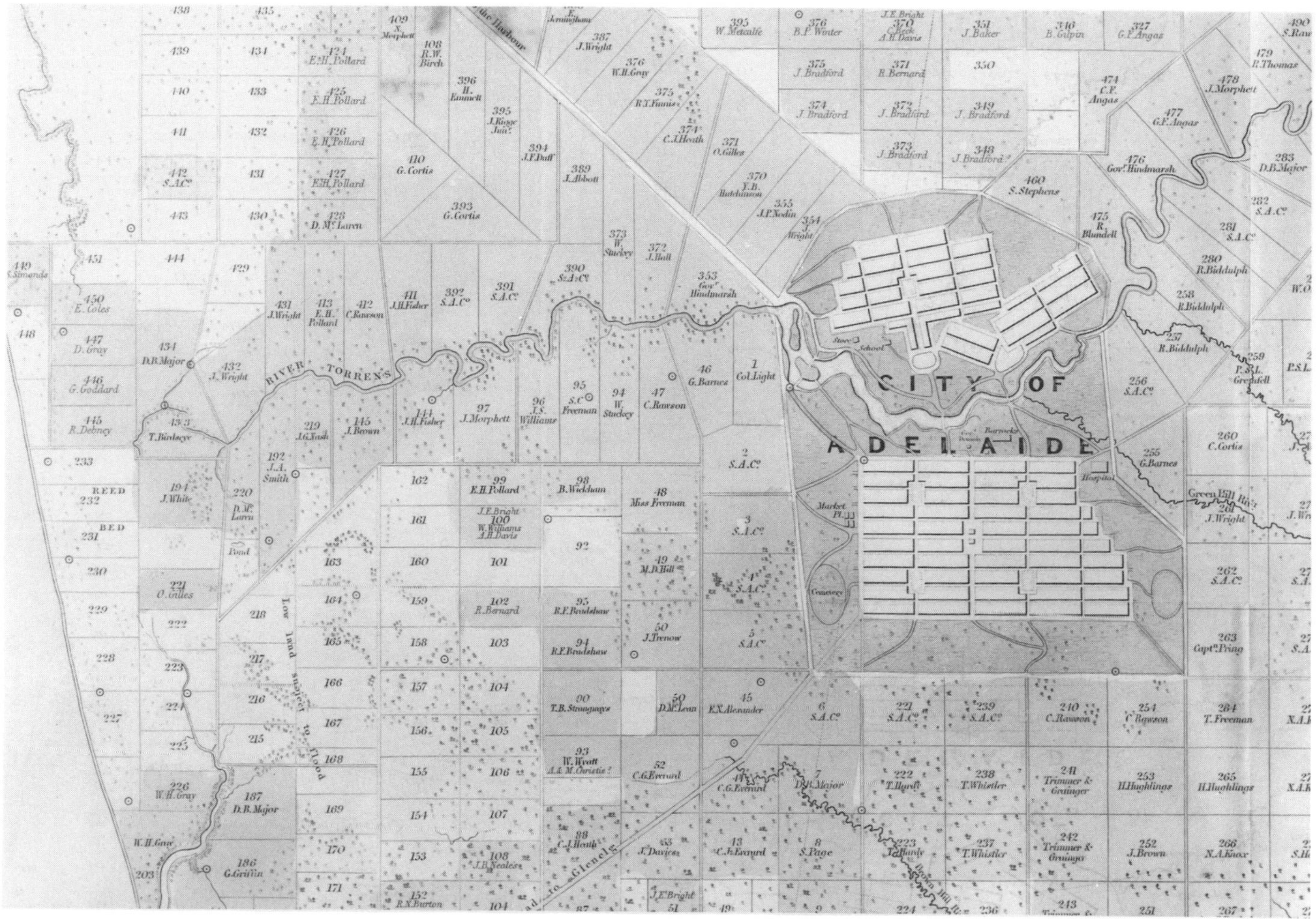

Fig. 8.28 Part of the *District of Adelaide, South Australia; as Divided into Country Sections from the Trigonometrical Surveys of Colonel Light late Survr. Genl.* (London: John Arrowsmith, 1839). The triangulation stations are indicated by dots enclosed in open circles. Source: Dawson 1841.

missioner in charge of tithe commutation surveys, Robert Kearsley Dawson (figure 8.29). Dawson's contribution was critical, both to the role of cadastral mapping in South Australia and for the Wakefield settlements of New Zealand. His ideas are expressed most fully in a report he made in 1840 on the most economical mode of effecting surveys in New Zealand "with general correctness but without minute accuracy"; his work in those provinces is discussed in section 8.14 below.[144]

In 1840 Surveyor-General Frome set down detailed instructions concerning the survey of

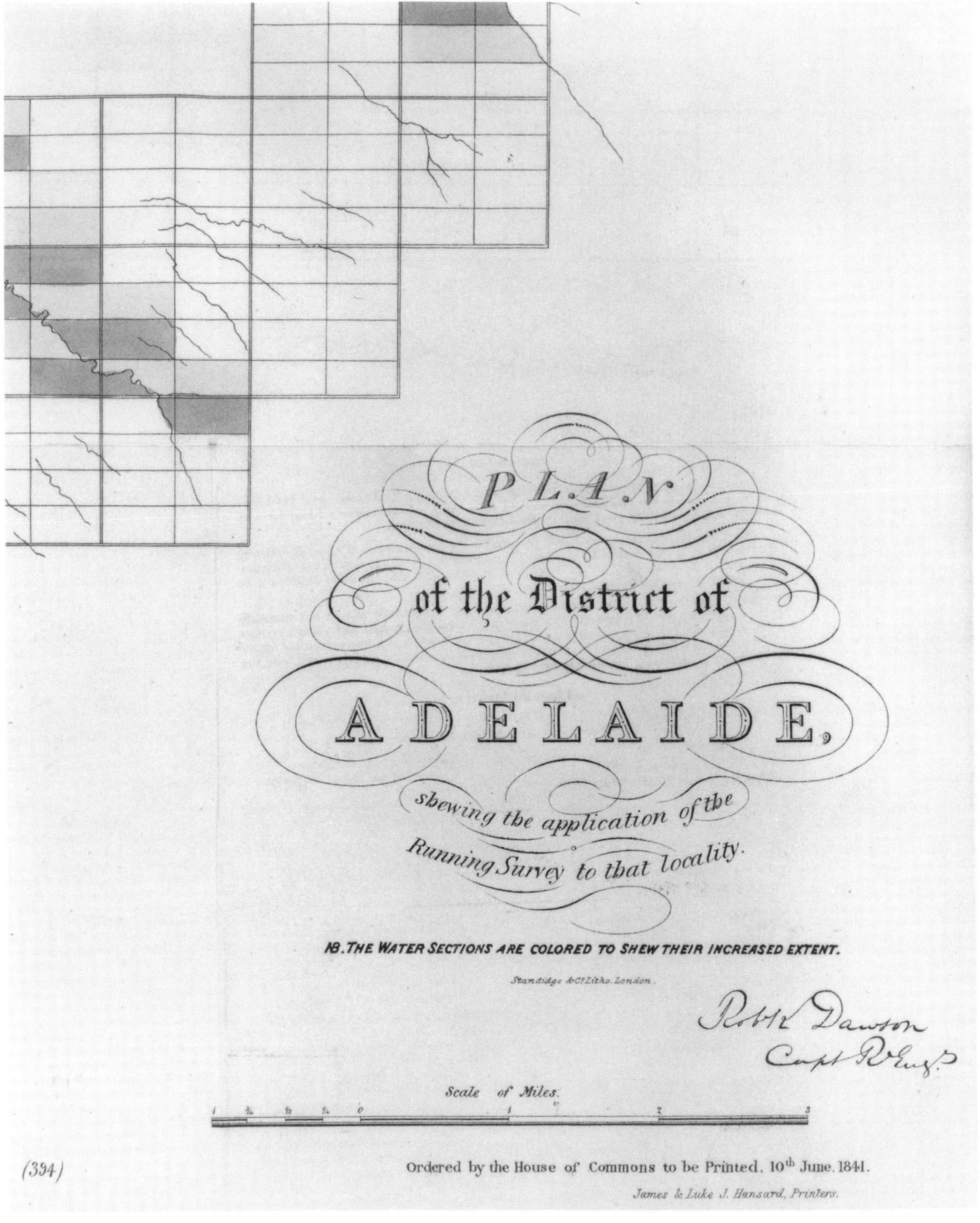

Fig. 8.29 Dawson illustrated the way in which a running survey could be used more quickly and at less cost than Colonel Light's trigonometric method to survey the same Adelaide district of the South Australia Wakefield Settlement. Source: Dawson 1841.

sections. These were to be square where terrain permitted and enclosed by occupation roads. For sections with river, coast, or main road frontages, the frontage was to be halved to increase the number of landowners with access to these advantages. In hilly areas size could be increased to take in wasteland, and the strict geometric grid of occupation roads enclosing sections could be varied to take terrain into account.

Land was divided into counties, hundreds, and sections for administrative purposes. Townships were not used (except that the name was applied to town settlements), and the hundred of approximately one hundred square miles became the unit by which the orderly disposal of land in sections was planned. By 1887, when the first *Handbook for Government Surveyors* was issued, this procedure had become formalized and stereotyped. Preliminary surveys of the hundreds were conducted to determine the general lie of the land and then returned to the controlling authorities, who prepared sketches of road lines. These general maps were passed to the cadastral surveyors, who set out sections within the hundreds.[145]

8.13 The Torrens System of Land Registration

In the years 1857–74, each of the Australian colonies (and New Zealand) adopted a system of land-title registration developed by Sir Robert Torrens and known thereafter as the Torrens System. Title to land under this system depended not on private deeds of transfer, as in England, but on registration of the land itself in an official register of titles and dealings which was open to public inspection. It was much simpler and cheaper than English practice and as such was a boon in the colonies with their hyperactive land markets.[146] Cadastral maps and plans deposited by licensed surveyors were an integral part of the registration process. Torrens statutes emanate from South Australia (1857–58), itself the colony with the most ordered system of land alienation at that date, and were adopted by the colonial parliaments of Queensland (1861); New South Wales, Tasmania, and Victoria (1862); and Western Australia (1874). These acts required that after a given date all land alienated from the Crown would follow the Torrens System, while land titles granted prior to the acts could be registered voluntarily.[147] Thus the Australian Torrens Acts (and the New Zealand Land Registry Act of 1860) established a precise and pivotal role for cadastral maps in the land registration process.[148]

As the process of land alienation proceeded in Australia, a spatial record of individual land titles was built up in colonial land offices by compiling and then periodically updating parish and county-scale cadastral maps. These are at scales of from one inch to between twenty and eighty chains (1:15,840 to 1:63,360) and record for each portion (an individual property) acreage, type of tenure, date, and name of original alienee or lessee. Most of these maps were first compiled in the latter part of the nineteenth century as counties were designated.[149]

In Victoria, the Grant Land Act was passed in 1865 to increase the agricultural as opposed to the pastoral occupation of the remaining public domain. It provided allotments of between 60 and 640 acres, but title was not granted until, at the end of a conditional lease, selectors had proved themselves by physically improving their allotments. By the end of 1865, almost 4 million acres in 157 agricultural areas had been opened for such settlement. Cadastral plans of each were lithographed and sold to prospective lessees to help them make their settlement decisions. At one and a third inches to a mile (1:47,520), they show the acreage, numbering, and boundary lines of allotments. Some 140,000 copies were sold by the end of that year.[150] For South Australia, there are state maps of alienated land for 1863, 1865, 1877–78, and 1894 onward.[151] With first-come, first-served "selection before survey" allowed in New South Wales under the Robertson Land Acts of 1861 to alienate more land for small-scale agriculture, there was not a uniform advance of survey but rather an uneven spread of selections with piecemeal surveys of individual properties to be fitted together.[152] This is all in stark contrast to the township plats of nineteenth-century United States of America, which were so easily brought together into county and state maps and atlases (section 8.10).

8.14 New Zealand

Following James Cook's landings of 1769, there was a gradual, small-scale immigration by missionaries. In 1840 New Zealand, which then contained but a few hundred European settlers, was annexed to Great Britain. Over the next half century, as in North America and Australia, surveyors and cadastral mappers played a pivotal role in the land alienation process.[153] Early surveys employed the running traverse and rectilinearity of other nineteenth-century colonies, but by midcentury, triangulation was being seriously discussed. In 1852 the country was divided into six provinces (later to become ten), and each province developed a separate survey and mapping system (figure 8.30).[154] When the provinces were abolished in 1876 with the establishment of one central government, triangulation was adopted for the whole country under a central Lands and Survey Department.[155] In this respect, the cadastral surveys and maps of New Zealand stand apart from those of the North American and Australian colonies.

From 1840 to 1876, government closely supervised settlement only in the Auckland district. The other main settlements of Wellington, Nelson, Taranaki (Plymouth), Otago, and Canterbury were controlled by the New Zealand Company and its associated settlement companies and were founded on Wakefield theory. The prime influence behind their initial adoption of running surveys as in New South Wales was Robert Kearsley Dawson of the English Tithe Commission.[156] Acknowledging the technical superiority of trigonometric surveys such as his own tithe surveys, Dawson considered that "by means of the running survey alone, sections may be set out with such

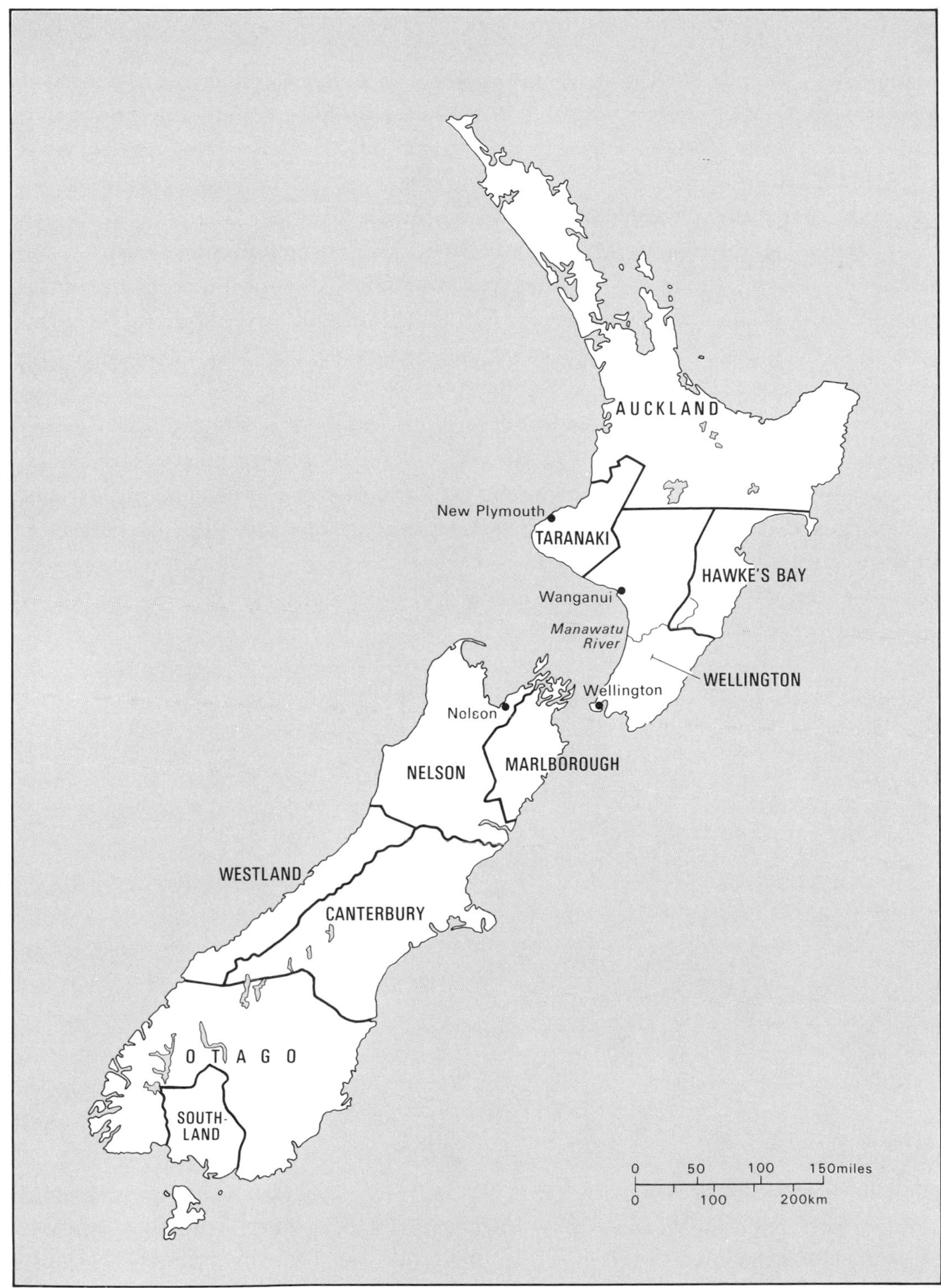

Fig. 8.30 New Zealand: districts, provinces, and places noted in the text.

accuracy as to secure the purchaser from loss, or, in other words, to assure him of the possession of, and clear title to, the full quantity of land he has paid for, it will be evident that the trigonometrical operation, may at least, be deferred until ulterior objects call for it, and the colony is better able to pay for it."[157]

In his report he makes explicit reference to the effectiveness of rectangular surveys in North America and also to their deficiencies, which he ascribed to "trusting entirely to the compass." He recommended the use of, and provided detailed specifications for, a simple four-inch theodolite rather more robust than that used at the time in India.[158] This type of survey was used for initial land alienation in Wellington, the Manawatu, Wanganui, New Plymouth, Nelson, and Otago (figure 8.31).[159]

Wellington was laid out in 1840 as the first of the New Zealand Company Wakefield settlements, though the role of the cadastral map was very different here from that in early South Australia. Many Wellington surveys were "paper surveys," with road and section lines marked out on the ground only after sale. Selection and sale *before* survey was usual in Wellington, and selections varied greatly in size and shape.[160] In Nelson (1841), Taranaki (1842), and over the first 150,000 acres of Otago (1848), in contrast, a more rigid system of rectangles was laid out as described in the satiric verse:

> Now the road through Michael's section,
> though it looked well on the map
> For the use it was intended
> wasn't really worth a rap.
> And at night was not unlikely
> to occasion some mishap.
>
> It was nicely planned on paper,
> and was ruled without remorse
> Over cliffs, and spurs and gullies,
> with a straight and even course
> Which precluded locomotion
> on part of man or horse.[161]

Canterbury (1850) was the fifth and last large-scale settlement to be established in New Zealand. Wakefield organized it as a special Church of England settlement, convinced that linking the Canterbury Association, which was to organize the settlement, to the church would reinvigorate the New Zealand Company's colonizing activity. It was also explicitly intended to avoid some of the mistakes of other settlements. Two years were set aside for exploration, assessment of agri-

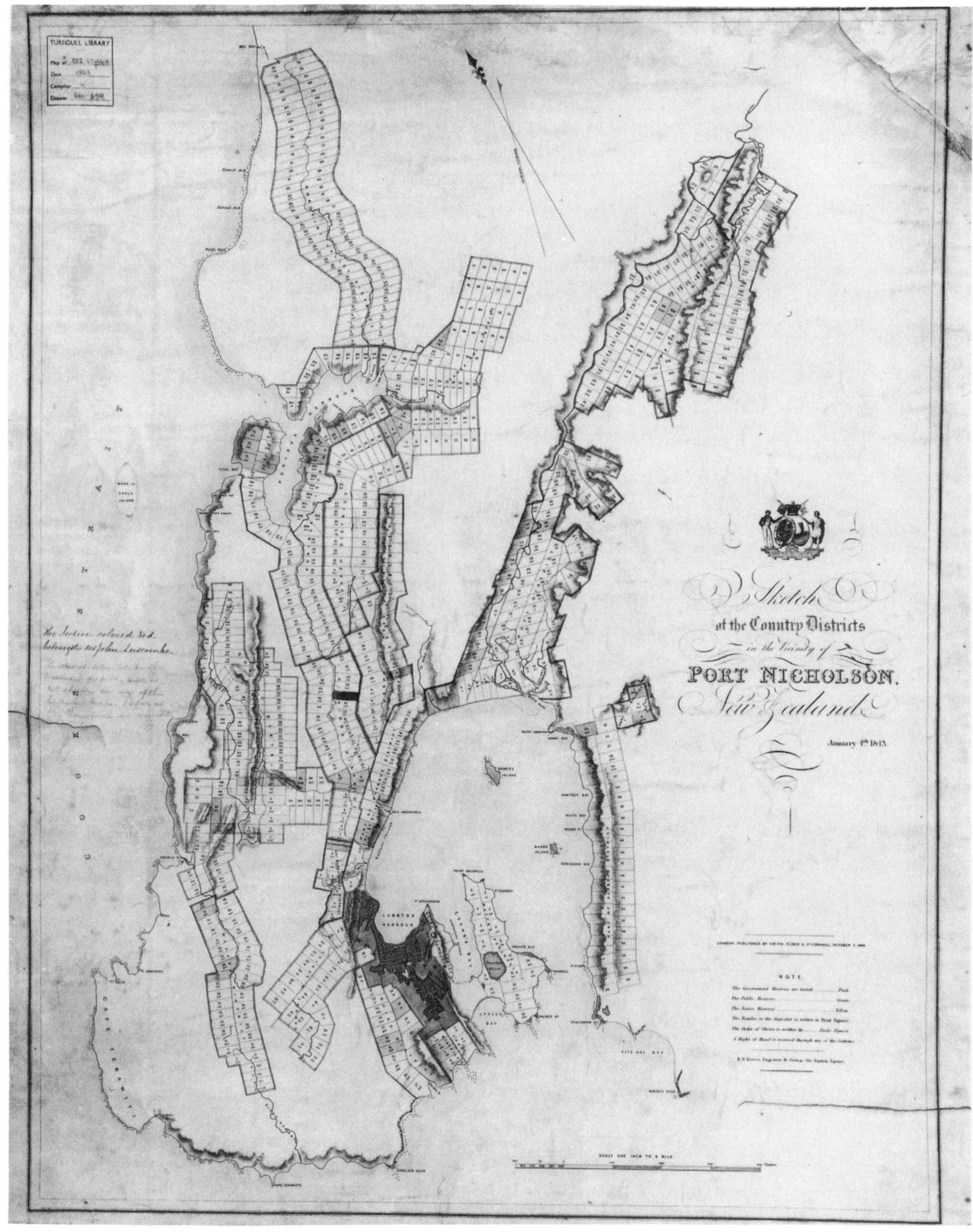

Fig. 8.31 "Sketch of the Country Districts in the Vicinity of Port Nicholson, New Zealand, 1843." Source: Alexander Turnbull Library, Wellington; a printed version of a manuscript plan held by the Wellington District Office of the Department of Lands and Survey, 8:32.47 gbbd/1843/Acc 457.

cultural areas, and settlement planning. Felix Wakefield, the younger brother of Edward Gibbon Wakefield, a man with practical experience of surveying in Van Diemen's Land and much influenced by the six-inch Ordnance Survey of Ireland, published his *Colonial Surveying with a View to the Disposal of Waste Land* . . . in 1849 on the very eve of the founding of this last Wakefield settlement. Felix Wakefield considered that the surveying and mapping of a new colony should serve four purposes: "first, that of facilitating intelligent selection by the buyer; secondly, that of designating selections both by the buyer in applying for the land, and by the seller in giving possession, to the end that every one should really obtain what he really intended to apply for; thirdly, that of recording selections . . . and fourthly, that of record as to title."[162]

The only way that Felix Wakefield considered all of these objects could be obtained was by returning to trigonometric survey as practiced by Colonel Light in the early days of South Australia. Wakefield was conscious that some temporary expedient was necessary to make land available before his trigonometric survey had been effected, otherwise those prevented from buying land would squat, as had happened in Australia when settlement pressures outpaced surveying. His solution was to permit selection before general survey but to require that:

> one boundary of every section is a natural mark so distinct and indelible, (a river bank, a stretch of coastline) as to be sure of always corresponding with the delineation of it on paper, and, therefore, of being always recognised as the mark by which the position of the section would be afterwards determined in the general survey. . . . the plan has been carried into effect with regard to a good deal of land in Van Dieman's Land, and successfully as a means of accomplishing its temporary and limited object.[163]

The New Zealand Company instructed its Canterbury surveyor, Captain Joseph Thomas, to conduct such a trigonometric survey. Instructions published in 1850 required that:

> Surveys of land intended for sale will not in future consist of arbitrary lines, dividing the surface into rectangular portions at the discretion of the surveyor, and liable to be obliterated by the action of the elements and growth of vegetation; but will comprise the natural and permanent features of the country; the crests of ranges of hills; the beds of watercourses; and the actual divisions of arable, pasture, forest, or marsh . . . and most valuable information as to the nature and quality of the soil will be placed permanently on record. . . . it is not the intention of the Association to divide the whole or any portion of the territory to be colonised (except the capital and other towns) into sections of uniform size and figure, which has been the system generally pursued in other settlements.[164]

Because much of the colony was in fact relatively flat land, rectangular forms were used for both town and country lots.

Settlement before survey was not, however, eliminated. Licenses to graze stock on pastoral "runs" were granted, and claimants submitted rough sketches of these properties for identification purposes prior to full survey. At this final stage, surveyors were able to impose some regularity by "scheming a section"—that is, taking the original application and then making its shape as rectangular as possible.[165]

In 1856, three years after Otago became a separate province, John Turnbull Thompson was appointed chief surveyor. He came with direct experience of the contemporaneous survey of India (section 8.15) and radically rethought the cadastral survey and mapping of New Zealand.[166] In his published report he considered that although the American rectangular survey was "admirably adapted to the country in which it is applied, it would be totally unsuitable here, on our high steep ridges and deep valleys, not to mention our high impassable mountains. Their arbitrary lines could never be taken across these."[167] He succinctly expressed what many thought about Dawson's system: It seemed a system devised at a desk in London by a man with firsthand experience of land-survey techniques in the settled world and knowledgeable about surveying in North America but quite ignorant of the very different physique of mountainous New Zealand. Thompson had worked in Bengal in the 1840s under Surveyor-General Colonel Thuillier (section 8.15), and this experience had convinced him of the merits of a minor triangulation to fix cadastral detail accurately. He transplanted the essentials of this "matured system" to Otago Province, New Zealand, and claimed that in so doing he had hybridized the rapid and inexpensive astronomical/geodesic methods of the United States Federal Land Survey which had so influenced Dawson, and the slower, more expensive but very accurate major and minor triangulation employed from the outset by the Ordnance Survey of Great Britain. Thompson considered that "the chief feature of the American system is its astronomical and geodesical character permeating all its processes, from primary positions to the remotest detail. The Otago survey is also astronomical and geodesical, but only in relation to primary and secondary geographical points. . . . Here we in Otago diverge and treat all subsequent processes as plane survey, which are effected by minor triangulation and by theodolite and chain traverse."[168]

Thompson surveyed and mapped the whole of Otago and Southland provinces in this manner and incorporated the rectangular blocks set out and granted earlier by the New Zealand Company. In Wellington Province Henry Jackson, another veteran of the Indian surveys, reformed the survey system in similar fashion (figure 8.32). By 1866, block and section surveys were tied into a primary and secondary triangulation.[169] Thompson's system was to become the basis for the whole of New Zealand after unification in 1876 and his appointment as surveyor-general

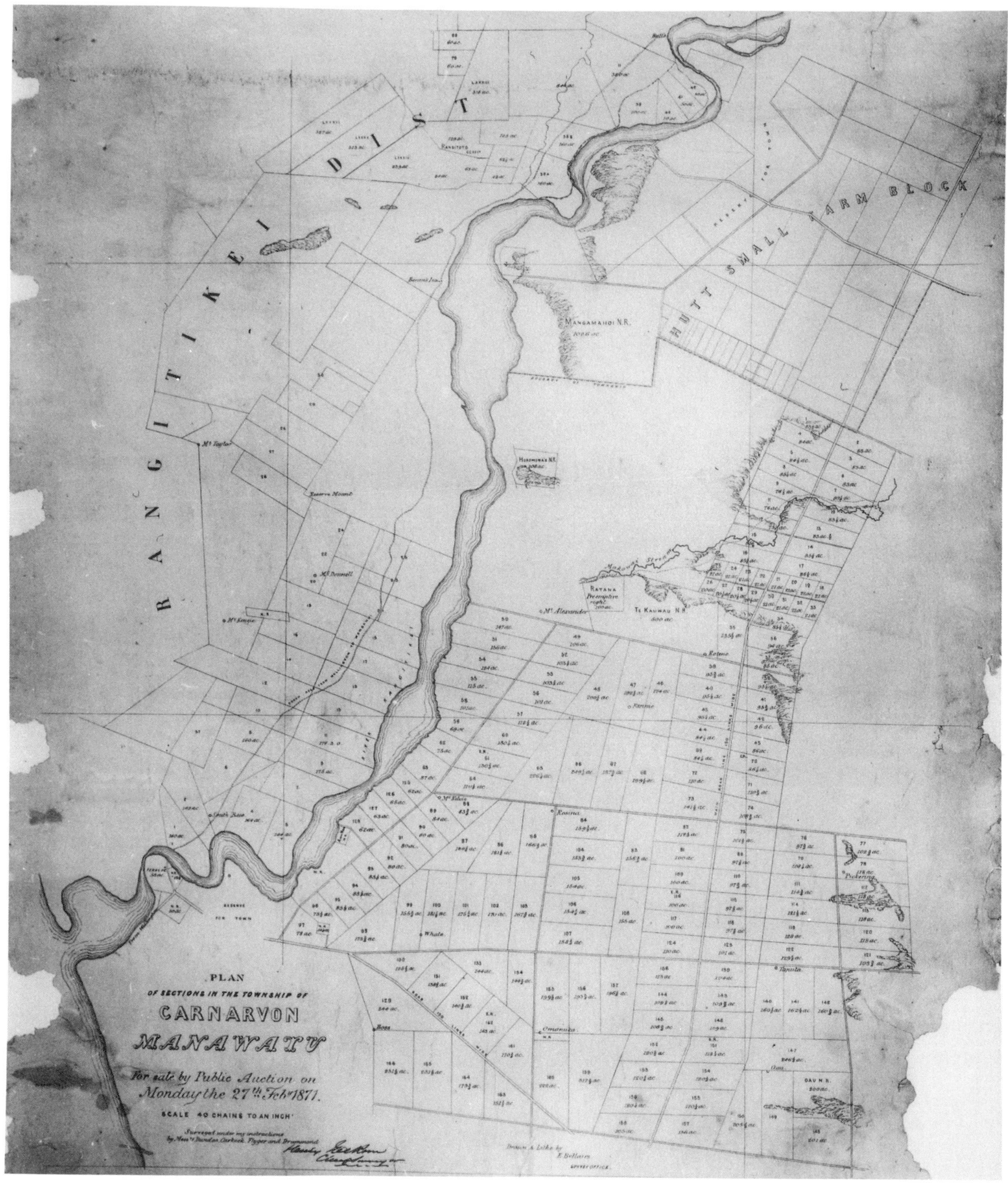

Fig. 8.32 "Plan of Sections in the Township of Carnavon, Manawatu, 1871." A lithograph produced for sale purposes from original cadastral plans. Source: Alexander Turnbull Library, Wellington, 8:32.43 gbbd/1871/Acc 9,271.

to the nation. Thompson's advocacy of minor triangulation in New Zealand as in India to establish cadastral detail suggests that Indian surveying practice might be considered a bridge between Old and New World mapping practices.

8.15 The Survey of India Revenue Surveys

James Rennell's massive eighteenth-century exploratory mapping in India based on routeway traverses with a perambulator to measure distances and a compass to establish directions and with positions fixed at intervals by astronomical observations was superseded from 1802 when William Lambton began the Great Trigonometrical Survey of India at Madras (figure 8.33).[170] As the triangulation was extended northward through the nineteenth century by Surveyor-General George Everest, the work of geodetic surveyors was augmented by plane-tablers constructing topographical maps and by the cadastral mappers of the Revenue Survey, who recorded details of field and village boundaries for tax purposes and plotted them on maps at four, sixteen, or thirty-two inches to a mile (1:15,840; 1:3,960; 1:1,980 respectively) or larger scales.[171] Though it was intended that these revenue maps would be integrated with the triangulation of the Great Trigonometrical Survey, in practice many village revenue surveys were not so controlled.[172] Nevertheless, Sir Clements Markham, the first historian of the Survey of India, comments that:

> The Revenue Surveys of India are one of the bases on which the whole fiscal administration of the country rests. By their means the wealth of the various provinces is ascertained, as well as their food-producing capabilities, and their power to bear taxation. The surveys furnish the information comprised in agricultural statistics, without which the statesman is deprived of the knowledge enabling him to improve the condition of the people, to increase their means of subsistence, to avert famines, to add to the wealth of the country, and to adjust taxation.[173]

Different types of revenue assessment were tried in various parts of the subcontinent. Most were based on land survey to settle boundaries, though not of land ownership per se, but of the right to the produce of the land.[174] Land, with but few exceptions, was not a commodity that could be sold or transferred in the market in pre-British India. Rather, the system centered on rights to shares in the produce of land.[175] So different were Mughal and British ideas of property ownership that the Revenue Surveys of British India can be viewed as an instrument by which the British government transplanted British social institutions and in particular concepts of absolute proprietary rights in the land to colonial India. Once land revenue "settlements" were completed

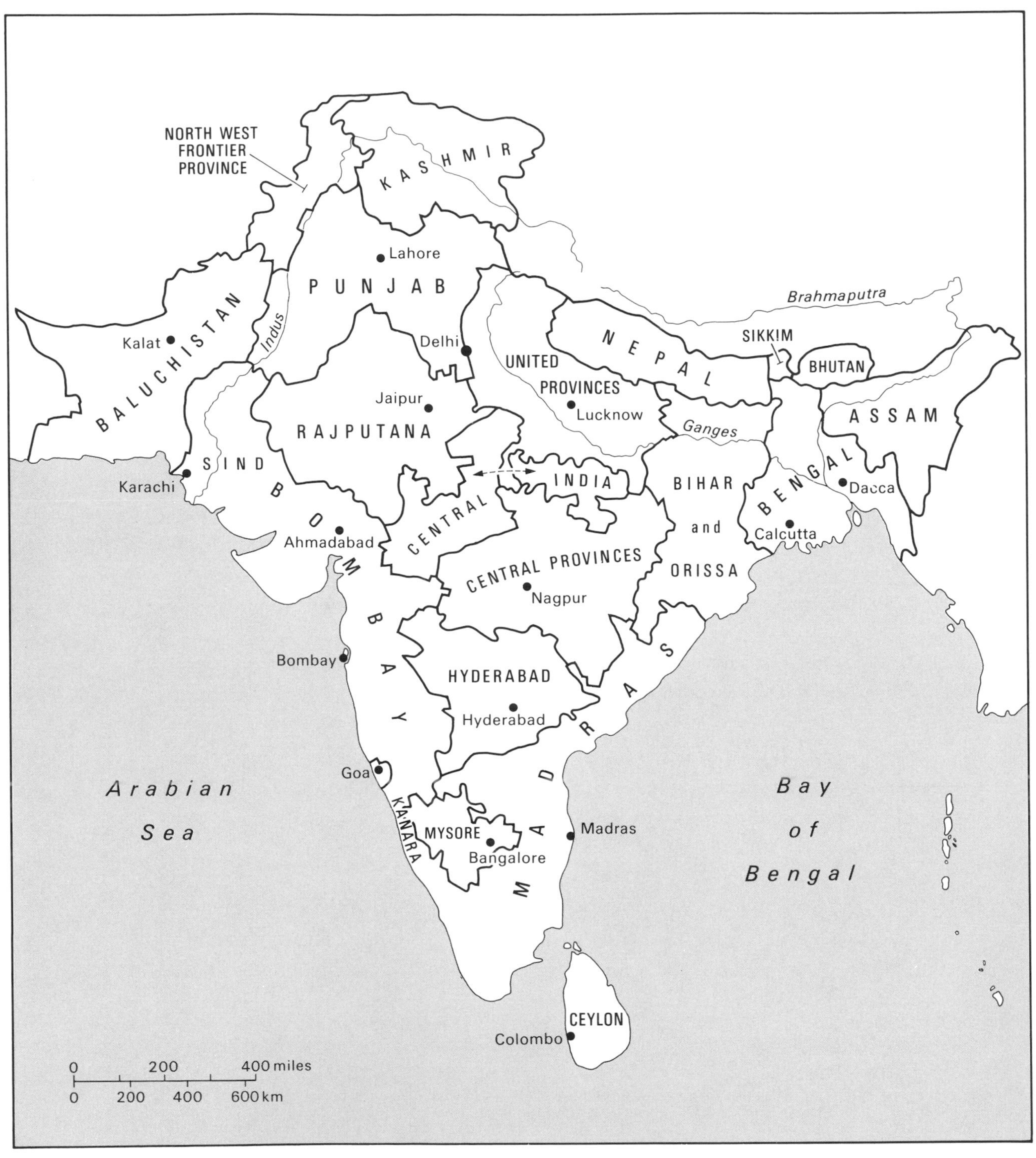

Fig. 8.33 The provinces of India. Sources: Droysens 1886; Muir and Philip 1927; Shepherd 1930; Kinder and Hilgermann 1974–78; Darby and Fullard 1978.

in a district, a detailed record of rights was drawn up, and those who paid the revenue were proclaimed the holders of heritable, alienable properties. The Uttar Pradesh Zamindari Abolition Committee summarized the effects of British revenue policy as follows: "Millions of people were, by these settlements, deprived of rights that they had enjoyed for well over two thousand years; hereditary cultivating proprietors of land were turned into rack-rented tenants at will."[176]

Before English colonial times, the Indian tradition had been that a village *patwari* (accountant) would prepare sketch diagrams of the relative positions of village fields to serve as indexes of village production rights and taxation records. As early as the middle of the sixteenth century, Raja Todarmull of the court of the Mughal emperor Akbar had introduced a system of land-tax assessment based on land measurement. In the early period of British rule, the three main presidencies of Madras, Bengal, and Bombay had their own survey departments and conducted taxation surveys.[177] A start had been made on cadastral mapping by East India Company officials. In Bombay, for example, Herman Blake had begun a detailed survey of properties in 1670, though illness prevented his completing it. Similar proposals were aired in 1679, 1710, 1747, and 1772. The earliest effective taxation surveys of Madras under British rule were made 1798–1815 and reduced to one inch to a mile (1:63,360) based on Lambton's triangulation.[178]

From 1837, the Survey of India began regular fiscal surveys plotted on maps at four inches to a mile (1:15,840), which became known as the Revenue Surveys. This relatively small scale proved inadequate, and the general scale adopted throughout India was ten chains to an inch (sixteen inches to a mile; 1:3,960), except in parts of Bengal and the United Provinces, where fields of just a few square yards required even larger scales. The method of construction varied from province to province, but Clements Markham considered the most efficient was that introduced into the North-West Provinces in 1871–72 and adopted later in Burma, Bengal, and elsewhere. Colonel Thuillier, the architect of these revenue surveys, intended that, not only would they prove "invaluable as a correct permanent record of the landed tenures for all purposes of revenue assessment, but an immense saving of expense will be effected in the end, by doing away with the constant necessity for partial remeasurements for irrigation, canals, railways, roads, and other purposes, which are now perpetually being made in an irregular, unsatisfactory, and expensive manner, for emergent engineering objects."[179]

Village boundaries were surveyed professionally by theodolite and chain, while internal boundaries were established and chained by a *patwari*. Every field, probably bounded by little more than a *mendh*, a ridge perhaps six inches high and a foot wide, was assigned a serial number to link the map to information recorded in field books. By 1873 there were four cadastral surveys underway in the North-West Provinces. Together they surveyed and computed the area of no fewer than 1,269,882 fields of an average size of 0.94 acres in that one year alone. In 1876 the seventeen revenue survey parties employed by the government of India covered more than 7 million acres (equivalent to about a quarter of all the tithe surveys of England and Wales), of which 1.4 million acres were mapped at scales of sixteen inches to a mile (1:3,960) or larger.[180] By the end of the nineteenth century, there were 1:3,960 cadastral sheets for much, but by no means all, the cultivated areas. To remedy this deficiency, an act in 1885 empowered local governments to make a cadastral survey and a record of rights to land in any area and to appoint settlement officers for the purpose.[181]

The Revenue Survey produced some twenty thousand volumes and maps covering the districts of Assam, Bengal, Bihar, Bombay, Central India, Central Provinces, North-West Frontier Province, Punjab, Rajputana, Sind, and United Provinces.[182] The compilation of these maps and records was a prodigious undertaking in terms of the area of land covered, in speed and economy of execution, in fitness for purpose (notwithstanding their very variable accuracy in cartometric terms), and in the detail of the cadastral mapping. If the Indian Revenue Surveys were not quite the "jewel in the crown" of Western cadastral mapping, at the very least they are a cartographic accomplishment which deserves fuller research in any history of land mapping.

8.16 Cadastral Maps and Colonial Settlement

The mapped cadastral survey was one of the most powerful instruments available to each of the royal colonies (in which the king appointed a governor), the proprietary colonies (where the proprietor assumed the role of appointing a governor), and the corporate colonies (governed by a corporation under a royal charter) for establishing their different political ideals by way of allocating land, their prime resource. The Wakefield settlements of South Australia and New Zealand used cadastral maps to reproduce overseas the English capitalist society based on landlord and wage laborer; the Indian Revenue Surveys established British absolute proprietary rights of landownership in place of the Mughal system of rights to the produce of land; the "New England method" of surveying rectangular townships in advance of settlement in North America was to ensure colonization by communities of predetermined size and character; while the "Virginia method" of facilitating the establishment of large, capitalist plantations founded on slave labor had quite the opposite social aim. To the individual shareholder, to land speculators, or to the migrants aspiring to become freeholders by perfecting title to a piece of land, the differences of political principle which resulted in land being surveyed either before or after settlement, and in either contiguous tracts or dispersed parcels, were incidental to their immediate concern of getting onto the land. In theory, until the surveyor had traversed the land with chain and compass, it could not be converted into private property. A cadastral survey and, from the middle of the seventeenth century onward, a cadastral survey recorded on a map were the legal means of achieving the status and security of a landed proprietor. Refinement of the Torrens System of land registration in the later nineteenth century further accentuated this role of cadastral maps. Nowhere was this more so than in the Cape Colony of South Africa, where every title deed is accompanied by a plan or diagram which must contain, in the words of a mid-nineteenth-century surveyor-general, "the lengths of boundary lines to two decimal places . . . and their angles of intersection of those boundaries to the nearest ten seconds; also the area; and no such plan or diagram must be used as the basis of a Title Deed before it has been tested as to the consistency of these elements by an Examining Officer."[183]

Whereas in Europe the cadastral surveyor was largely confined to demarcating and mapping long-settled lands, in the colonies the surveyor's prime responsibility was the imposition of a new economic and spatial order on "new territory," either erasing the precapitalist indigenous settlement or confining it to particular areas.[184] The surveyor was also explorer, resource appraiser, town planner, delineator of routeways, and the molder of rural cadasters and landscapes. As Paul Carter puts it in *The Road to Botany Bay*, "The survey, with its triple artillery of map, sketches and journal, was a strategy for translating space into a conceivable object, an object that the mind could possess long before the lowing herds."[185]

The differences among each of the new worlds was perhaps too great for us to expect any one, uniform set of cadastral procedures to emerge, but it is possible to identify three major groupings, each with its vocal advocates convinced of the superiority of their own approach. Two of these were for setting out landholdings in advance of settlement. The first is rectangulation, the imposition of a grid of land sections by running surveys. The scheming of sections was admired by some for its virtues of simplicity and cheapness, and its proponents could look to the successful alienation of public lands produced by such means in the United States, in Ontario, and in Upper Canada, as well as its recommendation by Robert Kearsley Dawson for Australia and New Zealand. The second approach to survey before alienation was promoted by the advocates of triangulation, who believed in survey by trigonometric methods with sections set out only after a regulatory framework was in place. They could look to the areally more limited surveys of Kanara and some other parts of India, Colonel Light's work in South Australia, and Henry Jackson's and J. Turnbull Thompson's transformation of the cadastral surveys of New Zealand. These two approaches to the scheming of sections before settlement were a purposive reaction to the perceived evils of the third method, that of surveying land by metes and bounds to prove title after selection. This "Virginia system," however, suited the southern colonies of America, as it enabled the acquisition of extensive plantations for farming with slave labor.

Once land in the colonies was alienated and settled by Europeans, cadastral mapping then played much the same role in the New World as private estate maps in the Old World. For example, William Godsoe, "surveyor for the town of Kittery" in coastal Maine, produced more than fifty maps to support or verify estate boundaries and disputed boundaries and to present as evidence in civil law suits.[186] These uses of property maps were long established in the Old World; the new role developed for cadastral maps in the colonies was as an instrument to effect the alienation of land.

Acknowledgments

We should like to acknowledge the help we have received from Prof. Anthony Christopher, Dr. Matthew Edney, Prof. Sam Hilliard, Prof. Josef Konvitz, Dr. Gordon Scurfield, and Dr. Michael Williams.

9

CADASTRAL MAPS IN THE SERVICE OF THE STATE

9.1 The Chronology of State Cadastral Mapping

More than 125 mapped cadastral surveys comprising many millions of individual cadastral maps have been reviewed in this book; some cadasters are maps of individual communities or parishes such as the English parliamentary enclosure maps and Dutch polder maps. Other maps, outnumbering by many times these individual maps, form part of systematic surveys of territories such that the whole cadaster may number many thousands or even hundreds of thousands of maps. Examples are the *ancien cadastre* of France, the Revenue Surveys of India, and, probably largest of all, the Federal Land Survey of the United States of America.

The reemergence of the state mapped cadaster was evident first in the Netherlands, where maps were used as early as the sixteenth century in association with the making of polders and disputes over tithes and boundaries. New mapping for drainage works peaked in the first half of the seventeenth century, by which time cadastral maps were also being used in many European countries for management of state resources, especially of forests, from the second half of the seventeenth century. Land colonization cadasters date from the second half of the sixteenth century (Ireland) and from then continued to be important, reflecting the successive opening of new lands for colonization. Enclosure mapping is a phenomenon of the eighteenth century, especially of the second half, but new enclosure surveys were begun as late as the second half of the nineteenth century in Norway. The largest single group of state-sponsored mapped cadasters discussed in this book are those which had tax reform as their primary objective. From beginnings in the early seventeenth century, again in the Netherlands, this mapping reached a peak in the first half of the nineteenth century.

This chronology of state cadastral mapping prompts two questions. First, What are the uses to which cadastral maps have been put by state agencies? Second, Why were mapped cadasters not axiomatic even by the end of the nineteenth century? In other words, why did some states or their agencies commission and undertake cadastral surveys, while others were either unaware of the potential of the mapped cadaster or consciously eschewed its use?

9.2 The Uses of Cadastral Maps

Cadastral Maps and Land Reclamation

Some of the earliest post medieval cadastral maps in Europe were produced in the course of late sixteenth- and seventeenth-century polder construction in the Netherlands, where the reclamation of coastal marshes and inland meres was something of a barometer of land values and economic conditions (chapter 2.1). The proposed cadastral patterns which they displayed were used to interest potential shareholders, to allot newly formed land plots, and then to serve as visual displays for those who had invested in the new land. To publicize the schemes as widely as possible, some of these Dutch polder maps were printed. Polder construction was also important in the coastal areas of East Friesland, and maps were associated with drainage activities in Brandenburg in northeastern Germany (chapter 4.3) and in the Fenland of eastern England. H. C. Darby describes the draining of the Fens as "one of the mighty themes in the story of Britain."[1] Though it is arguable how much of permanence was accomplished in the early period, sixteenth- and seventeenth-century administrators were conscious of the value of maps to help organize and administer their drainage undertakings. Skelton and Summerson cite a number of references in sixteenth-century documents to items such as "a platt of the country" or "plot of the drayne."[2]

In 1545 the Venetian government instituted the Officio dei Beni Inculti to supervise reclamation and drainage works in the valleys of the Veneto. Maps were important aids in the planning and execution of their schemes. For example, in 1570 the officio commissioned the cartographer Panfilo Piazzola to compile a map of the Menago River lowlands; at 1:15,000 it distinguishes land liable to flood and existing tilled lands, and it records the pattern of landownership.[3] Maps also played an essential part in the partition of naturally emergent land around the Gulf of Bothnia in Finland (chapter 3.7).

Cadastral Maps and the Evaluation and Management of State Land Resources

Some of both the earliest and the most recent European cadastral surveys were compiled as map inventories of state land resources to assist in their better and more productive exploitation. The precise purpose for which Gustav II Adolf of Sweden ordered the compilation of the *geometriska jordeböckerna* of his territories in 1628 is not clear (chapter 3.3). What is certain is that this,

Europe's first national mapped cadaster, had as one of its main objectives the mapping of the nation's land resources to identify potential improvements. Two hundred years later, many of the mapping programs to alienate public lands in the New World had as an important subsidiary purpose the gathering and dissemination of information about the resource potential of land. These components were made explicit in the official instructions for such as the Federal Land Survey of the United States of America (chapter 8.8), the Dominion Land Survey of Canada (chapter 8.9), and in New South Wales after 1864 (chapter 8.11). Between these dates occurred the flowering of the European Enlightenment, when rulers were occupied not only with questions of political modernization but also with the economic improvement of their lands.

In the seventeenth century deforestation of much of western and southern Europe by agriculture and by felling of trees for fuel and for building and construction timber was a matter of growing government concern. In England, John Evelyn lectured to the Royal Society on the economic role of forests and what he termed "this impolitick diminution of our timber."[4] Governments commissioned surveys of their dwindling forest resources, and some of these surveys had a cartographic base. In Russia the forests around Bolkhovsk were measured and mapped in 1647, and the program had been extended to other areas by the 1670s.[5] To the rulers of *ancien régime* France, the forest was at once of economic significance for building timber, of military importance for the construction of naval vessels, and of recreational value for hunting. In France a number of royal forests had been mapped by the second half of the sixteenth century; a map of the Forêt de Chantilly dating from the end of the fifteenth century is one of the earliest.[6] During the seventeenth century, woodland reserves were further diminished in France, and their precise extent was rendered uncertain by fires, military destruction, and illicit felling. An essential part of Louis XIV's late seventeenth-century forest reforms was the compilation of a complete cartographic inventory of the French royal forests. These maps—updated, copied, and recopied—continued to be used through the eighteenth century to regulate the felling and sale of timber (chapter 6.2). In eighteenth-century Norway, it was not so much forest depletion as its economic development that was the main concern. The Generalforstamt saw detailed mapping as the basis of systematic exploitation of the forests (chapter 3.19). Similarly in much of Germany, forests were a potentially valuable but underused resource at this time. In Baden-Württemberg, for example, forests which had passed out of church ownership in the Reformation were surveyed and mapped to help organize their more intensive exploitation (chapter 4.2).

This activity indicates that, by the seventeenth century, European governments and provincial rules not only were adopting maps for plotting national strategy and for organizing fortifications and warfare but also were using large-scale maps as land inventories. However, mapping did not inevitably accompany government-sponsored surveying of rural property, as can be seen from the parliamentary surveys of seventeenth-century England (chapter 7.1).

Cadastral Maps and Land Redistribution and Enclosure

Enclosure and redistribution of communally held land and of land divided into small strips was instituted to introduce more efficient farming, to expand the area under the plow, or to create improved pastures from moors or heaths. The reasons for enclosure are the subjects of long and continuing debate among agrarian and social historians, but seventeenth-century English surveying treatises promulgate a clear message that enclosure was a distinct agricultural improvement and one which could be better done with accurate measurement and maps. Surveyors were simultaneously the "great panegyrists of enclosing," the promoters of capitalist farming, and advocates of the cadastral map.[7] In England the use of maps both to determine the existing cadaster and to record the new cadaster was an integral part of enclosure by parliamentary act from about the middle of the eighteenth century onward (chapter 7.2). Cadastral maps were required by the legislation, but their production was entrusted to private surveyors with estate-mapping experience. In Sweden and Finland, however, enclosure mapping was carried out by the state surveying agency, although it was paid for by the farmers. Partition and redistribution began in the early eighteenth century, peaked about 1860, and continued into the present century (chapter 3.4). In Denmark, widespread enclosure got under way following the law of 1781, which was inspired by Physiocratic ideas and which provided a foundation for the comprehensive restructuring of Danish agriculture. Land parcels were mapped at three stages: before reallocation, when a provisional division was suggested, and when land divisions were confirmed (chapter 3.11). In Norway, as in the other Nordic countries, the enclosure movement was government directed, and maps were an integral part of the process, though here the fundamental legislation was in 1857 rather than in the eighteenth century, reflecting the much later penetration of agriculture by a market economy in this country (chapter 3.21). Together, Norway, Sweden, Finland, and Denmark represent the largest region of Europe subject to village-by-village land redistribution based on state-sponsored cadastral surveys and maps.

Agrarian reform in Germany in the eighteenth and nineteenth centuries mirrored that elsewhere in having two physical components: the common lands were divided into plots held in severalty, and strip fields were amalgamated into larger, enclosed parcels (chapter 4.4). In many German states, notably Brunswick, state-sponsored enclosure was either proposed or took place with maps as the base. Maps were used, as elsewhere in Europe, both in the reorganization process itself and to record its effects. By comparison with England and the Nordic countries, however, the physical aspects of agrarian reform were overshadowed by the overwhelming preoccupation in both Germany and the Austrian Habsburg lands with the emancipation of the peasantry. This was not the kind of reform in which mapping could play a significant role.

It is clear, however, that the cadastral map played a critical role in eighteenth- and nineteenth-century Europe in land reorganization and enclosure. It is rare to find village enclosure in Europe after about 1750 which does not use a map for devising and/or recording the new cadaster.

Cadastral Maps and Colonial Settlement

Cadastral maps have been used since the end of the Middle Ages by individuals and institutions to establish title to land. In the seventeenth, eighteenth, and nineteenth centuries, cadastral maps were used by state governments in the Old and New Worlds to organize, control, and record the settlement of "empty" lands, a process which in the New World often involved wresting control from indigenous peoples with very different concepts of landownership and exploitation.

Within Germany from the twelfth and thirteenth centuries there were important migrations eastward to occupy the territories east of the river Elbe and the underused heaths and forests throughout Germany. Some were individual settlers, but later others went to model colonies which were planned and set out on maps. Some, such as Carlsdorf in Hessen, bore the name of the enlightened prince who had established them (chapter 4.3). The settlement of the Alheden in Jutland in the eighteenth century (chapter 3.10), the extension of settlement in the eighteenth and nineteenth centuries over forested waste in Sweden (chapter 3.5), and the *allmenning* disposals in northern Norway from the 1750s (chapter 3.22) were each accompanied by systematic programs of mapmaking.

If cadastral maps were useful to governments wishing to promote the orderly settlement of underexploited parts of their home territory, they were soon established as a desideratum for settling overseas colonies. In Elizabethan Ireland one of the earliest experiments in the use of maps as models of the spatial organization of rural settlement was made in connection with the Munster plantation in 1585–86.[8] A diagram of the layout indicates that each Munster seigniory was initially intended to be about four miles square and divided into a range of different-sized holdings to create a balanced rural society.[9] In 1585 a tiny band of surveyors set about making the surveys which were to be used to allocate the confiscated lands to English settlers. In execution, the square seigniories disappeared, as the terrain was difficult, the countryside overgrown, and the local population generally hostile to the survey, which threatened their property rights: these were all problems encountered later by colonial surveyors across the Atlantic. In Ireland, as in North America, the advantage of a map to the state, in the words of Arthur Robins, one of the surveyors, was to show how the land "might most aptest be laid out in seignories to avoid their becoming excessively intermixed."[10] The Elizabethans achieved little in Munster but experienced much that was of use in establishing later plantations in Ireland; the method of plotting the results of a survey on maps was used when almost half the country was confiscated by the state after the 1641 rebellion.[11] This last produced the Down Survey, directed by William Petty between

1655 and 1659.[12] The maps of the Down Survey drawn by parishes at three or six inches to a mile (1:21,120 or 1:10,560) record the boundaries of forfeited town lands and contain an inventory of land as cultivable, bog, mountain, or wood.

In the New World the cadastral map was the instrument which enabled the settlement ideals of colonial governments to be realized. These were of all varieties, from one of encouraging large plantations as on the southern seaboard of North America, through one of individual proprietorship of holdings disposed with the regularity of the grid as with the federal land disposals of the United States of America, to one of strictly limiting land availability to establish a capitalist society with farmers and wage laborers, as in the Wakefield settlements of South Australia and New Zealand.

From the viewpoint of the individual settler, the cadastral map which defined the settler's claimed, granted, or purchased land was important for providing security of title. Indeed, the Torrens System of land registration (chapter 8.13) established a precise and pivotal role for cadastral maps in the land registration process as it used property maps rather than written deeds as the evidence of land title. In the Old World, the role of cadastral maps as evidence of landownership is somewhat equivocal. The *ancien cadastre* of early nineteenth-century France was expected to provide a legal record of ownership, but courts do not accept its evidence alone, as ownership was not verified at the time of its compilation (chapter 6.11). Exactly the same applies to the tithe surveys of mid-nineteenth-century England and Wales (chapter 7.3 and 7.4). While colonies in the New World were deemed "empty" by the European authorities, colonies in the Old World had a population which could not be defined out of existence. Imperial expansion within the Old World brought with it the desire not primarily for new settlements but for increased taxation revenues. Thus, in Sweden's Baltic lands in the seventeenth and eighteenth centuries, the most important projects were tax reform and the resumption by the Crown of lands alienated to the nobility. Maps were used, albeit with limited success, in these projects (chapter 3.7).

Cadastral Maps and Land Taxation

In the Old World more land was surveyed and mapped by the state for setting and recording land taxes than for any other single purpose. In antiquity cadastral survey and mapping enabled the Roman Empire to organize and control the levy of taxes in its far-flung dominions (chapter 1.1). During the seventeenth century, a number of European governments adopted maps as a means for assessing and recording the tax levied on land (cf. the Netherlands and the Swedish Baltic provinces in chapters 2 and 3.7 respectively). These cadastral maps can be distinguished from estate maps by the fact that they extend over all the properties in a parish, canton, or province, rather than the lands of a single manor or estate, or a number of manors belonging to the same lord. Such initiatives are related to the fact that, by the seventeenth century, taxes in some provinces were being more closely identified with the soil, which was held to generate wealth irrespective of the status of individuals who farmed the land or the communities which inhabited it. For example, a decree in 1585 in the Liberty of Bruges discontinued the tax on individual wealth and work and replaced this with one on land, with the result that the amount of land tax paid was related to land area.[13] Cadastral maps provided a concise and accurate method of both fairly assessing and permanently recording such charges on particular pieces of land.

In Renaissance Europe some of the earliest cadastral maps were produced as evidence in disputes over rights to levy tithe on particular tracts of land: for example, maps defining the lands tithable by particular appropriators were constructed from the late fifteenth century in the Netherlands (chapter 2.2) from the sixteenth century in France (chapter 6.1), and in some German territories (chapter 4.5). In nineteenth-century England and Wales a mapped cadastral survey was an integral part of one of the most celebrated of national tithe commutations, for which maps were used as a record of the rent-charges which replaced tithe and which were apportioned to particular fields or landholdings (chapter 7.3).

Examples of the use of cadastral maps to organize the collection of general land taxes can be found in all the territories discussed in this book with the exceptions of New World colonies, where landowners paid quitrents but not a land tax as such, and of England and Wales before 1910. Written cadasters had been used since the Middle Ages as the basis for collecting general land tax, and reference has been made in this book to, for example, the *compoix* of Languedoc which were renewed by survey in the thirteenth, fourteenth, and fifteenth centuries (chapter 6.3). The first uses of mapped surveys in the general realm of land dues occurred in the early map-conscious Netherlands (chapter 2.2), where in the second quarter of the sixteenth century, feudal dues of some *vroonlanden* (properties of the counts of Holland) were revised. After the Reformation (c. 1570–1600) mapped cadasters were commissioned to organize rent collection from newly secularized lands. Throughout the Netherlands in the seventeenth and early eighteenth centuries there was dissatisfaction with the inequity of land taxes, and in many rural communities surveying and mapping were spontaneously instigated to make the tax burden more equitable. Similar inequity in tax burden existed in Schleswig. In order to remedy this, Johannes Mejer was commissioned to survey and map fifty-one villages between 1639 and 1641 (chapter 3.15).

The Thirty Years' War brought financial as well as political dislocation to much of Europe. After the war, the practice of maintaining standing armies increased the burden of state expenditure; at the same time, the complexity of government was increasing, as were its costs. As the need for money increased, so states became ever more interested in accurately surveyed and mapped cadasters as instruments for increasing state revenues. Nowhere was this more evident than in the multifarious and mostly minute states which composed Germany at this time. Chapter 4.6 reviews initiatives in Hessen-Kassel, Fulda, Baden-Württemberg, Cleve, Schaumburg-Lippe, Osnabrück, Baden-Durlach, Oldenburg, and Hanover for the later seventeenth century and through the eighteenth century.

In Austrian Habsburg territories the Thirty Years' War brought such increased demand for tax revenue and such economic dislocation and devastation that the tax system was in almost total disarray by 1648 (chapter 5.2). Both Leopold I in the seventeenth century and Charles VI in the early eighteenth century gradually reformed or rectified the *Landeskontribution* (land tax), but their reforms were obstructed by vested interests in the various Diets of the Habsburgs' lands. By far the most significant early tax reform for the Austrian Habsburgs and subsequently for Spain, the Kingdom of Sardinia, and France was that instituted by Charles VI in the Duchy of Milan, one of the smallest but richest of the Habsburg lands. The *censimento* of Austrian Lombardy undertaken from 1718 was the earliest fully surveyed and mapped cadaster in Austrian Habsburg territory and is widely recognized as a pioneering example of modern taxation cadasters as it surveyed and mapped at a large scale all land, not just productive fields (chapter 5.4). It inspired parallel undertakings in eighteenth-century Spain, where tax reform was modeled directly on the Milanese method, and also in the neighboring Principality of Piedmont (chapter 5.5) and the Duchy of Savoy (chapter 6.4), both ruled at this time by the king of Sardinia. Not only was the method copied in Piedmont and Savoy, but also the same personnel were involved, as surveyors went from Milan to Piedmont and thence to Savoy. Meanwhile, in Austria itself, Maria Theresa's mid-eighteenth century revision of the land tax was undertaken without surveying and maps except in the Tyrol, where the taxation system was especially complicated (chapter 5.6).

Physiocracy provided a further spur to taxation reform and land survey in the later part of the eighteenth century. Physiocrats and other enlightened reformers held that there should be one tax, the *impôt unique,* equally apportioned, due to and collected by the central state. It was in France above all European states that the Physiocrats' views of taxation reform turned administrative attention to a cadaster-based land tax. To François Quesnay and the Physiocrats, land, rather than labor or some other factor of production, was the source of all value. Thus agriculture, not manufacturing or commerce, was the sole source of wealth. Agriculture was the only activity in their opinion to produce a *produit net* (surplus over all costs), and so the promotion of capitalist agriculture was, with free trade in grain and fiscal reform, one of their main practical aims. Fiscal reform would remove traditional feudal rights, notably those of exemption and local repartition. Once the taxes were repartioned and collected by the central state, precise information, including mapped information, was necessary, since the assessors and collectors no longer knew the locality in which they were collecting the taxes. It was seen as advantageous to have outsiders to do the assessment and collection, since they had no local vested interests. This meant, however, that they needed to be provided with far more background information than, for example, the lord's officials who had done the repartition and collection before. This is discussed further in chapter 5.4 to 5.7, where the attempts of the Austrian Habsburgs to wrest control of repartition from the lords are reviewed. The very idea of using a map-based cadaster to assess the land tax more accurately and levy it more equitably in the late eighteenth century was quintessentially Physiocratic, as were map-based agrarian reforms in, for example, Denmark (chapter 3.11 and 3.15), which aimed to improve the productivity of the land, the source of all wealth.

Joseph II of Austria's idea of one tax levied on land alone and to be paid by all, regardless of estate, was so contentious that it was not realized in his lifetime (chapter 5.7). His reforms were continued, but with the added improvement of a comprehensively mapped base, by the *Stabile* cadaster after 1806 (chapter 5.9). In France, after some tax reforms by *intendants* in particular *généralités* in the second half of the eighteenth century (chapter 6.5), the closing years of the *ancien régime* witnessed the publication of a plethora of didactic works advocating wholesale taxation reform. Most of these authors predicated reform on a mapped cadaster. A model and exemplar of a mapped cadaster was undertaken in Corsica (chapter 6.6), and the contribution of de Richeprey, the so-called father of the nineteenth-century French cadaster, in whose writings most of the components of the eventual cadaster were foreshadowed, is reviewed in chapter 6.7. After experimenting with surveys which defined only the boundaries of particular land-use types, from 1807 the French *ancien cadastre* embarked on a parcel-by-parcel survey to apportion global amounts, while large-scale maps which distinguished field boundaries and parcel ownership provided a permanent record of the apportionment and so discouraged future litigation.

Not only was this cadastral system extended over areas such as Belgium, the Netherlands, and Luxembourg, annexed to France by Napoleon (chapter 6.12), but it also provided a basic specification for other states in the nineteenth century. The Netherlands cadaster, for example, changed very little after the French retreated (chapter 2.4, 2.5 and 2.8). The Bavarian cadaster was conducted by the Topographisches Büro founded by Napoleon, and it first tried to emulate the method *par masses de cultures* before embarking on a parcel-by-parcel survey controlled by triangulation (chapter 4.7). Prussia, like Bavaria, gained much territory from the post-Napoleonic political settlements and inherited a great variety of taxation systems. Rhineland cadasters were begun under the Napoleonic Law of 1808, and when Prussia regained control, this work was continued and extended over all of Prussia's western provinces using the method of the French *Receuil méthodique*. This cadaster was revised and in turn extended to the eastern provinces between 1861 and 1865. In Hanover, however, the Franco-Prussian method of detailed, mapped surveys was eschewed for taxation reform. Hanover's English rulers seemed unmoved by advances in continental taxation reform based on cadastral maps and instead tried several methods, all more or less inaccurate, of gathering information to be used as a basis for tax collection. Landownership and tax burdens in Hanover were not mapped until the territory became part of Prussia in the 1870s. Nonetheless, by the later nineteenth century, in most areas of Germany the land tax was collected on the basis of a surveyed and mapped cadaster. Prussia stood at the head with its comprehensive and accurate mapping, arguably one of the more enduring of Napoleonic First Empire influences outside France.

In the realm of taxation reform, the cadastral map was established as an important tool for effective rule over wide areas of continental Europe by the late nineteenth century, but it was not a universal tool. In England (chapter 7.5) there was no systematic, surveyed, and mapped cadaster, perhaps because imposition of the land tax did not represent the power struggle which prompted reform by mapping elsewhere. In India, in contrast, the British government used its Revenue Surveys, first, as the base on which the whole fiscal administration of the country rested, and, second, as the instrument by which British social institutions and absolute proprietary rights to land were transplanted to Mughal India (chapter 8.15). This mapping represents one of the technical high points to which cadastral mapping was raised in the service of the state; more important, it shows how such technical knowledge cannot be neutral. The knowledge of how to draw a map and the information subsequently contained in it are instruments of power, and the map itself is a symbol of it.[14]

Cadastral Maps as Symbols of State Control over Land

The Swedish *geometriska jordebok* mapping of the 1630s onward was the first mapped cadaster of a Western nation (chapter 3.3). The practical purpose for which this survey was intended is not clear. However, this great project of mapping the Swedish lands and those which might be won for Sweden can be seen as part of the outburst of national creative energy which followed acceptance of *storgöticism*, the idea that the Swedes were descended from the ancient Gothic people and were regaining their rightful ascendancy in seventeenth-century Europe. The *geometriska jordeböckerna* were a symbol both of a nation fulfilling its historic destiny and of the power of the king. By 1630 Gustav II Adolf was a prince of European stature, and the national cadastral survey was a graphic reflection of his and the nation's glory.

At a regional scale, the polder maps of the Dutch polder authorities were objects of display as well as practical aids to administration (chapter 2.1). The cadastral base, combined with the coats of arms of the dike reeve and embellished with allegorical ornamentation, was a potent expression of an authority's power and status. Surviving examples of these spectacular wall maps testify to the emotive value as well as the monetary worth which the Netherlanders ascribed to their hard-won land.

For Napoleon Bonaparte, the cadaster of France was at once the means to finance the *grande armée* and a symbol of *égalité*. As with the architectural, engineering, and urban works, Napoleon's personal influence is clear. His new buildings and streets were to make Paris the most beautiful city the world would ever see; his cadaster, which was organized by his new administrative *départements* and which defined each individual land parcel and listed them by proprietor, symbolized the new order of his new legal code (chapter 6.9).

In colonial settlement, the cadastral map was a measure of an individual settler's stake in a new nation. The cadastral map as an icon of pride in individual landownership is probably nowhere better seen than in the published county and state landownership maps and atlases produced in such numbers in nineteenth-century United States and Canada (chapter 8.10). To the governments of imperial nations like Britain, France, and Spain, however, cadastral maps were the actual instruments of imperialism. These maps specified precise pieces of land in appropriated colonial territories; they enabled imperial states to alienate their colonial domains by sale, or by grant in return for a rent.[15]

Cadastral Maps as Tools of Rational Government

Enlightenment monarchs needed precise, detailed statistics as the basis for administrative reforms and mercantilist state direction of the economy. Where underused resources such as forests or "empty" colonial lands were under the direct control of the ruler, there was a further impetus to produce a map as an accurate medium for recording and displaying such information. By the nineteenth century, advocates of cadastral mapping were advancing a whole set of complementary roles which maps could play in addition to their primary purposes which might be for enclosure or tax reform. Almost immediately after its discovery, the process of lithography was taken up by those in charge of the Bavarian and the Austrian *Stabile* cadasters to make their maps available for a wide range of state and private uses (chapters 4.7 and 5.9). Dutch and Swedish cadasters were used as sources to compile later topographical and military maps respectively (chapters 2.5 and 3.6). The United States plats and notes constituted a basic reference source for county surveyors and for the United States Geological Survey and were compiled into the county maps for public sale (chapter 8.8). The plans and registers of the French *ancien cadastre* (chapter 6.11) have been widely used as a basis for property transactions and consulted by public bodies for purposes other than those concerned with taxation. Individual cadastral maps in all states have been used as evidence, with varying success, in countless thousands of property disputes.

All these complementary roles for cadastral maps were paraded by Robert Kearsley Dawson in 1836–37 when he argued that government ought to extend the tithe commutation mapping over that quarter of English parishes which fell outside the remit of the 1836 Tithe Commutation Act. In his papers to government he reviewed the general benefits which accrue from possession of large-scale property maps (chapter 7.3). The advantages he advanced included the resolution of boundary disputes, the easier transfer of property, identification of the best lines for canals and railways, the possibility of obtaining information about the "real capabilities of the country," and the opportunity to decide where investment might be most beneficial. Though the British government remained steadfastly unconvinced and unwilling to commit the additional expenditure, it would seem that by the mid-nineteenth century the British Parliament was something of an exception. By this time the practical value of a map-based cadaster to state governments was very much an accepted fact in Europe and the New World. A report from a commission on cadasters was a principal item on the agenda of the First International Statistical Congress held in 1853 in Brussels. After debating the report, the congress adopted a resolution which commended mapped cadasters as assets to government, not only as fiscal instruments, but also as inventories of a country's real property, the "fundamental book" in which individual proprietors could trace the title of their property. Cadasters were the basis for collection of agricultural statistics and for regulating mortgage and land credit systems. In short, the mapped cadaster was "the source of all information concerning properties. We consider, therefore, that a cadaster is one of the greatest benefits that a state can possibly procure." [16]

9.3 The Rediscovery of the Cadastral Map: Toward an Explanation

The wide range of practical uses and symbolic values of cadastral maps that had developed by the end of the nineteenth century has been summarized in section 9.2; for none of them, however, was the employment of a map absolutely indispensable. It was helpful but probably not essential to a court of law to have a map of a disputed boundary. In North America, land could be and was patented without a plat. In Europe, property taxes had long been collected without a cadastral map base, and they continued to be so levied in some countries. So it was also with the buying and selling of land, its drainage and improvement, its valuation and day-to-day management. Even enclosure and redistribution of land parcels, for which maps would be considered indispensable today, were conducted quite satisfactorily in England up to the third quarter of the eighteenth century with only a terrier to place a particular strip in the context of its furlong and field (chapter 7.2). In the nineteenth century, possession of a cadastral survey was not axiomatic. Even today there is no comprehensive map-based cadaster in England and Wales or Norway.

Though possession of the technical capability to survey and map individual properties at a large scale is an obvious prerequisite of state property mapping, there is no case reviewed in this book where mapping was prevented by lack of skilled personnel or equipment. The *agrimensores* were able to construct cadastral plans in the first century A.D. with simple equipment. The use of supervised, but untrained, peasant farmers on the Josephine cadaster in eighteenth-century Austria or *patwaris* in British India indicates how little formal training in surveying was necessary.

Triangulation control, about which there was so much debate in the nineteenth century, was not essential for successful survey, though clearly some purposes of cadastral mapping—for example, the registration of land titles in closely settled lands—were more dependent on planimetric accuracy than others. Technical knowledge itself, however, did not inevitably lead to mapping, as the techniques of cadastral survey were known in all those territories where governments did not commission cadastral maps, certainly after about 1650.

By definition, state mapping can be practiced only after the establishment of the state; until power was centralized, individual feudal estates would be the natural territorial units for mapping. In many instances, as for example in Milan and Denmark, reforms dependent on surveying and mapping had to await the establishment of a strong state. The existence of such a state, though, was not of itself enough to ensure that cadastral maps were constructed. Many nation-states, including England, did not commission comprehensive, mapped cadastral surveys.

Cadastral mapping is feasible, regardless of whether desirable or not, only when there is a physical division of land and not just a division of its value. In Norway, for example, until as recently as the late eighteenth century, it was *skyld,* or value, rather than land itself which was owned and leased (chapter 3.18). Owners of *skyld* could not point to any individual piece of land which was theirs. What they owned was a share in the value of the *gård* (cadastral farm). *Skyld* possessed no spatial coordinates and was thus impossible to map, just as it is today impossible to map the shareholdings in a joint stock company. The establishment of a system of private title to land in which individuals have absolute property rights over discrete plots of land is, however, not a sufficient condition for mapping. In some Austrian Habsburg territories before the nineteenth century and in England and Wales, absolute property rights existed but were not recorded by the state in a mapped cadaster. Nevertheless, in very many of the territories reviewed in this book, the association between cadastral mapping and the emergence of capitalist landowning and agrarian systems is clear.

Maps, notably those accurately surveyed, could provide the precise and complete information about land which came to be seen as necessary to effective, enlightened government in the eighteenth century. If such detail and precision were not perceived as important, or if, as with the land tax in England, there was judged to be little evasion of or opposition to the measure being effected, there was not much point in incurring the expense of mapping.

Conviction of the merits of mapping was a precondition for mapping itself. Not surprisingly, surveyors and cartographers everywhere were enthusiastic proponents of surveying and mapping; some, like Marinoni in the Milan surveys, got their way. Physiocratic convictions proved a spur to mapping throughout Europe. Their belief that land was the source of all value made French, Danish, and Austrian Physiocrats outspoken advocates of cadastral mapping as the scientific base for the improvement of the nation's fundamental asset. Yet conviction was not enough.

Joseph II was convinced of the merits of the Physiocratic position but could not bring his survey to fruition in the face of powerful opposition. Robert Kearsley Dawson was convinced that a national cadaster was a vital asset to government, but he lacked the power to sway the skeptical British government in 1836. It was not simply conviction that mattered, but the relative power of the convinced and unconvinced.

It is thus power—whether economic, social, or political—which lies at the heart of the history of cadastral mapping. The cadastral map is not simply an antiquarian curiosity, a cultural artefact, a useful source for the historian. It is an instrument of control which both reflects and consolidates the power of those who commission it. The cadastral map is *important*: it was recognized as such by both its advocates and its opponents throughout the four centuries reviewed in this book. The cadastral map is *partisan:* where knowledge is power, it provides comprehensive information to be used to the advantage of some and the detriment of others, as rulers and ruled were well aware in the tax struggles of the eighteenth and nineteenth centuries. Finally, the cadastral map is *active:* in portraying one reality, as in the settlement of the New World or in India, it helps obliterate the old. To look for one, all-sufficient precondition for its use would miss the point. As an instrument of power, the cadastral map can be understood only in the context of the balance of power and balance of interest in each area and in each period. The path to the acceptance of the cadastral map as an instrument in the service of the state was neither smooth nor uniform; it was as turbulent as the struggles for power themselves.

APPENDIX
A Regional Guide to German Cadastral Literature

Almost all of the cartographic and much historical and geographic literature in Germany, unlike most other countries, is concerned exclusively with individual places or regions. In Chapter 4 the subject matter has been treated thematically rather than regionally, and the chapter endnotes provide a consolidated bibliography. For ease of reference a review of the literature region by region is included in this appendix. It is not an exhaustive guide. The regions chosen are, in the main, cultural and geographic regions such as Lower Saxony, not political units, for example the Kingdom of Hanover. The exception is that of Brandenburg-Prussia. Of the vast literature on the development of cartography in Germany, only a very small proportion is concerned with cadastral maps. Only works which contain material directly relevant to the history of cadastral mapping are included. Most but not all the references cited below are referred to in chapter 4.[1]

BAVARIA

Katzenberger describes 175 years of cadastral surveying in Bavaria. His discussion includes the important early nineteenth-century cadaster, which is treated more fully in the works of Past and Ziegler.[2] Winschiers provides biographical sketches of cartographers and surveyors, including a few who worked on cadastral projects.[3]

SWABIA

Surveying in Swabia was given an impetus by its proximity to Basel, a notable early center of topographical and cadastral cartography. The Basel town council followed the lead of private estate owners and employed a succession of surveyors from the end of the sixteenth century onward to codify territorial and tenurial rights and also to put rights to tithes and taxes on a more scientific footing. Basel surveyors were also active in neighboring Swabia, and surveying there was aided by early manuals published in Basel.[4] Grenacher and Ulshöfer give more detailed accounts of Daniel Meyer's mapping in Swäbisch-Hall.[5]

Swabia was the scene of a number of important early mapped cadastral surveys. Religious authorities took the initiative, and there are many extant maps from the later seventeenth century; secular authorities followed at the turn of the eighteenth century.[6] Mapping in the region is also well documented. Oehme's pioneering work is the single most important account of cartographic development in southwestern Germany. In the second chapter he cites references to cadastral maps.[7] Bull-Reichenmiller gives a much briefer summary of map development in southwestern Germany, and her exhibition catalog is a useful, although short and unillustrated, description of some Swabian maps.[8] Musall et al. describe an interesting selection of historical maps of the area through four centuries. Their work and that of Bull-Reichenmiller include, though are not wholly devoted to, cadastral maps.[9] Individual maps are described by Grees (a map of the village of Obersteinach in 1717),[10] Miedel (two early maps by Hospinus),[11] Musall (a forest map of 1756–57),[12] and Scherzer (a picture map of 1547).[13] More general discussisons are Schäfer's treatment of the first eighteenth-century mapped survey of the Margravate of Baden and Scherzer's description of the development of maps in the Hochstift Würzburg.[14] Hoffman and Schumm provide inventories of maps in respectively the Staatsarchiv Wertheim and the Hohenlohe archive in Neuenstein.[15]

BRANDENBURG-PRUSSIA

Mapmaking in Brandenburg-Prussia was heavily dominated by the state.[16] As a result, the state archive, the Geheimes Staatsarchiv Preußische Kulturbesitz in Berlin, is one of the most important cartographic collections in Germany, despite very serious losses in the Second World War, and it is certainly among the best cataloged. Its extensive holdings of large-scale maps are made accessible by special inventories compiled by Bliß.[17] Each refers to an area of Prussia which was administered from 1723 to 1808 by a Kriegs- und Domänenkammer and after 1808 by a Regierung. These were important administrative bodies which had direct responsibility for taxation, excise, and the control and development of state lands in their area. They administered and improved forests, agricultural areas, heaths, and bogs with the aid of large-scale maps. Those which survived have been meticulously cataloged by Bliß. A selection of maps in the Geheimes Staatsarchiv Preußische Kulturbesitz is described in the well-illustrated exhibition catalog by Vogel and Zögner.[18] The selection includes some cadastral maps. Scharfe describes maps of Brandenburg, 1771–1821[19] and also provides a more general summary of mapping in Brandenburg.[20] Hanke and Degner discuss the development of state cartography, including cadastral and forest maps, in Brandenburg-Prussia.[21]

HESSEN

There is no consolidated account of cadastral mapping in the region, but an encyclopedia entry by Zögner and Zögner is a useful short summary.[22] There are several descriptions of individual cadastral maps: Eckhardt describes a tithe map of 1625, and Schäfer discusses a legal map of 1584 or 1587.[23] The work of individual cartographers is also investigated: Engel describes the life of the cartographer Joist Moers,[24] and Meers describes a mid-eighteenth-century cadaster of Korbach and its fields and forest.[25] Wolff and Engel present illustrated exhibition catalogs of maps, including cadastral maps, from archives in Hessen.[26]

LOWER SAXONY

Kleinn gives a brief summary of map development, including cadastral map development, in Lower Saxony.[27] Hake describes mapping in the environs of Hanover.[28] Kleinau, Penners, Pitz, and Vorthmann describe eighteenth-century maps of Brunswick.[29] Lang provides a discussion of mapping in East Friesland and describes large-scale property maps of a type more usually associated with the neighboring Netherlands.[30] Leerhoff presents a well-illustrated and annotated selection of maps, including several cadastral maps.[31] Jordan and Pitz discuss agricultural reform maps of Lower Saxony and Brunswick respectively.[32] Grossmann discusses eighteenth- and nineteenth-century maps of Lower Saxony, and v. d. Weiden describes the history of cadastral mapping in Lower Saxony.[33]

THE RHINELAND AND NORTH RHINE–WESTPHALIA

Kleinn gives a detailed survey of mapping in northwestern Germany in the last 250 years and a short account of mapping in North Rhine–Westphalia.[34] Sperling gives a similarly brief account of map development in the Rhineland.[35] The important Cleve cadaster carried out by the Prussians between 1731 and 1736 has been the object of sustained interest, and there are good accounts of its course and significance.[36] Grenacher reviews the development of Alsatian cartography. His main focus is on topographical maps, but there is some mention of maps used for a revision of the land tax and as a provincial inventory.[37] Maps from Osnabrück are described by Kleinn (the eighteenth-century survey of Osnabrück)[38] and Prinz (pre-nineteenth-century surveys of Osnabrück, particularly that of du Plat).[39] Meurer discusses maps of the Stift Heinsberg in the eighteenth century.[40] Prussian cadasters in the Rhineland are described by Müller-Wille (the surveys of 1822–35 and 1861–65)[41] and Osthoff (the cadaster of 1808–39).[42] Osthoff also gives an account of early agricultural reform in the Rhineland and the use of maps in that reform.[43]

RHINELAND-PALATINATE

There is no consolidated treatment of cadastral mapping in the region, but Hellwig and Werner provide a well-illustrated exhibition catalog which includes a few cadastral maps.[44] Bauer describes a forest map of 1767.[45]

SCHLESWIG-HOLSTEIN

The outstanding cartographic history of the Duchies of Schleswig, Holstein, and Lauenburg, which covered areas now in Germany and Denmark, is by Kahlfuß and consists of a detailed general history and a comprehensive catalog of maps of Schleswig-Holstein before 1864, when Prussian rule began.[46] Cadastral mapping is fully discussed. Unger's work complements Kahlfuß's in its discussion of post-1864 cadastral surveys.[47] Kahlfuß in a more recent work provides a brief general summary of cartography in the region, and Witt describes cartographic developments between 1475 and 1652.[48] Speiermann describes the cadastral mapping of lands belonging to the city of Lübeck since the 1870s.[49]

SAXONY AND THURINGIA (THÜRINGEN)

Stams gives a brief summary of mapping in the region.[50] Detailed accounts of early mapping in the Electorate of Saxony are provided by Beschorner (cartography before 1780)[51] and Bönisch (sixteenth- and seventeenth-century maps).[52] Richter describes Saxon cadastral mapping in the later period of 1835–41.[53] Emmerich discusses the holdings of rural cadastral maps in the Leipzig Stadtarchiv and their possible use.[54]

Two works which cover more than one region are by Vogel, who gives a brief region-by-region summary of German cadastral mapping with a useful bibliography, and Zögner, who provides further useful references.[55]

NOTES

Chapter One
Antiquity to Capitalism

1. Millard 1987:109–10.

2. Shore 1987:128–29.

3. Dilke (1985) considers that Byzantine "cadastral" ledgers known in Greek as *katastichon* ("line by line") are the likely etymological derivation for the modern Italian *catasto*, the French *cadastre*, and the English *cadaster*.

4. "Cadastral map consciousness" is individual and societal awareness of the value of map representations of landed properties.

5. Kish 1962.

6. Dilke 1971:88.

7. Dilke 1987:209–10.

8. Piganiol 1962; Pitte 1983:80–83; de Planhol 1988:66–67.

9. Translation by Dilke (1985:108).

10. Harley and Woodward 1987:507.

11. Kain forthcoming.

12. Fletcher 1990:8.

13. Thompson 1968:2.

14. Agas 1596:14–15.

15. Norden 1607:15–16.

16. Harvey 1987:464.

17. Beresford 1971:31–37.

18. Beresford 1971:25–62.

19. Darby 1933; Beresford 1971:46.

20. Emmison 1963.

21. De Dainville 1964:52.

22. Taylor 1947; Price 1955; Buisseret 1988.

23. Lynam 1947; Harvey 1980:102; Himley 1959; Grenacher 1964.

24. Roberts 1968.

25. Roberts 1968:36.

26. Anderson 1974:188.

27. Greslé-Bouignol 1973.

28. Price 1955; Skelton and Harvey 1969: 486–87; Harvey 1980:89–90.

29. Vannereau 1976.

30. De Dainville 1970:117.

31. L'Hoste 1629:129.

32. Norden 1607:30.

33. Eden 1983:77.

34. Leigh 1577:preface.

35. Harley 1983:37; cf. evidence from the Netherlands in chapter 2.

Chapter Two
The Northern and Southern Netherlands

1. Since 1987 the Werkgroep voor de Geschiedenis van de Kartografie at the University of Utrecht has been working to compile an exhaustive bibliography of the history of cartography of the Netherlands, defined as an area covering approximately present-day Belgium, Luxembourg, and

the Netherlands. A third edition was issued in 1991 (cf. Hameleers and van der Krogt 1991 as listed in the Bibliography), and a final edition is expected in 1993. Hameleers 1984 is a detailed catalog of more than a hundred Netherlands printed polder maps. The bibliography of the Netherlands chapter of this book is limited to books and maps which are directly referred to in the text. For more general and comprehensive guides, the reader is referred to the above-named works.

2. Place-names are given in English if a common form exists (e.g., Flanders, not Vlaanderen or Flandre). If a Flemish and a French form exist, the one most commonly used in English is given first, with the other in parentheses, thus Liège (Luik). In the notes and bibliography, where a title exists in Flemish and French, the Flemish is given first. Where a title exists in Flemish and German, the German is given first.

3. Kinder and Hilgermann 1974–78:vol. 1, pp. 244–45, 255, 269 and vol. 2; Scheffer 1977:17, 25.

4. Ristow 1974:138.

5. Zandvliet, personal communication.

6. Fockema Andreae 1954, 1982; van Iterson, letter of 12 April 1989; Lambert 1971:113, 213; van der Linden 1988.

7. Van der Gouw 1967.

8. Lambert 1971:14; Schama 1988:40–41, 51–93; J. de Vries 1974:28–29, 197–99.

9. Koeman gives details of map holdings of some *waterschappen* and *hoogheemraadschappen* in the Netherlands (1961:part 2, pp. 165–271), but his coverage is limited to the largest authorities. The Moll collection of maps in the library of the University of Utrecht is rich in polder maps. The published catalog records maps drawn for water regulatory bodies (Harms and Donkersloot–de Vrij 1977:31–64, nos. 131–289 inclusive). These are mostly large-scale maps (c. 1:5,000–1:15,000) from the seventeenth and eighteenth centuries. Most are copper engravings, though there are some manuscript maps (cf. Oddens 1989). Another important collection of maps is that of the Hoogheemraadschap van Rijnland. The maps were drawn by the official surveyors of the *hoogheemraadschap* for various official purposes, some of which demanded the drawing of maps with property boundaries. The collection includes 3,500 topographical maps dating from 1457–1857, of which 550 date from before 1700. There is a published inventory of these (Fockema Andreae 1933). There are also 137 atlases and 220 polder maps which date from 1720 to 1969 (van Iterson 1986, and letter of 12 April 1989). Other *waterschappen* have collections which are similar in composition but far smaller (Hameleers 1986:36).

10. Hameleers 1986; Mingroot 1989; de Vries 1983:3.

11. Koeman 1960; Koeman 1983:136.

12. Zandvliet 1985:137.

13. Lambert 1971:113–14; J. de Vries 1974: 197.

14. Teeling 1955:93.

15. Cf., e.g., Groenveld and Huussen 1975: 139 for a survey of Schieland by Jaspar Adriaenszoon in 1545.

16. Meeuwszoon had been involved in the count of Holland's survey in 1533 and with the *vroonland* survey immediately after that (see section 2.2).

17. Keuning 1952:53; Teeling 1955:94.

18. Fockema Andrea, with van 't Hoff 1947: 67; van der Gouw 1967:88, 123–26, 160–61.

19. Hameleers, personal communication.

20. Diebels 1986:2–3; Fockema Andreae, with van 't Hoff 1947:11–12 and pl. 2; van Iterson 1986:72.

21. Fockema Andreae, with van 't Hoff 1947: 19 and pl. 4; cf. Keuning 1952:41–43, 53.

22. Teeling 1981, 1983, 1984; Zandvliet 1985: 137 and personal communication 24 July 1991.

23. Baars 1979:34–35.

24. Lang 1952:81; Lang 1962:49–50; Westra 1983, 1986; cf. reproduction and discussion in Leerhoff 1985:78–79.

25. A list of some of the more important printed polder maps from the seventeenth to the

nineteenth centuries is given in Koeman 1983:143–48. Donkersloot–de Vrij catalogs over eight hundred manuscript and printed maps from before 1750 held in Dutch archives, including many polder and other cadastral maps (1981:51–174). See also Fockema Andreae, with van 't Hoff 1947:51.

26. Zandvliet, personal communication 24 July 1991.

27. Hameleers (1989) estimates that about two hundred copperplate maps were made for polder and water authorities; plates for about twenty maps survive. Lithography became the main means of reproduction from 1825 onward (Oddens 1989:31). Although the seventeenth and eighteenth centuries were the golden age of Dutch polder mapping, there was something of a revival in the nineteenth century, when state cadastral maps were used by engineers, cadastral officials, and employees of polder boards as a basis for maps of polder areas. They produced various maps as private, commercial enterprises, including the 1:10,000 maps of the South Holland islands (Fockema Andrea, with van 't Hoff 1947:93).

28. Ristow 1974:144; cf. reproductions in Donkersloot–de Vrij 1981:pl. 16; Fockema Andreae, with van 't Hoff 1947:pl. 13; Koeman 1983;139–44 and fig. 9.4; Zandvliet 1985:fig. 52, 56. A second map at a scale of 1:45,000 was printed in 1750 to try to overcome the illegibility of the first. (Hameleers 1984:79–84 and personal communication 2 September 1991).

29. Baars 1979:32; Muller and Zandvliet 1987a:5–7.

30. Lambert 1971:213–15; Wagret 1968: 78–85; Ristow 1974:139.

31. Ristow 1974:139; Lambert 1971:215.

32. Baars 1979:31–33; Hameleers, personal communication 2 September 1991.

33. A good example of this attitude would be Johan van Oldenbarneveldt, who invested in both the Dutch East India Company and lake drainage (Zandvliet, personal communication 24 July 1991).

34. Zandvliet 1985:138, 142–43.

35. Kamp 1971; Ristow 1974:139.

36. Koeman 1983:138, 148. Hameleers reports that two versions of the printed map exist. The map itself is the same, but the coats of arms surrounding it differ (personal communication 2 September 1991). Koeman suggests a date of 1660 for the map, but Hameleers (1984:147, 368) and Donkersloot–de Vrij (1981:no. 97) suggest a date of 1665.

37. Koeman 1983:138, 147; cf. reproduction in Donkersloot–de Vrij 1981:pl. 31.

38. Lambert 1971:215–16.

39. [Beemster] 1613.

40. De Vries 1983. Hameleers points out that the dates of publication of the maps are uncertain (personal communication 2 September 1991).

41. Zandvliet 1985:142; Lambert 1971:216–17; J. de Vries 1974:194.

42. Fockemae Andreae, with van 't Hoff 1947:36–37 and pl. 8.

43. De Vries 1983.

44. Lambert 1971:217–18.

45. Hameleers 1986:39, fig. 3.

46. J. de Vries 1974:200; Baars 1969:32; Lambert 1971:242; Hameleers 1984:43–48 and map 4.

47. Groenveld and Huussen 1975; Margry 1984; Muller and Zandvliet 1987a:7

48. Fockema Andreae, with van 't Hoff 1947:19–23; Muller and Zandvliet 1987a:10; Zandvliet 1989a:35 and personal communication 24 July 1991. Teeling (1981, 1983, 1984) gives an exhaustive list of Dutch surveyors from the fourteenth century onward.

49. Apprentices perhaps had simple books such as the manuscript textbook from the second half of the seventeenth century by Joannes Bollen of Limburg now in the Rijksarchief in Limburg (de la Haye 1988:46, pl. 10).

50. Huussen 1989; Margry 1984; Mingroot 1989; Schilder 1989; Pouls 1984; Muller and Zandvliet 1987a:15, 23; Muller and Zandvliet 1987b:3; Zandvliet 1985:137; Zandvliet 1989a:31–35. The surveyor was charged with accurate repre-

sentation of the "is," but it was the responsibility of judges to decide the "ought" in disputes (Hameleers, personal communication 2 September 1991).

51. Huussen 1976:148, 150; Keuning 1952: 41.

52. The copy and one original of the *vroonland* maps survive and are in the Algemeen Rijksarchief in The Hague.

53. Belonje 1935.

54. Biesbos was then in the County of Holland and now is in the province of North Brabant.

55. Zandvliet 1979; Zandvliet 1985:12–15; Zandvliet 1989b:20–21; the map of the Biesbos is now held in the Algemeen Rijksarchief in The Hague.

56. Ten Boom 1982:49–56.

57. Kok 1985.

58. J. de Vries 1974:210.

59. Written and mapped assessments of areas for the collection of tithes were also compiled, but many registers and maps have since been lost.

60. Roessingh 1968–69.

61. *Verponding,* a word that came into use around 1500, was originally a feudal levy, due in perpetuity to the feudal lord in return for the use of his land. After 1581, payment was due to the States, as the *verponding* had become a type of state land tax, composed of a fixed part, the *ordinaris,* and a part which fluctuated from year to year, the *extraordinaris* (Kok 1985:16–17; J. de Vries 1974: 283).

62. Koeman 1982:104; Koeman 1963:71; J. de Vries 1974:283. The Hollandse cadaster of 1795–1811 (see section 2.3) was never implemented.

63. Teeling 1955:95–96.

64. Teeling 1955:97.

65. Van der Steur 1985.

66. Teeling 1955:96.

67. Van Ham 1983, 1987.

68. Hessels 1891:3–6; Koeman 1982:104; Koeman 1963:71; Scheffer 1976.

69. Koeman 1982:104; Koeman 1963:71; Hessels 1891:4–5; Scheffer 1976:62.

70. Muller and Zandvliet 1987a, 1987b; Teeling 1981; Zandvliet, personal communication.

71. Hessels 1891:6–7; Scheffer 1977:17–18.

72. Teeling 1955:133–34.

73. Koeman 1963:72.

74. The bill was based on the laws of 1 May 1798 and the decrees of 25 March 1801 and 14 July 1805 (Hessels 1891:8).

75. Hessels 1891:9.

76. Scheffer 1977:20–23.

77. Muller and Zandvliet 1987a:46–48; Zandvliet 1985:156–57.

78. Scheffer 1977:23–24.

79. Hessels 1891:10; Koeman 1963:73; Scheffer 1977:25.

80. Koeman 1982:104.

81. Scheffer 1977:25.

82. Muller and Zandvliet 1987a:46–48.

83. Maps and registers from the cadaster are to be found in the Rijksarchieven of 's-Hertogenbosch, Arnhem, Haarlem, and Utrecht (Koeman 1982:104; Koeman 1983:226).

84. Hessels 1891:3, 11; Koeman 1982:104–5; Koeman 1983:226.

85. Muller 1981:178–80.

86. Keverling Buisman and Muller 1979:11; Koeman, 1983:227–28.

87. Koeman 1983:228.

88. De la Haye 1988:29; Muller 1979.

89. Muller 1979:253–54.

90. Koeman 1983:228–29.

91. Koeman 1983:187, 230–34. A useful discussion of the legal basis of the documents from Dutch nineteenth-century cadasters, their use in historical research, and their relation to land registration documents (*hypothecaires*) is that by Keverling Buisman and Muller (1979). A project to reproduce maps from the nineteenth-century Dutch cadaster is now underway in Gelderland, Drenthe, Friesland, and Zeeland (cf. *Kadastrale Atlas van Zeeland*).

92. Schonaerts 1976:67; Vuylsteke 1986.

93. Schonaerts 1976:36–37 and pl. 11 and 27; Vanhove 1986.

94. Duvosquel 1986; Minnen 1986.

95. Coppens 1968:598–99; Mosselmans 1976:xxxvi; Schonaerts 1976:57–59; van der Haegen 1986:6.

96. Coppens 1968:601; van der Haegen 1980: 115.

97. Ockeley 1986:51–58.

98. Bigwood 1898:395–96; Ockeley 1986: 51–58; Coppens 1968:599–601.

99. Bigwood 1898:397–98.

100. Ockeley 1986:54–55; the register is in the local archive in Opwijk, and the map is in the Rijksarchief.

101. Schonaerts 1976:75.

102. Mergaerts 1986.

103. Bigwood 1898:389–95; Coppens 1968: 598; van der Haegen 1980:114.

104. Van der Haegen 1980:114–16.

105. Bigwood 1898:398–402.

106. Bigwood 1898:403–10.

107. Bigwood 1906, 1912; Davis 1974:14–20.

108. Molemans 1988.

109. Coppens 1968:601–3; Hannes and Coppens 1967; Verhelst 1982.

110. Van der Haegen 1980:124–25; cf. Depuydt 1975 for discussion of cadastral maps for nineteenth-century topographical surveys.

111. Schama 1988; for the development of mapping in the Dutch colonies, cf., e.g., van der Brink 1982; Koeman 1973; Kok 1982; Loor 1973.

112. Cf. the disputed *vroonland* survey discussed in section 2.2.

Chapter Three
The Nordic Countries

1. Kinder and Hilgermann 1974–78:vol. 1, pp. 270–71.

2. Kinder and Hilgermann 1974:78:vol. 2, pp. 38–39.

3. Sporrong 1984a:12.

4. Lönborg 1901a, 1901b, 1903.

5. Örback 1978.

6. Peterson-Berger 1928:285.

7. Kungl. Instruktion, 4 April 1628, quoted in Ekstrand 1901–5:vol. 1, p. 1.

8. Bagger-Jörgensen 1922b:14–15.

9. Kungl. Instruktion, 4 April 1628, quoted in Ekstrand 1901–5:vol. 1, pp. 1–3.

10. Lönborg 1901a:114–15.

11. Kam. Koll. Instruktion, 20 April 1643, quoted in Ekstrand 1901–5:vol. 1, pp. 36–37.

12. Bagger-Jörgensen 1922b:12.

13. Kam. Koll. Fullmakt, 16 June 1633, quoted in Ekstrand 1901–5:vol. 1, p. 6.

14. Kam. Koll. Fullmakt, 19 June 1633, quoted in Ekstrand 1901–5:vol. 1, p. 6.

15. Kam. Koll. Fullmakt, 19 June 1633, quoted in Ekstrand 1901–5:vol. 1, p. 6.

16. Kam. Koll. Instruktion, 2 April 1634, quoted in Ekstrand 1901–5:vol. 1, p. 10.

17. Kungl. Instruktion, 30 April 1635, quoted in Ekstrand 1901–5:vol. 1, p. 14.

18. Johnsson 1965:16–17, 75.

19. Kam. Koll. Memorial, 19 May 1636, quoted in Ekstrand 1901–5:vol. 1, p. 21.

20. Cf. Lönborg 1901a:115 for 1643.

21. Johnsson 1965:18–19.

22. Ehrensvärd 1986a:123.

23. Peterson-Berger 1928:293.

24. Bagger-Jörgensen 1922b:26.

25. Sporrong 1984a:12; Bagger-Jörgensen 1922b:57.

26. Cf. Johnsson 1965:21, 28–29.

27. Johnsson 1965:61–70. The variation in the length of the *aln* together with inaccuracies of measurement and disagreement over what counted as cultivated land makes the calculation of the cultivated area in the seventeenth century difficult (Johnsson 1965:61–70).

28. Kam. Koll. Memorial, 19 May 1636, quoted in Ekstrand 1901–5:vol. 1, p. 20.

29. Peterson-Berger 1928:265; cf. Johnsson 1965:39–58.

30. Cf. Williams 1928:299–308; Curschmann 1935:52.

31. Bydén 1919.

32. Hedenstierna 1948:48–61.

33. [Statens Offentliga Utredning] 1986: 2–3.

34. Kam. Koll. Instruktion, 2 April 1634, quoted in Ekstrand 1901–5:vol. 1, p. 10.

35. Helmfrid 1959:225–27.

36. Kungl. Instruktion, 4 April 1628, quoted in Ekstrand 1901–5:vol. 1, p. 1; Peterson-Berger 1928.

37. Ehrensvärd 1984:556; Lönborg 1901a: 118, 137.

38. Kungl. Fullmakt, 30 April 1635, quoted in Ekstrand 1901–5:vol. 1, p. 15; Johnsson 1965: 11–12.

39. Johnsson 1965:11–12.

40. But cf. Helmfrid 1959:228 for an isolated incidence of an indirect use.

41. Ehrensvärd 1986a:111.

42. Cf. Wannerdt 1982.

43. Helmfrid 1959:228–29.

44. Roberts 1953:509–26.

45. Kungl. Instruktion, 4 April 1628, quoted in Ekstrand 1901–5:vol. 1, pp. 2–3.

46. Arrhenius 1945.

47. Johnsson 1965:36.

48. Sporrong 1984a:12.

49. Helmfrid 1959:224.

50. Lönborg 1901a:119–20; Ehrensvärd 1984:554.

51. Helmfrid 1959:224, 229–31. The *geometriska jordeböckerna* are in the main cadastral map archive in Sweden, Lantmäteriverkets Forskningsarkiv in Gävle, some two hundred kilometers north of Stockholm. This important archive houses about 200,000 maps, most of which have accompanying texts which include *protokoller,* or legal land registration documents signed by the surveyor and sometimes the parties concerned, and *beskrivninger,* which identify land plots and their owners and describe the use, quality, and extent of the land. There is some original material, notably the *geometriska jordeböckerna,* but chiefly the maps are *renovationer,* or clean copies of maps. The prick holes on them, where the maps were copied from originals, can clearly be seen. The *renovationer* are often beautifully illustrated and are easy to use but are in some respects incomplete. The *konceptkartor,* the original ground maps drawn up in the field and on which the *renovationer* are based, are to be found with their full notes and sometimes calculations and construction lines in the Länsstyrelsenslantmäterienheter, or county survey offices (Sporrong 1984a:7–8). They were reported in 1922 to be in rather poor condition and difficult to use (Bagger-Jörgensen 1922b:29). Since then, heavy use of certain archives, notably in Stockholm, has led to a deterioration in the state of the maps, some of which are now very fragile. When enclosure took place, a further copy of each map was made. This remained in the village in question and has in many cases been lost (Sporrong 1984a:18).

The Gävle archive is organized primarily by region: the maps are sorted by *län* (county), *härad* (hundred), *socken* (parish), and *by* (farm or village). For each *by,* the maps are then arranged chronologically; a brief description of the maps, the date each was drawn, and the date each was registered are given. Johnsson (1965:37) warns that the maps' dates refer not to the date on which they were drawn but to that on which they were delivered to the Räknerkammer (Exchequer). As well as the maps and descriptions which are in the map archive, Gävle includes the Adminstrativarkiv, which contains letters between the Lantmäteri administration and the surveyors out in the field. These reveal biographical details about the surveyors, not least about the difficulties they had in getting their salaries when they were working far from home.

It is unfortunate, given its unrivaled collection of seventeenth-century maps, that there is no published catalog to the Gävle archive and also unfortunate that calls for computerization of the indexing system (cf. Sporrong 1984a:22) have had no success. Most of the *geometriska jordebok* maps were microfilmed in 1954 by Uppsala University. Some were not, however, such as those for the island of Gotland, and in general the accompanying descriptions were not microfilmed ([Statens Lantmäteriverk] 1977:12). A new microfilming is now under way and is expected to be finished by the turn of the century, but films will unfortunately be only in black and white. Attempts have recently been made to bring the *geometriska jordeböckerna* to the attention of a wider public, as, for example, by Baigent (1990) and especially the first volume of the *National Atlas of Sweden,* a bilingual and well-illustrated volume entitled *Maps and Mapping* (Sporrong and Wennström 1990).

52. Kam. Koll. Instruktion, 17 May 1642, quoted in Ekstrand 1901–5:vol. 1, p. 30.

53. Sporrong 1984b:136, 140–42.

54. Sporrong 1984a:13–14.

55. The *rågångskartor* are mainly to be found in county survey offices (Sporrong 1984a:19).

56. Örback 1978:5.

57. Bagger-Jörgensen 1922a:5–6.

58. Cf., e.g., Faggot 1746.

59. Dahl 1961; Helmfrid 1961:115, 117–18.

60. Sporrong 1984a:14.

61. Sporrong 1984b:137; cf. Hoppe 1979 and 1982 for a critical account of the history of enclosure and of enclosure maps.

62. Wester 1960.

63. Bagger-Jörgensen 1922b:12, 29; Sporrong 1984a.

64. Peterson in Sporrong 1984a:15.

65. Hoppe 1979:57–58.

66. Bagger-Jörgensen 1922a:8.

67. Sporrong 1984a:17.

68. Peterson-Berger 1928:288; Sporrong 1984a:14–15.

69. Helmfrid 1961.

70. Helmfrid 1961; Hoppe, letter of 17 July 1989, adds a note of caution, however, in the interpretation of Helmfrid's 1961 maps, pointing out that the threshold of "50 percent of parish enclosed" could be reached in some parishes if just one farm was enclosed.

71. Bagger-Jörgensen 1922b:16, Örback 1978:5.

72. Åberg 1973.

73. [Statens Lantmäteriverk] 1977:9.

74. [Statens Lantmäteriverk] 1977:9.

75. [Statens Lantmäteriverk] 1977:11.

76. These *äldre avvittring* maps are in Gävle and come in particular from Gävleborg and Västernorrland *län* ([Statens Lantmäteriverk] 1977:12). The nineteenth-century *avvittring* maps are now in the Gävle archive.

77. Bagger-Jörgensen 1922a:9–10.

78. Peterson-Berger 1928:288–89.

79. Ehrensvärd 1984.

80. Ehrensvärd 1986a:109–13.

81. Helmfrid 1959:227–28; Lönborg 1901a: 128.

82. Ehrensvärd 1986a:113.

83. Ekstrand 1901–5:vol. 3, p. 246.

84. Jones 1977:160.

85. Kriesche 1944:262.

86. Niskanen 1963b:6.

87. Wallenius 1970:58.

88. Niskanen 1963b:6.

89. Jones 1977:93; Sporrong, letter of 1 February 1987; Niskanen 1963a:8.

90. These were private and not official land survey operations, but many of the documents are nonetheless preserved in the land survey archives (Jones 1977:118, 131, 157, 185–86).

91. Jones 1977:149–50.

92. Jones 1977:96.

93. Wallenius 1970:58. Finnish material is stored under a similar system to that used in Sweden: the Maanmittaushallitus, or Lantmäteristyrelsen (National Board of Land Survey), in Helsinki houses clean copies of the maps, while the originals are stored in the Länslantmäterikontor (county survey offices). Material for Åland, the islands between Sweden and Finland, is kept with that for Åbo and Björneborg *län* (Wallenius 1970:57).

94. Vasar 1930–31:pt. 2, p. ix. but cf. Soom 1954 for a more critical account of the "good Swedish times" and the assessment in Blumfeldt 1934.

95. Soom 1954.

96. Dunsdorfs 1981:208.

97. Kinder and Hilgermann 1974–78:vol. 1, pp. 270–71.

98. Schiemann 1882; Švābe 1933.

99. The singular is the Latin *uncus,* which corresponds to German *Haken,* Swedish *hake,* Latvian *arkls,* and English *plowland*

100. Liljedahl 1933:24–25.

101. There was also the *Bauernhaken,* on which a peasant's labor service to his lord was calculated (Dunsdorfs 1950:12).

102. Dunsdorfs 1981.

103. Ågren 1973: Tønnesson 1981.

104. Soom 1954:389.

105. Blumfeldt 1958:109–10; the resumption, however, was not linked to military reforms in the provinces (Dunsdorfs 1950:15).

106. Dunsdorfs, 1950:82.

107. The ethnic German nobles in Livonia should not be confused with ethnic Swedish nobles there.

108. Vasar 1930–31.

109. Dunsdorfs 1950:26–27.

110. Dunsdorfs 1950. Maps from the first Livonian cadaster have almost all been lost; those from the second cadaster (1688 onward) are to be found in the state archives at Riga (Latvia) and Tartu (Dorpat) (Estonia), as well as in the Riksarkiv in Stockholm (Dunsdorfs 1950:39, 193). Maps from the Estonian cadaster are in the Estonian State Archive (Eesti Riigi Kesarhiiv). There are c. 3,800 Swedish maps of Livonia and Estonia. Most are of Livonia, with relatively few from Estonia and Ösel (Saaremaa). The maps of Livonia are generally accompanied by books of descriptions and contain important information about the land, its cultivation, and its tax burden. The Estonian maps generally do not have accompanying descriptions (Liiv 1938; Treial 1931).

111. Kinder and Hilgermann 1974–78: vol. 1.

112. Švābe 1933:353.

113. Soom 1954.

114. Rubow-Kalähne 1954–55:647; Rubow-Kalähne 1959:663.

115. Pauli 1984:71; cf. Curschmann 1935: 54, which suggests that the scale of one quarter of a *Zoll* to eighteen thousand Swedish ells is equal to 1:8,333 and that the smaller-scale maps are thus 1:16,667.

116. Curschmann 1935:53–55.

117. Rubow 1954–55:644.

118. Pauli 1984.

119. Cf. Drolshagen 1920–23; Curschmann 1935, 1938, but cf. Rubow-Kalähne 1960; Hinkel, 1967. Maps from Pomerania are in the Staatsarchiv (National Archive) and University in Greifswald and in the Kongelige Bibliotek (Royal Library) in Copenhagen, as well as in Stockholm in the Riksarkiv. Maps from Mecklenburg are in Gävle (Pauli 1984:69).

120. Other Swedish archives with minor collections of cadastral maps are the Krigsarkiv (War Archive), the Kammararkiv (Treasury Archive), Riksarkiv (National Archive), the Kammarkollegium (Finance Department), and the Kunglige Bibliotek (Royal Library). All of these are in Stockholm.

121. Place-names are given in English if a common form exists (e.g., Zealand, not Sjælland). They are in German with the Danish equivalent in brackets if that is the form most commonly used in English, thus Schleswig (Slesvig). Otherwise place-names are in Danish with the German equivalent, if there is one, in parentheses, thus Ærø (Ärö). German names are used where these are undisputed.

122. Kinder and Hilgermann 1974–78.

123. Jensen 1975:13.

124. Fransden, letter of 3 June 1988; Hastrup 1964:259–63.

125. Rosén 1961:522–25.

126. Hansen and Steensberg 1951:423; Jensen 1975:30–31; Nordlund 1978:3–5.

127. Hansen and Steensberg 1951:424–25; Nordlund 1978:11; Jensen 1975:31.

128. Nordlund 1978:11 and cf. n. 166.

129. Fransden 1983:261.

130. Nordlund 1978:8–9; Jensen 1975:31–33; Thygesen 1915:545.

131. Hansen and Steensberg 1951:427.

132. Newcomb 1973:7.

133. Cf. Fransden 1983: Hansen 1964, 1981, 1985; Nordlund 1978. All Danish *matrikel* maps are held in the Matrikulsarkiv (Land Registry Archive), whose collection goes back to that of the Landmaalingsarkiv, founded in 1693 with the task of collecting and ordering the *protokoller* of Kristian V's cadaster. In 1953 all pre-1906 documents other than maps in the collection were transferred to the Rigsarkiv (National Archive) in Copenhagen, although they can be ordered through the Matrikulsarkiv (Balslev, Holst, and Rydahl 1986:70). The archive includes some 1,800 manuscript register books from the cadaster of 1681–83 (Newcomb 1973:1), although no maps from this *matrikel* survive, as well as the *specielle landmåling* maps (Jensen 1975:43) and more than 50,000 enclosure maps from 1790–1810 (Newcomb 1973:2), including those of the famous Bernstorff Estate survey of 1764 (Nissen 1944:59). Most of the maps in the Matrikulsarkiv were working documents to be altered and updated to form a current description of ownership. Thus many maps have crude lines superimposed on them; others were thrown out when they were no longer in use.

134. Nordlund 1978:18; Hansen and Steensberg 1951:438; Thygesen 1915:546.

135. The *rytterdistrikter* were Copenhagen-Fredriksborg-Kronborg, Antvorskov, Tryggevælde, Vordingborg, Skanderborg, and Dronningborg.

136. Surveying for the first *rytterdistrikt* mapping was carried out under Captain Abraham Christian Willars between 1720 and 1723. The maps were at scales of 1:33,000 to 1:70,000 and show which of the *ryttergårde* were under the various regiments. The maps show the boundaries of the *ryttergods* and show the land use of the fields and give details of the tithes and *hartkorn* attached to them. Much topographical information is also included, and the maps were used as the basis for Erik Pontoppidan's topographical atlas (Nørdlund 1943:57 and pl. 89–95; Brandt 1987a, 1987b; Jensen 1975:36). Maps from the first *rytterdistrikt* mapping with their accompanying texts are now in the Kongelige Bibliotek in Copenhagen.

137. Brandt 1987b:4–9; Hansen and Steensberg 1951:468; Nordlund 1978:20; Thygesen 1915:546.

138. Jensen 1975:40–42; cf. Balslev and Jensen 1975:vol. 2, map 5. Brandt 1987b considers the artistic development of the so-called special survey maps.

139. Jensen 1975:37–40; cf. Balslev and Jansen 1975:vol. 2, map 4.

140. Cf. Skrubbeltrang 1961 for an account of changes in land tenancy in Denmark in the eighteenth century.

141. Fink 1941:195.

142. Hatton 1957:342–43.

143. Hoffmeyer 1981:501–2.

144. Hoffmeyer 1981:496.

145. Balslev 1981:508; cf. Balslev and Jensen 1975:vol. 2, maps 6 and 7.

146. Hansen and Steensberg 1951:472–73; Jensen 1975:41; Nordlund 1978:20; cf., e.g., Bjørn 1981; Hansen 1981.

147. H. E. Jensen 1981:520; Nordlund 1978:20, 24.

148. Nordlund 1978:20–23; Hansen and Steensberg 1951:474.

149. From 1781 onward, surveyors were obliged to have a Danish decimal-scale rule with which to adjust their chains. The measures were to be adjusted with respect to the official measure which had been made by the Swede Johan Ahl, who was invited by the Videnskabernes Selskab to come to Denmark to make cartographic and mathematical instruments (Balslev 1981:508–12). It is interesting that Denmark looked to Sweden in this respect, as it had done earlier in the 1688 *matrikel*, which was led by a Swedish-Danish nobleman.

150. H. E. Jensen 1981:519–20.

151. H. E. Jensen 1981:519–20; Nordlund 1978:20–23; Staunskjær 1981:546.

152. Jensen 1975:43. Both original and copy enclosure maps are today to be found in the central collection at the Matrikulsarkiv (Jensen 1975:43). These maps are extremely accurate and even today provide the best large-scale coverage of rural Denmark (cf. Balslev and Jensen 1975:vol. 2, map 8)

and form the basis for numerous historical and geographic studies (cf. Hastrup 1964; Newcomb 1973).

153. Derry 1965:482; cf. Bjørn 1977 for an attempt to review agricultural reform from the point of view of the peasantry.

154. Jensen 1975:44; Nordlund 1978:24–25; Thygesen 1915:547.

155. Jensen 1975:45–46; Brandt 1987a: 360; Nissen 1944:53–55; Thygesen 1915:548; cf. Balslev and Jensen 1975:vol. 2, map 9.

156. Nordlund 1978:27–29.

157. The *protokoller* from the 1844 *matrikel* are in the Matrikulsarkiv.

158. Pedersen 1944:47–49.

159. Nissen 1944:55–57; Pedersen 1944: 49–50.

160. Jensen 1975:47.

161. Thygesen 1915:550–51; Heering 1958:1.

162. Balslev, Holst, and Rydahl 1986:34–35. Difficulties with the maps include the following: techniques at the time of survey were rather primitive, the paper on which the maps were drawn has often either shrunk or stretched, and the scale has become variable from one part of the map to another as copies were made by pricking through the originals.

163. Fransden 1984a:11–13, 18 and cf. Kruse 1953 and Skrubbeltrang 1961.

164. Cf. Bugge 1795–98.

165. Jensen 1975:24–30; Nissen 1944: 58–59.

166. Jensen 1975:24, 27, 62–63; cf. Balslev and Jensen 1975:vol. 2, map 2. The standardization in style and training was matched by increasing uniformity in units of measurements used. At the beginning of the period the German, Lübeck, Gotland, Norwegian, Jutland, and Zealand ells were all in use. In 1521 the Zealand ell was declared the official national measure, but for his *matrikel* Kristian V declared in 1683 that the Rhineland foot was to be the standard measure. Two Rhineland feet were to equal one Danish or Norwegian ell. This unit remained in use until the introduction of the metric system in 1907. Surveyors divided the *fod* (foot) into ten *decimaltommer* (decimal inches), while the normal division was into twelve *duodecimaltommer* (duodecimal inches): this paralleled the situation in Sweden (cf. section 3.3). The unit of areal measurement was the *tønde land* (acre): one *tønde land* equaled fourteen thousand *kvatratal* (square ells) (Jensen 1975:15–16).

167. Kinder and Hilgermann 1974–78:vol. 2, pp. 198–99; Balslev, Holst, and Rydahl 1986:14–17.

168. Balslev, Holst, and Rydahl 1986:17.

169. Kahlfuß 1969.

170. Cf. Nørdlund 1942:vols. 1–2.

171. Kahlfuß 1969:22; Jensen 1975:57; Nørdlund 1942:vol. 3, pp. v–vi; cf. facsimile reproduction in Nørdlund 1942:vol. 3. The *Erdt Buch des Ambts Apenrade* is now in the Kongelige Bibliotek in Copenhagen.

172. Kahlfuß 1969:24–25; Jensen 1975:57–58. Maps and accompanying material from Samuel Gries's survey are to be found in the Rigsarkiv in Copenhagen.

173. Jensen 1975:58–59. The maps and other material from Pape's survey are in the Landsarkiv (County Record Office) in Odense.

174. Hoffmeyer 1981:502–6; Grau Møller 1984: Jensen 1975:59; Kahlfuß 1969:26, 147.

175. Kahlfuß 1969:22–23.

176. Grau Møller 1984:27–29; Kahlfuß 1969:87–90, 148. Material from this survey is in the Slesvig-Holstenske Landkommission's documents in the Rentekammer's archive in the Rigsarkiv in Copenhagen. The remarkably complete collection includes surveying documents and calculations, reports and letters, as well as details of enclosure and taxation. The maps form part of the Kieleraflevering (Kiel accession) in the Rigsarkiv.

177. Kahlfuß 1969:90. Kahlfuß points to the likely existence of large numbers of enclosure maps which were in the keeping of the farmers in newly enclosed villages as well as the better-documented existence of maps from large landed

estates which were mapped in connection with ownership changes or, more often, enclosure and agricultural restructuring on the estate.

178. Balslev, Holst, and Rydahl 1986:85–86; Jensen 1975:59–60.

179. Jensen 1975:57–60.

180. Thygesen 1915:549.

181. Pálsson 1986:327–28; Pálsson and Edwards 1972; Dronke, personal communication.

182. Sigurdsson, letter of 15 November 1986.

183. Cf. Sigurdsson 1978.

184. Nørdlund 1944:44.

185. Balslev, Horst, and Rydahl 1986:18–19.

186. Balslev, Horst, and Rydahl 1986:18–19.

187. Kinder and Hilgermann 1974–78:vol. 1, p. 199, and vol. 2, p. 84.

188. Fladby, Imsen, and Winge 1974:221, 225.

189. Fladby, Imsen, and Winge 1974:225, 227.

190. Sevatdal, letter of 14 June 1988.

191. Fladby, Imsen, and Winge 1974:227–28; Michael Jones, letter of 23 June 1986.

192. Sevatdal 1988:7.

193. Andressen 1982; Borgedal 1959:101–33; Fladby 1969; Holmsen 1958, 1961, 1979, 1982; Martinsen 1982; Oakley 1986; Sevatdal 1982, 1988; Winge 1982.

194. Sevatdal 1988:10.

195. Sevatdal 1988:10–11.

196. Sevatdal 1988:11.

197. Sevatdal 1988:12–14.

198. Evers 1943; Fladby, Imsen, and Winge 1974:300–301; Nissen 1937a, 1937b.

199. Aanrud 1977.

200. Evers 1943.

201. Nissen 1959; Opsal 1958:28.

202. Aanrud 1977:101.

203. Engelstad 1981:30–36.

204. Engelstad 1981:36–42. Maps drawn up in property disputes together with the accompanying descriptions and legal material for the period 1752–1801 are held in the Generalkonduktørarkiv in the Riksarkiv in Oslo. The collection is not complete (Engelstad 1981:34), but there are nearly four hundred maps with their accompanying descriptions, mostly from Akerhus, Hedmark, Oppland, and Buskerud. They are mostly from the first half of the fifty-year period during which the office existed; many come from Christopher Hammer's home area. They are often rather primitive in technical terms, but they are full of detail (Johannessen 1972:609).

205. Engelstad 1981; Johannessen 1972: 607–9; Jones 1985:73–74; Tønnesson 1981.

206. Jones 1982b.

207. Jones 1982a:135; cf. section 3.15.

208. Sevatdal, letter of 14 June 1988.

209. Hovstad 1981:19.

210. Björkvik 1956:34n.

211. Hovstad 1981:19.

212. Jones 1982b:177–78.

213. Hovstad (1981) has investigated the extant enclosure maps. The number of maps bear only a very crude relationship to the number of *gårder* which underwent enclosure, as one map might cover a number of *gårder,* or more commonly, several maps might be needed to cover one *gård:* one map was needed for its *innmark* (cultivated fields and some meadowland), and one or more for its *utmark* (wastes).

214. Hovstad 1981; Björkvik 1956:54.

215. Jones 1982a:136; Jones 1985; Grendahl 1959; Sevatdal 1982.

216. Holmsen 1956:25–29.

217. Frimannslund 1956:64; Holmsen 1961: 158. Enclosure maps are housed in Jordskifteverkets Kartarkiv (the Enclosure Board's Map Archive) in Ås.

218. Sevatdal 1985.

219. It operated under the title of Norges Grændsers Oppmaaling.

220. Eggen 1987; Johannessen 1972:613; Jones 1987.

221. Aanrud 1981:75–76.

222. Great efforts have been made by the Norsk Lokalhistorisk Institutt to encourage the use of the long-neglected Norwegian maps. A collected catalog of unpublished maps has been produced in the *Gamle norsk kart* series of eighteen volumes, one for each *fylke* (county) (Fladby et al. 1979–85). This series covers all unpublished maps in public institutions, but not in communes or private collections, drawn before 1900 and *gårdskarter,* maps of one or part of a *gård* or several *gårder* until the present (Fladby 1979:7–8; Fladby 1981:18). Marthinsen 1971 is a guide to other unpublished maps. Some earlier maps are to be found at Norges Riksarkiv (National Archive) in Oslo or at Statens Kartverk, until recently Norges Geografiske Oppmåling, at Hønefoss. One difficulty with Norwegian maps is that of orthography. When Norway was united with Denmark, Danish was the language of government, and the struggle for the reform and codification of the Norwegian language has continued ever since Norway became fully independent. Since substantial changes in orthography have been made, maps can be difficult to compare, especially in northern regions, in which Lapp and Finnish place-names create a further complication (Evers 1943; Michael Jones, letter of 23 June 1988).

223. Roberts 1953:112–37.

224. Hopkins 1987, 1989.

Chapter Four
Germany

1. Place-names are given in English if a universally recognized form exists (e.g., Germany, not Deutschland). Other places are given in English form (e.g., Hanover, not Hannover), except where this practice seems antiquated (e.g., Cleve, not Cleves; Hessen, not Hesse).

2. Heer 1968; Kinder and Hilgermann 1974–78; Koch 1984; Vierhaus 1988.

3. Musall et al. 1986:151.

4. Schäfer 1979:126; Werner 1985:147; Wolff and Engel 1988:15.

5. Grossmann 1955:24; Witt 1982:26–27; Hellwig 1985:47; Scherzer 1970, 1977; Werner 1985:147.

6. Aymans, letter of 2 January 1989.

7. Grees 1979:2–3.

8. Hellwig 1985; Werner 1985; cf. Müller-Wille 1940:49 and Hoffmann 1983:45; Aymans 1985; Schäfer and Weber 1971:viii. See Faber 1988b and Jung 1988 for maps in archives generally.

9. Aymans, letter of 2 January 1989; Wolff and Engel 1988:22.

10. Cf. Hoffmann 1983; Wolff and Engel 1988:28; Vollmer 1982; Witt 1982:27–32.

11. Wolff and Engel 1988:35.

12. Schäfer 1979:151–52.

13. Wolff and Engel 1988:22–23.

14. Cf. Witt 1982:34–35 for a 1588 map of Schauenburg, and Wolff and Engel 1988:15–21 for maps of Hessen's disputed borders used in boundary negotiations in the sixteenth century.

15. Cf. Emmerich 1962:113.

16. Bönisch 1970:1–33, 51; Stams 1986:693. Öder's maps of the Electorate of Saxony are held in the Staatsarchiv Dresden.

17. Witt 1982:18–19.

18. Werner 1985:147; Wolff and Engel 1988: 36–39.

19. Wolff and Engel 1988:39–45.

20. Wolff and Engel 1988:28, 38; cf. Witt 1982:20–25 for maps of the changing course of the Elbe and territorial adjustments resulting therefrom.

21. Hellwig 1985:47; Werner 1985:148.

22. Grees 1979.

23. Kahlfuß 1969:22.

24. Emmerich 1962:117–20. The atlas of the estate of Pfaffendorf is now in the Leipzig Stadtarchiv.

25. Leerhoff 1985:158–59.

26. Meurer 1981. The remaining maps of Heinsberg are now in the Hauptstaatsarchiv in Düsseldorf.

27. Werner 1985:132, 152.

28. Vierhaus 1988:16.

29. Leerhoff 1985:150–51.

30. Bull-Reichenmiller 1971:35.

31. Bauer 1971.

32. Cf. Schäfer 1979, map of 1585; Werner 1955:157–58, for eighteenth- and nineteenth-century forest maps in the province of Hanover, including areas of the Harz Mountains; Wolff and Engel 1988:22–23, for forest maps from Hessen; cf. Bliß 1978a, 1978b, 1979, 1981, 1982 for Prussian collections and Hofmann, with Semmler 1983, for the Löwenstein-Wertheim-Rosenberg map collections, which contain many forest maps.

33. Oehme 1961:43–45. Kieser's registers of Württemberg forests still survive; unhappily, the original maps were destroyed in the Second World War, although photographic copies of them survive.

34. Hanke and Degner 1935:171–81; cf. Bliß 1978a, 1978b, 1979, 1981, 1982; Vogel and Zögner 1979:61–63. Prussian Forestry Department maps are now in the Geheimes Staatsarchiv Preußische Kulturbesitz in Berlin.

35. Musall 1983.

36. Vogel and Zögner 1979:63–64.

37. Aymans 1985:39 and personal communication. The maps of the Asperd heath colony are now in the Archiv Gaesdonk, Grafenthaler Akten.

38. Wolff and Engel 1988:34.

39. Bruford 1971:116; Wolff 1987:41, 46–47.

40. Lang 1962:49–53.

41. Henderson 1963:80–83; 128–30; Scharfe 1972:41, 66, 74.

42. Cf. Farr 1986; Hagen 1986; Harnisch 1986.

43. Conze 1969; Lee 1972:xiii; Hagen 1986; Harnisch 1986; Moeller 1986; Schissler 1986.

44. Bruford 1971:107–11; Conze 1969:57; Haines 1982; Lee 1972:xiii; Harnisch 1986:40.

45. Harnisch 1986; Kriedtke, Medick, and Schlumbohm 1977.

46. Conze 1969:56–57; Clapham 1928: 29–34.

47. Conze 1969:63–64.

48. These two decrees of 1802 and 1808 were intended as first steps in a program to reduce the rigid legal barriers between the *Stände* (the three Estates: nobility, bourgeoisie, and peasantry) and to improve the lot of the peasants. Stein, the prime mover behind the reform, was forced by the French to resign in 1808, and the project foundered.

49. Bruford 1965:377–78; Clapham 1928: 42–44; Hagen 1986; Harnisch 1986; Schissler 1986.

50. Clapham 1928:48–49; Conze 1969:65–66.

51. Clapham 1928:202; Harnisch 1986; Schissler 1986.

52. Local variation was more pronounced in the west than in the east. Mooser (1986:55–56), for example, describes two neighboring areas in Westphalia. In Minden-Ravensberg the peasants, who were involved in proto-industry, initiated agricultural reform; by 1800 almost all the commons had been divided. In adjoining Paderborn, with more noble estates and lack of proto-industry, division of the commons did not really start until the 1840s and was bitterly resisted by the peasants, who depended for their survival on access to the wastes.

53. Conze 1969:58–62.

54. Bruford 1965:368, 391–92; Kitchen 1978:14; Clapham 1928:42, 49, 202; Lee 1972:xvii, 188–98, 219–22, 254, 257.

55. Conze 1969:59; Lee 1972:202; Clapham 1928:46–47, 202–3, 215.

56. Cf. map in Borchardt 1978:129.

57. Clapham 1928:202–3.

58. Reform of land taxes inevitably accompanied any agricultural reform which redistributed land or extended the privately owned area; tithes, too, were often converted from kind to cash, commuted or abolished when landholding was reformed (Leerhoff 1985:154). Tax and/or tithe re-

forms were entailed in most of the agricultural mapped surveys described below, but their main purpose was not fiscal (cf. Pitz 1967).

59. Grossmann 1955:24–25; Jordan 1955: 149–50; Kleinau 1968; Kleinau and Penners 1956; Leerhoff 1985:162–63; Pitz 1957, 1967; Vorthmann 1956.

60. Wolff and Engel 1988:26.

61. Jordan 1955:141–44. Hanoverian enclosure maps from the 1770s are now in the Niedersächsiches Landeskulturamt.

62. Clapham 1928:50; Conze 1969:59.

63. Jordan 1955:144–48; Grossmann 1955: 25, 40–41.

64. Jordan 1955:153.

65. Kitchen 1978:11–14.

66. Osthoff 1956:8–22.

67. Kahlfuß 1969:23–24.

68. Kahlfuß 1969:58–59. Peter Barber (personal communication) suggests that dynastic links and the consequent exchanges of personnel and ideas among Holstein-Gothorp, Sweden, and Russia may show the Holstein-Gothorp mapping to have been influenced by developments in Swedish cartography and in turn to have influenced Russian cartographic development. No evidence has been found on this point, but it remains an intriguing thought.

69. Bruford 1965:368; Clapham 1928:32; Conze 1969:68; Kitchen 1978:14; Kahlfuß 1969:90–91.

70. Farr 1986:1–6.

71. Schön 1970:66.

72. Lee 1975:158.

73. Lee 1975:159–61.

74. Lee 1975:160.

75. Cf. Hoffmann 1983:48, 50, for tithe maps in Staatsarchiv Wertheim; Meurer 1981, for tithe maps of the Stift Heinsberg in the Hauptstaatsarchiv in Düsseldorf.

76. Eckhardt 1961; Wolff 1987:41, 44, 45; Wolff and Engel 1988:22, 30–31. Wilhelm Dilich's tithe map of Niederzwehren is now in the Stiftsarchiv Kaufungen.

77. Leerhoff 1985:156–57.

78. Leerhoff 1985:154–55.

79. Aymans, personal communication; Ketter 1929:47; Leerhoff 1985:153.

80. Leerhoff 1985:152–53.

81. Krüger 1982:113–14; Wolff 1987:42–43; Wolff and Engel 1988:24–25.

82. Grossmann 1955:25–26; Krüger 1982: 114–15. Some taxation maps of Hessen-Kassel dating from the 1750s and produced in the survey were still in use in 1955 at the cadastral office in Rinteln.

83. Wolff 1987:43; Wolff and Engel 1988:24.

84. Wolff and Engel 1988:26.

85. Grenacher 1960; Oehme 1961:73, 112, and pl. 8; Ulshöfer 1968.

86. Schäfer 1968:141–43.

87. Some of the material from 1705 was used in the successful cadaster of 1731–36.

88. Ketter 1929:33.

89. Aymans 1987:168–69.

90. Aymans 1987; Aymans, Burggraaff, and Jansen 1988; Ketter 1929:18–42, 62, 73–75; Schulte 1984.

91. Ketter 1929:59–60; Aymans 1986: 16, 24.

92. Cf. Vollmer 1982.

93. Aymans 1986:17; Ketter 1929:68.

94. Aymans 1986; Ketter 1929:68–72. Some 1,500 maps are known to be extant, and they are widely scattered in public and private archives in Germany and the Netherlands. Aymans (1985:39) provides a key map to the known extant maps. A project at the University of Bonn to construct a small-scale map by joining all known original maps is well advanced (cf. Aymans 1986 and Aymans, Burggraaff, and Jansen 1988).

95. Schulte 1984:193, 200.

96. Leerhoff 1985:161–62.

97. Prinz 1950:111–18.

98. Grossmann 1955:26; Kleinn 1976–77; Leerhoff 1985:164–65; Prinz 1950. The maps from the Osnabrück survey have survived well and

are mainly in the Staatsarchiv Osnabrück (422 maps), with others in the Staatsarchiv Münster and the Katasteramt Wiedenbrück. A project to reproduce a simplified version of the maps on a smaller scale was completed in 1972. The bishop of Osnabrück at the time of mapping was Frederick, duke of York, George IV of England's younger brother. This dynastic link probably explains the employment of du Plat (Peter Barber, personal communication).

99. Jordan 1955:152; Leerhoff 1985:168–70.

100. Kahlfuß 1969:22–23. The documents from the eighteenth-century Lauenburg survey survive, as do finished maps from Lauenburg and draft maps from Neuhaus. The finished maps from Neuhaus, however, were destroyed by fire.

101. Hanke and Degner 1935:136.

102. Vogel 1929:225–26.

103. Cf. Lee 1975; Vogel 1929.

104. Lee 1972:30; Lee 1975:159.

105. Cf. maps in Ziegler 1976:9, 10, 12, 19.

106. Katzenberger 1977; Past 1978; Ziegler 1976.

107. Lee 1972:325–27.

108. Müller-Wille 1940:49–59; Osthoff 1950.

109. Müller-Wille 1940:59–63.

110. V. d. Weiden 1955:156–66; Kahlfuß 1986:710; Speiermann 1974:338–39.

111. Grossmann 1955:37–38; v. d. Weiden 1955:160–61.

112. Grossmann 1955:36–37, 45.

113. Grossmann 1955:45–48; Hake 1978:62; Jordan 1955:148; v. d. Weiden 1955:157–59.

114. Richter 1921.

115. Lee 1975:159.

116. Ketter 1929:15–16; Krüger 1982:113; Schön 1970:66; Grossman 1955:25; Past 1978:9.

117. Leerhoff 1985:166.

118. Grenacher 1971; Schäfer 1968:152–61.

119. Emmerich 1962:111–13; Hellwig 1985:47; Leerhoff 1985:149; Schäfer 1979:126; Musall et al. 1986:151, 153.

120. Bauer 1971; Scherzer 1970:161; Past 1978:14; Wolff and Engel 1988:22.

121. Grees 1979:4–5; Grossmann 1955:36–37; Past 1978.

122. Prinz 1950:124, 126, 136; Past 1978:10–11; Grees 1979:6; cf. Jordan 1955:142; Harnisch 1986.

123. Cf. appendix for the influence of Swiss cartography on Swabia.

124. Grossmann 1955:37–40.

125. Müller-Wille 1940:50.

126. Meers 1958.

127. Schäfer 1968:141–43.

128. Ziegler 1976:5; Past 1978:12.

129. Ketter 1929:34.

130. Hanke and Degner 1935:137; cf. n. 68 above.

131. Ketter 1929:47–53; Lang 1962:44–53.

132. Jordan 1955:152; Leerhoff 1985:186–270; Grossmann 1955:25.

133. Leerhoff 1985:166–67; Jordan 1955:141, 145; Grossmann 1955:36–37.

134. Scharfe 1972:41, 64–68.

Chapter Five
The Austrian Habsburg Lands

1. Place-names are given in English alone if a universally accepted form exists (e.g., Austria, not Österreich). Otherwise, the most commonly used English form is given with the German in parentheses, thus Transylvania (Siebenbürgen). Czech, Slovak, Hungarian, and other place-names are not given except where the German place-name is not well known in English—for example, Ljubljana (Laibach).

2. Cf. Bernleithner 1959:211–14; Dörflinger

1979:59–60; cf. Dörflinger 1984; Grüll 1952; Ulbrich 1967c:169–70 for further examples of privately sponsored large-scale mapped surveys.

3. The maps remain in Schloß Stetteldorf in the private ownership of the Hardegg family.

4. Oberhummer 1934–35.

5. The emperor's personal atlas is now in the Nationalbibliothek (National Library) in Vienna: the second set is in the Staatsarchiv in Vienna (Oberhummer 1932).

6. Oberhummer 1932; Bernleithner 1959: 213.

7. Dörflinger 1979:59.

8. The map by Cornelius in 1715 is in the Niederösterreichische Landesbibliothek, Vienna, ref. B IV 154 (Dr. G. König, personal communication); Bernleithner 1959:211, 213–14; Oberhummer 1932:154–55.

9. Dickson 1987:vol. 2, pp. 12, 185, 205–6.

10. Regele 1955:15.

11. Spielman 1977:39–40.

12. Spielman 1977:141, 164. The money, however, was formally voted each year by the Diet (Dickson, personal communication).

13. Dickson 1987:vol. 1, p. 297, and vol. 2, pp. 211–19, and personal communication.

14. Peter Barber, personal communication.

15. Zangheri 1980:107; [Archivio di Stato di Milano], 1988.

16. Klang 1977:5.

17. Ulbrich 1967c:170; Zangheri 1980:109.

18. Lego 1968a:89; Lego 1968b:1; Zangheri 1980:110.

19. Kretschmer 1968:63; Zangheri 1980: 121.

20. Zangheri 1980:114.

21. Lego 1968a:93; Lego 1968b:2, 7.

22. Lego 1968a:94–95; Lego 1968b:7, 10.

23. Kretschmer 1968:63.

24. Lego 1968a:96; Lego 1968b:11.

25. Lego 1968a:97–98; Lego 1968b:12.

26. Smith 1976:835–36.

27. Kretschmer, letter of 1 March 1989.

28. Cf. Lego 1968a:98; Lego 1968b:13.

29. Zangheri 1980:109.

30. Grab 1989; Zangheri 1980:109.

31. Klang, 1977:11.

32. Grab 1983–84; Grab 1977:45–65; Zangheri 1980:120–24.

33. Peter Barber, personal communication.

34. Vivoli 1987.

35. Ricci 1983:187.

36. Ricci 1983:187.

37. Ricci 1983:187.

38. Ricci 1983:188.

39. A *giornata* equaled 3,810 square meters, which represented the area plowable by a team of oxen in one day.

40. Ricci 1983:188–89.

41. Ricci 1983:189.

42. Zangheri 1980:95–96.

43. Zangheri 1980:95.

44. Ricci 1983:189.

45. The areas between Piedmont and Milan passed in stages from the control of the Duchy of Milan to that of the Principality of Piedmont between 1713 and 1748. These areas had been included in the Milanese cadaster, which continued in operation until the Piedmont cadaster was introduced there after an edict of 5 December 1775. The Duchy of Aosta was brought into the cadaster on 7 October 1783, up to which point it had had its own taxation system. The area was not mapped, however, until the nineteenth-century Napoleonic cadaster (Ricci 1983:189–90).

46. Ricci 1983:189–90.

47. Ricci 1983:190–91.

48. Scaraffia and Sereno 1976:515, 518, and figs. 48 and 49.

49. Scheider 1968:para. 21; Matilla Tascón 1947.

50. Dickson 1987:vol. 2, pp. 185–210.

51. Dickson, personal communication; Dickson 1987:vol. 2, pp. 211–16, 380–81.

52. Kretschmer 1968:63; Dickson (1987:vol. 2, p. 241 n. 118), however, points to isolated refer-

ences to trained surveyors using compasses in the Bohemian lands.

53. Cf. Grüll 1952:47–54 for further examples of private mapping in connection with the Theresian cadaster.

54. Lego 1968b:23–25; Norz 1967.

55. Dickson 1987:vol. 2, pp. 186, 190.

56. Cf. Wangermann 1969.

57. Wright 1966:131.

58. Wright 1966:132–33.

59. Wright 1966:131–37.

60. Kretschmer 1968:63–64; Lego 1968b; Wright 1966:136–37.

61. Wright 1966:137–42.

62. Cited in Lego 1968b:18.

63. Lego 1968b:15–18.

64. Ulbrich 1967c:171 reports the existence of some extant *Brouillons* in the Esterhàzy-Planarchiv in Eisenstadt bound into the parish cadastral books; Grüll 1952:54 reports the survival of high-quality maps in some parishes, such as those of Gleink and Hargelsberg, drawn in 1788.

65. Grüll 1952:54.

66. Lego 1968b:15; Wright 1966:142.

67. Wright 1966:143–44.

68. Wright 1966:146–48.

69. The main documents from the Josephine cadaster are the *Grundsteuerkatasteroperate,* which are housed in Landesarchive (regional archives). The *Summarien* (taxation figures) are in the Hofkammerarchiv in Vienna (Dickson, personal communication).

70. Kretschmer 1968:63–64; Lego 1968b: 22; but cf. Dickson 1987:vol. 2, p. 230 n. 75 for a discussion of the later use of parts of the Josephine cadaster.

71. Cited in Roskievicz 1873:249–50.

72. Von Nischer-Falkenhof 1937:84; Regele 1955:18; Stavenhagen 1904:19.

73. Von Nischer-Falkenhof 1937:84; Regele 1955:18; Stavenhagen 1904:19.

74. Lego 1968b:25–27.

75. Böhm 1967.

76. Lego 1968b:23, 25.

77. Lego 1968b:29–38.

78. Moritsch 1970:439.

79. Until 1861, lithography involved wetting the original maps, which caused them to shrink and distort.

80. Kretschmer 1968:64; Lahr 1967:Lego 1968b:40; Moritsch 1970:439; Ulbrich 1967a.

81. Moritsch 1970:440–41.

82. Moritsch 1970:445–47.

83. Kretschmer 1968:65; Kamenik 1967: 83–84.

84. Cf. Hruda 1967; Messner n.d.:101.

85. Lego 1968b:45–48.

86. Kretschmer, letter of 1 March 1989.

87. The Franciscan cadastral documents are housed in Landesarchive (regional archives) in Austria. Those for Lower Austria are incomplete, as the *Steuerschätzungsoperate* appear to have been destroyed. The Franciscan cadaster for the Burgenland is said to be in Budapest and to be rather incomplete. The *Steuerschätzungsoperate* and *Parzellenprotokolle* for the Austrian coastland and for Dalmatia are in the cadastral office in Trieste. The maps are for the most part in local cadastral offices. For Carniola the maps and some documents are accessible in the Slovenian Staatsarchiv (National Archive) in Ljubljana (Laibach), but the *Steuerschätzungsoperate,* although they appear to be in the archive, are not accessible. All material for the Lower Styria (Untersteiermark) is in Graz (Moritsch 1969:15–16, 149; Moritsch 1970:447).

88. Moritsch 1970:43; Lego 1968b:38.

89. Moritsch 1970:440.

90. Moritsch 1970:440; cf., e.g., Thomas 1988.

91. Kretschmer 1968:65–66.

92. Hruda 1967; Messner n.d.:101.

93. Stavenhagen 1904:37.

94. Stavenhagen 1904:35.

95. Eddie 1989:844–51.

96. Quoted in Eddie 1989:849.

97. Bernleithner 1971, 1978.

Chapter Six
France

1. Gillispie 1980:3.

2. Gillispie 1980:4.

3. Bloch 1929:392.

4. Konvitz 1987:41.

5. [Bibliothèque Nationale] 1939.

6. Neveux, Jacquart, and Le Roy Ladurie 1975: 215.

7. Bloch 1929.

8. De Dainville 1961; Le Roy Ladurie 1966: vol. 1.

9. Frêche 1971.

10. Fougères 1945:55.

11. Greslé-Bouignol 1973:11.

12. Frêche 1971:331.

13. Pitte 1978.

14. De Dainville 1970; Mousnier 1986.

15. Mousnier 1979:493–94.

16. [Comité des Travaux Historiques et Scientifiques] 1975:7; Guerout 1987:15.

17. Guerout 1987:15; Bloch 1929:66–67.

18. Fréminville 1752:vol. 1, p. 106, cited in Bloch 1929:66.

19. Cited in Desreumaux 1979:1.

20. Guerout 1987:15.

21. Bloch 1929:70.

22. Babeuf 1789:54.

23. Bloch 1929:66.

24. De Dainville 1964:50; Fournioux 1982.

25. Broc 1980:135.

26. De Froidour 1759.

27. Hervé 1960; Pelletier 1988.

28. Very many of the original plans are to be found in *archives départmentales;* cf. Boissiere 1986.

29. Dubois 1978:102.

30. De Coincy 1922–23; de Dainville 1961.

31. [Bibliothèque Nationale] 1668; [Comité des Travaux Historiques et Scientifiques] 1975.

32. De Poorter 1980:12.

33. Marstboom, Bourlon, and Jacobs 1956: 24.

34. The maps of the Savoy cadaster were first brought to the attention of modern historians at the end of the nineteenth century (Bruchet 1896) and were also discussed by Marc Bloch (1929:390–92) in his survey of European cadasters for the *Annales.* The fullest discussion of the antecedents, production, and content of the *cadastre sarde* remains the seminal paper by Paul Guichonnet (1955), which is drawn on extensively in this section. The Musée Savoisien (1980) has published a detailed analysis of the cartographic characteristics of the survey, and Bruchet's pioneering study (1896) was reprinted in 1988 with an introduction by Paul Guichonnet.

35. Guichonnet 1955:262.

36. Jones and Siddle 1982:32.

37. Guichonnet 1955:264–68.

38. Guichonnet 1955:268–70.

39. Dufournet 1978.

40. Guichonnet 1955:274; see Bruchet 1896: xviii–59.

41. Fougères 1945:56–58.

42. Fougères 1945:58.

43. Fougères 1945:61.

44. Fougères 1945:62.

45. Fougères 1945:64.

46. Fournier 1941; Fougères 1945:67–68.

47. Fougères 1945:69–70; Clout 1967, 1969.

48. Huguenin 1970:123–27

49. Willis 1980:328–29.

50. Cited in Albitreccia 1942:29.

51. Cited in Albitreccia 1942:32.

52. Cited in Albitreccia 1942:77.

53. Albitreccia 1942:93.

54. Huguenin 1970:130.

55. Huguenin 1962–63.

56. De Richeprey's journal of observations in this part of France has been published with an introduction by Guilhamon (1952).

57. Cited in Goubert 1954:382–83

58. Munier 1779; Dutillet de Villars 1781; Lamy 1789; de Richeprey 1782 (see n. 56).

59. Aubry-Dubrochet 1790:2.

60. Herbin and Pebereau 1953:16.

61. [Direction Générale des Impôts] 1987:31.

62. [Direction Générale des Impôts] 1987:31.

63. Konvitz 1987:49.

64. [Direction Générale des Impôts] 1987:33.

65. Perpillou 1935:11.

66. Quoted in [Direction Générale des Impôts] 1987:33.

67. Quoted in [Direction Générale des Impôts] 1987:33.

68. Herbin and Pebereau 1953:24.

69. [Direction Générale des Impôts] 1987:33.

70. [Ministre des Finances] 1811.

71. Herbin and Pebereau 1953:23.

72. Herbin and Pebereau 1953:26–27.

73. Bloch 1935; Herbin and Pebereau 1953:30–52.

74. De Poorter 1980; Jouanne 1933; Levron 1937. This section on compiling the *ancien cadastre* is based largely on these accounts and on the general review of Herbin and Pebereau (1953).

75. Herbin and Pebereau 1953:68.

76. Clout and Sutton 1969:217.

77. Lewison 1987:71.

78. Herbin and Pebereau 1953:77.

79. Konvitz 1987:60.

80. Most notably by the geographer Aimé Perpillou; cf. Solle 1987.

81. Herbin and Pebereau 1953:84–85.

82. Guichonnet 1963; Walter 1980.

83. Challe 1973; Mosselmans and Schonaerts 1976:80.

84. Hannes 1967.

85. De Smet 1969.

86. Hannes 1968:137–38.

87. Marstboom, Bourlon, and Jacobs 1956:119–20.

88. Hannes 1968:146; Depuydt 1975; Marstboom, Bourlon, and Jacobs 1956:130–31; see also chapters 2.4, 2.5, and 2.8.

Chapter Seven
England and Wales

1. Lawrence 1985:54.

2. Madge 1938:133–40.

3. Newton 1968:51.

4. Sandell 1971:1.

5. Chambers and Mingay 1966:79–80.

6. Chapman 1987:27.

7. Turner 1980:63–93; Turner 1984a:16–17.

8. Turner 1978:28–32.

9. [House of Commons] 1800:vol. 27, p. 7.

10. Emmison 1937; Harley 1972:34; [Byways and Bridleways Trust] 1988.

11. Turner 1978:39.

12. Turner 1984b:143.

13. Marshall 1804:29.

14. Marshall 1804:345.

15. Marshall 1801:51.

16. Marshall 1801:47.

17. Marshall 1801:47.

18. Marshall 1801:61.

19. 41 Geo III, c. 109 (1801):IV and V.

20. Homer 1766:43–46.

21. Homer 1766:53–54.

22. Beresford 1979; Wordie 1983; Chapman 1984.

23. Chapman 1978.

24. Fletcher 1990.

25. Kain 1986.

26. Kain and Prince 1985:6–15.

27. Cobbett 1912:vol. 2, p. 124.

28. Rudé 1967; Hobsbawm and Rudé 1969; Charlesworth 1979, 1983.

29. [House of Commons] 1837a:vol. 41, pp. 11–16.

30. [House of Commons] 1837a:vol. 41, p. 3.

31. Cited in Whalley 1838:240.

32. [House of Commons] 1837a:vol. 41, p. 10.

33. [House of Commons] 1837a:vol. 41, p. 10.

34. [House of Commons] 1837a:vol. 41, p. 16.

35. Thompson 1968:104–6; Kain 1975a; Kain and Prince 1985:69–86; Beech, 1985.

36. [House of Commons] 1837a:vol. 41, p. 4.

37. [House of Commons] 1837b:vol. 6, pp. 15, 24, 27.

38. Thompson 1968:105.

39. Whalley 1838:194–98.

40. [Tithe Commissioners] 1837.

41. Dowson and Shepherd 1952:12–13.

42. Cited in Dowson and Shepherd 1952:14.

43. [House of Commons] 1850:vol. 32, p. 15.

44. [House of Lords] 1857, 2d sess.: vol. 25.

45. Dowson and Shepherd 1952:37–39.

46. Sweeney and Simson 1967:11–12.

47. Dowson and Shepherd 1952:40.

48. Cited in Dowson and Shepherd 1952:41.

49. Beech 1988.

50. Beech 1988:191.

51. Dowson and Shepherd 1952:15.

52. Dowson and Shepherd 1952:15.

53. In some areas the aristocracy and the church lost their right to exemption from tax on their feudal lands. In other areas they retained their right to exemption, but administrative changes meant that they could no longer evade tax on their nonfeudal lands.

54. Recent research suggests that the tax was collected in various ways in different parts of the country (cf. Beckett 1985:292).

55. Beckett 1980; Baigent 1985:95–107, Baigent 1988:31–32; Gibson and Mills 1983; Turner and Mills 1986.

56. The most famous incidence of this was their defeat in 1733 of Walpole's plan for a general excise and the abolition of the land tax (cf. Beckett 1985:303).

57. Cf. Berg 1985:27 for a table showing the growth of government in the period.

58. The wars and certainly war debts and hence the land tax were in fact fairly continuous (cf. Beckett 1985:299–300).

59. Samuel Pepys, diary entry for 5 October 1666, in Latham 1985:688.

60. The best account of the financial changes of the period is Dickson 1967.

61. The land tax was progressive, but excise duties were highly regressive, as they fell mainly on items of mass consumption like beer, soap, salt, tea, and sugar. By resisting them, the landowners thus increased their own tax burden. However, this was not necessarily clear to them at the time. Many contemporaries thought of customs and excise duties as taxes on luxury and hence on the rich, since items such as silk and brandy were charged. Others thought that all taxes, whether land or excise, were ultimately borne by landowners (cf. Beckett 1985).

62. In particular, landowners from the south wanted those from the north to contribute more (cf. Beckett 1985:290–92).

63. Mathias and O'Brien 1976:614 report that the land tax produced £1.5 million in 1770–79.

64. Mathias and O'Brien 1976:601–50.

65. For a good new general account of the politics of the period, see Williams and Ramsden 1990.

66. Brooks 1974.

67. This was the case, for example, in Bristol, cf. Baigent 1985:92–94 and Baigent 1988:34.

68. Brooks 1974 and Mathias and O'Brien 1976 both contain interesting comparisons between the taxation systems of England and France.

69. [House of Lords] 1854–55:vol. 8, question 128.

70. [House of Commons] 1862:vol. 6.

71. Harley 1979.

72. [House of Commons] 1854:vol. 41, p. 187.

73. [Congrès Général de Statistique] 1853: 138.

74. Oliver 1986:162–251.

75. Harley 1964:18.

76. Kain 1975a.

77. Short 1989; the documents are preserved in the Public Record Office and in local repositories.

Chapter Eight
Colonial Settlement from Europe

1. Christopher 1976a:37–39.

2. Hughes 1979:11.

3. Hughes 1976:68.

4. Hilliard 1987:156.

5. Folkingham 1610:preface.

6. Norwood 1637; Love 1970:15–19.

7. Love 1688:preface.

8. Earle 1975:182–202; Christopher 1976a.

9. Wall 1788; Clendinin 1793; Moore 1796; Dewey 1799; Jess 1799; Little 1799; Flint 1804; Conway 1807; Eliot 1807; Gummere 1814; Anthony 1817; Day 1817; Fairlamb 1818; Hanna 1818.

10. Hayward, in Craven and Hayward 1945: xxxvii.

11. Craven, in Craven and Hayward 1945: xxviii.

12. Cited in Robinson 1957:33.

13. Quoted in Hughes 1979:48.

14. Earle 1975:184.

15. Earle 1975:192.

16. Hughes 1979:52; Westover was the most prosperous plantation in Virginia at this time. Its owner, William Byrd, F.R.S., was an exceptional man.

17. Hughes 1979:52.

18. Cited in Wacker 1975:368.

19. Wacker 1975:370–71.

20. Hughes 1979:72–105.

21. Hughes 1979:120.

22. Hilliard 1982:417.

23. Cited in Merrens 1964:24.

24. Cited in Hilliard 1982:417.

25. Cited in Merrens 1964:25.

26. Pett-Conklin 1986:28–31.

27. Roome 1883:45, cited in Wacker 1975: 369–70.

28. Hilliard 1982.

29. Cited in Marschner 1959:14.

30. Hammon 1980.

31. Sherman 1925:31.

32. Clark 1968:79–91.

33. Harris 1966:3–4.

34. Thomson 1966:48–49.

35. Thomson 1966:76.

36. Harris 1987:115.

37. Harris 1966:23.

38. Harris 1966:127.

39. Gaudet 1927.

40. Clark 1968:198.

41. Cited in Clark 1968:199.

42. Clark 1968:341–43.

43. Thomson 1966:120.

44. Eaton 1981:16.

45. Hall 1970:6–47.

46. Hilliard 1973.

47. Hall 1970:47.

48. Jordan 1974:82.

49. Hall 1970:98–110.

50. "Parish" is the local terminology for county in this area.

51. French 1978:32–56.

52. Hébert 1987.

53. McManis 1975:42–46; Wood 1978:17–22, 58–153.

54. Cronon 1983:70–72.

55. Marschner 1960:25.

56. Thrower 1966:10.

57. Garvin 1980.

58. Quoted in Marschner 1959:32.

59. Lemon 1972:50.

60. Marschner 1959:31–33; Lemon 1972: 50–59.

61. Cited in Love 1970:9–10.

62. Cited in Love 1970:81.

63. Wacker 1975:221–72.

64. Quoted in Wacker 1975:300.
65. Thrower 1966:11.
66. Freund 1946.
67. Marschner 1960:42.
68. Cited in Marschner 1959:15.
69. White 1983:11.
70. Meinig 1986:342.
71. Cited in Pattison 1961:339.
72. Pattison 1961:339–45.
73. Cited in Johnson 1976:43.
74. Pattison 1957:75–76.
75. Cited in Rohrbough 1968:5.
76. Cited in Johnson 1976:44.
77. Johnson 1976:44–45. The full text of the 1785 ordinance is reprinted in White 1983:11–14.
78. Ernst 1958:33–78; Pattison 1959.
79. Cazier 1976:35–103.
80. Coles 1957.
81. Jordan 1974, 1982.
82. Uzes 1977:145–54.
83. Moore 1796:iv.
84. Hilliard 1987:157–58.
85. Grim 1985.
86. The instructions are described and many are reprinted in White 1983; see Stewart 1935.
87. Carstensen 1963:xvi.
88. Rohrbough 1968:295.
89. Johnson 1957.
90. [U.S. Bureau of Land Management] 1947:10.
91. Johnson 1976:78.
92. Meinig 1986:412.
93. Carstensen 1963:xvi.
94. Pattison 1957:133.
95. Pattison 1956:11.
96. See, for example, Sears 1925–26; Kenoyer 1933; Dick 1936; Schafer 1940; Hughes 1950; Pattison 1956; and for a comparative Canadian example, Clarke and Finnegan 1984.
97. Cited in Thomson 1966:99.
98. Glendinning 1934:232–37.
99. Thomson 1966:115–23.
100. Gentilcore and Donkin 1973.
101. Cited in Thomson 1966:221.
102. Sebert 1980:69.
103. Cited in Gentilcore and Head 1984:78.
104. Gentilcore and Donkin 1973:20–21.
105. Cited in Sebert 1980:80.
106. Harris and Warkentin 1974:247.
107. Warkentin and Ruggles 1970:183.
108. Lester 1963:20–28.
109. MacGregor 1981.
110. Sebert 1980:88–90.
111. Lester 1963:26–27.
112. Wacker 1975:316.
113. Ristow 1977:14; Stephenson 1967:ix; Conzen 1984a:11.
114. Edmonds 1986. Clara le Gear (1950–53) lists some 4,000 different United States county atlases, and Richard Stephenson (1967) some 1,449 county landownership maps in the Library of Congress alone. These numbers are being revised upward by Michael Conzen's research (Conzen 1984b).
115. Conzen 1984a:17; Stephenson 1967:xvi; Ristow 1966; and Thrower 1961, who both draw on Harrington 1879.
116. Thrower 1972:49.
117. Sebert 1980:98–99; Gentilcore and Head 1984:79, 102–5.
118. Quoted in Thrower 1972:50.
119. Ristow 1965:314.
120. Williams 1975:61; Wynn 1977:255.
121. Toms and Plunkett 1982.
122. Williamson 1984b:3.
123. Jeans 1966a:121; Jeans 1972:106; Jeans 1975:7; Powell 1970a:32–33.
124. Jeans 1966a.
125. Cited in Jeans 1966–67:245.
126. Roberts 1924:39–50.
127. Cited in Perry 1963:44.
128. Jeans 1966–67:245.
129. Jeans 1981:227.
130. Jeans 1966b.
131. Roberts 1924:80.
132. Jeans 1978:94; Williamson 1984c.
133. Cited in Williamson 1982:252; the text of the regulations appears in full on pp. 251–53.

134. Cited in Jeans 1978:94–95.
135. Powell 1970a:47–48; Williamson 1984a: 290–91.
136. Jeans 1966a.
137. Heathcote 1965:33–34.
138. Cited in Mitchell 1979:31.
139. Williams 1975.
140. Roberts 1924:83–85.
141. Wakefield 1841:333.
142. Williams 1974:67–68.
143. Williams 1974:99.
144. Dawson's report of 1840 was published in 1841.
145. Williams 1974:82–90.
146. Powell 1972.
147. Toms 1976:205–8.
148. Hogg 1905.
149. Smailes 1966–67; Mitchell 1979.
150. Powell 1970a:119–23.
151. Williams 1966:347.
152. Robinson 1974; Powell 1977:75–76.
153. Jourdain 1925.
154. Barton 1980:34–35.
155. B. de Vries 1974:510.
156. Patterson 1984:712–23.
157. Dawson 1841:342.
158. Dawson 1841:342.
159. De Vries 1966:23–36.
160. Marais 1927:109–12; de Vries 1966:38; Patterson 1980.
161. An anonymous outback poet quoted in de Vries 1966:45.
162. Wakefield 1849:50.
163. Wakefield 1849:82.
164. Quoted in de Vries 1966:48–49.
165. De Vries 1966:49–50.
166. Thompson 1875.
167. Thompson 1875:8.
168. Thompson 1875:8.
169. Patterson 1980; Patterson 1984:748–64.
170. Heaney 1957.
171. Binns 1953:16–19; Gole 1983:98–99; Heaney 1968.
172. Edney 1990.
173. Markham 1878:180.
174. Baden-Powell 1907:148–230.
175. Beaglehole 1966:35–54; Cohn 1969; Edney 1990.
176. Cited in Cohn 1969:53.
177. Dowson and Shepherd 1952:24–26; Oswal and Singh 1975:156–58.
178. Phillimore 1945–50:vol. 1, pp. 133–47, and vol. 2, pp. 180–84; Phillimore 1947:203.
179. Cited in Markham 1878:183.
180. Markham 1878:183–88.
181. Barron 1885; Dowson and Shepherd 1952:32.
182. Madan 1971. The Revenue Surveys are held in the National Archives in New Delhi.
183. De Smidt 1970:9; cf. Christopher 1984: 9, which reveals that all this precision stands in stark contrast to the Cape Dutch system of the eighteenth century, where farm claims were marked not by survey but by the settler riding his horse from the centroid of his intended farm for half an hour in various directions to establish boundaries and in so doing laying claim to an area of an average of 3,000 *morgens* (6,750 acres).
184. Christopher 1988:194–95.
185. Carter 1987:113.
186. Candee 1980.

Chapter Nine
Cadastral Maps in the Service of the State

1. Darby 1940:28.
2. Skelton and Summerson 1971.
3. Salgaro 1980.
4. Evelyn 1664:1.
5. Bagrow 1975.
6. De Dainville 1964:50.
7. Cited in Darby 1933:530; see also Harley 1988:284–86.

8. Andrews 1970.

9. Andrews 1985:28–51.

10. Cited in Andrews 1980:189.

11. Andrews 1985:52–82.

12. O'Domhnaill 1943; Andrews 1961, 1962; Baker and Butlin 1973:12.

13. Mosselmans and Schonaerts 1976.

14. Cf. Morgan 1979; Helgerson 1986; and Harley 1989:12, which notes, "Especially where maps are ordered by government . . . it can be seen how they extend and reinforce the legal statutes, territorial imperatives, and values stemming from the exercise of political power."

15. Cf. Stone 1988.

16. [Congrès Général de Statistique] 1853: 138.

Appendix

1. The Deutsche Forschungsgemeinschaft is currently funding a project for the publication of systematic catalogs to many German libraries and some archives (for Passau, e.g., see Haversath and Struck 1986).

2. Katzenberger 1977; Past 1978; Ziegler 1976.

3. Winschiers 1982.

4. Grenacher 1971, 1975.

5. Grenacher 1960; Ulshöfer 1968.

6. Grees 1979:5–8; Bull-Reichenmiller 1971: 35–44.

7. Oehme 1961, especially 33–108.

8. Bull-Reichenmiller 1986, 1971 (the catalog).

9. Musall et al. 1986; Bull-Reichenmiller 1971.

10. Grees 1979.

11. Miedel 1975–76.

12. Musall 1983.

13. Scherzer 1977.

14. Schäfer 1968; Scherzer 1970.

15. Hoffmann 1983; Schumm 1961.

16. Hanke and Degner 1934:5.

17. Bliß 1978a, maps of Bromberg, 1772–1912, Bliß 1978b, maps of Frankfurt an der Oder, 1670–1870; Bliß 1981, maps of Potsdam, 1651–1850; Bliß 1982, maps of Marienwerder, 1670–1919.

18. Vogel and Zögner 1979.

19. Scharfe 1972.

20. Scharfe 1986.

21. Hanke and Degner 1935.

22. Zögner and Zögner 1986.

23. Eckhardt 1961; Schäfer 1979.

24. Engel 1982.

25. Meers 1958.

26. Wolff 1987; Wolff and Engel 1988.

27. Kleinn 1986.

28. Hake 1978.

29. Kleinau 1968; Kleinau and Penners 1956; Pitz 1957; Vorthmann 1956.

30. Lang 1962.

31. Leerhoff 1985.

32. Jordan 1955; Pitz 1967.

33. Grossmann 1955; v. d. Weiden 1955.

34. Kleinn 1964–65, 1986.

35. Sperling 1986.

36. Aymans 1986, 1987; Aymans, Burggraaff and Jansen 1988; Ketter 1929; Schulte 1984.

37. Grenacher 1964:70–71.

38. Kleinn 1976–77.

39. Prinz 1950.

40. Meurer 1981.

41. Müller-Wille 1940.

42. Osthoff 1950.

43. Osthoff 1956.

44. Hellwig 1985; Werner 1985.
45. Bauer 1971.
46. Kahlfuß 1969.
47. Unger 1964.
48. Kahlfuß 1986; Witt 1982.
49. Speiermann 1974.
50. Stams 1986.
51. Beschorner 1921.
52. Bönisch 1970.
53. Richter 1921.
54. Emmerich 1962.
55. Vogel 1929; Zögner 1984.

BIBLIOGRAPHY

Aanrud, Roald. 1977. "Generalforstamtet og norsk kartografi. Et 200 års-minne om Johann Georg von Langen." *Norsk Geografisk Tidsskrift* 31:97–101.

———. 1981. "Gammel kartleggingskunst." In *Våre gamle kart,* edited by Rolf Fladby and Leif T. Andressen, 69–80. Oslo: Norsk Lokalhistorisk Institutt and Universitetsforlaget.

Abel, Thomas. 1761. *Subtensial Plain Trigonometry . . . And this Method Apply'd to Navigation and Surveying.* Philadelphia: Andrew Steuart.

Åberg, Alf. 1973. "The Swedish Army, from Lützen to Narva." In *Sweden's Age of Greatness, 1632–1718,* edited by Michael Roberts, 265–87. London: Macmillan.

Agas, Radulph. 1596. *A Preparative to Platting of Landes and Tenements for Surueigh.* London: Thomas Scarlet.

Ågren, Kurt. 1973. "The *Reduktion.*" In *Sweden's Age of Greatness, 1632–1718,* edited by Michael Roberts, 237–64. London: Macmillan.

Albitreccia, Antoine. 1942. *Le Plan terrier de la Corse au 18e siècle: Etude d'un document géographique.* Paris: Presses Universitaires de la France.

[Alexander Turnbull Library, Wellington]. 1980. *The Surveying and Mapping of Wellington Province, 1840–76: An Exhibition of Manuscript Maps, Field Books. . . .* Wellington: Alexander Turnbull Library.

Anderson, Perry. 1974. *Passages from Antiquity to Feudalism.* London: New Left Books.

Andressen, Leif T. 1982. "Om bruksstørrelse i lys av skyld og skatt." In *Den eldeste matrikkelen: En infallsport til historien: Skattematrikkelen 1647,* edited by Rolf Fladby and Harald Winge, 50–58. Oslo: Norsk Lokalhistorisk Institutt and Universitetsforlaget.

Andrews, John H. 1961. *Ireland in Maps.* Dublin: Dolmen Press.

———. 1962. "Ireland in Maps: A Bibliographical Postscript." *Irish Geography* 4:234–43.

———. 1970. "Geography and Government in Elizabethan Ireland." In *Irish Geographical Studies in Honour of E. Estyn Evans,* edited by Nicholas Stephens and Robin E. Glasscock, 178–91. Belfast: Department of Geography, The Queen's University.

———. 1980. "Henry Pratt, Surveyor of Kerry Estates." *Journal of the Kerry Archaeological and Historical Society* 13:5–31.

———. 1985. *Plantation Acres: An Historical Study of the Irish Land Surveyor and His Maps.* Ulster Historical Foundation.

Anthony, Daniel. 1817. *The Guaging Inspector, and Measurer's Assistant. . . .* Providence: Miller and Hutchens.

[Archives de France]. 1987. *Espace français: Vision et aménagement, XVIe–XIXe siècles.* Paris: Archives Nationales.

[Archivio di Stato di Milano]. 1988. *L'immagine interessata: Territorio e cartografia in Lombardia tra 500 e 800.* Como: New Press.

Arrhenius, Olof. 1945. "Fördelning av 1600-talets geometriska kartor över Sverige." *Globen* 24, no. 2:29–32.

Aubry-Dubrochet, Pierre F. 1790. *Exécution du cadastre général de la France et d'un cadastre provisoire pour la répartition des impôts en 1791.* Paris: Imprimerie Nationale.

Aymans, Gerhard. 1985. "Die handschriftliche Karte als Quelle geographischer Studien." *Landkarten als Geschichtsquellen.* Landschaftsverband Rheinland–Archivberatungsstelle Rheinland 16:21–46. Cologne: Rheinland Verlag.

———. 1986. "Die preußische Katasteraufnahme im Herzogtum Kleve der Jahre 1731–38." *Erdkunde* 40:14–28.

———. 1987. "Das Kataster im Herzogtum Kleve der Jahre 1731–38 und der Fall Gennep." *Publications*:162–84.

Aymans, Gerhard; Burggraaff, Peter; and Jansen, Wolfgang. 1988. *De regio Gennep aan de Ketting: Gennep, Heijen, Milsbeek, Oeffelt, Ottersum, Ven-Zelderheide in katasterkaarten (1731–1732).* Venray: Gemeente Gennap, Gemeente Oeffelt and Stichting Historie Peel-Maas-Niersgebied.

Baars, C. 1979. "Geschiedenis van de bedijking van het deltagebied." *Landbouwkundig Tijdschrift* 91, no. 2:29–36.

Babeuf, François N., and Audiffred, J. P. 1789. *Cadastre perpetuel.* Paris: Garnery and Volland.

Baden-Powell, Baden H. 1907. *A Short Account of the Land Revenue and Its Administration in British India, with a Sketch of the Land Tenures.* 2d ed. Oxford: Clarendon Press.

Bagger-Jörgensen, L. Olof. 1922a. "Lantmäteriet." In *Sveriges kartläggning: En översikt,* by L. Olof Bagger-Jörgensen et al., 5–10. Stockholm: Stockholm Kartografiska Sällskapet.

———. 1922b. "Lantmäteriets organisation." In *Sveriges kartläggning: En översikt,* by L. Olof Bagger-Jörgensen et al., 12–62. Stockholm: Stockholm Kartografiska Sällskapet.

Bagrow, Leo. 1975. *A History of Russian Cartography up to 1800.* Edited by H. W. Castner. Wolfe Island, Ontario: Walker Press.

Baigent, Elizabeth. 1985. "Bristol Society in the Later Eighteenth Century, with Special Reference to the Handling by Computer of Fragmentary Historical Sources." University of Oxford D.Phil. thesis.

———. 1988. "Assessed Taxes as Sources for the Study of Urban Wealth: Bristol in the Later Eighteenth Century." *Urban History Yearbook,* 31–48.

———. 1990. "Swedish Cadastral Mapping, 1628–1700: A Neglected Legacy." *Geographical Journal* 156:62–69.

Baker, Alan R. H., and Butlin, Robin A., eds. 1973. *Studies of Field Systems in the British Isles.* Cambridge: Cambridge University Press.

Balslev, Svend. 1981. "Nogle landinspektører og landmålere bag udskiftningerne." In *Udskiftningsforordning 200 år = Landinspektøren* 30, no. 8), 508–16.

Balslev, Svend; Holst, Anne; and Rydahl, Steen, eds. 1986. *Håndbog om matrikelvæsen: Matrikeldirektoratets opgaver og organisation.* Copenhagen: Landbrugsministeriet.

Balslev, Svend, and Jensen, Hans Ejner. 1975. *Landmåling og landmålere: Danmarks økonomiske opmåling.* Vol. 1, text; vol. 2, maps. Copenhagen: Den Danske Landinspektørforening.

Barron, W. 1885. "The Cadastral Survey of India." *Royal Geographical Society Supplementary Papers* 1, part 4:597–618.

Barton, Phil L. 1980. "The History of the Mapping of New Zealand." *Map Collector* 11:28–35.

Bateman, John. 1883. *The Great Landowners of Great Britain and Ireland.* London: Harrison.

Baudot, Marcel. 1937. *Archives départementales de l'Eure, répertoire numérique des cartes et plans.* Evreux: Imprimerie Hérissey.

Bauer, Erich. 1971. "Siegmund Jakob Haeckher als Architekt und Landmesser in der Herrschaft Trippstadt." *Pfälzer Heimat* 22, no. 2: 68–73.

Beaglehole, Timothy H. 1966. *Thomas Munro and*

the Development of Administrative Policy in Madras, 1792–1818. Cambridge: Cambridge University Press.

Beckett, John V. 1980. *Local Taxation: National Legislation and the Problems of Enforcement.* London: Bedford Square Press.

———. 1985. "Land Tax or Excise: The Levying of Taxation in Seventeenth- and Eighteenth-Century England." *English Historical Review* 100:285–308.

Beech, Geraldine. 1985. "Tithe Maps." *Map Collector* 33:20–25.

———. 1988. "Cartography and the State: The British Land Registry Experience." *Journal of the Society of Archivists* 9:190–96.

[Beemster]. 1613. *Extract uyt het Octroy vande Beemster met de Cavel-Condition en de Caerten van Dien. 't Register vande Participanten. . . .* Pascart: Geeritszoon.

Belonje, J. 1935. "Twee bijzondere landkaarten." *West Friesland Oud en Nieuw,* 70–76.

Bendall, A. Sarah. 1989. "The Mapping of Rural Estates: A Case-Study of Cambridgeshire *c.* 1600–1836." University of Cambridge Ph.D. thesis.

Beresford, Maurice W. 1971. *History on the Ground: Six Studies in Maps and Landscapes.* Rev. ed. London: Methuen. Orig. pub. London: Lutterworth Press, 1957.

———. 1979. "The Decree Rolls of Chancery as a Source for Economic History." *Economic History Review,* 2d ser., 32:1–10.

Berg, Maxine. 1985. *The Age of Manufactures: Industry, Innovation, and Work in Britain, 1700–1820.* Totowa, N.J.: Barnes and Noble.

Bernleithner, Ernst. 1949. "Die Entwicklung der österreichischen Ländekunde an der Wende des 18. und 19. Jahrhunderts." University of Vienna Ph.D. thesis.

———. 1953. "Niederösterreich im Kartenbild der Zeiten." *Unsere Heimat* 24:188–97.

———. 1959. "Die Entwicklung der Kartographie in Österreich." *Berichte zur Deutschen Landeskunde* 22, no. 2:191–224.

———. 1971. "Austria's Share in World Cartography." *Imago Mundi* 25:65–73.

———. 1978. "Österreichs Kartographie zur Zeit des Grafen Ferraris." In *De cartografie in de 18de eeuw en het werk van Graaf de Ferraris (1726–1814) / La Cartographie au XVIIIe siècle et l'œuvre du Comte de Ferraris (1726–1814),* 129–48. Gemeentekrediet van Belgie. Historische Uitgaven Pro Civitate / Crédit Communal de Belgique. Collection Histoire Pro Civitate 8° Series, no. 54.

Beschorner, Hans. 1921. "Landvermessung und Kartenwesen Kursachsens bis 1780." In *Beiträger zur deutschen Kartographie,* edited by Hans Beschorner, 32–46. Leipzig: Akademische Verlagsgesellschaft.

[Bibliothèque Nationale]. 1668. *Plans des forêts . . . de la grande maîtrise . . . de l'Isle de France . . . Louis XIV.* Paris: Bibliothèque Nationale.

———. 1939. *Les Travaux et les jours dans l'ancienne France. Exposition organisée . . . pour commémorer le IVe centenaire d'Olivier de Serres.* Paris: Bibliothèque Nationale.

Bigwood, Georges. 1898. "Matricules et cadastres: Aperçu sur l'organisation du cadastre en Flandre, Brabant, Limbourg, et Luxembourg avant la domination française." *Annales de la Société d'Archéologie de Bruxelles* 12:388–411.

———. 1906. "Les Emprunts à lots aux Pays-Bas autrichiens." *Annales de la Société d'Archéologie de Bruxelles* 20:439–56.

———. 1912. "La Loterie aux Pays-Bas autrichiens." *Annales de la Société Royale d'Archéologie de Bruxelles* 26:53–134.

Billingsley, John. 1802. "An Essay on the Best Method of Inclosing, Dividing, and Cultivating Waste Lands." *Letters to the Bath and West of England Agricultural Society* 11:1–93.

Binns, Sir Bernard. 1953. *Cadastral Surveys and Records of Rights in Land.* Rome: Food and Agriculture Organization of the United Nations.

Björkvik, Halvard. 1956. "The Farm Territories, Habitation and Field Systems, Boundaries, and Common Ownership in the Old Norwegian Peasant Community." *Scandinavian Economic History Review* 4, no. 1:33–61.

Bjørn, Claus. 1977. "The Peasantry and Agrarian Reform in Denmark." *Scandinavian Economic History Review* 25, no. 1:117–37.

———. 1981. "Udskiftningsforordningen af 23. april 1781 og dens plads i dansk landbrugs historie." In *Anledning at 200 året: 1781–23. april—1981.* 5–30. Copenhagen: Landbrugsministeriet: Matrikeldirektorat.

Blakemore, Michael J., and Harley, J. Brian. 1980. *Concepts in the History of Cartography: A Review and Perspective.* Monograph 26, *Cartographica* 17.

Bliß, Winfried. 1978a. *Die Plankammer der Regierung Bromberg: Spezialinventar 1772 bis 1912.* Veröffentlichungen aus den Archiven Preußischer Kulturbesitz 16. Cologne, Vienna: Böhlau Verlag.

———. 1978b. *Die Plankammer der Regierung Frankfurt an der Oder: Spezialinventar 1670 bis 1870.* Veröffentlichungen aus den Archiven Preußischer Kulturbesitz 15. Cologne, Vienna: Böhlau Verlag.

———. 1979. "Die Kartenabteilung des Geheimen Staatsarchivs Preußischer Kulturbesitz." In *Preußen im Kartenbild,* edited by Werner Vogel and Lothar Zögner, 8–13. Berlin: Geheimes Staatsarchiv Preußische Kulturbesitz.

———. 1981. *Die Plankammer der Regierung Potsdam: Spezialinventar 1651 bis 1850.* Veröffentlichungen aus den Archiven Preußischer Kulturbesitz 18. Cologne, Vienna: Böhlau Verlag.

———. 1982. *Die Plankammer der Regierung Marienwerder: Spezialinventar 1670 bis 1919.* Veröffentlichungen aus den Archiven Preußischer Kulturbesitz 19. Cologne, Vienna: Böhlau Verlag.

Bloch, Marc. 1929. "Les Plans parcellaires." *Annales d'Histoire Economique et Sociale* 1:60–70, 390–98.

———. 1935. "Une Nouvelle Image de nos terroirs: La mise à jour du cadastre." *Annales d'Histoire Economique et Sociale* 7:156–59.

Blumfeldt, Evald. 1934. "Den svenska tiden i estnisk historieforskning." *Äratrükk Akad. Rootsi-Eesti Seltsi Aastaraamatust,* 124–38.

———. 1958. "Reduktionen på Ösel." *Svio-Estonica* 14 (n.s. 5): 109–71.

Böhm, Josef. 1967. "Katastrale Lage- und Höhenaufnahmen." In *150 Jahre österreichischer Grundkataster, 1817–1967: Ausstellungskatalog,* edited by Robert Messner, 35–53. Vienna: Bundesamt für Eich- und Vermessungswesen.

Böhme, Helmut. 1976. *An Introduction to the Social and Economic History of Germany: Politics and Economic Change in the Nineteenth and Twentieth Centuries.* Oxford: Basil Blackwell.

Boissiere, Jean. 1986. "Espaces et paysages forestiers à travers la cartographie et les archives forestières de la période moderne: Introduction." *Hommes et Terres du Nord* 2–3:200–203.

Bönisch, Fritz. 1970. *Genauigkeitsuntersuchungen am öderschen Kartenwerk von Kursachsen.* Abhandlungen der sächsischen Akademie der Wissenschaften zu Leipzig. Philologisch-historische Klasse 61, 3. Berlin: Akademie-Verlag.

Boom, H. ten. 1982. "Geschiedenis van de kerkelijke instellingen in Nederland." The Hague: Rijksarchiefschool. Typescript.

Borchardt, Knut. 1978. "Germany, 1700–1914." In *The Fontana Economic History of Europe*; vol. 1, *The Emergence of Industrial Societies,* edited by Carlo M. Cipolla, 76–160. Glasgow: William Collins.

Borgedal, Paul. 1959. "Jordeiendommenes historie i Norge." In *Jordskifteverk gjennom 100 år, 1859–1958,* edited by Torleif Grendahl, 9–166. Oslo: Det Kgl. Landbruksdepartement.

Brandt, Marie Louise. 1987a. "Kartografiske kilder til information om landskabsmønsteret før udskiftningen." *Landinspektøren* 33, no. 7:358–66.

———. 1987b. "Signaturer i eldre økonomiske kort. I: Den specielleste opmaaling, 1768–1776." Typescript.

Brink, Paul P. W. J. van den. 1982. "De kaartenverzameling van de Holland Land Company (1789–1869)." *Caert-Thresoor* 1, no. 4:54–56.

Broc, Numa. 1980. *La Géographie de la Renaissance, 1420–1620.* Paris: Bibliothèque Nationale.

Brooks, Colin. 1974. "Public Finance and Political Stability: The Administration of the Land Tax, 1688–1720." *Historical Journal* 17:281–300.

Bruchet, Max. 1896. *Notice sur l'ancien cadastre de Savoie.* Annecy: Abry. Reprinted with an introduction by Paul Guichonnet. Annecy: Archives de la Haute-Savoie, 1988.

Bruford, Walter H. 1965. "German Constitutional and Social Development, 1795–1830." In *The New Cambridge Modern History;* vol. 9, *War and Peace in an Age of Upheaval, 1793–1830,* edited by Charles W. Crawley, 367–94. Cambridge, Cambridge University Press.

———. 1971. *Germany in the Eighteenth Century: The Social Background of the Literary Revival.* Cambridge: Cambridge University Press.

Bugge, Thomas. 1795–98. *Mathematiske forelæsningar, I–II deel. De første grunde til regning. Geometri plantrigonometri og landmaaling.* Copenhagen.

Buisseret, David. 1988. *Rural Images: The Estate Plan in the Old and New Worlds.* Chicago: Newberry Library.

Bull-Reichenmiller, Margareta. 1971. "Schwäbisches Land in alten Karten und Plänen." Typescript.

———. 1986. "Südwestdeutschland." In *Lexikon zur Geschichte der Kartographie von den Anfängen bis zum Ersten Weltkrieg,* edited by Ingrid Kretschmer, Johannes Dörflinger, and Franz Wawrik, 796–801. Vienna: Franz Deuticke.

Bydén, Artur. 1919. "Några anteckningar rörande 1600-talets svenska kartor." *Ymer* 39, no. 2:100–122.

[Byways and Bridleways Trust]. 1988. *Inclosure Acts and Awards.* London: Byways and Bridleways Trust.

Candee, Richard, M. 1980. "Land Surveys of William and John Godsoe of Kittery, Maine: 1689–1769." In *New England Prospect: Maps, Place Names, and the Historical Landscape,* edited by Peter Benes, 9–46. Proceedings of the Dublin Seminar for New England Folklife, 1980. Boston: Boston University.

Carstensen, Vernon, ed. 1963. *The Public Lands: Studies in the History of the Public Domain.* Madison: University of Wisconsin Press. Reprinted 1968.

Carter, John. 1774. *The Young Surveyor's Instructor; or, An Introduction to the Art of Surveying.* Philadelphia: W. and T. Bradford.

Carter, Paul. 1987. *The Road to Botany Bay: An Essay in Spatial History.* London: Faber and Faber.

Cazier, Lola. 1976. *Surveys and Surveyors of the Public Domain, 1785–1975.* Washington, D.C.: U.S. Government Printing Office.

Challe, J. P. 1973. "Le Cadastre primitif et son utilisation." *Bulletin Trimestriel du Crédit Communal de Belgique* 105:149–64.

Chambers, John D., and Mingay, Gordon E. 1966. *The Agricultural Revolution, 1750–1880.* London: Batsford.

Chapman, John. 1978. "Some Problems in the Interpretation of Enclosure Awards." *Agricultural History Review* 26:108–14.

———. 1984. "The Chronology of English Enclosure." *Economic History Review* 37:557–59.

———. 1987. "The Extent and Nature of Parliamentary Enclosure." *Agricultural History Review* 35:25–35.

Charlesworth, Andrew. 1979. *Social Protest in a Rural Society.* Research Series no. 1. Norwich: Historical Geography Research Group.

———, ed. 1983. *An Atlas of Rural Protest in Britain, 1548–1900.* London: Croom Helm.

Christopher, Anthony J. 1976a. *Southern Africa.* Folkestone: William Dawson.

———. 1976b. "The Variability of the South African Standard Farm." *South African Geographical Journal* 58:107–17.

———. 1984. *Crown Lands of British South Africa, 1853–1914.* Kingston, Ontario: Limehouse Press.

———. 1988. *The British Empire at Its Zenith.* London: Croom Helm.

Clapham, John H. 1928. *The Economic Development of France and Germany, 1815–1914.* Cambridge: Cambridge University Press.

Clark, Andrew H. 1968. *Acadia: The Geography of Early Nova Scotia to 1700.* Madison: University of Wisconsin Press.

Clarke, John, and Finnegan, Gregory F. 1984. "Colonial Survey Records and the Vegegation of Essex County, Ontario." *Journal of Historical Geography* 10:119–38.

Clendinin, John. 1793. *The Practical Surveyor's Assistant. . . .* Philadelphia: Benjamin Johnson.

Clout, Hugh D. 1967. "The Pays de Brays: A Study of Land Use Change, 1750–1965." University of London M.Phil. thesis.

———. 1969. "Structures agraires et utilisations agricoles du sol dans le Bray au XVIIIe siècle." *Acta Geographica,* jan–mars, 13–22.

Clout, Hugh D., and Sutton, Keith. 1969. "The *Cadastre* as a Source for French Rural Studies." *Agricultural History* 43:215–23.

Cobbett, William. 1912. *Rural Rides.* London: Dent.

Cohn, Bernard S. 1969. "Structural Change in Indian Rural Society, 1596–1885." In *Land Control and Social Structure in Indian History,* edited by Robert E. Frykenberg, 53–121. Madison: University of Wisconsin Press.

Coincy, H. de. 1922–23. "Les Archives toulousaines de la réformation générale des eaux et forêts." *Bibliographe Moderne* 21:89–182.

Coles, Harry L. 1957. "Applicability of the Public Land System to Louisiana." *Mississippi Valley Historical Review* 43:39–58.

[Comité des Travaux Historiques et Scientifiques, 100e Congrès National des Sociétés Savantes, Paris, 1975]. 1975. *Exposition: De l'Isle de France rurale à la grande ville.* Paris: Bibliothèque Nationale.

[Congrès Général de Statistique]. 1853. *Compte rendu des travaux du Congrès Général de Statistique réuni à Bruxelles les 19, 20, 21, et 22 septembre 1853.* Brussels: M. Hayez.

Conway, Miles W. 1807. *Geodaesia; or, A Treatise of Practical Surveying . . . Made for the Use of the Western Surveyors in Particular. . . .* Lexington, Ky.: Daniel Bradford.

Conze, Werner. 1969. "The Effects of Nineteenth-Century Liberal Agrarian Reforms on Social Structure in Central Europe." In *Essays in European Economic History, 1789–1914,* edited by Francois M.-J. Crouzet, William H. Chaloner, and Walter M. Stern, 53–81. London: Edward Arnold.

Conzen, Michael P. 1984a. "The County Landownership Map in America, Its Commercial Development and Social Transformation, 1814–1939." *Imago Mundi* 36:9–31.

———. 1984b. "Landownership Maps and County Atlases." *Agricultural History* 58:118–22.

Coppens, H. 1968. "Oud en nieuw kadaster." *Spiegel Historiael* 3, no. 11:597–603.

Craven, Wesley F., and Hayward, Walter B., eds. 1945. *The Journal of Richard Norwood, Surveyor of Bermuda.* New York: Scholar's Facsimiles and Reprints for Bermuda Historical Monuments Trust.

Cronon, William. 1983. *Changes in the Land: Indians, Colonists, and the Ecology of New England.* New York: Hill and Wang.

Curschmann, Fritz. 1935. "Die schwedischen Matrikelkarten von Vorpommern und ihre wissenschaftliche Auswertung." *Imago Mundi* 1:52–57.

———. 1938. "Die schwedische Matrikelkarten von Vorpommern und ihre Bedeutung für die Siedlungs-, Sozial-, und Wirtschaftsgeschichte des Landes." *Beiträge zur Raumforschung und Raumordnung* 1:165–75.

Dahl, Sven. 1961. "Strip Fields and Enclosure in Sweden." *Scandinavian Economic History Review* 9, no. 1:56–67.

Dainville, François de. 1961. *Cartes anciennes du Languedoc, XVIe–XVIIIe siècles.* Montpellier: Société Languedocienne de Géographie.

———. 1964. *Le Langage de géographes: Termes, signes, couleurs des cartes anciennes, 1500–1800.* Paris: Editions A. and J. Picard.

———. 1970. "Cartes et contestations au XVe siècle." *Imago Mundi* 24:99–121.

Darby, Sir H. Clifford. 1933. "The Agrarian Contribution to Surveying in England." *Geographical Journal* 82:529–35.

———. 1940. *The Draining of the Fens.* Cambridge: Cambridge University Press.

Darby, Sir H. Clifford, and Fullard, Harold, eds. 1978. *The New Cambridge Modern History.* Vol. 14, *Atlas.* Cambridge: Cambridge University Press.

Davis, Walter W. 1974. *Joseph II: An Imperial Reformer for the Austrian Netherlands.* The Hague: Martinus Nijhoff.

Dawson, Robert K. 1841. "Report on Surveying; Considered with reference to New Zealand, and Applicable to the Colonies Generally." *British Parliamentary Papers, House of Commons,* 1st sess., vol. 4, Appendix to "Report from the Select Committee on South Australia."

Day, Jeremiah. 1817. *The Mathematical Principles of Navigation and Surveying, with the Mensuration of Heights and Distances.* New Haven, Conn.: Steele and Gray.

Depuydt, F. 1975. "The Large Scale Mapping of Belgium, 1800–1850." *Imago Mundi* 27:23–26.

Derry, T. K. 1965. "Scandinavia." In *The New Cambridge Modern History;* vol. 9, *War and Peace in an Age of Upheaval, 1793–1830,* edited by Charles W. Crawley, 480–94. Cambridge: Cambridge University Press.

Desreumaux, Roger. 1979. "Relations entre les arpenteurs et leurs employeurs à propos de plans terriers." Paper presented at the Eighth International Conference on the History of Cartography, Berlin.

Dewey, Solomon. 1799. *A Short and Easy Method of Surveying . . . to Which is Added the Square Root.* Hartford: Elisha Babcock.

Dick, W. Bruce. 1936. "A Study of the Original Vegetation of Wayne County, Michigan." *Papers of the Michigan Academy of Science, Arts, and Letters* 22:329–34.

Dickson, Peter G. M. 1967. *The Financial Revolution in England: A Study in the Development of Public Credit, 1688–1756.* London: Macmillan.

———. 1987. *Finance and Government under Maria Theresia, 1740–1780.* Vol. 1, *Society and Government;* vol. 2, *Finance and Credit.* Oxford: Oxford University Press.

Diebels, Peter. 1986. *Beschrijving van de oudste kaarten, 1457–1580.* Leiden: Het Hooheemraadschap van Rijnland.

Dilke, Oswald A. W. 1971. *The Roman Land Surveyors: An Introduction to the Agrimensores.* Newton Abbot: David and Charles.

———. 1985. *Greek and Roman Maps.* London: Thames and Hudson.

———. 1987. "Maps in the Service of the State: Roman Cartography to the End of the Augustan Era." In *The History of Cartography;* vol. 1, *Cartography in Prehistoric, Ancient, and Medieval Europe and the Mediterranean,* edited by J. Brian Harley and David Woodward, 201–11. Chicago: University of Chicago Press.

[Direction Générale des Impôts]. 1987. "Le Cadastre moderne." In *Espace français: Vision et aménagement, XVIe–XIXe siècles,* 31–34. Paris: Archives Nationales.

Donkersloot–de Vrij, Ypkje Marijke. 1981. *Topografische kaarten van Nederland vóór 1750: Handgetekende en gedrukte kaarten, aanwezig in de Nederlandse Rijksarchieven.* Groningen: Wolters-Noordhoff / Bouma's Boekhuis.

Dörflinger, Johannes. 1979. "Die österreichische Kartographie vom spanischen Erbfolgerkrieg bis nach dem Wiener Kongress unter besonderer Berücksichtigung der Privatkartographie zwischen 1780 and 1820." 3 vols. University of Vienna *Habilitation* thesis.

———. 1984. *Die österreichische Kartograhie im 18. und bis zum Beginn des 19. Jahrhunderts unter besonderer Berücksichtigung der Privatkartographie zwischen 1780 und 1820.* Vol. 1, *Österreichische Karten des 18. Jahrhunderts.* Vienna: Österreichische Akademie der Wissenschaften.

Dowson, Sir Ernest, and Sheppard, Vivian L. O.

1952. *Land Registration.* London: Her Majesty's Stationery Office.

Drolshagen, Carl. 1920–23. *Die schwedische Landesaufnahme und Hufenmatrikel von Vorpommern als ältestes deutsches Kataster.* 2 vols. Beiheft zum 37./38. und 40./41. *Jahresbericht der Geographischen Gesellschaft Greifswald.* Greifswald.

Droysens, G. 1886. *Allgemeine historische Handatlas.* Bielefeld and Leipzig: Verlag von Velhagen and Klasing.

Dubois, Jean-Jacques. 1978. "Les Plans des forêts de la Région du Nord au XVIIe–XVIIIe siècles: Quelques remarques sur leur utilisation." *Bulletin de la Section de Géographie* (Comité des Travaux Historiques et Scientifiques) 82:101–26.

Dufournet, P. 1978. *Une Communauté agraire secrète et organise son territoire, à Bassy (Province de Genevois, Haute Savoie): Contribution à la connaissance du paysage historique.* Paris: Picard.

Duhamel du Monceau, Henri L. 1750–61. *Traité de la culture des terres, suivant le principes de M. Tull.* 6 vols. Paris: H.-L. Guérin and L.-F. Delatour.

Dunsdorfs, Edgars. 1950. *Der große schwedische Kataster in Livland, 1681–1710.* Kungl. Vitterhets Historie och Antikvitets Akademiens Handlingar 72. Stockholm: Almqvist and Wiksell.

———. 1974. *Der große schwedische Kataster in Livland, 1681–1710.* Part 2, *Kartenband.* Stockholm: Almqvist and Wiksell.

———. 1981. *The Livonian Estates of Axel Oxenstierna.* Stockholm: Almqvist and Wiksell.

Dutillet de Villars. 1781. *Précis d'un projet d'établissement du cadastre dans le royaume.* Paris: Imprimerie de Clousier.

Duvosquel, Jean-Marie. 1986. "La Correspondance du Duc Charles de Croÿ avec son arpenteur Pierre de Bersacques (1598–1606) à propos de la confection des cadastres de ses seigneuries." In *Oude kaarten en plattegronde: Bronnen voor de historische geografie van de zuidelijke Nederlanden (16de–18de eeuw) / Cartes et plans anciens: Sources pour la géographie historique de Pays-Bas méridionaux (XVIe–XVIIIe siècles),* edited by H. van der Haegen, F. Daelemans, and Eduard van Ermen, 11–50. Archief- en Bibliotheekwezen in België / Archives et Bibliothèques de Belgique, special no. 31. Brussels.

Earle, Carville V. 1975. *The Evolution of a Tidewater Settlement System: All Hallow's Parish, Maryland, 1650–1783.* Department of Geography, Research Paper no. 170. Chicago: University of Chicago.

Eaton, E. L. 1981. "The Survey Plan of Cornwallis Township, Kings County." *Nova Scotia Historical Review* 1:16–33.

Eckhardt, Wilhelm Alfred. 1961. "Wilhelm Dilichs Zehntkarte von Niederzwehren." *Zeitschrift des Vereins für hessische Geschichte und Landeskunde* 72:99–121.

Eddie, Scott M. 1989. "Economic Policy and Economic Development in Austria-Hungary, 1867–1913." In *The Cambridge Economic History of Modern Europe;* vol. 8, *The Industrial Economies: The Development of Economic and Social Policies,* edited by Peter Mathias and Sidney Pollard, 814–86. Cambridge: Cambridge University Press.

Eden, Peter. 1983. "Three Elizabethan Estate Surveyors: Peter Kempe, Thomas Clerke, and Thomas Langdon." In *English Map-Making, 1500–1650,* edited by Sarah Tyacke, 68–84. London: British Library.

Edmonds, Michael. 1986. "The U.S. General Land Office and Commercial Map Making." *Government Publications Review* 13:571–80.

Edney, Matthew. 1989. "The Ordnance Survey and British Surveys in India." *Sheetlines* 26:3–8.

———. 1990. "Mapping and Empire: British Trigonometrical Surveys in India and the European Concept of Systematic Survey, 1799–1843." University of Wisconsin-Madison Ph.D. thesis.

Eggen, Brynjulf. 1987. "Statistikk på gamle kart." *A La Kart: Bedriftsavis for Statens Kartverk* 1, no. 1:10–11.

Ehrensvärd, Ulla. 1984. "Cartographical Representation of the Scandinavian Arctic Regions." *Arctic* 37, no. 4:552–61.

———. 1986a. "Fortifikationsofficeren som kartograf." In *Fortifikationen: 350 år, 1635–1985,* edited by Bertil Runnberg, with Sten Carlsson, 109–24. Stockholm: Fortifikationenskåren.

———. 1986b. "Sockenkartverket och rikets ekonomiska kartverk." In *Ekonomisk karta över Skaraborgs Län, 1877–1882,* 1–16. Skara: Skaraborgslänsmuseum, supplement to new edition.

Ekstrand, K. Victor. 1901–5. *Samlingar i landtmäteri.* Vol. 1, *Instruktioner och bref, 1628–1699;* vol. 2, *Bilder ur landtmäternes lif;* vol. 3, *Förteckining öfver landtmäteri författningar, 1628–1904.* Stockholm: Isaac Marcus.

Elliot, William. 1807. *A Traverse Table for the Use of Surveyors.* New Brunswick: the author.

Embrey, Ainslie T. 1969. "Landholding in India and British Institutions." In *Land Control and Social Structure in Indian History,* edited by Robert E. Frykenberg, 33–52. Madison: University of Wisconsin Press.

Emmerich, Werner. 1962. "Flurpläne aus der Zeit des sächsischen Kurfürsten Friedrich August I im Leipziger Stadtarchiv. Ihre Bedeutung als kartographische Leistung und als isedlungsgeschichtliche Quelle." *Jahrbuch für die Geschichte Mittel- und Ostdeutschlands* 11:111–35.

Emmison, Frederick G. 1937. *Types of Open Field Parishes in the Midlands.* Pamphlet no. 108. London: Historical Association.

———. 1963. "Estate Maps and Surveys." *History* 48:34–37.

Engel, Werner. 1982. "Joist Moers im Dienste des Landgrafen Moritz von Hessen. Ein Beitrag zu seiner späten Landmesstätigkeit und zugleich zur Schiffahrtsgeschichte der Fulda." *Hessisches Jahrbuch für Landesgeschichte* 32:165–73.

Engelstad, Sigurd. 1981. "Landmalingskonduktørene." In *Våre gamle kart,* edited by Rolf Fladby and Leif T. Andressen, 30–44. Oslo: Norsk Lokalhistorisk Institutt and Universitetsforlaget.

Ernst, Joseph W. 1958. "With Compass and Chain: Federal Land Surveyors in the Old Northwest." Columbia University (New York) Ph.D. thesis. Pub. New York: Arno Press, 1979.

Estienne, Charles, and Liébault, Jean. 1564. *L'Agriculture et maison rustique. . . .* Paris: J. de Puys. English translation in 1600 by Richard Surflet and in 1616 by Gervase Markham as *Maison Rustique; or, The Countrey Farme.* London: John Bill, 1616.

Evans, Eric J. 1976. *The Contentious Tithe: The Tithe Problem and English Agriculture, 1750–1850.* London: Routledge and Kegan Paul.

Evans, Richard J., and Lee, William R., eds. 1986. *The German Peasantry.* Beckenham: Croom Helm.

Evelyn, John. 1664. *Sylva; or, a Discourse of Forest-trees, and the Propagation of Timber. . . .* London: Royal Society of London for Improving of Natural Knowledge.

Evers, W. 1943. "Entwicklung und Stand der Kartographie Norwegens." *Petermanns Geographische Mitteilungen* 89:355–62.

Faber, Thomas F. 1988a. "Ausgewählte Karten aus dem Schloßarchiv Wissen—eine Arbeitsausstellung." In *Erschließung und Auswertung historischer Landkarten / Ontsluiting en gebruik van historische landkaarten.* 79–92. Landschaftsverband Rheinland—Archivberatungsstelle Rheinland. Archivhefte 18 / Werken uitgegeven door Limburgs Geschied- en Oudheidkundig Genootschap gevestigd te Maastricht 10. Cologne: Rheinland Verlag and Betriebsgesellschaft des Landschaftverbandes Rheinland.

———. 1988b. "Kartensammlung und archivischer Kartenbestand: Erfahrungsbericht über die Erschließungsarbeiten in drei Projekten." In *Erschließung und Auswertung historischer Landkarten / Ontsluiting en gebruik van historische landkaarten,* 51–77. Landschaftsverband Rheinland—Archivberatungsstelle Rheinland. Archivhefte 18 / Werken uitgegeven door Limburgs Geschied- en Oudheidkundig Genootschap gevestigd te Maastricht 10. Cologne: Rheinland Verlag and Betriebsgesellschaft des Landschaftverbandes Rheinland.

Faggot, Jacob. 1746. *Svenska landbrukets hinder och hjelp.* Stockholm.

Fairlamb, Jonas Preston. 1818. *A New and Concise Method of Completely Obviating the Difficulty Occasioned in Surveying by Local Attractions of the Magnetic Needle. . . .* Wilmington, Del.: William A. Miller.

Fallenbüchl, Zoltàn. 1983. "Vermessungsinstrumente und Vermessungswesen in Ungarn bis zum Ausgang des 18. Jahrhunderts." *Globusfreund* 31–32:158–72.

Farr, Ian. 1986. "'Tradition' and the Peasantry: On the Modern Historiography of Rural Germany." In *The German Peasantry,* edited by Richard J. Evans and William R. Lee, 1–36. Beckenham: Croom Helm.

Fink, Troels. 1941. *Udskiftningen i Sønderjylland indtil 1770.* Copenhagen: Ejnar Munksgaard.

Fitzherbert, Sir John Anthony. 1539. *The Boke of Surveyinge and Improvements.* London: Berthelet edition.

Fladby, Rolf. 1981. "Kartets plass i lokalhistorisk forskning." In *Våre gamle kart,* edited by Rolf Fladby and Leif T. Andressen, 9–18. Oslo: Norsk Lokalhistorisk Institutt and Universitetsforlaget.

———, ed. 1969. *Skattematrikkelen 1647.* Vol. 1, *Østfold Fylke.* Oslo: Norsk Lokalhistorisk Institutt and Universitetsforlaget.

———. 1969–78. *Skattematrikkelen 1647.* 17 vols. Oslo: Norsk Lokalhistorisk Institutt and Universitetsforlaget.

Fladby, Rolf, and Andressen, Leif T. 1981. "Forord." In *Våre gamle kart,* edited by Rolf Fladby and Leif T. Andressen, 7. Oslo: Norsk Lokalhistorisk Institutt and Universitetsforlaget.

Fladby, Rolf; Andressen, Leif T.; Nakken, Afhild; Schou, Libaek; and Schou, Terje, eds. 1979–85. *De gamle norske kart: Samkatolog over utrykte kart for de siste 300 år.* 18 vols. Oslo: Norsk Lokalhistorisk Institutt and Universitetsforlaget.

Fladby, Rolf; Imsen, Steinar; and Winge, Harald, eds. 1974. *Norsk historisk leksikon: Næringsliv, rettsvesen, administrasjon, mynt, mål og vekt, militære forhold, byggeskikk, m m 1500–1850.* Oslo: Norsk Lokalhistorisk Institutt and Cappelen.

Fletcher, David H. 1990. "Estate Maps of Christ Church, Oxford: The Emergence of Map Consciousness, *c.* 1600–*c.* 1840." University of Exeter Ph.D. thesis.

Flint, Abel. 1804. *A System of Geometry and Trigonometry together with a Treatise on Surveying.* Hartford: Oliver Cooke. 3d ed., 1813.

Fockema Andreaea, Sybrandus Johannes. 1933. *Beknopte inventaris van de oude archieven van het hoogheemraadschapp van Rijnland, 1255–1857.*

———. 1954. *Willem I, graaf van Holland, 1203–1222, en de Hollandse hoogheemraadschappen.* Wormerveer: Iris Pers.

———. 1982. *Het hoogheenraadschap van Rijnland: Zijn recht en zijn bestuur van den vroegsten tijd tot 175.* Alphen aan den Rijn: Canaletto. Orig. pub. Leiden: IJdo, 1934.

Fockema Andreae, Sybrandus Johannes, with van 't Hoff, B. 1947. *Geschiedenis der kartografie van Nederland van den Romeinschen tijd tot het midden der 19de eeuw.* The Hague: Martinus Nijhoff.

Folkingham, William. 1610. *Feudigraphia: The Synopsis or Epitome of Surveying Methodized.* London: Richard Moore.

Fougères, M. 1945. "Les Plans cadastraux de l'ancien régime." *Mélanges d'Histoire Sociale, Annales d'Histoire Sociale* 3:54–69.

Fournier. P.-F. 1941. "Lieux dits et cadastres." *Bulletin Historique et Scientifique de l'Auvergne* 61:126-.

Fournioux, Bernard. 1982. "Sur un plan figuré de la fin du XVe siècle d'une forêt vicomtale en Périgord." *Annales du Midi* 94:197–207.

Frandsen, Karl-Erik. 1983. "Vang og tægt: Studier over dyrkningssystemer og agrarstrukturer i Danmarks landsbyer, 1682–1683." University of Copenhagen doctoral thesis, Esbjerg: Bygd.

———, ed. 1984. *Atlas over Danmarks administrative inddeling efter 1660.* Vol. 1, maps; vol. 2, text. Dansk Historisk Fællesforening. Vojens: P. J. Schmidt.

Fransson, R., et al. 1956. "Örtugadelning och ägoinnehav före storskiftet." *Upplands Forminnesförenings Tidskrift* 48, no. 1.

Frêche, Georges. 1971. "Compoix, propriété foncière, fiscalité, et demographie historique en pays de taille réelle (XVIe–XVIIIe siècles)." *Revue d'Histoire Moderne et Contemporaine* 18:321–53.

Fréminville, Edme de la Poix de. 1752. *La Pratique universelle pour la rénovation des terriers.* Paris: Gissy.

French, Carolyn O. 1978. "Cadastral Patterns in Louisiana: A Colonial Legacy." Louisiana State University Ph.D. thesis.

Freund, Rudolf. 1946. "Military Bounty Lands and the Origin of the Public Domain." *Agricultural History* 20:8–18.

Frimanslund, Rigmor. 1956. "Farm Community and Neighbourhood Community." *Scandinavian Economic History Review* 4, no. 1:62–81.

Froidour, Louis de. 1759. *L'Instruction pour les ventes des bois du roy.* Paris: Berrier. Orig. pub. Toulouse, 1668.

Frykenberg, Robert E., ed. 1969. *Land Control and Social Structure in Indian History.* Madison: University of Wisconsin Press.

Garvin, James L. 1980. "The Range Township in Eighteenth-Century New Hampshire." In *New England Prospect: Maps, Place Names, and the Historical Landscape,* edited by Peter Benes, 47–68. Proceedings of the Dublin Seminar for New England Folklife, 1980. Boston: Boston University.

Gaudet, Placide. 1927. "Les Seigneuries de l'ancienne Acadie." *Bulletin des Recherches Historiques* 33:343–47.

Gentilcore, R. Louis, and Donkin, Kate. 1973. *Land Surveys of Southern Ontario: An Introduction and Index to the Field Notebooks of the Ontario Surveyors, 1784–1859.* Cartographica Monographs no. 8. Toronto.

Gentilcore, R. Louis, and Head, C. Grant. 1984. *Ontario's History in Maps.* Toronto: University of Toronto Press.

Gibson, Jeremy, and Mills, Dennis, eds. 1983. *Land Tax Assessments* c. *1690*–c. *1950.* Plymouth: Federation of Local History Societies.

Gibson, Robert. 1785. *A Treatise of Practical Surveying* . . . .4th ed., "Adapted to the Use of American Surveyors," Philadelphia: Joseph Crukshank.

Gillispie, Charles Coulston. 1980. *Science and Polity in France at the End of the Old Regime.* Princeton: Princeton University Press.

Glendinning, Robert M. 1934. "The Distribution of Population in the Lake St. John Lowland, Quebec." *Geographical Review* 24:232–37.

Gole, Susan. 1983. *India within the Ganges.* New Delhi: Jayaprints.

Goubert, Pierre. 1954. "En Rouergue: Structures agraires et cadastre au XVIIIe siècle." *Annales. Economies. Sociétés. Civilisations* 9:382–86.

Gouw, Jacobus Leonardus van der. 1967. *De Ring van Putten: Onderzoekingen over een hoogheemraadschap in het deltagebied.* The Hague: Albani.

Grab, Alexander. 1983–84. "Enlightened Despotism and State Building: The Case of Austrian Lombardy." *Austrian History Yearbook* 19–20, no. 2:43–72.

———. 1989. "Enlightened Despotism and Commons Enclosure: The Case of Austrian Lombardy." *Agricultural History* 63, no. 1:49–72.

Grau Møller, Per. 1984. *Udskiftningen og dens økonomiske og sociale følger i Sønderjylland circa 1730–1830: En analyse af et udvalgt område på Nordals.* Studier utgivet af Historisk Samfund for Sønderjylland, no. 1. Åbenrå.

Grees, Hermann. 1979. "Dorfgemarkung Obersteinach 1717 von Johann Mattäus Beck(er)." In *Historischer Atlas von Baden-Württemberg, Er-*

läuterungen. Beiwort zur Karte 1.7. Stuttgart: Kommission für geschichtliche Landeskunde in Baden-Württemberg.

Grenacher, Franz. 1960. "Daniel Meyer, ein unbekannter schweizerischer Kartograph und der Kataster seiner Zeit." *Geographica Helvetica* 15:8–16.

———. 1964. "Current Knowledge of Alsatian Cartography." *Imago Mundi* 18:60–77.

———. 1971. "'Cartographia Wiesenthalensis' im 17. und 18. Jahrhundert." *Regio Basiliensis* 12:147–73.

———. 1975. "Standortbestimmung der Basler Kartographie des 17. Jahrhunderts." *Regio Basiliensis* 16:1–27.

Grendahl, Torleif, ed. 1959. *Jordskifteverket gjennom 100 år, 1859–1958.* Oslo: Det Kgl. Landbruksdepartement.

Greslé-Bouignol, Maurice. 1973. *Les Plans de villes et du villages notables du Département du Tarn conservés dans divers depôts.* Albi: Archives Départementales du Tarn.

Grim, Ronald E. 1985. "Mapping Kansas and Nebraska: The Role of the General Land Office." *Great Plains Quarterly* 5:177–97. Reprinted in *Mapping the North American Plains: Essays in the History of Cartography,* edited by F. E. Luebke, Francis W. Kaye, and Gary E. Moulton, 127–44. Norman: University of Oklahoma Press, 1987.

Groenveld, S., and Huusssen, A. H., Jr. 1975. "De zestiende-eeuwse landmeter Jaspar Adriaensz. en zijn kartografisch werk." *Hollandse Studien* 8:131–77.

Grossman, W. 1955. "Niedersächsische Vermessungsgeschichte im 18. und 19. Jahrhundert." In *C. F. Gauss und die Landesvermessung in Niedersachsen,* by W. Gronwald et al., 17–59. Hanover: Niedersächsisches Landesvermessungsamt.

Grüll, Georg. 1952. "Die Ingenieure Knittel im Rahmen der oberösterreichischen Mappierungen im 17. und 18. Jahrhundert." *Mitteilungen des Oberösterreichischen Landesarchivs* 2:43–76.

Guerout, Jean. 1987. "Origine et evolution des plans domaniaux." In *Espace français: Vision et aménagement, XVIe–XIXe siècles,* 15–16. Paris: Archives Nationales.

Guichonnet, Paul. 1955. "Le Cadastre savoyard de 1738 et son utilisation pour les recherches d'histoire et de géographie sociales." *Revue de Géographie Alpine* 43:255–98.

———. 1963. "Les Cadastres genevois du XVIIIe siècle et de la période française." *Genava* 2:519–40.

Guilhamon, H. 1952. *Journal des voyages en Haute-Guyenne de J.-F. Henry de Richeprey.* Vol. 1, *Rouergue.* Rodez: Archives Historiques de Rouergue.

Gummere, John. 1814. *A Treatise on Surveying, Containing the Theory and Practice. . . .* Philadelphia: W. Brown.

Habakkuk, Hrothgar J. 1968. "Population, Commerce, and Economic Ideas." In *The New Cambridge Modern History;* vol. 8, *The American and French Revolutions, 1763–93,* edited by A. Goodwin, 25–54. Cambridge: Cambridge University Press.

Haegen, H. van der. 1980. "Bronnen voor de reconstructie van de agrarische structuur: Landschap, bedrijven, huur- en eigendomsverhoudingen in Vlaanderen." In *Bronnen voor de historische geografie van België / Sources de la géographie historique de la Belgique,* edited by Jacques Mertens, 113–28. Report from the conference held 25–27 April 1979. Brussels: Algemeen Rijksarchief en Rijksarchief in de Provincien / Archives Générales du Royaume et Archives de l'Etat dans les Provinces.

———. 1986. "Oude kaarten en plannen en hun bruikbaarheid voor de historische geograaf." In *Oude kaarten en plattegronde: Bronnen voor de historische geografie van de zuidelijke Nederlanden (16de–18de eeuw) / Cartes et plans anciens: Sources pour la géographie historique de Pays-Bas méridionaux (XVIe–XVIIIe siècles),* edited by H. van der Haegen, F. Daelemans, and Eduard van Ermen, 5–9. Archief- en Bibliotheekwezen in België / Archives et Bibliothèques de Belgique, special no. 31. Brussels.

Hagen, William M. 1986. "The Junkers' Faithless Servants: Peasant Insubordination and the Breakdown of Serfdom in Brandenburg-Prussia, 1763–1811." In *The German Peasantry,* edited by Richard J. Evans and William R. Lee, 71–101. Beckenham: Croom Helm.

Haines, Michael R. 1982. "Agriculture and Development in Prussian Upper Silesia, 1846–1913." *Journal of Economic History* 42:355–84.

Hake, Gunter. 1978. "Historische Entwicklung des Kartenwesens im Raum Hannover." In *Hannover und sein Umland: Festschrift zur Feier des 100 jährigen Bestehens des Geographischen Gesellschaft zu Hannover, 1878–1978,* edited by Wolfgang Eriksen and Adolf Arnold, 50–67. Hanover: Geographischen Gesellschaft zu Hannover.

Hall, John W. 1970. "Louisiana Survey Systems: Their Antecedents, Distribution, and Characteristics." Louisiana State University Ph.D. thesis.

Ham, Willem A. van. 1983. "Een late overzichtskaart van het Markiezaat van Bergen op Zoom." *Caert-Thresoor* 2, no. 3:34–35.

———. 1987. "De landmeters uit de Familie Adan en hun betekenis voor de kartografie van westelijk Noord-Brabant." *Caert-Thresoor* 6, no. 2:23–26.

Hameleers, Marc M. 1984. "Biografie van Nederlandse gedrukte polderkaarten." University of Utrecht doctoral thesis.

———. 1986. "Representativiteit en functionaliteit van polder-en waterschapskaarten toegelicht med de kaart van Rijnland door Floris Balthasars (1615)." In "Polder-, waterschaps-, en rivierkartografie," typescript edited by Marc Hameleers and Peter van der Krogt, 33–56. Utrecht: Werkgroep voor de Geschiedenis van de Kartografie van de Nederlandse Vereniging voor Kartografie and Geographisch Institut.

———. 1989. "Copperplates in the Northern Netherlands." *Map Collector* 47:36–39.

Hameleers, Marc M., and van der Krogt, Peter. 1991. "Bibliografie van de geschiedenis van de kartografie in Nederland." 3d. ed. Typescript.

Hammon, Neal O. 1980. "Land Acquisition on the Kentucky Frontier." *Register of the Kentucky Historical Society* 78:297–321.

Hanke, Max, and Degner, Hermann. 1934. *Die Pflege der Kartographie bei der königlich preußichen Akademie der Wissenschaften unter der Regierung Friedrichs der Großen.* Abhandlungen der preußischen Akademie der Wissenschaften, 33 (Physikalisch-mathematische Klasse no 2). Berlin: Verlag der Akademie der Wissenschaften.

———. 1935. *Geschichte der amtlichen Kartographie Brandenburg-Preußens bis zum Ausgang der friderizianischen Zeit.* Geografische Abhandlungen Series 3, vol. 7. Stuttgart: Verlag J. Engelhorns Nachf.

Hanna, James. 1818. *The American Instructor; or, Everyone his Own Teacher; Comprising . . . Surveying. . . .* Trenton, N.J.: the author.

Hannes, J. 1967. "La Constitution du cadastre parcellaire—étude des sources." *Bulletin Trimestriel du Crédit Communal de Belgique* 80:79–88.

———. 1968. "L'Atlas cadastral parcellaire de la Belgique de P. C. Popp." *BulletinTrimestriel du Crédit Communal de Belgique* 85:137–46.

Hannes, J., and Coppens, H. 1967. "Mededeling over de oudste parcellaire plans Provincie Antwerpen." *Tijdschrift van de Belgische Verening voor Aardrijkskundige Studies / Bulletin Société Belge Géographique* 36:182–84.

Hansen, Carl Rise, and Steensberg, Axel. 1951. *Jordfordeling og udskiftning; Undersøgelse i tre sjællandske landsbyer.* Det Kongelige Danske Videnskabernes Selskab: Historisk-Filologiske Skrifter 2, no. 1. Copenhagen.

Hansen, Viggo. 1964. *Landskab og bebyggelse i Vendsyssel. Studier over landebebyggelsens udvikling indtil slutningen af 1600-talet.* Kulturgeografiske Skrifter no. 7. Copenhagen: C. A. Reitzels Forlag.

———. 1981. "Landskabsændringer i årtierne efter udskiftningen." In *Udskiftningsforordning 200 år (= Landinspektøren* 30, no. 8), 525–40.

———. 1985. "Økologiske vilkår i en vestjysk landbrugsegn belyst gennem 1688-matriklen." *Fortid og Nutid* 33:271–82.

Harley, J. Brian. 1964. *The Historian's Guide to Ordnance Survey Maps.* London: National Council for Social Service.

———. 1972. *Maps for the Local Historian.* London: Standing Conference for Local History and National Council for Social Service.

———. 1979. *The Ordnance Survey and Land-Use Mapping: Parish Books of Reference and the County Series 1:2500 Maps, 1855–1918.* Research Series no. 2. Norwich: Historical-Geography Research Group.

———. 1983. "Meaning and Ambiguity in Tudor Cartography." In *English Map-Making, 1500–1650,* edited by Sarah Tyacke, 22–45. London: British Library.

———. 1988. "Maps, Knowledge, and Power." In *The Iconography of Landscape: Essays on the Symbolic Representation, Design, and Use of Past Environments,* edited by Denis Cosgrove and Stephen Daniels, 277–312. Cambridge: Cambridge University Press.

———. 1989. "Deconstructing the Map." *Cartographica* 26:1–20.

Harley, J. Brian, and Woodward, David, eds. 1987. *The History of Cartography.* Vol. 1, *Cartography in Prehistoric, Ancient, and Medieval Europe and the Mediterranean.* Chicago; University of Chicago Press.

Harms, J., and Donkersloot–de Vrij, Ypkje Marijke. 1977. *Catalogus van de kaartencollectie Moll.* Utrecht: Universiteits Bibliotheek.

Harnisch, Harmut. 1986. "Peasants and Markets: The Background to the Agrarian Reforms in Feudal Prussia East of the Elbe, 1760–1807." In *The German Peasantry,* edited by Richard J. Evans and William R. Lee, 37–70. Beckenham: Croom Helm.

Harrington, Bates. 1879. *How 'Tis Done: A Thorough Ventilation of the Numerous Schemes Conducted by Wandering Canvassers together with the Various Advertising Dodges for the Swindling of the Public.* Chicago: Fidelity Publishing Company.

Harris, R. Cole. 1966. *The Seigneurial System in Early Canada: A Geographical Study.* Madison: University of Wisconsin Press. 2d ed., Montreal: McGill-Queen's University Press, 1984.

———, ed. 1987. *Historical Atlas of Canada.* Vol. 1, *From the Beginning to 1800.* Toronto: University of Toronto Press.

Harris, R. Cole, and Warkentin, John. 1974. *Canada before Confederation: A Study in Historical Geography.* New York: Oxford University Press.

Harvey, Paul D. A. 1980. *The History of Topographical Maps: Symbols, Pictures, and Surveys.* London: Thames and Hudson.

———. 1987. "Local and Regional Cartography in Medieval Europe." In *The History of Cartography;* vol. 1, *Cartography in Prehistoric, Ancient, and Medieval Europe and the Mediterranean,* edited by J. Brian Harley and David Woodward, 464–501. Chicago: University of Chicago Press.

Hastrup, Frits. 1964. *Danske landsbytyper: En geografisk analyse.* 2 vols. Århus: Clemenstrykkeriet.

Hatton, R. M. 1957. "Scandinavia and the Baltic." In *The New Cambridge Modern History;* vol. 8, *The Old Regime, 1713–63,* edited by Jean O. Lindsay, 339–64. Cambridge: Cambridge University Press.

Haversath, J.-B., and Struck, E. 1986. *Passau und das Land der Abtei in historischen Karten und Plänen.* Passauer Schriften zur Geographie 3. Passau: Passavia Universitätsverlag.

Haye, Régis de la. 1988. "Kaartenverzamelingen in de Nederlandse Provincie Limburg." In *Erschließung und Auswertung historischer Landkarten / Ontsluiting en gebruik van historische landkaarten,* 29–50. Landschaftsverband Rheinland—Archivberatungsstelle Rheinland. Archivhefte 18 / Werken uitgegeven door Limburgs Geschied- en Oudheidkundig Genootschap gevestigd te Maastricht 10. Cologne: Rheinland Verlag and Betriebsgesellschaft des Landschaftverbandes Rheinland.

Heaney, George F. 1957. "The Story of the Survey of India." *Geographical Magazine* 30:182–90.

———. 1968. "Rennell and the Surveyors of India." *Geographical Journal* 134:318–27.

Heathcote, Ronald L. 1965. *Back of Bourke: A Study of Land Appraisal and Settlement in Semi-arid Australia.* Melbourne: Melbourne University Press.

Hébert, John R. 1987. "Vicente Sebastien Pintado, Surveyor General of Spanish West Florida, 1805–17: The Man and His Maps." *Imago Mundi* 39:50–72.

Hedenstierna, B. 1948. "Stockholms skärgård: Kulturgeografiska studier i Värmdö gamla skeppslag." *Geografiska Annaler* 30:1–444.

Heer, Friedrich. 1968. *The Holy Roman Empire.* London: Weidenfeld and Nicholson.

Heering, H. T. 1958. "The Danish Matricel (Cadastre)." Report to the Ninth International Congress of Surveyors. Typescript.

Helgerson, Richard. 1986. "The Land Speaks: Cartography, Chorography, and Subversion in Renaissance England." *Representations* 16:51–85.

Hellwig, Fritz. 1985. "Zur Geschichte der älteren Kartographie vom Mittelrhein und Mosselland." In *Mittelrhein und Moselland im Bild alter Karten,* edited by Fritz Hellwig, 9–52. Koblenz: Landesarchivverwaltung Rheinland-Pfalz.

Helmfrid, Staffan. 1959. "De geometriska jordeböckerna—'skattläggningskartor?'" *Ymer* 79, no. 3:224–31.

———. 1961. "The *storskifte, enskifte,* and *laga skifte* in Sweden—General Features." *Geografiska Annaler* 43:114–29.

Henderson, William O. 1963. *Studies in the Economic Policy of Frederick the Great.* London: Frank Cass.

Herbin, R., and Pebereau, A. 1953. *Le Cadastre français.* Paris: Les Editions Francis Lefebvre.

Hervé, Roger. 1960. "Les Plans de forêts de la grande réformation colbertienne, 1661–90." *Bulletin de la Section de Géographie* (Ministère de l'Education Nationale, Comité des Travaux Historiques) 73:143–71.

Hessels, J. T. 1891. "Het kadaster tijdens de Bataafsche Republiek en het Koninkrijk der Nederlanden." *Tijdschrift voor Kadaster en Landmeetkunde,* 3–11.

Hilliard, Sam B. 1973. "An Introduction to Land Survey Systems in the Southeast." *West Georgia College Studies in the Social Sciences* 12:1–15.

———. 1982. "Headright Grants and Surveying in Northeastern Georgia." *Geographical Review* 72:416–29.

———. 1987. "A Robust New Nation, 1783–1820." In *North America: The Historical Geography of a Changing Continent,* edited by Robert D. Mitchell and Paul A. Groves, 149–71. London: Hutchinson.

Himley, François-J. 1959. *Catalogue des cartes et plans manuscrits antérieurs à 1790.* Strasbourg: Archives Départmentales du Bas-Rhin. Supplement, 1978.

Hindle, Paul. 1988. *Maps for Local History.* London: Batsford.

Hinkel, Heinz. 1967. "Pommerschen Karten in der Staatsbibliothek des Stiftung Preußischer Kulturbesitz." *Zeitschrift für Ostforschung* 16:342–53.

Hobsbawm, Eric J., and Rudé, George F. E. 1969. *Captain Swing.* London: Lawrence and Wishart.

Hoffmeyer, Johs. 1981. "Den historiske baggrund for udskiftningsreformen i 1781." In *Udskiftningsforordning 200 år* (= *Landinspektøren* 30, no. 8), 496–507.

Hofmann, Norbert, with Semmler, Hans. 1983. *Inventar des löwenstein-wertheim-rosenbergschen Karten- und Planselekts im Staatsarchiv Wertheim, 1725–1835.* Veröffentlichungen der staatlichen Archivverwaltung Baden-Württemberg 43. Stuttgart: Landesarchivdirektion Baden-Würtemberg and Verlag W. Kohlhammer.

Hogg, James E. 1905. *The Australian Torrens System.* London: Wm. Clowes.

Holmsen, Andreas. 1956. "General Survey and Historical Introduction." *Scandinavian Economic History Review* 4, no. 1:17–32.

———. 1958. "Landowners and Tenants in Norway." *Scandinavian Economic History Review* 6, no. 2:121–31.

———. 1961. "The Transition from Tenancy to Freehold Peasant Ownership in Norway." *Scandinavian Economic History Review* 9, no. 2:152–64.

———. 1979. *Gård, skatt, og skyld.* Oslo: Universitetsforlaget.

———. 1982. "Skattematrikkelen 1647." In *Den eldeste matrikkelen: En infallsport til historien: Skattematrikkelen 1647,* edited by Rolf Fladby and Harald Winge, 9–12. Oslo: Norsk Lokalhistorisk Institutt and Universitetsforlaget.

Homer, Henry S. 1766. *An Essay on the Nature and Method of Ascertaining the Specifick Shares of Proprietors upon the Inclosure of Common Fields. . . .* Oxford: S. Parker.

Hopkins, Daniel. 1987. "The Danish Cadastral Survey of St. Croix, 1733–1734." Louisiana State University Ph.D. thesis.

———. 1989. "An Extraordinary Eighteenth-Century Map of the Danish Sugar-Plantation Island St. Croix." *Imago Mundi* 41:44–58.

Hoppe, Göran. 1979. *Enclosure in Sweden: Background and Consequences.* Kulturgeografiskt Seminarium 9.

———. 1982. *"At the Ventilation of the Suggested Redistribution much Controversy was Disclosed . . .": Enclosure in Väversunda Village, Östergötland.* Kulturgeografiskt Seminarium 5.

Hoste, Jean L'. 1629. *Sommaire de la sphère artificielle et de l'usage d'icelle.* Nancy: the author.

[House of Commons]. 1795–1800. "Three Reports from the Select Committee Appointed to Take Into Consideration the Means of Promoting the Cultivation and Improvement of the Waste, Uninclosed, and Unproductive Lands in the Kingdom." *British Parliamentary Papers (House of Commons),* First Series of Reports, 9.

———. 1800. "Report of the Select Committee on Bills of Inclosure." *British Parliamentary Papers (House of Commons)* 27.

———. 1837a. "Copy of Papers Respecting the Proposed Survey of Lands under the Tithe Act." *British Parliamentary Papers (House of Commons)* 41.

———. 1837b. "Report from the Select Committee on Survey of Parishes (Tithe Commutation Act), with the Minutes of Evidence." *British Parliamentary Papers (House of Commons)* 6.

———. 1850. "First Report from the Registration and Conveyancing Commission." *British Parliamentary Papers (House of Commons)* 32.

———. 1854. "Correspondence Respecting the Scale of the Ordnance Survey and upon Contouring and Hill Delineation." *British Parliamentary Papers (House of Commons)* 41.

———. 1856. "Report from the Select Committee on Ordnance Survey of Scotland together with the Proceedings of the Committee, Minutes of Evidence, Appendix, and Index." *British Parliamentary Papers (House of Commons)* 14.

———. 1862. "Report from the Select Committee on the Cadastral Survey." *British Parliamentary Papers (House of Commons)* 6.

[House of Lords]. 1830–31. "Second Report of the Commissioners on the Law of Real Property." *British Parliamentary Papers (House of Lords)* 284.

———. 1854–55. "Report of the Select Committee of the House of Lords . . . to Inquire into . . . Agricultural Statistics." *British Parliamentary Papers (House of Lords)* 8.

———. 1857. "Report of the Commissioners Appointed to Consider the Subject of the Registration of Title with Reference to the Sale and Transfer of Land." *British Parliamentary Papers (House of Lords),* 2d sess., 25.

Hovstad, Håkon. 1981. "Hva jordskiftekart kan fortelle." In *Våre gamle kart,* edited by Rolf Fladby and Leif T. Andressen, 19–29. Oslo: Norsk Lokalhistorisk Institutt and Universitetsforlaget.

Hruda, Hans. 1967. "Die Entwicklung der agrarischen Operationen und deren Auswirkungen auf den österreichischen Grundkataster." In *150 Jahre österreichischer Grundkataster,* edited by Robert Messner, Josef Mitter, and Manfred Schenk, 51–63. Vienna: Bundesamt für Eich- und Vermessungswesen.

Hughes, Jonathan R. T. 1976. *Social Control in the Colonial Economy.* Charlottesville: University Press of Virginia.

Hughes, Leslie. 1950. "Some Features of Early Woodland and Prairie Settlement in a Central Iowa County." *Annals of the Association of American Geographers* 40:40–57.

Hughes, Sarah S. 1979. *Surveyors and Statesmen: Land Measuring in Colonial Virginia.* Richmond: Virginia Surveyors' Foundation and the Virginia Association of Surveyors.

Huguenin, Marcel. 1962–63. "La Cartographie ancienne de la Corse." *Bulletin d'Information de l'Association des Ingénieurs Géographes* 23:85–98; 26:33–55.

———. 1970. "French Cartography of Corsica." *Imago Mundi* 24:123–37.

Huussen, A. H., Jr. 1976. "Kartografie en rechtspraak: Een verkenning." *Spiegel Historiael* 11, no. 3:148–55.

———. 1989. "Polemische kaarten." In *Kaarten met geschiedenis 1550–1800: Een selectie van oude getekende kaarten van Nederland uit de collectie Bodel Nijenhuis,* edited by Dirk de Vries, 31–41. Utrecht: H. and S. Publishers.

Iterson, Rob van. 1986. "De verzameling kaarten, tekeningen, atlassen, en kaartboeken van het Hoogheemraadschap van Rijnland." In "Polder-, waterschaps-, en rivierkartografie," typescript edited by Marc Hameleers and Peter van der Krogt, 69–88. Utrecht: Werkgroep voor de Geschiedenis van de Kartografie van de Nederlandse Vereniging voor Kartografie and Geographisch Institut.

Jäger, Eckhard. 1978. *Bibliographie zur Kartengeschichte von Deutschland und Osteuropa: Eine Auswahl des kartographischen Schriftums mit einem Exkurs über Landkarten Preise im 18. Jahrhundert im Vergleich zu anderen Kosten.* Lüneburg: Norddeutsches Kulturwerk.

Jeans, Dennis N. 1966a. "The Breakdown of Australia's First Rectangular Grid Survey." *Australian Geographical Studies* 4:119–28.

———. 1966b. "Crown Land Sales and the Accommodation of the Small Settler in N.S.W., 1825–1842." *Historical Studies Australia and New Zealand* 12:205–12.

———. 1966–67. "Territorial Divisions and the Locations of Towns in New South Wales, 1826–1842." *Australian Geographer* 10:243–55.

———. 1972. *An Historical Geography of New South Wales to 1901.* Sydney: Reed Education.

———. 1975. "The Impress of Central Authority upon the Landscape: South-eastern Australia, 1788–1850." In *Australian Space Australian Time: Geographical Perspectives,* edited by Joseph M. Powell and Michael Williams, 1–17. Melbourne: Oxford University Press.

———. 1978. "Use of Historical Evidence for Vegetation Mapping in N.S.W." *Australian Geographer* 14:93–97.

———. 1981. "Official Town-founding Procedures in New South Wales, 1828–1842." *Journal of the Royal Australian Historical Society* 67:227–37.

Jensen, Hans Ejner. 1975. "Danmarks økonomiske opmåling." In *Landmåling og landmålere: Danmarks økonomiske opmåling;* vol. 1, text, edited by Svend Balslev and Hans Ejner Jensen, 11–64, 323–31. Copenhagen: Den Danske Landinspektørforening.

———. 1981. Landinspektøren under landboreformerne." In *Udskiftningsforordning 200 år* (= *Landinspektøren* 30, no. 8), 517–24.

Jensen, Peter Flint. 1981. "Jordfællesskabets ophævelse." In *Udskiftningsforordning 200 år* (= *Landinspektøren* 30, no. 8), 478–95.

Jess, Zachariah. 1799. *A Compendious System of Practical Surveying and Dividing of Land. . . .* Wilmington: Bonsal and Niles.

Johannessen, Knut. 1972. "NLI registrerer eldre kart." *Heimen* 15:607–14.

Johnson, Hildegard Binder. 1957. "Rational and Ecological Aspects of the Quarter Section: An

Example from Minnesota." *Geographical Review* 47:330–48.

———. 1976. *Order upon the Land: The U.S. Rectangular Land Survey and the Upper Mississippi Country.* New York: Oxford University Press.

Johnsson, Bruno. 1965. "Synpynkter på 1600-talets tidiga geometriska kartering med särskild hänsyn till Västmanlands län." *Ymer* 85:9–84.

Jones, Anne M., and Siddle, David J. 1982. *Sources for the Reconstruction of Peasant Systems in an Upland Area of Europe, 1561–1735.* Liverpool Papers in Human Geography no. 3. Liverpool: Liverpool University, Department of Geography.

Jones, Michael. 1977. *Finland: Daughter of the Sea.* Folkestone: Dawson.

———. 1982a. "Inovasjonsstudier i historisk-geografisk perspektiv: Eksemplifisert ved spredning av jordskifte i Norden." In *Geografi som samfunnsvitenskap: Filosofi metode, anvendbarhet,* edited by Sverre Strand, 134–42, 200–201, 218. Ad Novas—Norwegian Geographical Studies 19. Bergen-Oslo-Trondheim: Universitetsforlaget.

———. 1982b. "Land Consolidation in Theory and Practice: Examples from Aurland Commune in Sogn, Western Norway, 1860–1975." In *Proceedings of Historical Geography—Twenty-fourth International Geographical Congress, Section,* 9, edited by T. Tanoika and T. Ukita, 176–80. Tokyo.

———. 1985. "Datakilder, datainnsamling, og verdisyn." In *Metode på tvers: Samfunnsvitenskapelige forskningsstrategier som kombinerer metod og analysenivåer,* edited by Britt Dale, Michael Jones, and Willy Martinnssen, 57–84. Trondheim: Tapir.

———. 1987. "Summarisk oversikt over Norges kartverk." Typescript.

Jordan, G. 1955. "Die alten Teilungs- und Verkoppelungskarten im Raume Niedersachsen." In *C. F. Gauss und die Landesvermessung in Niedersachsen,* by W. Gronwald et al., 141–54. Hanover: Niedersächsisches Landesvermessungsamt.

Jordan, Terry G. 1974. "Antecedents of the Long-Lot in Texas." *Annals of the Association of American Geographers* 64:70–86.

———. 1982. "Division of the Land." In *This Remarkable Continent: An Atlas of United States and Canadian Society and Cultures,* edited by John F. Rooney, Wilbur Zelinsky, and Dean R. Louder, 54–70. College Station: Texas A&M University Press, for The Society for the North American Cultural Survey.

Jouanne, René. 1933. *Les Origines du cadastre ornais.* Alençon: Imprimerie Alençonnaise.

Jourdain, William R. 1925. *Land Legislation and Settlement in New Zealand.* Wellington: Government Printer.

Jung, R. 1988. "Die Erschließung von Landkarten im Bibliotheken." In *Erschleißung und Auswertung historischer Landkarten / Ontsluiting en gebruik van historische landkaarten.* 93–113. Landschaftsverband Rheinland–Archivberatungsstelle Rheinland. Archivhefte 18 / Werken uitgegeven door Limburgs Geschied- en Oudheidkundig Genootschap gevestigd te Maastricht 10. Cologne: Rheinland Verlag and Betriebsgesellschaft des Landschaftverbandes Rheinland.

Kadastrale Atlas van Zeeland, 1832. 1988. Middleburg: Stichting Kadastrale Atlas van Zeeland.

Kahlfuß, Hans-Jürgen. 1969. *Landesaufnahme und Flurvermessung in den Herzogtümern Schleswig, Holstein, und Lauenburg vor 1864.* Beiträge zur Geschichte der Kartographie Nordalbingiens. Neumünster: Karl Wachholtz Verlag.

———. 1986. "Schleswig-Holstein." In *Lexikon zur Geschichte der Kartographie von den Anfängen bis zum Ersten Weltkrieg,* edited by Ingrid Kretschmer, Johannes Dörflinger, and Franz Wawrik, 709–10. Vienna: Franz Deuticke.

Kain, Roger J. P. 1975a. "R. K. Dawson's Proposals in 1836 for a Cadastral Survey of England and Wales." *Cartographic Journal* 12:81–88.

———. 1975b. "Tithe Surveys and Landownership." *Journal of Historical Geography* 1:39–48.

———. 1986. *An Atlas and Index of the Tithe Files of Mid-Nineteenth-Century England and Wales.* Cambridge: Cambridge University Press.

———. Forthcoming. "Maps and Rural Land Management in the Renaissance." In *The History of Cartography;* vol. 3, *Cartography in Renaissance Europe, 1460–1670,* edited by J. Brian Harley and David Woodward. Chicago: University of Chicago Press.

Kain, Roger J. P., and Prince, Hugh C. 1985. *The Tithe Surveys of England and Wales.* Cambridge: Cambridge University Press.

Kamenik, Walter. 1967. "Katastralneuvermessung, historische Kontinuität und zeitgenössische Aspekte." In *150 Jahre österreichischer Grundkataster,* edited by Robert Messner, Josef Mitter, and Manfred Schenk, 81–89. Vienna: Bundesamt für Eich- und Vermessungswesen.

Kamp, A. F., ed. 1971. *Caerte vant Houtbos ende Zijplant.* Alkmaar: Hoogheemraadschap Noordhollands Nordkwartier te Alkmaar.

Katzenberger, Ludwig. 1977. "175 Jahre bayerische Landesvermessung in kartographischer Sicht." *International Yearbook of Cartography* 17:104–12.

Kenoyer, Leslie. 1933. "Forest Distribution in Southwestern Michigan as Interpreted from the Original Land Survey (1826–32)." *Papers of the Michigan Academy of Arts, Science, and Letters* 19:107–11.

Ketter, Kurt. 1929. *Der Versuch einer Katasterreform in Cleve unter Friedrich Wilhelm I.* Rheinisches Archiv 9. Bonn: Ludwig Röhrscheid Verlag.

Keuning, Johannes. 1952. "Sixteenth-Century Cartography in the Netherlands." *Imago Mundi* 9:35–63.

Keverling Buisman, F., and Muller, E. 1979. *Kadaster-gids: Gids voor de raadpleging van hypothecaire en kadastrale archieven uit de 19e en de eerste helft van de 20e eeuw.* The Hague: Rijksarchiefdienst.

Kinder, Hermann, and Hilgermann, Werner. 1974–78. *The Penguin Atlas of World History.* Vol. 1, *From the Beginning to the Eve of the French Revolution;* vol. 2, *From the French Revolution to the Present.* Harmondsworth: Penguin.

Kish, George. 1962. "Centuriato: The Roman Rectangular Land Survey." *Surveying and Mapping* 22:233–44.

———. 1978. *The Discovery and Settlement of North America, 1500–1865: A Cartographic Perspective.* New York: Harper and Row.

Kitchen, Martin 1978. *The Political Economy of Germany, 1815–1914.* London: Croom Helm.

Klang, Daniel M. 1977. *Tax Reform in Eighteenth-Century Lombardy.* East European Quarterly. East European Monographs 27.

Kleinau, Hermann. 1968. "Die Karte des Landes Braunschweig im 18. Jahrhundert." *Braunschweigisches Jahrbuch* 49:202–8.

Kleinau, Hermann, and Penners, Theodor. 1956. "Historische Karte des Landes Braunschweig im 18. Jahrhundert. I: Erläuterungen in historischer Sicht." *Niedersächsisches Jahrbuch für Landesgeschichte* 28:1–8.

Kleinn, Hans. 1964–65. "Nordwestdeutschland in der exakten Kartographie der letzten 250 Jahre." *Westfälische Forschungen* 17:28–82; 18:43–74.

———. 1976–77. "Die Reproduktion der Landesvermessung des Fürstbistums Osnabrück (1784–1790) von J. W. Du Plat." *Westfälische Forschungen* 28:181–84.

———. 1986. "Niedersachsen." In *Lexikon zur Geschichte der Kartographie von den Anfängen bis zum Ersten Weltkrieg,* edited by Ingrid Kretschmer, Johannes Dörflinger, and Franz Wawrik, 531–32. Vienna: Franz Deuticke.

Kloiber, Otto, and Schwarzinger, Karl. 1986. *1883–1983: 100 Jahre Führung des Katasters.* 2d ed. Vienna: Bundesamt für Eich- und Vermessungswesen.

Koch, Hannsjoachim W. 1984. *A Constitutional History of Germany in the Nineteenth and Twentieth Centuries.* London: Longman.

Koeman, Cornelis. 1960. "De betekenis van de waterschapskaarten van Delfland voor de geschiedenis van de kartografie van Nederland." *Tijdschrift van het Koniklijk Nederlandsch Aardrijkskundig Genootschap,* 2d ser., 77:132–34.

———. 1961. *Collections of Maps and Atlases in the Netherlands: Their History and Present State.* Leiden: Brill.

———. 1963. "De administratie van de grondbelasting als bron voor de historisch-topografische documentatie." In *Handleiding van en studie van en topografische kaarten van Nederland, 1750–1850,* by Cornelis Koeman, 71–76. Groningen: Wolters-Noordhoff / Bouma's Boekhuis.

———. 1973. *Bibliography of Printed Maps of Suriname, 1671–1971.* Amsterdam: Theatrum Orbis Terrarum.

———. 1982. "Bijdragen van het kadaster aan de kartografie van Nederland." in *Op goede gronden: Een bundel opstellen ter gelegenheid van het 150-jarig bestaan van de dienst van het kadaster en de openbare registers,* 104–29. The Hague: Staatsuitgeverij.

———. 1983. *Geschiedenis van de kartografie van Nederland. Zes eeuwen land- en zeekaarten en stadsplattegronden.* Alphen aan den Rijn: Canaletto.

Kok, Marjanne. 1982. "Johan Christoph Heneman: Kartograaf van Suriname en Guyana van 1780–1806." *Caert-Thresoor* 1, no. 1:4–12.

———. 1985. *Kaartboek van het Baljuwschap van Naaldwijk: Kaartboek van de landerijen gelegen in de Heerlijkheden Naaldwijk, Honselersdijk en het Honderland, omstreeks 1620 door Floris Jacobsz. vervaardigd.* Alphen aan den Rijn: Canaletto.

Konvitz, Joseph. 1987. *Cartography in France, 1660–1848.* Chicago: University of Chicago Press.

Kretschmer, Ingrid. 1968. "150 Jahre österreichischer Grundkataster." *Mitteilungen der Österreichischen Geographischen Gesellschaft* 110, no. 1:62–71.

Kretschmer, Ingrid; Dörflinger, Johannes; and Wawrik, Frans, eds. 1986. *Lexikon zur Geschichte der Kartographie von den Anfängen bis zum Ersten Weltkrieg.* 2 vols. Vienna: Deuticke.

Kriedte, Peter; Medick, Hans; and Schlumbohm, Jürgen. 1977. *Industrialisierung vor der Industrialisierung: Gewerbliche Warenproduktion auf dem Land in der Formationsperiode des Kapitalismus.* Veröffentlichungen des Max Planck Instituts für Geschichte 53. Göttingen: Vandenhoeck and Ruprecht.

Kriesche, Ulrich. 1944. "Entwicklung und Stand der Kartographie Finlands." *Petermanns Geographische Mitteilungen* 90:261–68.

Krüger, Kersten. 1982. "Entstehung und Ausbau des hessischen Steuerstaates vom 16. bis zum 18. Jahrhundert—Akten der Finanzverwaltung als frühenzeitlicher Gesellschaftsspiegel." *Hessisches Jahrbuch für Landesgeschichte* 32:103–25.

Kruse, Louis Frederick Vinding. 1953. *The Right of Property.* 2 vols. Oxford: Oxford University Press. (Danish edition, 1939.)

Lahr, Walter. 1967. "Alois Senefelder und die Reproduktion der österreichischen Katastralmappe." In *150 Jahre österreichischer Grundkataster,* edited by Robert Messner, Josef Mitter, and Manfred Schenk, 91–97. Vienna: Bundesamt für Eich- und Vermessungswesen.

Lambert, Audrey H. 1971. *The Making of the Dutch Landscape: An Historical Geography of the Netherlands.* London: Seminar Press.

Lamy. 1789. *Cadastre universel.*

Lang, Arend. 1962. "Kleine Kartengeschichte Frieslands zwischen Ems und Jade." *Hier Büst Du to Huus* 6. Norden.

Lang, Wilhelm. 1952. "Martin Faber's Map of the Ems Mouth." *Imago Mundi* 9:79–82.

Latham, Robert, ed. 1985. *The Shorter Pepys.* Berkeley and Los Angeles: University of California Press.

Lawrence, Heather. 1985. "John Norden and His Colleagues: Surveyors of Crown Lands." *Cartographic Journal* 22:54–56.

Le Gear, Clara E. 1950–53. *United States Atlases.* Vol. 1, *A List of National, State, County, City, and*

Regional Atlases in the Library of Congress; vol. 2, *A Catalog of National, State, County, City and Regional Atlases in the Library of Congress and Cooperating Libraries.* Washington, D.C.: Library of Congress.

Le Roy Ladurie, Emmanuel. 1966. *Les paysans de Languedoc.* Paris: SEVPEN.

Lee, William R. 1972. "Some Economic and Demographic Aspects of Peasant Society in Oberbayern (1752–1855), with Special Reference to Certain Estates in the Former Landgericht Kranzberg." University of Oxford D.Phil. thesis.

———. 1975. "Tax Structure and Economic Growth in Germany (1750–1850)." *Journal of European Economic History* 4, no. 1:153–78.

Leerhoff, Heiko. 1985. *Niedersachsen in alten Karten.* Neumünster: Karl Wachholtz Verlag.

Lego, Karl. 1968a. *Geschichte des österreichischen Grundkatasters.* Vienna: Bundesamt für Eich- und Vermessungswesen. Pp. 1–7 and 10–13 of this book reprinted in Lego 1968b.

———. 1968b. "Der Mailänder Kataster—Vorbild der Katastralvermessungen des 19. Jahrhunderts." *Nachrichten der niedersächsischen Vermessungs- und Katasterverwaltung* 18, no. 3:89–100.

Leigh, Valentine. 1577. *The Moste Profitable and Commendable Science, of Surueying of Landes, Tenementes, and Heriditamentes.* London: Andrew Maunsell.

Lemon, James T. 1972. *The Best Poor Man's Country: A Geographical Study of Early Southeastern Pennsylvania.* Baltimore: Johns Hopkins University Press. Reprinted New York: Norton, 1976.

———. 1987. "Colonial America in the Eighteenth Century." In *North America: The Historical Geography of a Changing Continent,* edited by Robert D. Mitchell and Paul A. Groves, 121–46. London: Hutchinson.

Lester, Carl. 1963. "Dominion Land Surveys." *Alberta Historical Review* 11:20–28.

Levron, Jacques. 1937. *La Confection du cadastre dans le Département du Maine-et-Loire.* Angers: Archives du Maine-et-Loire.

Lewison, Anthony. 1987. "Creating a Cadastre in the Alpes-Maritimes, 1831–1842." *Landscape History* 9:65–76.

Liiv, Otto. 1938. "The Aspect of the Estonian Landscape at the End of the Seventeenth Century Based on the Maps of the Swedish Times in the Central State Archives." *Äratrükk Õpetatud Eesti Seltsi Toimetustest* 30:370–83.

Liljedahl, Ragnar. 1933. *Svensk förvaltning i Livland, 1617–1634.* Uppsala: Almqvist and Wiksell.

Linden, H. van der. 1988. "De Nederlandse waterhuishouding en waterstaatsorganisatie tot aan de moderne tijd." *Bijdragen en Mededelingen Betreffende de Geschiedenis der Nederlanden* 103, no. 4:534–53.

Little, Ezekiel. 1799. *The Usher. Comprising . . . Surveying; The Surveyor's Pocket Companion. . . .* Exeter, N.H.: H. Ranlet.

Lönborg, Sven E. 1901a. "Geografiska och kartografiska arbeten i Sverige under 1600-talet." *Ymer* 21, no. 2:59–78.

———. 1901b. "Om de äldsta kartorna över Sverige." *Ymer* 21, no. 1:113–44.

———. 1903. *Sveriges kartor: Tiden till omkring 1850.* Arbeten Utgifna med Understöd af Vilhelm Ekmans Universitetsfond 2. Uppsala.

Loor, A. H. 1973. "Cartography in the Use of Agriculture." In *Schakels met het verleden: Geschiedenis van de kartografie van Suriname / Links with the Past: The History of the Cartography of Suriname,* edited by Cornelis Koeman, 36–73. Amsterdam: Theatrum Orbis Terrarum.

Love, John. 1688. *Geodaesia; or, The Art of Surveying and Measuring of Land Made Easie.* London: John Taylor. 12th ed., "Adapted to American Surveyors," New York: Samuel Campbell, 1793.

Love, John B. 1970. "The Colonial Surveyor in Pennsylvania." University of Pennsylvania Ph.D. thesis.

Luka, K. 1980. "Pamje mbi ekonomine dhe topgraffine e visevi te Hofit" (summary in French). *Studime Hist* 34, no. 4:219–47.

Lynam, Edward. 1947. *British Maps and Map-Makers.* 3d rev. impression. London: W. Collins.

MacGregor, James G. 1981. *Vision of an Ordered Land: The Story of the Dominion Land Survey.* Saskatoon, Saskatchewan: Western Producer Prairie Books.

McManis, Douglas. 1975. *Colonial New England: A Historical Geography.* New York: Oxford University Press.

Madan, P. L. 1971. "Cartographic Records in the National Archives of India (1700–1900)." *Imago Mundi* 25:79–80.

Madge, Sidney J. 1938. *The Domesday of Crown Lands: A Study of the Legislation, Surveys, and Sales of Royal Estates under the Commonwealth.* London: Routledge.

Marais, Johannes S. 1927. *The Colonisation of New Zealand.* London: Oxford University Press.

Margry, P. J. 1984. "Drei proceskaarten (Geertruidenberg versus Standhazen) uit 1448." *Caert-Thresoor* 3, no. 2:27–33.

Marinoni, Joannes Jacobus. 1751. *De Re Ichnographica, cujus hodierna praxis exponitur et propriis exemplis Pluribus illustratur. . . .* Vienna: Kaliwoda.

Markham, Clements R. 1878. *A Memoir on the Indian Surveys.* 2d ed. London: Her Majesty's Stationery Office.

Marschner, Francis J. 1959. *Land Use and Its Patterns in the United States.* Agricultural Handbook no. 153. Washington, D.C.: U.S. Department of Agriculture.

———. 1960. *Boundaries and Records in the Territory of Early Settlement from Canada to Florida, with Historical Notes on the Cadaster and Its Potential Value in the Area.* Washington, D.C.: U.S. Department of Agriculture.

Marshall, William. 1801. *On the Appropriation and Inclosure of Commonable and Intermixed Lands. . . .* London: W. Bulmer.

———. 1804. *On the Landed Property of England. . . .* London: G. and W. Nichol.

Marstboom, Léon; Bourlon, Roger; and Jacobs, Albert E. 1956. *Le Cadastre et l'impôt foncier.* Brussels: Lielens.

Marthinsen, Jörgen H. 1971. "Oversikt over en del samlinger av utrykte Kart." *Geografen* 5, no. 1:54–58.

Marthinsen, Liv. 1982. "Landskyld og landskyldvarer." In *Den eldeste matrikkelen: En infallsport til historien: Skattematrikkelen 1647,* edited by Rolf Fladby and Harald Winge, 13–18. Oslo: Norsk Lokalhistorisk Institutt and Universitetsforlaget.

Mathias, Peter, and O'Brien, Patrick. 1976. "Taxation in Britain and France, 1715–1810: A Comparison of the Social and Economic Incidence of Taxes Collected for the Central Governments." *Journal of European Economic History* 5:601–50.

Matilla Tascón, A. 1947. *La única contribución y el catastro de la ensenada.* Servicio de Estudios de la Inspección General del Ministerio de Hacienda. Madrid: Sanchez Ocaña.

Mead, William R. 1975. "An Atlas of Settlement in Sixteenth-Century Finland." *Journal of Historical Geography* 1:17–20.

———. 1981. *An Historical Geography of Scandinavia.* London: Academic Press.

Meers, Karl. 1958. "Die erste Korbacher Katasterkarte aus den Jahren 1749–1756." *Vermessungstechnische Rundschau*:421–24.

Meinig, Donald W. 1986. *The Shaping of America: A Geographic Perspective on Five Hundred Years of History.* Vol. 1, *Atlantic America, 1492–1800.* New Haven, Conn.: Yale University Press.

Mergaerts, S. 1986. "Het kaartboek van Grimbergen van 1696." In *Oude kaarten en plattegronde: Bronnen voor de historische geografie van de zuidelijke Nederlanden (16de–18de eeuw) / Cartes et plans anciens: Sources pour la géographie historique de Pays-Bas méridionaux (XVIe–XVIIIe siècles),* edited by H. van der Haegen, F. Daelemans, and Eduard van Ermen, 205–31. Archief- en Bibliotheekwezen in België / Archives et Bibliothèques de Belgique, special no. 31. Brussels.

Merrens, H. Roy. 1964. *Colonial North Carolina in the Eighteenth Century: A Study in Historical Geography.* Chapel Hill: University of North Carolina Press.

Messner, Robert. 1967. "Dokumentation zur Geschichte des österreichischen Grundkatasters." In *150 Jahre österreichischer Grundkataster, 1817–1967: Ausstellungskatalog,* edited by Robert Messner, 69–157. Vienna: Bundesamt für Eich- und Vermessungswesen.

———. N.d. "Die katastrale Bearbeitung agrarische Operationen." In *50 Jahre Bundesamt für Eich- und Vermessungswesen,* edited by Robert Messner, 101–6. Vienna: Bundesamt für Eich- und Vermessungswesen.

Meurer, Peter H. 1981. "Zur Kartierung Heinsberger Stiftsbesitzes im 18. Jahrhundert." *Heimatkalender des Kreises Heinsberg,* 32–37.

Miedel, Hilde. 1975–76. "Zwei unbekannte Landkarten von Michael Hospinus." *Zeitschrift für Württembergische Landesgeschichte* 34–35:358–62.

Millard, A. R. 1987. "Cartography in the Ancient Near East." In *The History of Cartography;* vol. 1, *Cartography in Prehistoric, Ancient, and Medieval Europe and the Mediterranean,* edited by J. Brian Harley and David Woodward, 107–16. Chicago: University of Chicago Press.

Mingroot, E. van. 1989. "De oude kaart als historische bron." In *Kaarten met geschiedenis 1550–1800: Een selectie van oude getekende kaarten van Nederland uit de collectie Bodel Nijenhuis,* edited by Dirk de Vries, 16–30. Utrecht: H. and S. Publishers.

[Ministre des Finances]. 1811. *Le Recueil méthodique des lois, décrets, règlements, instructions, et décisions sur le cadastre de la France.* Paris: Imprimerie Impériale.

Minnen, Bart. 1986. "Achter de kaleidoscoop: En kritische kijk op de prekadastrale kaartboeken van Karel van Croÿ (†1612) voor het Hertogdom Aarschot." In *Oude kaarten en plattegronde: Bronnen voor de historische geografie van de zuidelijke Nederlanden (16de–18de Eeuw) / Cartes et plans anciens: Sources pour la géographie historique de Pays-Bas méridionaux (XVIe–XVIIIe siècles),* edited by H. van der Haegen, F. Daelemans, and Eduard van Ermen, 163–203. Archief- en Bibliotheekwezen in België / Archives et Bibliothèques de Belgique. special no. 31. Brussels.

Mitchell, J. E. 1979. "Victoria's Maps and Plan System: The Evolutionary Process." *Globe: Journal of the Australian Map Collectors Circle* 11:22–37.

Mitchell, Robert D. 1987. "The Colonial Origins of Anglo-America." In *North America: The Historical Geography of a Changing Continent,* edited by Robert D. Mitchell and Paul A. Groves, 93–120. London: Hutchinson.

Mitchell, Robert D., and Groves, Paul A., eds. 1987. *North America: The Historical Geography of a Changing Continent.* London: Hutchinson.

Moeller, Robert G. 1986. "Locating Peasants and Lords in Modern German Historiography." In *Peasants and Lords in Modern Germany: Recent Studies in Agricultural History,* edited by Robert G. Moeller, 1–23. Winchester, Mass.: Allen and Unwin.

Molemans, Jos. 1988. "Achtergronden van het ontstaan van het 18de-eeuwse kadaster in de Limburgse Kempen." In *Erschließung und Auswertung historischer Landkarten / Ontsluiting en gebruik van historische landkaarten,* 223–53. Landschaftsverband Rheinland—Archivberatungsstelle Rheinland. Archivhefte 18 / Werken uitgegeven door Limburgs Geschied- en Oudheidkundig Genootschap gevestigd te Maastricht 10. Cologne: Rheinland Verlag and Betriebsgesellschaft des Landschaftverbandes Rheinland.

Moore, Samuel. 1796. *An Accurate System of Surveying. . . .* Litchfield, Conn.: T. Collier.

Mooser, Josef. 1986. "Property and Wood Theft: Agrarian Capitalism and Social Conflict in Rural Society, 1800–50: A Westphalian Case Study." In *Peasants and Lords in Modern Germany: Recent Studies in Agricultural History,* edited by Robert G. Moeller, 52–80. Winchester, Mass.: Allen and Unwin.

Morgan, Victor. 1979. "The Cartographic Image of 'The Country' in Early Modern England." *Transactions of the Royal Historical Society* 29:129–54.

Moritsch, Andreas. 1969. *Das nahe triester Hinterland. Zur wirtschaftlichen und sozialen Entwicklung vom Beginn des 19. Jahrhunderts bis zur Gegenwart.* Wiener Archiv für Geschichte des Slaventums und Osteuropas 7. Vienna: Böhlau.

———. 1970. "Der franziszeische Kataster und die dazugehörigen Steuerschätzungsoperate als wirtschafts- und sozialhistorische Quellen." *East European Quarterly* 3, no. 4:438–48.

Mosselmans, Jean. 1976. "Introduction." In *De landt-meeters van de XVIe tot de XVIIIe eeuw in onze provincies / Les Géomètres-Arpenteurs du XVIe au XVIIIe siècle dans nos provinces,* by Jean Mosselmans and Roger Schonaerts, vii–lxiv. Brussels: Koninklijke Bibliotheek Albert I / Bibliothèque Royale Albert Ier.

Mosselmans, Jean, and Schonaerts, Roger. 1976. *De landt-meeters van de XVIe tot de XVIIIe eeuw in onze provincies / Les Géomètres-Arpenteurs du XVIe au XVIIIe siècles dans nos provinces.* Brussels: Koninklijke Bibliotheek Albert I / Bibliothèque Royale Albert Ier.

Mousnier, Mireille. 1986. "A propos d'un plan figuré de 1521: Paysages agraires et passages sur la Garonne." *Annales du Midi* 98:517–28.

Mousnier, Roland. 1979. *The Institutions of France under the Absolute Monarchy, 1598–1789.* Chicago: University of Chicago Press.

Muir, Ramsay, and Philip, George. 1927. *Philip's Historical Atlas Medieval and Modern.* 6th ed. London: London Geographical Institute.

Muller, E. 1979. "De invoering van het kadaster in Limburg." *Geodesia* 21, no. 7/8:253–57.

———. 1981. "Kadastrale plans vóór 1832." *Geodesia* 23, no. 5:178–82.

Muller, E., and Zandvliet, Kees. 1987a. "Historisch Overzicht." In *Admissies als landmeter in Nederland vóór 1811: Bronnen voor de geschiedenis van de landmeetkunde en haar toepassing in administratie, architectuur, kartografie, en vesting- en waterbouwkunde,* edited by E. Muller and Kees Zandvliet, 5–59. Alphen aan den Rijn: Canaletto.

———. 1987b. "Introduction." In *Admissies als landmeter in Nederland vóór 1811: Bronnen voor de geschiedenis van de landmeetkunde en haar toepassing in administratie, architectuur, kartografie, en vesting- en waterbouwkunde,* edited by E. Muller and Kees Zandvliet, 3–4. Alphen aan den Rijn: Canaletto.

Müller-Wille, Wilhelm. 1940. "Die Akten der Katastralabschätzung 1822–35 und der Grundsteuerregelung 1861–65 in ihrer Bedeutung für die Landesforschung in Westfalen." *Westfälische Forschungen* 3, no. 1:48–64.

Munier, Etienne. 1779. *Essai d'une méthode générale propre à étendre les connaissances, ou recueil d'observations.* Paris: Moutard.

Musall, Heinz. 1983. "Eine Forstkarte des Hardtwaldes nördlich Karlsruhe von 1756–57." *Beiträge zur Kartographie,* 51–57.

Musall, Heinz; Neumann, Joachim; Reinhard, Eugen; Salaba, Marie; and Schwarzmaier, Hansmartin. 1986. *Landkarten aus vier Jahrhunderten.* Karlsruher geowissenschaftliche Schriften Series A, 3. Karlsruhe: Fachhochschule Karlsruhe Fachbereich Vermessungswesen und Kartographie.

[Musée Savoisien]. 1980. *Le Cadastre sarde de 1730 en Savoie.* Chambéry: Musée Savoisien.

Neveux, Hugues; Jacquart, Jean; and Le Roy Ladurie, Emmanuel. 1975. *Histoire de la France rurale.* Vol. 2, *L'Age classique des paysannes, 1340–1789.* Paris: du Seuil.

Newcomb, Robert M. 1973. "Exploiting a Danish Domesday in the Study of Ridge and Furrow." Report to Occasional Discussions in Historical Geography, University of Cambridge.

Newton, S. C. 1968. "Short Guides to Records. No. 17, Parliamentary Surveys." *History* 53:51–54.

Nischer-Falkenhof, Ernst von. 1937. "The Survey by the Austrian General Staff under the Empress Maria Theresa and the Emperor Joseph

II, and the Subsequent Initial Surveys of Neighbouring Territories during the Years 1749–1854." *Imago Mundi* 2:83–88.

Niskanen, Viljo. 1963a. "Land Survey in the Past Thirty Years." In *Maanmittaushallitus: Lantmäteristyrelsen, 1812–1962,* by Viljo Niskanen et al., 8–33. Maanmittaushallituksen Julkaisu vol. 38. Helsinki.

———. 1963b. "Preface." In *Maanmittaushallitus: Lantmäteristyrelsen, 1812–1962,* by Viljo Niskanen et al., 6. Maanmittaushallituksen Julkaisu vol. 38. Helsinki.

Nissen, Kristian. 1937a. "Brødrene Franz Philip von Langen og Johann Georg von Langen." In *Norsk biografisk leksikon,* 8:185–90. Oslo.

———. 1937b. "Brødrene von Langen og deres virksomhet i Norge, 1737–1747." *Tidsskrift for Skogbruk* 45, no. 4:379–87.

———. 1959. "Skogbruk, sagbruk, og trelasthandel i Risørs Oppland 1732 på bakgrunn av Andreas Heitmans anonyme og udaterte karter over området." *Tidsskrift for Skogbruk* 67, no. 4:167–95.

Nissen, Poul. 1944. "Matrikulens kortgrundlag, dets oprindelse og vedligeholdelse." *Tidskrift for Opmaalings- og Matrikulsvæsen* 17, no. 1:53–59.

Norden, John. 1607. *The Surueyors Dialogue. . . .* London: H. Astley.

Nordlund, H. O. 1978. *Matriklen og matrikelkortene som hjælpekilder ved lokalhistoriske undersøgelser.* Lynby: Lokalhistoriske arkiver i Storkøbnhavn.

Nørdlund, Niels Erik. 1942. *Johannes Mejers kort over det danske rige.* Vol. 1, *Sjælland, Bornholm, Skaane, Halland, Blekinge, Gotland, og Færøerne;* vol. 2, *Jylland og Fyn;* vol. 3, *Åbenrå amt.* Geodætisk Instituts Publikationer 1–3. Copenhagen: Ejnar Munksgaard.

———. 1943. *Danmarks kortlægning. En historisk fremstilling,* 1 *Tiden till afslutningen af videnskabernes selskabs opmaaling.* Geodætisk Instituts Publikationer 4. Copenhagen: Ejnar Munksgaard.

———. 1944. *Færøernes kortlægning. En historisk fremstilling.* Geodætisk Instituts Publikationer 6. Copenhagen: Ejnar Munksgaard.

Norwood, Richard. 1637. *The Sea-mans practice . . . with Certaine Tables and Other Rules Usefull in Navigation: As also in the Plotting and Surveying of Places. . . .* London: George Hurlock.

Norz, Rudolf. 1967. "Katasterartige Vermessungen in Tirol vor der allgemeinen Katastralaufnahme." In *150 Jahre österreichischer Grundkataster,* edited by Robert Messner, Josef Mitter, and Manfred Schenk, 125–37. Vienna: Bundesamt für Eich- und Vermessungswesen.

Oakley, Stewart P. 1986. "Reconstructing Scandinavian Farms, 1660–1860: Sources in Denmark, Iceland, Norway, and Sweden." *Scandinavian Economic History Review* 34, no. 3: 181–203.

Oberhummer, Eugen. 1932. "Ein Jagdatlas Kaiser Karl VI." *Unsere Heimat* 6, no. 1:152–59.

———. 1934–35. "Die Herrschaft der Grafen von Hardegg im 18. Jahrhundert nach der Aufnahme von J. Marinoni, 1715–27." *Unsere Heimat* 7, no. 3:77–85; 8, no. 1:21–22, 152–59.

Ockeley, J. 1986. "De Westbrabantse landmeters en hun bijdrage tot de meting- en kaartboeken van gemeenten en abdijen in de streek tussen Zenne en Dender einde 16de eeuw–18de eeuw." In *Oude kaarten en plattegronde: Bronnen voor de historische geografie van de zuidelijke Nederlanden (16de–18de eeuw) / Cartes et plans anciens: Sources pour la géographie historique de Pays-Bas méridionaux (XVIe–XVIIIe siècles),* edited by H. van der Haegen, F. Daelemans, and Eduard van Ermen, 51–74. Archief- en bibliotheekwezen in België / Archives et Bibliothèques de Belgique, special no. 31. Brussels.

Oddens, Roelof P. 1989. "Maps in the Faculty of Geographical Sciences at the University of Utrecht." *Map Collector* 47:28–32.

O'Domhnaill, S. 1943. "The Maps of the Down Survey." *Irish Historical Studies* 3:381–92.

Oehme, Ruthardt. 1961. *Die Geschichte der Kartographie des deutschen Südwestens. Arbeiten zum Historischen Atlas von Südwestdeutschlands.* Vol. 3, Kommission für geschichtliche Landeskunde in Baden-Württemberg. Konstanz and Stuttgart: Jan Thorbecke.

Oliver, Richard R. 1986. "The Ordnance Survey in Great Britain, 1835–1870." University of Sussex D.Phil. thesis.

Opsal, Arnt. 1958. "I holzførsternes fotspor. IV: Johann Fridrich Lange." *Tidsskrift for Skogbruk* 66:16–30.

Örback, Alfred. 1978. "En återblick genom seklerna." *Lantmäteriverket Information* 6:2–5.

Osthoff, Friedrich. 1950. "Die Entstehung des rheinisch-westfälischen Katasters, 1808–1839." University of Bonn Landwirtschaftlichen Fakultät dissertation.

———. 1956. *Die älteren Flurbereinigungen im Rheinland und die Notwendigkeit von Zweitbereinigungen.* Schriftenreihe für Flurbereinigung 11. Bundesministerium für Ernährung, Landwirtschaft und Forsten.

Oswal, H. L., and Singh, Vijay. 1975. "Cadastral Surveys in India." *Survey Review* 3:156–65.

Padover, Saul K. 1967. *The Revolutionary Emperor: Joseph II of Austria.* London: Eyre and Spottiswoode.

Pálsson, Hermann. 1986. "Landnámabók." In *Dictionary of the Middle Ages,* edited by J. R. Strayer, 7:327–28. New York: American Council of Learned Societies.

Pálsson, Hermann, and Edwards, Paul, trans. 1972. *The Book of Settlements. Landnámabók.* Manitoba: University of Manitoba Press.

Past, Franz. 1978. "Das Teifelszeug." *Kultur und Technik* 1:8–14.

Patterson, Bradford R. 1980. "Introduction." In *The Surveying and Mapping of Wellington Province, 1840–76.* Wellington: Alexander Turnbull Library.

———. 1984. "Reading between the Lines: People, Politics, and the Conduct of Surveys in the Southern North Island, New Zealand, 1840–1876." Victoria University of Wellington Ph.D. thesis.

Pattison, William D. 1956. "Use of the U.S. Public Land Survey Plats and Notes as Descriptive Sources." *Professional Geographer* 8:10–14.

———. 1957. *Beginnings of the American Rectangular Land Survey System.* Department of Geography, Research Paper no. 50. Chicago: University of Chicago.

———. 1961. "The Original Plan for an American Rectangular Land Survey." *Surveying and Mapping* 21:339–45.

Pauli, Ulf. 1984. "Den geometriska uppmätning av Pommern: En pionjärinsats av svenska lantmätare under karolinsk tid." *Karolinska Förbundets Årsbok,* 63–79.

Pedersen, V. E. 1944. "Matriklernes anlæg og vedligeholdelse." *Tidskrift for Opmaalings- og Matrikulsvæsen* 17, no. 1:47–52.

Pelletier, Monique. 1988. "De nouveaux plans de forêts à la Bibliothèque Nationale." *Revue de la Bibliothèque Nationale* 29:56–62.

Perpillou, Aimé. 1935. "Les Documents cadastraux dans les études de géographie économique." *Bulletin de l'Association de Géographes Français* 84:10–18.

Perry, Thomas M. 1963. *Australia's First Frontier: The Spread of Settlement in New South Wales, 1788–1829.* Melbourne: Melbourne University Press.

Peterson-Berger, Einar. 1928. "Lantmäteriets kartografiska verksamhet: Resultat och betydelse." In *Svenska Lantmäteriet, 1628–1928,* vol. 1, 257–97. Stockholm: Sällskapet för Utgifvande av Lantmäteriets Historia.

Pett-Conklin, Linda M. 1986. "Cadastral Surveying in South Carolina: A Historical Geography." Louisiana State University and Agricultural and Mechanical College Ph.D. thesis.

Phillimore, Reginald H. 1945–50. *Historical Records of the Survey of India.* Vol. 1, *Eighteenth Century;* vol. 2, *1800 to 1815.* Dehra Dun, United Provinces: Surveyor-General of India.

———. 1947. "Historical Maps of the Survey of India." *Indian Archives* 1:198–205.

Piganiol, André. 1962. *Les Documents cadastraux de la colonie romaine d'Orange.* Paris: Centre National de la Recherche Scientifique.

Pitte, Jean-Robert. 1978. "Les Origines et l'évolution de la châtaigneraie vivaraise à travers un document cadastral du XVIIIe siècle." *Bulletin de la Section de Géographie* (Comité des Travaux Historiques et Scientifiques) 82:165–78.

———. 1983. *Histoire du paysage français.* Vol. 1, *Le Sacré: De la préhistoire au 15e siècle.* Paris: Tallandier.

Pitz, Ernst. 1957. "Die historische Karte des Landes Braunschweig im 18. Jahrhundert." *Braunschweigisches Jahrbuch* 38:141–49.

———. 1967. *Landeskulturtechnik, Markscheide-, und Vermessungswesen im Herzogtum Braunschweig bis zum Ende des 18. Jahrhunderts.* Veröffentlichungen der Niedersächsische Archivverwaltung 23.

Planhol, Xavier de. 1988. *Géographie historique de la France.* Paris: Fayard.

Plaul, Hainer. 1986. "The Rural Proletariat: The Everyday Life of Rural Labourers in the Magdeburg Region, 1830–80." In *The German Peasantry,* edited by Richard J. Evans and William R. Lee, 102–28. Beckenham: Croom Helm.

Poorter, Serge de. 1980. "Introduction." In *Répertoire numérique de la sous-série 3P,* by Serge de Poorter, Jocelyne Girot, and Noëlle Riva, 11–97. Caen: Archives Départementales du Calvados.

Postma, C. 1977. "Nicolaes Kruikius en zijn werk." Introduction to *Kruikius kaart van Delfland 1712,* edited by C. Postma. Alphen aan den Rijn: Canaletto.

Pouls, H. C. 1984. "Landmeetkundige methoden en instrumenten tot 1800." In *Stad in kart: Voordrachten gehouden op het Congres "De historische stadsplattegrond—Spiegel van wens en werklijheid,"* edited by F. J. Bakker, E. A. J. Boiten, and W. K. van der Veen, 13–28. Alphen aan den Rijn: Canaletto.

Pounds, Norman J. G. 1979. *An Historical Geography of Europe, 1500–1840.* Cambridge: Cambridge University Press.

Powell, Joseph M. 1970a. *The Public Lands of Australia Felix: Settlement and Land Appraisal in Victoria, 1834–91, with Special Reference to the Western Plains.* Melbourne: Oxford University Press.

———. 1970b. "The Victorian Survey System, 1837–1860." *New Zealand Geographer* 26:50–69.

———. 1972. "The Records of the New Zealand Lands and Survey Department." *New Zealand Geographer* 28:72–77.

———. 1977. *Mirrors of the New World: Images and Image-Makers in the Settlement Process.* Folkestone: Dawson.

Powell, Joseph M., and Williams, Michael, eds. 1975. *Australian Space Australian Time: Geographical Perspectives.* Melbourne: Oxford University Press.

Price, D. J. 1955. "Medieval Land Surveying and Topographical Maps." *Geographical Journal* 121:1–10.

Prinz, Joseph. 1950. "Die ältesten Landkarten, Kataster-, und Landesaufnahmen des Fürstentums Osnabrück. III." *Mitteilungen des Vereins für Geschichte und Landeskude von Osnabrück* 64:110–45.

Raam, Pedeer Nilsson. 1670. *Ortuga deelo-bok.* Strangnäs.

Rålamb, Åke Claesson. 1690. *Adelig öfning.* Vol. 1.

Ratsma, P. 1975. "De landmeter Jan Jansz. Potter: De topografie van Rotterdam en omgeving in de tweede helft van de zestiende eeuw." *Holland: Regionaal-Historische Tijdschrift* 7:300–321.

Regele, Oskar. 1955. *Beiträge zur Geschichte der staatlichen Landesaufnahme und Kartographie in Österreich bis zum Jahre 1918.* Vienna: Verlag des Notringes der Wissenschaftlichen Verbände Österreichs.

Ricci Massabò, Isabella. 1983. "Conoscenza, memoria, gestione della terra nella rapresentazione catastale." In *Arte e scienza per il disegno del mondo,* edited by Carlo Pirovano, 187–94. Milan: Electra.

Richter, Alfred. 1921. "Die sächsische Landesvermessung (Katastervermessung)." In *Beiträger zur deutschen Kartographie,* edited by Hans Praesent, 61–63. Leipzig: Akademische Verlagsgesellschaft.

Ristow, Walter W. 1965. "Nineteenth-Century Cadastral Maps in Ohio." *Papers of the Bibliographical Society of America* 59:306–15.

———. 1966. "Alfred T. Andreas and His Minnesota Atlas." *Minnesota History* 40:120–29.

———. 1974. "Dutch Polder Maps." *Quarterly Journal of the Library of Congress* 31, no. 3:136–49.

———. 1977. *Maps for an Emerging Nation: Commercial Cartography in Nineteenth-Century America.* Washington, D.C.: Library of Congress.

Roberts, Brian K. 1968. "An Early Tudor Sketch Map: Its Context and Implications." *History Studies* 1:33–38.

Roberts, Michael. 1953. *Gustavus Adolphus: A History of Sweden, 1611–1632.* Vol. 1. London: Longmans.

Roberts, Stephen H. 1924. *History of Australian Land Settlement, 1788–1920.* Melbourne: Macmillan. Reprinted 1968.

Robinson, M. E. 1974. "The Robertson Land Acts in New South Wales, 1861–84." *Transactions of the Institute of British Geographers* 61:17–33.

Robinson, W. Stitt. 1957. *Mother Earth: Land Grants in Virginia, 1607–1699.* Williamsburg, Va.: Virginia 350th Anniversary Celebration Corporation.

Roessingh, H. K. 1968–69. "Hoe zijn de gelderse verpondingskohieren uit het midden van de 17e eeuw ingericht? Bronnen voor het lokaalhistorisch onderzoek." *Gelre Bijdragen en Mededelingen* 63:61–71.

Rohrbough, Malcolm J. 1968. *The Land Office Business: The Settlement and Administration of American Public Lands, 1789–1837.* New York: Oxford University Press.

Roome, William. 1883. *The Early Days and Early Surveys of New Jersey.* Morristown, N.J.: Jerseyman Press.

Rosén, Jerker. 1961. "Scandinavia and the Baltic." In *The New Cambridge Modern History;* vol. 5, *The Ascendancy of France, 1648–88,* edited by Francis L. Carsten, 519–42. Cambridge: Cambridge University Press.

Roskievicz, J. 1873. "Zur Geschichte der Kartographie in Österreich." *Mitteilungen der k. und k. Geographischen Gesellschaft* 16:248–63, 289–98.

———. 1875. *Die Kartographie in Österreich vom Jahre 1750 bis zum Jahre 1873.* Vienna: Siedel.

Rouselle, Joseph. 1770. *Instructions pour les seigneurs et leurs gens d'affaires.* Paris: Lotin.

Rubow, Ernst. 1954–55. "Die historische Geographie in Greifswald und die Arbeiten am schwedischen Matrikelwerk (1692–1698)." *Wissenschaftliche Zeitschrift der Ernst Moritz Arndt-Universität Greifswald.* Mathematisch-Naturwissenschaftliche Reihe 6–7:643–46.

Rubow-Kalähne, Marianne. 1954–55. "Die Revision der schwedischen Landesaufnahme von Vorpommern und Rügen und ihr geographischer Wert." *Wissenschaftliche Zeitschrift der Ernst Moritz Arndt-Universität Greifswald.* Mathematisch-Naturwissenschaftliche Reihe 6–7:647–51.

———. 1959. "Langstreifenfluren in Neu-Vorpommern: Eine Auswertung der schwedischen Matrikelkarten." *Wissenschaftliche Zeitschrift der Martin-Luther-Universität Halle-Wittenberg.* Mathematisch-Naturwissenschaftliche Reihe 8:663–76.

———. 1960. *Matrikelkarten von Vorpommern 1692–1698 nach der schwedischen Landesaufnahmen: Eine kurze Erläuterung zu den kartenblättern Neuenkirchen, Griefswald, Wusterhusen, und Hanshagen.* Leipzig: Verlag Enzyklopädie.

Rudé, George F. E. 1967. "English Rural Disturbances on the Eve of the First Reform Bill, 1830–1." *Past and Present* 37:87–102.

Salgaro, Silvino. 1980. "Il governo delle acque nella pianura Veronese da una carta del XVI secolo." *Bollettino della Società Geografica Italiana,* ser. 10, no. 9:327–50.

Sandell, Richard E., ed. 1971. *Abstracts of Wiltshire Inclosure Awards and Agreements.* Devizes: Wiltshire Record Society.

Scaraffia, Lucetta, and Sereno, Paulo. 1976. "L'area Piedmontese." In *Storia d'Italia;* vol. 6, *Atlante,* edited by Ruggiero Romano and Corrado Vivanti, 506–19. Turin: Einaudi.

Schäfer, Alfons. 1968. "Die erste amtliche Vermessung und Landesaufnahme in der Markgrafschaft Baden im 18. Jahrhundert." In *Beiträge zur Geschichtliche Landeskunde. Geographie, Geschichte, Kartographie. Festgabe für Ruthardt Oehme zur Vollendung des 65. Lebensjahres,* 141–65. Veröffentlichungen der Kommission für geschichtliche Landeskunde in Baden-Württemberg, Series B 46. Stuttgart: W. Kohlhammer.

Schäfer, Alfons, and Weber, H. 1971. *Inventar der handgezeichneten Karten und Plänen zur europäischen Kriegesgeschichte des 16.-19. Jahrhunderts im Generallandesarchiv Karlsruhe.* Veröffentlichungen der Staatliche Archivverwaltung Stuttgart 25.

Schafer, Joseph. 1940. "The Wisconsin Domesday Book: A Method of Research for Agricultural Historians." *Agricultural History* 14:23–32.

Schäfer, Karl. 1979. "Leben und Werk des korbacher Kartographen Joist Moers." *Geschichtsblätter für Waldeck* 67:123–77.

Schama, Simon. 1988. *The Embarrassment of Riches: An Interpretation of Dutch Culture in the Golden Age.* London: Fontana.

Scharfe, Wolfgang. 1972. *Abriss der Kartographie Brandenburgs, 1771–1821.* Veröffentlichungen der historischen Kommission zu Berlin 35. Berlin: Walter de Gruyter.

———. 1986. "Brandenburg". In *Lexikon zur Geschichte der Kartographie von den Anfängen bis zum Ersten Weltkrieg,* edited by Ingrid Kretschmer, Johannes Dörflinger, and Franz Wawrik, 105–7. Vienna: Franz Deuticke.

Scheffer, A. 1976. "De landmeting in de Meijerij anno 1792–1793." *Nederlands Geodetisch Tijdschrift* 6, no. 4:59–64.

———. 1977. "Het 'Hollandse kadaster.'" *Nederlands Geodetisch Tijdschrift* 7, no. 2:17–26.

Scheider, Theodor, ed. 1968. *Handbuch der europäischen Geschichte.* Vol. 4. Stuttgart: Union-Verlag.

Scherzer, Walter. 1970. "Die Entwicklung der Kartographie im Hochstift Würzburg." In *Volkskultur und Geschichte: Festgabe für Josef Dünninger zum 65. Geburtstag,* edited by Dieter Harmening, Gerhard Lutz, Bernhard Schemmel, and Erich Wimmer, 153–69. Berlin: Erich Schmidt Verlag.

———. 1977. "Die Darstellung spätmittelalterlicher—frühneuzeitlicher Siedlungsentwicklung auf einer zeitgenössichen Karte." *Mainfränkisches Jahrbuch für Geschichte und Kunst* 29:66–80.

Schiemann, Theodor. 1882. *Der älteste schwedische Kataster Livlands und Estlands: Eine Ergänzung zu der baltischen Güterchroniken.* Reval (Tallin): Franz Kluge.

Schilder, Gunter. 1989. "Historische Kartografie: Quo Vadis?" *Kartografisch Tijdschrift* 15, no. 2:23–30.

Schissler, Hanna. 1986. "The Junkers: Notes on the Social and Historical Significance of the Agrarian Elite in Prussia." In *Peasants and Lords in Modern Germany: Recent Studies in Agricultural History,* edited by Robert G. Moeller, 24–51. Winchester, Mass.: Allen and Unwin.

Schön, Berthold. 1970. "Die ersten Flurkarten." *Heimatkalender Jahrbuch Kreis Dinslaken* 27:66–72.

Schonaerts, Roger. 1976. "Catalogue." In *De landtmeeters van de XVIe tot de XVIIIe eeuw in onze provincies / Les Géomètres-Arpenteurs du XVIe au XVIIIe siècle dans nos provinces,* by Jean Mosselmans and Roger Schonaerts, 1–153 Brussels: Koninklijke Bibliotheek Albert I / Bibliothèque Royale Albert Ier.

Schulte, Paul-Günter. 1984. "Vom klevischen Grundsteuerregister zur Katasterkarte." *Nachrichten aus dem öffentlichen Vermessungsdienst Nordrhein-Westfalen* 17, no. 3:185–201.

Schumm, Karl. 1961. *Inventar der handschriftlichen Karten im Hohenlohe-Zentralarchiv in Neuenstein.* Inventare der Nichtstaatlichen Archive Baden-Württemberg 8.

Schwartz, Seymour I., and Ehrenberg, Ralph H. 1980. *The Mapping of America.* New York: H. M. Abrams.

Sears, Paul B. 1925–26. "The Natural Vegetation of Ohio. I: Map of the Virgin Forest; II: The Prairies." *Ohio Journal of Science* 25 (1925): 139–49; 26:128–46.

Sebert, L. M. 1980. "The Land Surveys of Ontario, 1750–1850." *Cartographica* 17:65–106.

Selling, Gösta. 1981. "Hässelby-karta återförd till Sverige." *Spånga Byden. Spånga Fornminnes- och Hembygdsgille* 1:5–10.

Sems, Johan, and Dou, Jan Pieterszoon. 1600. *Practijck des landmeterns.* Leiden: Bouwens.

Serres, Olivier de. 1600. *Le Théatre d'agriculture et mesnage des champs.* Paris: Jamet-Métayer.

Sevatdal, Hans. 1982. "Forelesningar og studiemateriale til kurs i eigedomshistorie." Ås: Norges Landbrukshøgskole. Typescript.

———. 1985. *Offentlig grunn og bygdeallmenninger. Nasjonalatlas for Norge.* Hovedtema 8. Jord- og skogbruk." Oslo: Norges Geografiske Oppmåling.

———. 1988. "Innleiing til matrikkellære: Forelesningsnotat i kurset E2/E3 1988." Ås: Norges Landbrukshøgskole. Typescript.

Shanks, Royal E. 1953. "Forest Composition and Species Association of the Beech-Maple Forest Region of Western Ohio." *Ecology* 34:455–66.

Shepherd, William R. 1930. *Historical Atlas.* London: University of London Press.

Sherman, Christopher E. 1925. *Final Report of the Ohio Cooperative Topographic Survey.* Vol. 3, *Original Ohio Land Subdivisions.* Columbus: Press of the Ohio State Reformatory.

Shore, A. F. 1987. "Egyptian Cartography" In *The History of Cartography;* Vol. 1, *Cartography in Prehistoric, Ancient, and Medieval Europe and the Mediterranean,* edited by J. Brian Harley and David Woodward, 117–29. Chicago: University of Chicago Press.

Short, Brian. 1989. *Lloyd George's Domesday.* Research Series no. 22. Cheltenham: Historical Geography Research Group.

Sigurdsson, Harald. 1978. *Kortasaga Islands frá lokum 16.* Reykjavík: Bókaútgáfa Menningarsjods og Pjodvinafélagsins.

Skelton, Raleigh A., and Harvey, Paul D. A. 1969. "Local Maps and Plans before 1500." *Journal of the Society of Archivists* 3:496–97.

Skelton, Raleigh A., and Summerson, John. 1971. *The Maps and Architectural Drawings at Hatfield House.* Oxford: Roxburghe Club.

Skrubbeltrang, Fridlev. 1961. "Developments in Tenancy in Eighteenth-Century Denmark as a Move towards Peasant Proprietorship." *Scandinavian Economic History Review* 9, no. 1:167–75.

Smailes, P. J. 1966–67. "The Large Scale Cadastral Map Coverage in Australia, and the Parish Maps of New South Wales." *Australian Geographer* 10:81–94.

Smet, Antoine de. 1969. *Philippe Vandermaelen, 1795–1869.* Brussels: Koninklijke Bibliotheek Albert I.

Smidt, A. de. 1970. "A Brief History of the Surveys and Cartography of the Colony of the Cape of Good Hope." *South Africa Survey Journal* 74:3–11.

Smith, Adam. 1976. *An Inquiry into the Nature and Causes of the Wealth of Nations.* Edited by R. H. Campbell, Andrew S. Skinner, and W. B. Todd. 2 vols. Oxford: Oxford University Press.

Smith, Robert E. F. 1977. *Peasant Farming in Moscovy.* Cambridge: Cambridge University Press.

Solle, Henriette. 1987. "L'Utilisation agricole du sol en France: Les cartes Aimé Perpillou." *Acta Geographica* 45:1–25.

Soom, Arnold. 1954. *Der Herrenhof in Estland im 17. Jahrhundert.* Lund: Skånska Centraltrykeriet.

Speiermann, F. 1974. "Ein Jahrhundert Lübecker

Kataster." *Allgemeine Vermessungsnachrichtung* 81:338–41.

Sperling, Walter. 1986. "Rheinland." In *Lexikon zur Geschichte der Kartographie von den Anfängen bis zum Ersten Weltkrieg,* edited by Ingrid Kretschmer, Johannes Dörflinger, and Frans Wawrik, 670–73. Vienna: Franz Deuticke.

Spielman, John P. 1977. *Leopold I of Austria.* London: Thames and Hudson.

Sporrong, Ulf. 1984a. "Bebyggelsehistoria genom kartstudier." *Bebyggelsehistorisk Tidskrift* 7:6–24.

———. 1984b. "Samhällshistoria genom våre äldre lantmäteriakter: Några reflexioner med utgångspunkt från tidiga radikala jordaskifte i Sörmland under 1700-talets förra hälft." *Ymer* 104:134–46.

Sporrong, Ulf, and Wennström, Hans-Frederik, eds. 1990. *National Atlas of Sweden.* Vol. 1, *Maps and Mapping.* Stockholm: National Atlas of Sweden.

Stams, Werner. 1986. "Sachsen und Thüringen." In *Lexikon zur Geschichte der Kartographie von den Anfängen bis zum Ersten Weltkrieg,* edited by Ingrid Kretschmer, Johannes Dörflinger, and Franz Wawrik, 692–96. Vienna: Franz Deuticke.

[Statens Lantmäteriverk]. 1977. "Utredningsrapport: Forskningsarkiv och kartlagning." Typescript.

[Statens Offentliga Utredningar]. 1986. *Rätten till jord—utveckling och reformkrav.* Statens Offentliga Utredningar 52, no. 2.

Staunskjær, Jørgen. 1981. "Udskiftningen ved Brahetrolleborg." In *Udskiftningsforordning 200 år* (= *Landinspektøren* 30, no. 8), 541–48.

Stavenhagen, W. 1904. *Skizze der Entwicklung und des Standes des Kartenwesens des außerdeutschen Europa.* Ergänzungsheft no. 148 to *Petermanns Mitteilungen.* Gotha.

Stephenson, Richard W. 1967. *Landownership Maps: A Checklist of Nineteenth Century United States County Maps in the Library of Congress.* Washington, D.C.: Library of Congress.

Steur, A. G. van der, ed. 1985. *Chaerte vande vrye Heerlickheydt Warmondt: Een pre-kadastrale kaarte uit 1667 vervaardigd door Johan Dou(w).* Alphen aan den Rijn: Canaletto.

Stewart, Lowell O. 1935. *Public Land Surveys: History, Instructions, Methods.* Ames, Iowa: Collegiate Press.

Stokman, J. 1986. "Hendrik Verhees." *Caert-Thresoor* 5, no. 1:12–17.

Stone, Jeffrey C. 1988. "Imperialism, Colonialism, and Cartography." *Transactions of the Institute of British Geographers,* n.s., 13:57–64.

Styffe, Carl G. 1856. *Samlingar af instructioner rörande den civila förvaltningen i Sverige och Finnland.* Föranstaltad av Kongl. Samfundet för utgifwande af handskrifter rörande Skandinaviens Historia 1. Stockholm: Hörberga Boktrykeriet.

Švābe, Arveds. 1933. "Die älteste schwedische Landrevision Livlands." *Latvijas Universitates Raksti (Acta Universitatis Latviensis). Tautsaimniecìbas uu Tiesìbu Zinàtuu Fakultàtes* 2, no. 3:337–596.

Sweeney, C. J., and Simson, J. A. 1967. "The Ordnance Survey and Land Registration." *Geographical Journal* 133:10–23.

Taylor, Eva G. R. 1947. "The Surveyor." *Economic History Review* 17:121–33.

Teeling, P. S. 1955. "Van verpondingsadminstratie tot grondslag van de ruimtelijke ordening." Parts 1–3. *Orgaan der Vereninging van technische Ambtenaren van het Kadaster* 13, no. 3:92–98; no. 4:131–36; no. 6:208–15.

———. 1981. *Repertorium van oud-Nederlandse landmeters 14e tot 18e eeuw.* Part 1, *Bibliography;* part 2, *Surveyors.* Apeldoorn: Hoofdirectie van de Dienst van het Kadaster en de Openbare Registers.

———. 1983. *Repertorium van oud-Nederlandse landmeters 14e tot 18e Eeuw: Index.* Compiled by P. C. J. van der Krogt. Apeldoorn: Hoofdirectie van de Dienst van het Kadaster en de Openbare Registers.

———. 1984. *Landmeters van de kadastrering van Nederland.* Apeldoorn: Hoofdirectie van de

Dienst van het Kadaster en de Openbare Registers.

Thirsk, Joan, ed. 1984. *The Agrarian History of England and Wales, 1640–1750.* Cambridge: Cambridge University Press.

Thomas, Colin. 1988. "Industrial Development in Agrarian Communities: Case Studies from Slovenia." *Geografiska Annaler* 70, B no. 1:227–37.

Thompson, Francis M. L. 1968. *Chartered Surveyors: The Growth of a Profession.* London: Routledge and Kegan Paul.

Thompson, John T. 1875. *An Exposition of Processes and Results of the Survey System of Otago.* Dunedin: Henry Wise.

Thomson, Don W. 1966. *Men and Meridians: The History of Surveying and Mapping in Canada.* Vol. 1, *Prior to 1867.* Ottawa: The Queen's Printer. Vols. 2 and 3 published 1967 and 1969.

Thrower, Norman J. W. 1957. "Cadastral Surveys and Roads in Ohio." *Annals of the Association of American Geographers* 47:181–82.

———. 1961. "The County Atlas of the United States." *Surveying and Mapping* 21:365–73.

———. 1966. *Original Survey and Land Subdivision: A Comparative Study of the Form and Effect of Contrasting Cadastral Surveys.* Chicago: Association of American Geographers.

———. 1972. "Cadastral Survey and County Atlases in the United States." *Cartographic Journal* 9:43–51.

Thygesen, P. 1915. "Matrikulsmaalinger og andre økonomiske maalinger." In *Forelæsninger over økonomisk landmaaling,* paragraph 44. 2d ed. Copenhagen: Det Privat Ingeniørfonds Foranstaltning and Philipsen.

[Tithe Commissioners for England and Wales]. 1837. *Instructions as to Forms of Apportionment and Maps.* London: Tithe Commission, 31 July.

Toms, K. N. 1976. "The Dimensions of Cadastre: A Historical Approach." *Australian Surveyor* 28:187–216.

Toms, K. N., and Plunkett, P. M. 1982. "Crown Land Survey Administration in Van Dieman's Land: A Historical Indicator for Integrationists." *Australian Surveyor* 31:72–83.

Tønnesson, Kåre. 1981. "Tenancy, Freehold, and Enclosure in Scandinavia from the Seventeenth to the Nineteenth Century." *Scandinavian Journal of History* 6:191–206.

Treial, Helene. 1931. "Einige Daten über die Landkartensammlung des estnischen Staatszentralarchivs." *Õpetatud Eesti Seltsi Aastaraamat / Sitzungsberichte der Gelehrten Estnischen Gesellschaft,* 158–75.

Turner, Michael E. 1977. "Enclosure Commissioners and Buckinghamshire Parliamentary Enclosure." *Agricultural History Review* 25:120–29.

———. 1980. *English Parliamentary Enclosure: Its Historical Geography and Economic History.* Folkestone: Dawson.

———. 1984a. *Enclosures in Britain, 1750–1830.* London: Macmillan.

———. 1984b. "The Landscape of Parliamentary Enclosure." In *Discovering Past Landscapes,* edited by Michael Reed, 132–66. London: Croom Helm.

———, ed. 1978. *W. E. Tate, a Domesday of English Enclosure Acts and Awards.* Reading: University of Reading.

Turner, Michael E., and Mills, Dennis, eds. 1986. *Land and Property: The English Land Tax, 1692–1832.* Gloucester: Alan Sutton.

Ulbrich, Karl. 1961. "Genauigkeit der ersten Messtisch-Katastralvermessung in Österreich." *Österreichische Zeitschrift für Vermessungswesen* 49, no. 2:44–53.

———. 1967a. "Die Entwicklung des Zeichenschlüssels der österreichischen Katastralvermessung." In *150 Jahre österreichischer Grundkataster,* edited by Robert Messner, Josef Mitter, and Manfred Schenk, 159–66. Vienna: Bundesamt für Eich- und Vermessungswesen.

———. 1967b. "Vorläufer des Stabilen franziszeischen Katasters." In *150 Jahre österreichischer Grundkataster,* 1817–1967: *Ausstellungskatalog,* edited by Robert Messner, 7–21. Vienna: Bundesamt für Eich- und Vermessungswesen.

———. 1967c. "Zeittafel zur historischen Entwicklung der österreichischen Katastralvermessung." In *150 Jahre österreichischer Grundkataster,* edited by Robert Messner, Josef Mitter, and Manfred Schenk, 167–96. Vienna: Bundesamt für Eich- und Vermessungswesen.

Ulshöfer, Kuno. 1968. "Der Feldmesser Daniel Meyer aus Basel und die Landrenovatur in Hall um 1700." *Der Hallquell: Blätter für Heimatkunde des Haller Landes* 20 no. 13:49–56.

Unger, Horst. 1964. *Das Kataster- und Vermessungswesen Schleswig-Holsteins in Jahrzehnten.* Kiel: Landesvermessungsamt Schleswig-Holstein.

[United States Bureau of Land Management]. 1947. *Manual of Instructions for the Survey of the Public Lands of the United States.* Washington, D.C.: U.S. Department of the Interior.

[United States General Land Office]. 1855. *Instructions to the Surveyors General of Public Lands of the United States for those Surveying Districts Established in and since the Year 1850; Containing also a Manual of Instructions to Regulate the Field Operations of Deputy Surveyors, Illustrated by Diagrams.* Washington, D.C.: A. O. P. Nicholson.

Uzes, Francois D. 1977. *Chaining the Land: A History of Surveying in California.* Fort Sutter Station, Calif.: Landmark Enterprises.

Vanhove, A. Luc. 1986."De kaartboeken van de abdij van Park-Heverlee." In *Oude kaarten en plattegronde: Bronnen voor de historische geografie van de zuidelijke Nederlanden (16de–18de Eeuw) / Cartes et plans anciens: Sources pour la géographie historique de Pays-Bas méridionaux (XVIe–XVIIIe siècles),* edited by H. van der Haegen, F. Daelemans, and Eduard van Ermen, 121–61. Archief- en Bibliotheekwezen in België / Archives et Bibliothèques de Belgique, special no. 31. Brussels.

Vannereau, Marie-Antoinette. 1976. *Places et provinces disputés: Exposition des cartes et plans du XVe au XIXe siècle.* Comité des Travaux Historiques et Scientifiques 101e Congrès National des Sociétés Savantes, Lille, 1976. Paris: Bibliothèque Nationale.

Vasar, Juhan. 1930–31. *Die große livländische Güterreduktion. Die Entstehung des Konflikts zwischen Karl XI und der livländischen Ritter- und Landschaft, 1678–1684.* Eesti vabariigi tartu ülikooli Toimetused (Acta et Commentationes Universitatis Tartuensis [Dorpatensis]) B Humaniora pt. 1, vol. 20; pt. 2, vol. 22.

Verhelst, J. 1982. *De Documenten uit de ontstaansperiode van het moderne kadaster en van de grondbelasting (1790–1835).* Miscellanea Archivistica 31. Brussels: Algemeen Rijksarchief.

Verhoeve, Antoon. 1980. "Bodemkaart en landboek in een landschapsgenetisch perspectief." In *Bronnen voor de historische geografie van België / Sources de la géographie historique de la Belgique,* edited by Jacques Mertens, 241–54. Report from the conference held 25–27 April 1979. Brussels: Algemeen Rijksarchief en Rijksarchief in de Provincien / Archive Général du Royaume et Archives de l'Etat dans les Provinces.

———. 1986. "Het landboek als cartografische bron voor cultuur-historisch landschapsonderzoek." In *Oude kaarten en plattegronde: Bronnen voor de historische geografie van de zuidelijke Nederlanden (16de–18de eeuw) / Cartes et plans anciens: Sources pour la géographie historique de Pays-Bas méridionaux (XVIe–XVIIIe siècles),* edited by H. van der Haegen, F. Daelemans, and Eduard van Ermen, 253–74. Archief- en Bibliotheekwezen in België / Archives et Bibliothèques de Belgique, special no. 31. Brussels.

Vierhaus, Rudolf. 1988. *Germany in the Age of Absolutism.* Translated by Jonathan B. Knudsen. Cambridge: Cambridge University Press.

Vivoli, Carlo. 1987. "Cartographie et institutions: Les plans des fiefs de la Toscane du XVIIIe siècle." Paper presented at the Thirteenth International Conference on the History of Cartography, Paris.

Vogel, Walther. 1929. "La Documentation de l'histoire économique: Les plans parcellaires: Allemagne." *Annales d'Histoire Economique et Sociale* 1:225–29.

Vogel, Werner, and Zögner, Lothar, eds. 1979. *Preußen im Kartenbild.* Berlin; Geheimes Staatsarchiv Preußische Kulturbesitz.

Vollmer, Giesela. 1982. "Eine Wasserburg." In *Zeugnisse rheinischer Geschichte,* edited by Frans-Josef Heyen and Wilhelm Janssen, 236–40, 442. Rheinischer Verein für Denkmalpflege und Landschafftschutz, Jahrbuch 1982–83. Neuss: Verlag Gesellschaft für Buchdruckerei AG.

Vorthmann, Albert. 1956. "Historische Karte des Landes Braunschweig im 18. Jahrhundert. II: Erläuterungen in technischer Sicht." *Niedersächsiches Jahrbuch für Landesgeschichte* 28:8–14.

Vries, Barbara de. 1966. "The Role of the Land Surveyor in the Development of New Zealand, 1840–76." Victoria University of Wellington M.A. thesis.

———. 1974. "The Colonial Surveyors of New Zealand, 1840–76: An Overview." *New Zealand Surveyor* 27:509–17.

Vries, Dirk de. 1983. *Beemsterlants caerten: Een beredeneerde lijst van oude gedrukte kaarten.* Alphen aan den Rijn: Canaletto.

Vries, Jan de. 1974. *The Dutch Rural Economy in the Golden Age, 1500–1700.* New Haven, Conn. Yale University Press.

Vuylsteke, G. 1986. "De betrouwbaarheid van landmeterskaarten van de Parochie Bavikhove." In *Oude kaarten en plattegronde: Bronnen voor de historische geografie van de zuidelijke Nederlanden (16de–18de eeuw) / Cartes et plans anciens: Sources pour la géographie historique de Pays-Bas méridionaux (XIVe–XVIIIe siècles),* edited by H. van der Haegen, F. Daelemans, and Eduard van Ermen, 233–51. Archief- en Bibliotheekwezen in België / Archives et Bibliothèques de Belgique, special no. 31. Brussels.

Wacker, Peter O. 1975. *Land and People. A Cultural Geography of Preindustrial New Jersey: Origins and Settlement Patterns.* New Brunswick, N.J.: Rutgers University Press.

Wagret, Paul. 1968. *Polderlands.* London: Methuen.

Wakefield, Edward G. 1841. "Letter to the Colonization Commissioners for South Australia, 2 June 1835." *British Parliamentary Papers, House of Commons,* 1st sess., vol. 4, Appendix, 332–39.

Wakefield, Felix. 1849. *Colonial Surveying with a View to the Disposal of Waste Land: In a Report to the New Zealand Company.* London: John W. Parker.

Wall, George, Jnr. 1788. *A Description with Instructions for the Use of a Newly Invented Surveying Instrument, Called the Trigonometer.* Philadelphia: Zachariah Poulson Jnr.

Wallenius, Helmer. 1970. "Lantmäteriarkiven i Finland." *Kart og Plan* 30, no. 2:57–62.

Walter, François. 1980. "Cadastre et histoire rurale: Contribution à l'étude de la petite propriété familiale en pays de Fribourg (milieu du XIXe siècle), questions et méthode." *Schweitzerische Zeitschrift für Geschichte* 30:29–58.

Wangermann, Ernst. 1969. *From Joseph II to the Jacobin Trials: Government Policy and Public Opinion in the Habsburgh Dominions in the Period of the French Revolution.* Oxford: Oxford University Press 2nd edition.

Wannerdt, Arvid. 1982. *Den svenska folkbokföringens historia under tre sekler.* Riksskatteverkets utbildning och information. Stockholm: Riksskatteverket.

Warkentin, John, and Ruggles, Richard I. 1970. *Manitoba Historical Atlas: A Selection of Facsimile Maps, Plans, and Sketches from 1612 to 1969.* Winnipeg: Historical and Scientific Society of Manitoba.

Weiden, A. v. d. 1955. "Die Urkataster und die Entwicklung sowie Neugestaltung der katasteramtlichen Messungs- und Kartenwerke in Niedersachsen." In *C. F. Gauss und die Landesvermessung in Niedersachsen,* by W. Gronwald et al., 155–84. Hanover: Niedersächsisches Landesvermessungsamt.

Werner, Kristine. 1985. "Archivische Karten als

historische Quelle." In *Mittelrhein und Mosel-land im Bild alter Karten*, edited by Fritz Hellwig, 145–61. Koblenz: Landesarchivverwaltung Rheinland-Pfalz.

Wester, Ethel. 1960. *Några skånska byar enligt lantmäterikartorna.* Meddelande från Lund Universitets Geografiska Institution 386. Lund: Geografiska Institutionen.

Westermanns Atlas zur Weltgeschichte. 1956. Brunswick: Westermann.

Westra, F. 1983. "De landmeter-ingenieur Johan Sems en de kaarten van Leewarden (1600–1603) en Franeker." *Caert-Thresoor* 2, no. 1:2–5.

———. 1986. "Bunderneuland: De kaart van Johan Sems (1628) beschouwd in kartografische en historische samenhang." *Caert-Thresoor* 5, no. 3:53–58.

Whalley, George H. 1838. *The Tithe Act and the Tithe Act Amendment Act.* London: Shaw and Sons.

White, C. Albert. 1983. *A History of the Rectangular Survey System.* Washington, D.C.: U.S. Government Printing Office.

Williams, Eberhard. 1928. "Skattläggningsväsendet och lantmätarna." In *Svenska lantmäteriet, 1628–1928,* vol. 1, pp. 299–363. Stockholm: Sällskapet för Utgifvande av Landmäteriets Historia.

Williams, Glyn, and Ramsden, John. 1990. *Ruling Britannia: A Political History of Britain, 1688–1988.* London: Longman.

Williams, Michael. 1966. "Delimiting the Spread of Settlement: An Examination of Evidence in South Australia." *Economic Geography* 42:336–55.

———. 1974. *The Making of the South Australian Landscape: A Study in the Historical Geography of Australia.* London: Academic Press.

———. 1975. "More and Smaller Is Better: Australian Rural Settlement, 1788–1914." In *Australian Space Australian Time: Geographical Perspectives,* edited by Joseph M. Powell and Michael Williams, 61–103. Melbourne: Oxford University Press.

Williamson, I. P. 1982. "Early Regulations for the Survey and Marking of Crown Land in New South Wales." *Australian Surveyor* 31:246–54.

———. 1984a. "Coordination of Cadastral Surveys in New South Wales." *Australian Surveyor* 32, no. 4:274–92.

———. 1984b. "The Development of the Cadastral Survey System in New South Wales." *Australian Surveyor* 32, no. 1:2–20.

———. 1984c. "A Historical Review of Measurement and Marking Techniques for Cadastral Surveying in New South Wales." *Australian Surveyor* 32, no. 2:106–12.

Willis, F. R. 1980. "Development Planning in Eighteenth-Century France: Corsica's Plan Terrier." *French Historical Studies* 11:328–51.

Winge, Harald. 1982. "Skatt, gård, og skyld." In *Den eldeste matrikkelen: En infallsport til historien: Skattematrikkelen 1647,* edited by Rolf Fladby and Harald Winge, 24–28 Oslo: Norsk Lokalhistorisk Institutt and Universitetsforlaget.

Winschiers, Kurt. 1982. *500 Jahre Vermessung und Karte in Bayern: Ein Überblick in 60 biographischen Skizzen.* Mitteilungsblatt Deutscher Verein für Vermessungswesen 34. Sonderheft 2.

Witt, Reimer. 1982. *Die Anfänge von Kartographie und Topographie Schleswig-Holsteins, 1475–1652.* Heide: Verlag Boyens.

Wolff, Fritz. 1987. *Karten im Archiv.* Veröffentlichungen der Archivschule Marburg—Institut für Archivwissenschaft 13. Marburg.

Wolff, Fritz, and Engel, Werner. 1988. *Hessen im Bild alter Landkarten. Ausstellung der Hessischen Staatsarchiv 1988.* Marburg: Hessische Staatsarchiv zu Marburg.

Wood, Joseph S. 1978. "The Origin of the New England Village." Pennsylvania State University Ph.D. thesis.

Wordie, J. R. 1983. "The Chronology of English Enclosure, 1500–1914." *Economic History Review* 36:483–505.

Wright, William E. 1966. *Serf, Seigneur, and Sovereign: Agrarian Reform in Eighteenth-Century Bohemia.* Minneapolis: University of Minnesota Press.

Wyn, Graeme. 1977. "Discovering the Antipodes: A Review of Historical Geography in Australia and New Zealand, 1969–1975." *Journal of Historical Geography* 3:251–65.

Yelling, James A. 1977. *Common Field and Enclosure in England, 1450–1850.* London: Macmillan.

Zandvliet, Kees. 1979. *The Biesbos near Geertruidenberg by Pieter and Jacob Sluyter, 1562 / De Biesbos bij Geertruidenberg door Pieter en Jacob Sluyter, 1562.* The Hague: Algemeen Rijksarchief.

———. 1985. *De groote waereld in 't kleen geschildert: Nederlandse kartografie tussen de Middeleeuwen en de industriële revolutie.* Alphen aan den Rijn: Canaletto.

———. 1989a. "Kaart en bewijs: Enkele opmerkingen over de 'Kaart als akte' naar aanleiding van de weergave van de landscheiding op de kaart van Rijnland 1614." *Kartografisch Tijdschrift* 15, no. 2:31–36.

———. 1989b. "Not for Sale: The Map Collection of the Dutch General State Archives." Map Collector 47: 20–24.

Zangheri, Renato. 1980. *Catasti e storia della proprietà terriera.* Turin: Einaudi.

Ziegler, T. 1976. *Die Entstehung des bayerischen Katasterwerks.* Sonderheft des Deutschen Vereins für Vermessungswesen. Munich: Landesverein Bayern.

Zögner, Gudrun K., and Zögner, Lothar. 1986. "Hessen." In *Lexikon zur Geschichte der Kartographie von den Anfängen bis zum Ersten Weltkrieg,* edited by Ingrid Kretschmer, Johannes Dörflinger, and Franz Wawrik, 291–93. Vienna: Franz Deuticke.

Zögner, Lothar. 1984. *Bibliographie zur Geschichte der deutschen Kartographie.* Sonderheft 2, *Bibliographia Cartographia.* Munich.

INDEX

Page references to illustrations are in **bold** type.